Geothermal Energy Systems

Geothermal Energy Systems

Ibrahim Dincer
Ontario Tech. University, Oshawa, Ontario, Canada

Murat Ozturk
Isparta University of Applied Sciences, Isparta, Turkey

ELSEVIER

Elsevier
Radarweg 29, PO Box 211, 1000 AE Amsterdam, Netherlands
The Boulevard, Langford Lane, Kidlington, Oxford OX5 1GB, United Kingdom
50 Hampshire Street, 5th Floor, Cambridge, MA 02139, United States

British Library Cataloguing-in-Publication Data
A catalogue record for this book is available from the British Library

Library of Congress Cataloging-in-Publication Data
A catalog record for this book is available from the Library of Congress

ISBN: 978-0-12-820775-8

Publisher: Brian Romer
Acquisitions Editor: Lisa Reading
Editorial Project Manager: Allison Hill
Production Project Manager: Poulouse Joseph
Cover Designer: Mark Rogers

Typeset by MPS Limited, Chennai, India

Contents

Preface

This book aims to cover the essentials of geothermal energy systems and applications in an innovative way. Its spectrum is quite diverse, focusing on almost every kind of geothermal system, ranging from basic to advanced and integrated systems for various multigenerational purposes. Diversified applications of geothermal energy systems for power, heating-cooling, hot and freshwater, drying, hydrogen, and ammonia production, are considered and illustrated by ranging from traditional to innovative systems for the generation of multiple useful outputs. The book is also intended to supply a unique perspective for researchers, scientists, engineers, technologists, students, policy makers, and others who wish to learn more about geothermal energy–based systems and applications.

This book is divided into nine chapters. In this regard, Chapter 1 dwells on thermodynamic fundamentals, with a brief presentation of the concept of thermodynamic analysis for studies in the follow-up chapters when considering heat and its transfer in geothermal energy systems. The types of thermodynamic systems are categorized and defined, and the balance equations, namely mass, energy, entropy, and exergy for different systems and their components are specifically written. The chapter ends with a presentation of the different ways of defining energy and exergy efficiencies. Chapter 2 primarily focuses on energy, the environment and sustainable development aspects of geothermal energy sources, systems, and implementation practices. In Chapter 3, geothermal energy sources are introduced and discussed, along with some information about the ten largest geothermal energy–deploying countries. Geothermal energy classifications by resource temperature and application areas are clearly presented in this chapter. Geothermal energy utilization in the world, particularly for heating, cooling, power, hydrogen, and ammonia production options are discussed in Chapter 4. Next, in Chapter 5, basic geothermal energy systems along with comprehensive case studies are presented to cover both energy and exergy analyses for geothermal energy–based power generation systems, such as direct steam power generation, single flash steam power generation, double flash power generation, triple flash power generation, quadruple flash power generation, binary cycle power generation, and combined power generation. In addition, Chapter 6 presents an overview of advanced geothermal energy systems based on thermodynamic analysis and assessments to cover various cases where the advanced geothermal energy systems are classified, designed, analyzed and evaluated both energetically and exergetically. In Chapter 7, multigenerational geothermal energy systems are presented to emphasize the need and importance of multiple useful outputs, and a detailed case study of a geothermal energy–based integrated system is given to provide a better understanding of geothermal energy–based multigeneration is given. Some detailed background and application methods on geothermal energy–based district systems and plants are described and discussed with examples in Chapter 8. Finally, the future directions, along with potential development opportunities for geothermal energy–based integrated plants for multigeneration, are discussed in Chapter 9. Furthermore, this chapter provides some closing remarks and recommendations.

This book, in closing, offers unique perspectives on the fundamentals, processes and applications of geothermal energy–based systems. Following the International System of Units (SI), the book presents different examples and case studies in each chapter through thermodynamic analyses of geothermal energy–based systems. Also, the environmental effects and some design indicators on the performance of the basic and advanced systems are evaluated for more efficient system design goals. At the end of each chapter, there are useful references for readers who seek further information. Moreover, chapter-end questions and problems help instructors adopt this publication as a textbook.

Ibrahim Dincer and Murat Ozturk

Acknowledgments

In this book on *Geothermal Energy Systems*, we gratefully acknowledge the assistance provided by Dr. Yunus Emre Yuksel, Dr. Fatih Yilmaz, and PhD student Nejat Tukenmez in reviewing and revising several chapters and preparing figures, tables, questions/problems, etc.

Prof. Dr. Dincer also acknowledges the support provided by the Turkish Academy of Sciences.

Last but not least, we warmly thank our families for being a great source of support and motivation, patience, and understanding.

Ibrahim Dincer and Murat Ozturk

Chapter 1

Thermodynamic fundamentals

1.1 Introduction

Whenever or wherever energy systems and application-related matters come up, thermodynamics is always needed as a main instrument for properly designing, analyzing, assessing, and improving such systems for practical applications. The terminological meaning of "thermodynamics" is heat power, and it originally comes from the Greek words *therme* and *dynamis*, which could be stated as the conversion of heat into power.

The basic definition of thermodynamics is commonly the science of energy and entropy. However, Dincer [1] has proposed the following definition: the discipline of energy (which comes from the first law of thermodynamics) and exergy (which comes from the second law of thermodynamics). This is a clear indication that thermodynamics stands on the two pillars of the first and second laws of thermodynamics as the governing laws of thermodynamics. When a thermodynamic analysis is needed, it begins with writing the balance equations for mass, energy, entropy, and exergy under these governing laws. So the approach in this chapter is to identify the type of system, pinpoint all inputs and outputs (along with generations and destructions), and write the balance equations accordingly. The next step is to define the efficiency in order to be able to determine the performance of the thermodynamic system. This can be done in terms of either energy efficiency (referring to energy analysis under the first law) or exergy efficiency (referring to exergy analysis under the second law). Of course, these will be clearly outlined with the examples throughout the chapter as well the book.

The primary purpose of this chapter is to introduce thermodynamic systems in terms of closed system (CS) and open system (OS), define the key thermodynamic concepts and fundamentals, and discuss them for various types of systems and their applications. In addition, numerous examples and case studies are presented to better illustrate all such systems for design, analysis, and assessment.

Furthermore, in this chapter, the primary ideas of thermodynamics are investigated, with a focus on proposed geothermal energy—based power production plants in the ensuing chapters. In developed and underdeveloped countries, there is a continuously growing need for electric energy as a basic commodity to meet a lot of societal demands. Thermodynamic analysis is

Geothermal Energy Systems. DOI: https://doi.org/10.1016/B978-0-12-820775-8.00010-6

the fundamental discipline concerned with the conversion of thermal energy, fuel energy, nuclear energy, or other forms of energy into beneficial motive energetic content. Since sources are not unlimited, it is essential to produce useful outputs with the highest effectiveness. Another aspect relates to the environment. Generally, fossil energy sources—based power production systems, without clean technologies such as carbon capture and storage, intensely pollute the environment by emitting greenhouse gaseous and other pollutants. New and advanced plants for generating power and other useful products are in development worldwide, such as geothermal energy—based integrated plants for multigeneration aims, capable of less pollution and higher efficiency. Regardless of the geothermal power plant model, thermodynamic assessment is the most fundamental way to conduct plant analysis, modeling, design, irreversibility rate, pollution reduction, etc. The general thematic thermodynamic concepts are submitted at the beginning of the chapter. Thermodynamic analysis based on the first and second laws is a modern approach to thermal design and optimization that utilizes four kinds of balance equalities: mass balance equation (MBE), energy balance equation (EBE), entropy balance equation (EnBE), and exergy balance equation (ExBE). Both the energetic and exergetic performance viewpoints and equations are defined. Finally, the balance and performance formulations are written for most important subplants, such as the turbine, pump, compressor, valve, flashing, three-way valve, separator, ejector, mixer, purifier, HEX, condenser, boiler, preheater, superheater, evaporator, and reverse osmosis.

1.2 Thermodynamic systems

In this subsection, we first need to define the thermodynamic system as a system having inputs and outputs, resulting in changes in thermodynamic properties and hence states. Here, one should remember several things:

- *Property*, such as pressure or temperature, is a characteristic of the system.
- *State* is a condition defined by at least two properties, such as inlet and exit states.
- *Process* is a change from one state to another, such as isothermal (constant temperature) and isobaric (constant pressure).
- *Cycle* is to go through processes depending on the changes at the state points and to come back to the original starting point, such as Carnot cycle, Rankine cycle, Brayton cycle, etc.

The prime idea behind a thermodynamic analysis is to analyze a thermodynamic system defined by Carnot in 1824. By description, the thermodynamic system is isolated by an actual or unreal boundary from the rest of the universe, stated as the *surroundings*. Based on this definition, for the aim of thermodynamic assessment, the universe can be divided into two sections, the thermodynamic system and its surroundings. Also, a thermodynamic system can be described as the quantity of matter or the field in space selected for study as defined by Cengel and Boles [2]. In addition, the mass or field outside the system can be defined as the surroundings. The actual or unreal surface that separates the system from its surroundings can be described as the *boundary*. For better visualization, these terms are depicted in Fig. 1.1.

A thermodynamic system can be separated into two types, a closed system (CS) and an open system (OS) based on whether a fixed mass or a fixed volume in space is selected for the study. Any thermodynamic system that can transfer energy in the form of work and heat but does not permit mass to enter or exit is a *closed* type of thermodynamic system. In other words, the heat and work transfers are possible for the CS, but the mass is not transferred to or from the system, as illustrated in Fig. 1.2.

In addition, if the energetic exchange is allowed only in the form of work, the CS can then be defined as *adiabatic*, as illustrated in Fig. 1.3, where no heat transfer is coming in or going out.

Furthermore, some special cases may be attributed to the various types and operations. One of these cases is known as an *isolated* system where no transfer of mass and energy (in terms of heat and work) is coming in and leaving the system, as shown in Fig. 1.4.

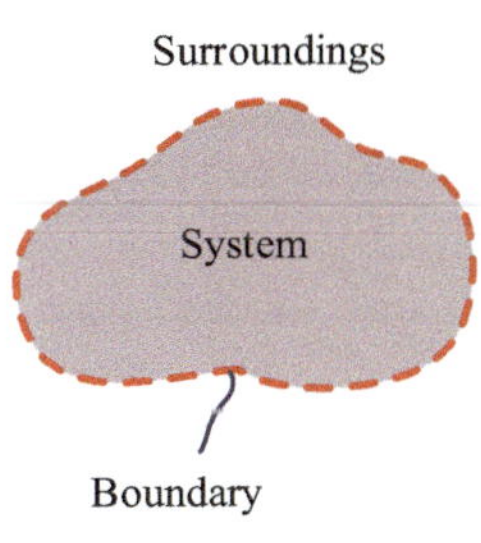

FIGURE 1.1 Expression of surroundings, system, and boundary of any plant or process.

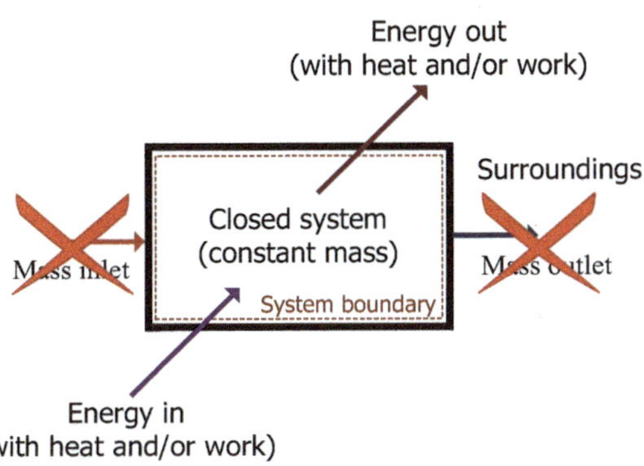

FIGURE 1.2 Schematic diagram of a closed thermodynamic system.

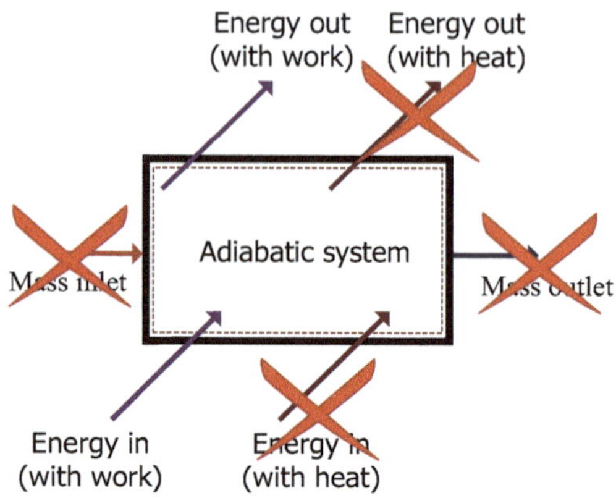

FIGURE 1.3 Schematic diagram of an adiabatic system.

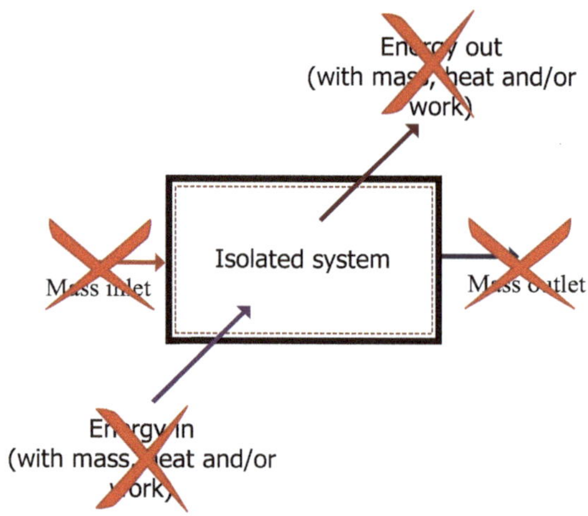

FIGURE 1.4 Schematic diagram of an isolated system.

If the thermodynamic system can interact with its surroundings by means of mass and energy in the form of heat or work transfers, in that case the system is an *open* system or control volume; this type of system is illustrated in Fig. 1.5. The OS generally encloses a component that involves mass flow such as a compressor, pump, turbine, nozzle, etc. The flow through these components is best studied by choosing the field within the component as the control volume. But it should not be forgotten that, in the OS, both mass and energy can cross the boundary of the control volume of the investigated study.

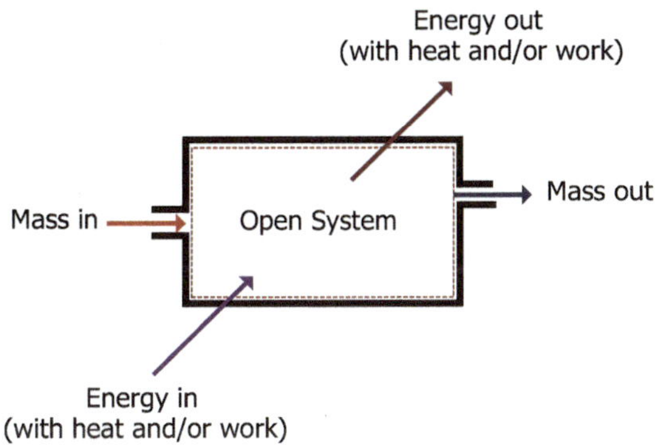

FIGURE 1.5 Schematic diagram of an open thermodynamic system.

1.3 Energy and exergy analyses

The energetic and exergetic assessments based on the first and second laws of thermodynamics have been extensively utilized as a powerful procedure for the design and optimization of power production plants and for many other engineered plants. Additionally, researchers and scientists have benefited greatly from the minimum entropy generation procedure for plant design and operations [3]. This methodology, however, requires a simultaneous consideration of thermodynamic laws and heat transfer viewpoints while predicting the indicators of a preferable model with decreased irreversibilities (or minimized produced entropy rates).

In the engineering of power production plants, a strong effort has always been made toward making preferable devices, cycles, and plants. This view point has a complex form and is expanded toward the principle of sustainability. It is generally agreed that, to achieve such a requirement, one needs sustainable power production decisions. Dincer and Zamfirescu [4] previously introduced some key targets for sustainable development:

- better efficiency
- better cost-effectiveness
- better utilization of resources
- better environment
- better energy security
- better management

In each of these requirements, exergy analysis and assessment are required for thermodynamic systems in addition to the energy analysis that is traditionally done. When an exergy analysis is conducted, it is also equally important to study the interactions with the surroundings. That is why Dincer and Rosen [3] defined exergy as the confluence of energy, environment, and sustainable development.

Exergetic assessment offers a determination of the maximum reversible work, which is the work produced or consumed for the plant to achieve the dead state per the indicators of the reference surroundings. For the exergetic assessment, the reference surroundings are assumed to be a thermodynamic system at a dead state. In order to determine the plant irreversibility or exergy destruction, the reversible work of the plant must be compared with its actual work [5]. The determination of different plants' or subsystems' exergetic destructions is one of the primary aims of the exergetic assessment. Another significant purpose of this assessment is to calculate the real performance of these plants or subsystems.

1.3.1 Balance equations

Generally, any balance equation for a quantity in a process should be defined as follows:

$$\text{Input} + \text{Generation} - \text{Output} - \text{Consumption} = \text{Accumulation} \tag{1.1}$$

This relationship is referred to as the quantity balance, and it should be expressed as the quantity accumulated in a process during a cycle that is equal to the difference between the net quantity transfer through the process boundary plus the quantity produced and the quantity consumed within the plant boundaries.

- *Mass balance equation*: The conservation of mass is a fundamental rule in investigating any thermodynamic process. As given in Fig. 1.6, the MBE for a control volume for non-steady-state process conditions can be defined mathematically:

$$\text{MBE:} \sum \dot{m}_i = \sum \dot{m}_e + \frac{dm_{CV}}{dt} \tag{1.2}$$

Here, m and $\dot{m}$ show the mass and mass flow rate, subscripts i and e show the inlet and outlet flows, t is time, and CV is the control volume. For a steady-flow process, MBE can be revised:

$$\text{MBE}_{\text{sf}}: \sum \dot{m}_i = \sum \dot{m}_e \tag{1.3}$$

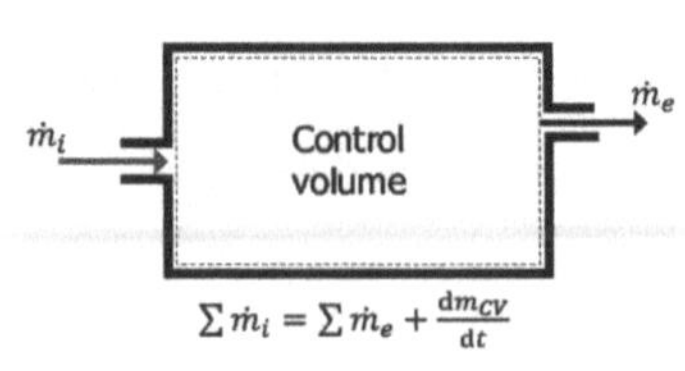

FIGURE 1.6 Schematic diagram of control volume mass balance equation.

- *Energy balance equation*: Based on the first law of thermodynamics, the EBE, which shows the variation of process energetic condition between inlet and outlet flows, can be defined:

$$\Delta E = m\Delta e = m\left[\left(u_e + 0.5\vartheta_e^2 + gz_e\right) - \left(u_i + 0.5\vartheta_i^2 + gz_i\right)\right] \tag{1.4}$$

Here, u is the internal energy; ϑ, g, and z show velocity, gravitational acceleration, and elevation. Also, e is the total specific energy for a nonflowing thermodynamically process and is given mathematically as follows:

$$e = u + 0.5\vartheta^2 + gz \tag{1.5}$$

The EBE for a CS can be expressed in the rate form:

$$\text{EBE}_{\text{CS}}: \sum \dot{q}_i + \sum \dot{w}_i = \sum \dot{q}_e + \sum \dot{w}_e + \frac{\mathrm{d}\dot{e}}{\mathrm{d}t} \tag{1.6}$$

where $\dot{q}$ and $\dot{w}$ show specific heat transfer rate and specific power. As shown in Fig. 1.7, if the flow and boundary work exit for the OS control volume, the EBE for this system becomes:

$$\text{EBE}_{\text{OS}}: \sum \dot{m}_i\theta_i + \sum \dot{Q}_i + \sum \dot{W}_i = \sum \dot{m}_e\theta_e + \sum \dot{Q}_e + \sum \dot{W}_e \frac{\mathrm{d}(me)}{\mathrm{d}t} \tag{1.7}$$

Here, $\dot{Q}$ and $\dot{W}$ show the heat transfer rate and power, and θ shows the total energy of flowing materials and can be defined as:

$$\theta = \text{internal energy} + \text{flow work} + \text{kinetic energy} + \text{potential energy} \tag{1.8}$$

or

$$\theta = u + Pv + 0.5\vartheta^2 + gz = h + 0.5\vartheta^2 + gz \tag{1.9}$$

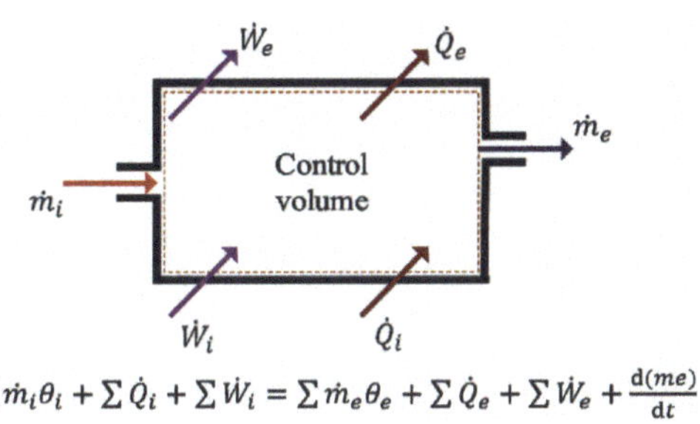

FIGURE 1.7 Schematic diagram of control volume energy balance equation.

The MBE for a steady flow process (mass flow rate, pressure, temperature, etc. do not change in time) is described as follows:

$$\begin{aligned}\mathrm{EBE_{sf}}: \sum \dot{m}_i e_i + \sum \dot{m}_i h_i + \sum \dot{Q}_i + \sum \dot{W}_i \\ = \sum \dot{m}_e e_e + \sum \dot{m}_e h_e + \sum \dot{Q}_e + \sum \dot{W}_e\end{aligned} \tag{1.10}$$

where h gives the specific enthalpy. When kinetic and potential energy changes between inlet and outlet conditions are unimportant, this balance equation can be revised:

$$\mathrm{EBE_{sf}}: \sum \dot{m}_i h_i + \sum \dot{Q}_i + \sum \dot{W}_i = \sum \dot{m}_e h_e + \sum \dot{Q}_e + \sum \dot{W}_e \tag{1.11}$$

- *Entropy balance equation*: The second law of thermodynamics should be defined in the form of the EnBE: For thermodynamic process, the entropy inlet plus generated entropy is equal to entropy exit plus entropy change within the process. Or it can be defined as entropy balance; that is, the entropy change of a thermodynamic process is equal to the generated entropy in the process boundary plus the net entropy transferred to the process across its boundary. It must be stated that entropy can be transferred outward from a process as heat energy, and it cannot be transferred as work energy [6]. In real processes, entropy exiting from the process is always higher than entropy going into the process, where the difference based on the internal irreversibilities is considered entropy generation. Only in the ideal (reversible) process is the amount of irreversibility equal to zero; therefore, there is no entropy generation. The general EnBE can be expressed as:

$$\mathrm{EnBE}: \sum \dot{S}_i + \dot{S}_{gen} = \sum \dot{S}_e + \frac{\mathrm{d}S_{sys}}{\mathrm{d}t} \tag{1.12}$$

where $\dot{S}$ gives the entropy flow or generation rate, and S and s show the entropy and specific entropy, respectively. The entropy transfer equation for CS is defined as follows:

$$\mathrm{EnBE_{CS}}: \sum \left(\int \frac{\mathrm{d}\dot{Q}_i}{T} \right) + \dot{S}_{gen} = \sum \left(\int \frac{\mathrm{d}\dot{Q}_e}{T} \right) + \frac{\mathrm{d}S_{sys}}{\mathrm{d}t} \tag{1.13}$$

This equation can be modified for the closed adiabatic system:

$$\dot{S}_{sys} = \dot{S}_{gen}. \tag{1.14}$$

As illustrated in Fig. 1.8, the entropy transfer equation for an OS is described in the rate form as:

$$\mathrm{EnBE_{OS}}: \sum \int \frac{\mathrm{d}\dot{Q}_i}{T} + \sum \dot{m}_i s_i + \dot{S}_{gen} = \sum \int \frac{\mathrm{d}\dot{Q}_e}{T} + \sum \dot{m}_e s_e + \frac{\mathrm{d}S_{sys}}{\mathrm{d}t} \tag{1.15}$$

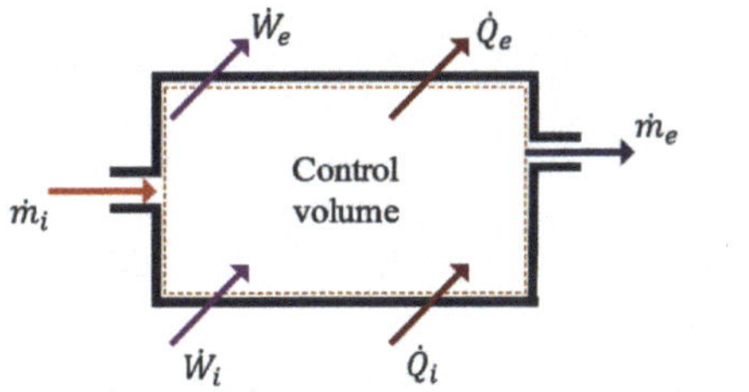

FIGURE 1.8 Schematic diagram of control volume entropy balance equation.

For the steady flow, Eq. (1.15) is rewritten as

$$\mathrm{EnBE_{sf}}: \sum\left(\int \frac{d\dot{Q}_i}{T}\right) + \sum \dot{m}_i s_i + \dot{S}_{gen} = \sum\left(\int \frac{d\dot{Q}_e}{T}\right) + \sum \dot{m}_e s_e \quad (1.16)$$

When the mass flow rate between inlet and outlet conditions of the process is constant ($\dot{m}_i = \dot{m}_e = \dot{m}$), Eq. (1.16) is simplified as:

$$\sum\left(\int \frac{d\dot{Q}_i}{T}\right) + \dot{m}(s_i - s_e) + \dot{S}_{gen} = \sum\left(\int \frac{d\dot{Q}_e}{T}\right) \quad (1.17)$$

Finally, the EnBE for an open adiabatic system can be formulated:

$$\sum \dot{m}_i s_i + \dot{S}_{gen} = \sum \dot{m}_e s_e \quad (1.18)$$

- *Exergy balance equation:* Exergetic analysis is a methodology for the investigation of the performance of components and processes and includes investigating the exergetic performance at different points in a series of energetic conversion stages. "Maximum theoretical work," "reference environment," and "energy quality thermodynamic" are terms that can be used to describe the exergy. When energy content is converted into a different, less beneficial form, the lost part of the beneficial content cannot be recovered again; it is part of the energy that is not conserved as the total energetic content of the process [7]. The quantity of beneficial work is described as exergetic content. Exergy (also called available energy or availability) of a process is the "maximum shaft work that can be made by the composite of the process and a specified dead-state environment." In each thermal process, heat energy transfer either within the process or between the process and reference environment happens at a finite temperature difference, which is an essential contributor to irreversibility rates for the process.

Exergetic analysis takes into account the several thermodynamic amounts of dissimilar energetic forms and quantities. The exergetic transfer associated with the shaft work is equal to the shaft work. The exergetic transfer associated with heat energy, however, depends on the temperature at which it occurs in relation to the dead state temperature of the surroundings [8]. A number of significant characteristics of the exergetic viewpoint are listed here:

- A plant in complete equilibrium with its surroundings does not have any exergetic content.
- The exergetic content of a plant rises the more it deviates from the dead state conditions.
- When the energetic loses its quality or is degraded, exergetic content is destroyed.
- Exergetic content, by description, depends not only on the state of a plant or stream but also on the condition of the dead-state surroundings.
- Exergetic performances are the measurement of the system's approach to ideality (or reversibility). This statement is not necessarily right for energetic performances, which are often misleading.
- Energetic conditions with high exergetic contents are typically more valued and beneficial than energetic conditions with low exergetic content.

The ExBE presents the destroyed exergetic term, which symbolizes the maximum work potential that cannot be recovered for beneficial aims due to irreversibility rates. For the reversible process, there is no exergetic destruction since all of the work produced in the process boundary can be made beneficial. Also, the exergy destruction and entropy generation terms can be connected:

$$\dot{E}x_d = T_o \Delta \dot{S}_{gen} \tag{1.19}$$

Here, T_o is the dead-state temperature, and $\dot{E}x_d$ shows the exergy destruction rate. There are three states based on the $\dot{E}x_D$ condition:

$\dot{E}x_d > 0 \Rightarrow$ system is irreversible.
$\dot{E}x_d = 0 \Rightarrow$ system is reversible.
$\dot{E}x_d < 0 \Rightarrow$ system is impossible.

The net exergy entering a thermodynamic process must be balanced by the net exergy exiting the process plus the change of exergetic content of the process plus the exergy destruction. Exergetic variables can be transferred to or from the process in three ways: as work, heat, and mass. Based on these definitions, as shown in Fig. 1.9, the general ExBE can be written in rate form [9]:

$$\text{ExBE:} \sum \left(\dot{m}_i ex_i + \dot{W}_i + \dot{E}x_{Q,i}\right) = \sum \left(\dot{m}_e ex_e + \dot{W}_e + \dot{E}x_{Q,e}\right) + \frac{dE_x}{dt} + \dot{E}x_d \tag{1.20}$$

Here, ex shows the specific exergy and can be described as:

$$ex = ex_{ph} + ex_{kn} + ex_{pt} + ex_{ch} \tag{1.21}$$

The physical or flow exergy is associated with the deviation of temperature and pressure relative to the dead-state ambient and can be defined as:

$$ex_{ph} = (h - h_o) + T_o(s - s_o) \tag{1.22}$$

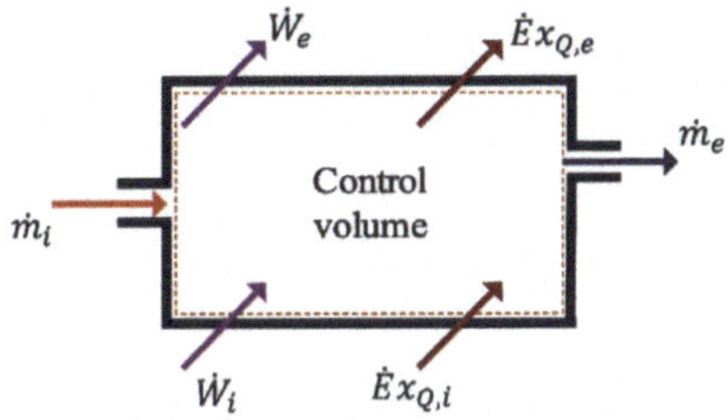

$$\Sigma(\dot{m}_i ex_i + \dot{W}_i + \dot{E}x_{Q,i}) = \Sigma(\dot{m}_e ex_e + \dot{W}_e + \dot{E}x_{Q,e}) + \frac{dE_x}{dt} + \dot{E}x_d$$

FIGURE 1.9 Schematic diagram of control volume exergy balance equation.

The kinetic exergy is related with the process velocity, measured relative to a chosen reference level, and is written as:

$$ex_{kn} = \frac{1}{2}\vartheta^2 \tag{1.23}$$

The potential exergy is related with the process height, measured relative to a chosen reference level, and can be given as:

$$ex_{pt} = g(z - z_o) \tag{1.24}$$

The chemical energy is related to the deviation of the chemical combination of the process relative to a chosen dead-state condition and can be written as:

$$ex_{ch} = \Delta_f G^o + \sum_{elements} vex^{ch}_{element} \tag{1.25}$$

Here, $\Delta_f G^o$ shows the Gibbs free energy of formation and can be calculated as:

$$\Delta_f G^o = \Delta_f H^o - T_o \Delta_f S^o \tag{1.26}$$

where $\Delta_f H^o$ is the formation enthalpy, and $\Delta_f S^o$ is the formation entropy and can be described as follows:

$$\Delta_f S^o = s^o - \sum_{elements} vs^o_{f,\,element} \tag{1.27}$$

In Eq. (1.25), v shows the number of moles of the element in the chemical equation of formation, $ex^{ch}_{element}$ is the chemical exergy of elements, subscript f shows the reactants. The exergetic variables due to heat transfer rate can be written as:

$$\dot{E}x_Q = \left(1 - \frac{T_o}{T}\right)\dot{Q} \tag{1.28}$$

The exergy destruction is directly involved in the entropy generation within the process control volume and is described as follows:

$$\dot{E}x_d = T_0 \dot{S}_{gen} \tag{1.29}$$

For a thermodynamic process at a steady flow, the exergetic balance equation can be defined as:

$$\text{ExBE}_{\text{sf}}\text{:}\sum\left(\dot{m}_i ex_i + \dot{W}_i + \dot{E}x_{Q,i}\right) = \sum\left(\dot{m}_e ex_e + \dot{W}_e + \dot{E}x_{Q,e}\right) + \dot{E}x_d \quad (1.30)$$

1.3.2 Energy and exergy efficiencies

Energetic performance presents lower energy consumption costs and investment costs in energy systems, more independence from imported fossil energy sources, and the reduction of greenhouse gas emissions and local air pollution. The efficient utilization of energy can be adopted in buildings, appliances, transport, industry, and lighting by achieving a system's full potential [9]. Developed, developing, and underdeveloped countries promote energetic performance in both power supply and demand in many fields such as transport, buildings, industry, and home devices. They also take action by using renewable energy resources, including geothermal, wind, solar, and biomass. In other words, energy plays a necessary role in the lives of people, communities, countries, and the world. The efficient utilization of energy resources by means of considering the useful outlets, as well as cost and environmental impact, can be a key solution.

Efficiency can be defined as a measure of the effectiveness and/or performance of a plant, and it can be defined in terms of the useful output and the total input:

$$\text{Efficiency} = \frac{\text{Useful output}}{\text{Total input}} \quad (1.31)$$

As the measure of energy-related performance, a process energy efficiency (η) can be defined, based on the first law of thermodynamics, as the ratio of useful generation from the process boundary to the energy input to the process:

$$\eta = \frac{\text{Useful energy output}}{\text{Total energy input}} \quad (1.32)$$

or in the rate form:

$$\eta = \frac{\dot{E}n_{out}}{\dot{E}n_{in}} \quad (1.33)$$

All real systems, including natural processes, are irreversible, and the plant efficiency is reduced as a result of these exergy destructions in each thermodynamic process making up the plant. The availability quantity is decreased by the irreversibilities, and based on this definition, the related amount of energy content turns into unusable conditions.

Entropy generation evaluates the impact of these irreversibilities in a plant during a cycle and assists in assessing each part in the plant based on how much they contribute to the operational inefficiencies of the whole plant. Hence, entropy generation related to each cycle has to be examined to determine the whole plant performance. Although the energetic assessment is the most usually utilized way for investigating thermal plants, this method is concerned only with the conservation of energy, which neither takes the corresponding environmental cases into calculation nor identifies how, where, and why the plant efficiency is reduced. As a result, energetic assessment only calculates the quantity of energy and does not reveal the full performances of the plant. Therefore, investigated plants must be analyzed with respect to exergetic assessment in order to better understand the true performances of the system parts by analyzing the irreversibilities in each process, as well as in the whole plant, and how nearly the respective efficiencies approach optimal cases.

By investigating both the quality (usefulness) and the quantity of the energy, the true magnitude of losses and their causes and locations are identified by analyzing the locations of exergetic destruction in order to increase individual system parts and the whole system. According to the second law of thermodynamics, the exergy efficiency (ψ) equation for the steady-state operation should be described as:

$$\psi = \frac{\text{Exergy output}}{\text{Total exergy input}} \tag{1.34}$$

or in the rate form:

$$\psi = \frac{\dot{E}x_{out}}{\dot{E}x_{in}} \tag{1.35}$$

The expressions for the mass, energy, entropy, and ExBEs, as well as for the energy and exergy efficiency equations for some selected components, are defined in Table 1.1.

General fundamental thermodynamic balance, energy and exergy efficiency equations as presented in this chapter will be used for the energy and exergy analyses of the basic geothermal energy systems (Chapter 5), advanced geothermal energy systems (Chapter 6), multigenerational geothermal energy systems (Chapter 7) and geothermal district energy systems (Chapter 8).

Example 1.1: An air compressor, as illustrated in Fig. 1.10, compresses the air entering it at the pressure and temperature of 100 kPa and 15°C, respectively, and the air exits the device at the pressure and temperature of 500 kPa and 112°C, respectively. The air flows at a rate of 0.5 kg/s. The heat loss during the compressor process is assumed to equal an amount of 45% of the total work rate running this device. Calculate the following:

1. the power required to drive the air compressor,
2. both energy and exergy efficiencies of the air compressor,
3. the variation of energy and exergy efficiencies of the air compressor, when the reference temperature increases from 0°C to 40°C.

TABLE 1.1 Mass, energy, entropy, and exergy balance, energy and exergy efficiency equations of the different significant components.

Components	Equations
1. Turbine 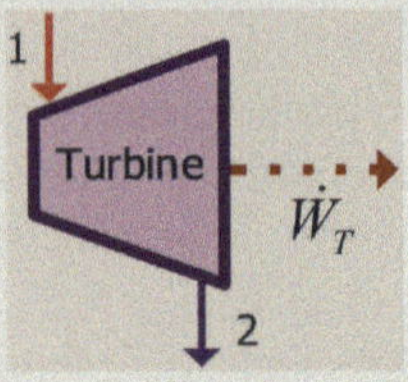	Balance equations: MB:$\dot{m}_1 = \dot{m}_2$ EBE:$\dot{m}_1 h_1 = \dot{m}_2 h_2 + \dot{W}_T$ EnBE:$\dot{m}_1 s_1 + \dot{S}_{g,T} = \dot{m}_2 s_2$ ExBE:$\dot{m}_1 ex_1 = \dot{m}_2 ex_2 + \dot{W}_T + \dot{E}x_{d,T}$ Efficiency equations: $\eta_T = \frac{\dot{W}_T}{\dot{m}_1 h_1}$ $\psi_T = \frac{\dot{W}_T}{\dot{m}_1 ex_1}$
2. Pump 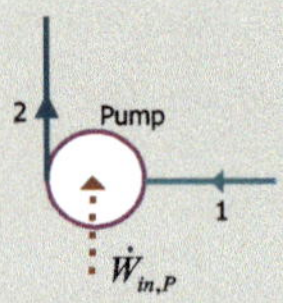	Balance equations: MB:$\dot{m}_1 = \dot{m}_2$ EBE:$\dot{m}_1 h_1 + \dot{W}_P = \dot{m}_2 h_2$ EnBE:$\dot{m}_1 s_1 + \dot{S}_{g,P} = \dot{m}_2 s_2$ ExBE:$\dot{m}_1 ex_1 + \dot{W}_P = \dot{m}_2 ex_2 + \dot{E}x_{d,P}$ Efficiency equations: $\eta_P = \frac{(\dot{m}_2 h_2 - \dot{m}_1 h_1)}{\dot{W}_P}$ $\psi_P = \frac{(\dot{m}_2 ex_2 - \dot{m}_1 ex_1)}{\dot{W}_P}$
3. Compressor 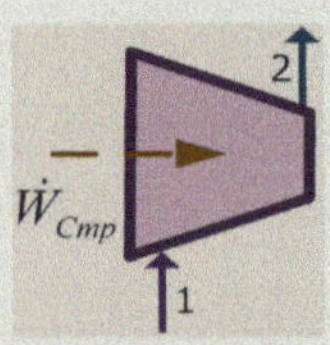	Balance equations: MB:$\dot{m}_1 = \dot{m}_2$ EBE:$\dot{m}_1 h_1 + \dot{W}_{Cmp} = \dot{m}_2 h_2$ EnBE:$\dot{m}_1 s_1 + \dot{S}_{g,Cmp} = \dot{m}_2 s_2$ ExBE:$\dot{m}_1 ex_1 + \dot{W}_{Cmp} = \dot{m}_2 ex_2 + \dot{E}x_{d,Cmp}$ Efficiency equations: $\eta_{Cmp} = \frac{(\dot{m}_2 h_2 - \dot{m}_1 h_1)}{\dot{W}_{Cmp}}$ $\psi_{Cmp} = \frac{(\dot{m}_2 ex_2 - \dot{m}_1 ex_1)}{\dot{W}_{Cmp}}$
4. Valve 	Balance equations: MB:$\dot{m}_1 = \dot{m}_2$ EBE:$\dot{m}_1 h_1 = \dot{m}_2 h_2$ EnBE:$\dot{m}_1 s_1 + \dot{S}_{g,Val} = \dot{m}_2 s_2$ ExBE:$\dot{m}_1 ex_1 = \dot{m}_2 ex_2 + \dot{E}x_{d,Val}$ Efficiency equations: $\eta_{Val} = \frac{\dot{m}_2 h_2}{\dot{m}_1 h_1}$ $\psi_{Val} = \frac{\dot{m}_2 ex_2}{\dot{m}_1 ex_1}$
5. Three-way valve 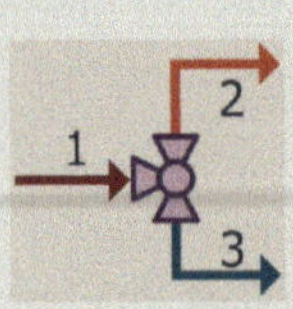	Balance equations: MB:$\dot{m}_1 = \dot{m}_2 + \dot{m}_3$ EBE:$\dot{m}_1 h_1 = \dot{m}_2 h_2 + \dot{m}_3 h_3$ EnBE:$\dot{m}_1 s_1 + \dot{S}_{g,3WV} = \dot{m}_2 s_2 + \dot{m}_3 s_3$ ExBE:$\dot{m}_1 ex_1 = \dot{m}_2 ex_2 + \dot{m}_3 ex_3 + \dot{E}x_{D,3WV}$ Efficiency equations: $\eta_{3WV} = \frac{(\dot{m}_2 h_2 + \dot{m}_3 h_3)}{\dot{m}_1 h_1}$ $\psi_{3WV} = \frac{(\dot{m}_2 ex_2 + \dot{m}_3 ex_3)}{\dot{m}_1 ex_1}$

(Continued)

TABLE 1.1 (Continued)

Components	Equations
6. Separator 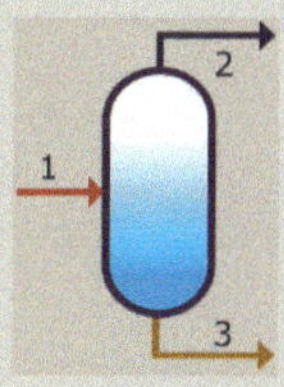	Balance equations: MB:$\dot{m}_1 = \dot{m}_2 + \dot{m}_3 s$ EBE:$\dot{m}_1 h_1 = \dot{m}_2 h_2 + \dot{m}_3 h_3$ EnBE:$\dot{m}_1 s_1 + \dot{S}_{g,Sp} = \dot{m}_2 s_2 + \dot{m}_3 s_3$ ExBE:$\dot{m}_1 ex_1 = \dot{m}_2 ex_2 + \dot{m}_3 ex_3 + \dot{E}x_{d,Sp}$ Efficiency equations: $\eta_{Sp} = \frac{(\dot{m}_2 h_2 + \dot{m}_3 h_3)}{\dot{m}_1 h_1}$ $\psi_{Sp} = \frac{(\dot{m}_2 ex_2 + \dot{m}_3 ex_3)}{\dot{m}_1 ex_1}$
7. Ejector 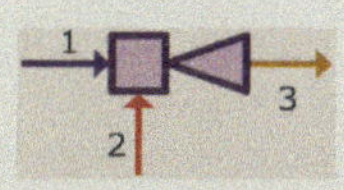	Balance equations: MB:$\dot{m}_1 + \dot{m}_2 = \dot{m}_3$ EBE:$\dot{m}_1 h_1 + \dot{m}_2 h_2 = \dot{m}_3 h_3$ EnBE:$\dot{m}_1 s_1 + \dot{m}_2 s_2 + \dot{S}_{g,Ej} = \dot{m}_3 s_3$ ExBE:$\dot{m}_1 ex_1 + \dot{m}_2 ex_2 = \dot{m}_3 ex_3 + \dot{E}x_{d,Ej}$ Efficiency equations: $\eta_{Ej} = \frac{\dot{m}_3 h_3}{(\dot{m}_1 h_1 + \dot{m}_2 h_2)}$ $\psi_{Ej} = \frac{\dot{m}_3 ex_3}{(\dot{m}_1 ex_1 + \dot{m}_2 ex_2)}$
8. Open feedwater heater 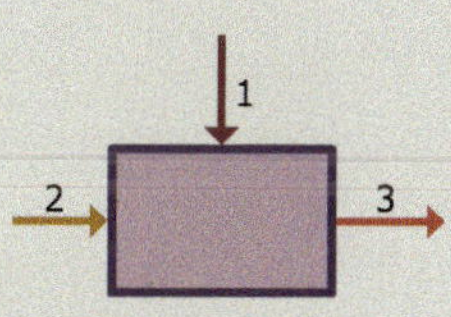	Balance equations: MB:$\dot{m}_1 + \dot{m}_2 = \dot{m}_3$ EBE:$\dot{m}_1 h_1 + \dot{m}_2 h_2 = \dot{m}_3 h_3$ EnBE:$\dot{m}_1 s_1 + \dot{m}_2 s_2 + \dot{S}_{g,ofh} = \dot{m}_3 s_3$ ExBE:$\dot{m}_1 ex_1 + \dot{m}_2 ex_2 = \dot{m}_3 ex_3 + \dot{E}x_{d,ofh}$ Efficiency equations: $\eta_{ofh} = \frac{\dot{m}_3 h_3}{(\dot{m}_1 h_1 + \dot{m}_2 h_2)}$ $\psi_{ofh} = \frac{\dot{m}_3 ex_3}{(\dot{m}_1 ex_1 + \dot{m}_2 ex_2)}$
9. Dehumidifier 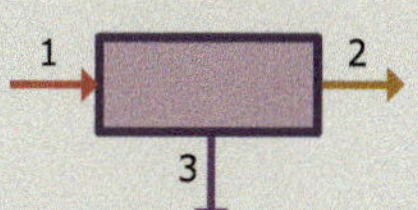	Balance equations: MB:$\dot{m}_1 = \dot{m}_2 + \dot{m}_3$ EBE:$\dot{m}_1 h_1 = \dot{m}_2 h_2 + \dot{m}_3 h_3$ EnBE:$\dot{m}_1 s_1 + \dot{S}_{g,dh} = \dot{m}_2 s_2 + \dot{m}_3 s_3$ ExBE:$\dot{m}_1 ex_1 = \dot{m}_2 ex_2 + \dot{m}_3 ex_3 + \dot{E}x_{d,dh}$ Efficiency equations: $\eta_{dh} = \frac{\dot{m}_2 h_2 + \dot{m}_3 h_3}{\dot{m}_1 h_1}$ $\psi_{dh} = \frac{\dot{m}_2 ex_2 + \dot{m}_3 ex_3}{\dot{m}_1 ex_1}$
10. Heat exchanger 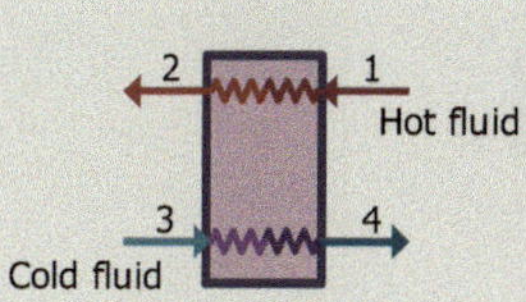	Balance equations: MB:$\dot{m}_1 = \dot{m}_2$ and $\dot{m}_3 = \dot{m}_4$ EBE:$\dot{m}_1 h_1 + \dot{m}_3 h_3 = \dot{m}_2 h_2 + \dot{m}_4 h_4$ EnBE:$\dot{m}_1 s_1 + \dot{m}_3 s_3 + \dot{S}_{g,HEX} = \dot{m}_2 s_2 + \dot{m}_4 s_4$ ExBE:$\dot{m}_1 ex_1 + \dot{m}_3 ex_3 = \dot{m}_2 ex_2 + \dot{m}_4 ex_4$ $+ \dot{E}x_{d,HEX}$ Efficiency equations: $\eta_{HEX} = \frac{(\dot{m}_4 h_4 - \dot{m}_3 h_3)}{(\dot{m}_1 h_1 - \dot{m}_2 h_2)}$ $\psi_{HEX} = \frac{(\dot{m}_4 ex_4 - \dot{m}_3 ex_3)}{(\dot{m}_1 ex_1 - \dot{m}_2 ex_2)}$

(Continued)

TABLE 1.1 (Continued)

Components	Equations
11. Reverse osmosis 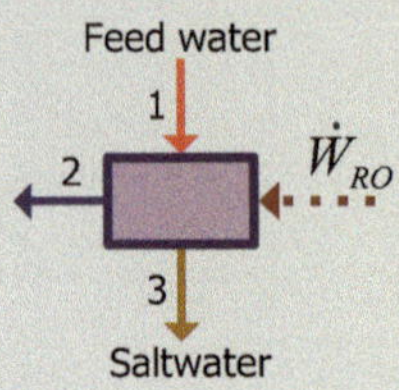 	Balance equations: MB: $\dot{m}_1 = \dot{m}_2 + \dot{m}_3$ EBE: $\dot{m}_1 h_1 = \dot{m}_2 h_2 + \dot{m}_3 h_3$ EnBE: $\dot{m}_1 s_1 + \dot{S}_{g,RO} + \dot{W}_{RO} = \dot{m}_2 s_2 + \dot{m}_3 s_3$ ExBE: $\dot{m}_1 ex_1 + \dot{W}_{RO} = \dot{m}_2 ex_2 + \dot{m}_3 ex_3$ $+ \dot{E}x_{d,RO}$ Efficiency equations: $\eta_{RO} = \frac{\dot{m}_2 h_2}{\dot{m}_1 h_1 + \dot{W}_{RO}}$ $\psi_{RO} = \frac{\dot{m}_2 ex_2}{\dot{m}_1 ex_1 + \dot{W}_{RO}}$
12. Nozzle 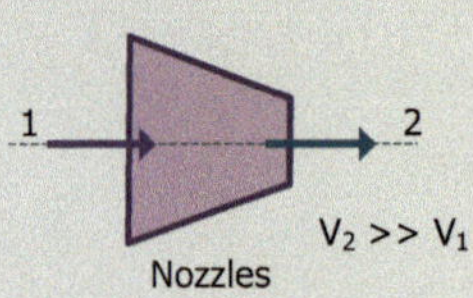 	Balance equations: MB: $\dot{m}_1 = \dot{m}_2$ EBE: $\dot{m}_1 h_1 + \frac{1}{2}\dot{m}_1 v_1 = \dot{m}_2 h_2 + \frac{1}{2}\dot{m}_2 v_2$ EnBE: $\dot{m}_1 s_1 + \dot{S}_{g,nz} = \dot{m}_2 s_2$ ExBE: $\dot{m}_1 ex_1 + \frac{1}{2}\dot{m}_1 v_1 = \dot{m}_2 ex_2 + \frac{1}{2}\dot{m}_2 v_2$ $+ \dot{E}x_{d,nz}$ Efficiency equations: $\eta_{nz} = \frac{\dot{m}_2 (h_2 + \frac{1}{2} v_2)}{\dot{m}_1 (h_1 + \frac{1}{2} v_1)}$ $\psi_{nz} = \frac{\dot{m}_2 (ex_2 + \frac{1}{2} v_2)}{\dot{m}_1 (ex_1 + \frac{1}{2} v_1)}$
13. Diffuser 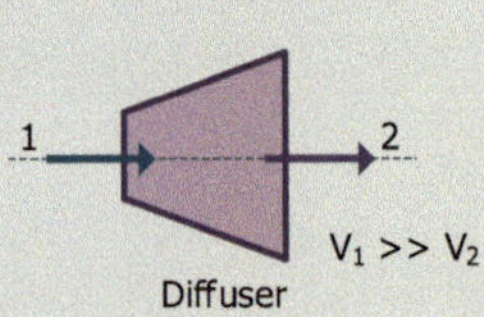 	Balance equations: MB: $\dot{m}_1 = \dot{m}_2$ EBE: $\dot{m}_1 h_1 + \frac{1}{2}\dot{m}_1 v_1 = \dot{m}_2 h_2 + \frac{1}{2}\dot{m}_2 v_2$ EnBE: $\dot{m}_1 s_1 + \dot{S}_{g,df} = \dot{m}_2 s_2$ ExBE: $\dot{m}_1 ex_1 + \frac{1}{2}\dot{m}_1 v_1 = \dot{m}_2 ex_2 + \frac{1}{2}\dot{m}_2 v_2$ $+ \dot{E}x_{d,df}$ Efficiency equations: $\eta_{df} = \frac{\dot{m}_2 (h_2 + \frac{1}{2} v_2)}{\dot{m}_1 (h_1 + \frac{1}{2} v_1)}$ $\psi_{df} = \frac{\dot{m}_2 (ex_2 + \frac{1}{2} v_2)}{\dot{m}_1 (ex_1 + \frac{1}{2} v_1)}$

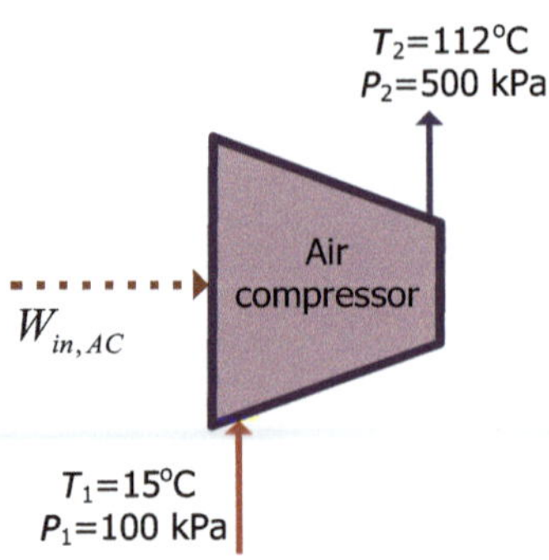

FIGURE 1.10 A schematic diagram of Example 1.1 for an air compressor.

Solution: The compression process is steady, and power input is required.

Assumptions:

- The reference temperature and pressure are taken as 101.3 kPa and 25°C, respectively.
- The air is treated as an ideal gas.
- The changes in the kinetic and potential energies and exergies are neglected.
- The ideal gas properties of air are taken from the Engineering Equation Solver (EES) program, and also the parametric study is made using this software program.

Analysis:

The first step in solving this example is to write the mass, energy, entropy, and ExBEs:

$$\text{Mass balance: } \dot{m}_1 = \dot{m}_2$$

$$\text{Energy balance: } \dot{m}_1 h_1 + \dot{W}_{in,AC} = \dot{m}_2 h_2 + \dot{Q}_{L,AC}$$

$$\text{Entropy balance: } \dot{m}_1 s_1 + \dot{S}_{g,AC} = \dot{m}_2 s_2 + \frac{\dot{Q}_{L,AC}}{T_{AC}}$$

$$\text{Exergy balance: } \dot{m}_1 ex_1 + \dot{W}_{in,AC} = \dot{m}_2 ex_2 + \dot{Ex}^{Q}_{L,AC} + \dot{Ex}_{d,AC}$$

1. The power consumed throughout the compression process needs to be computed in this subsection. Utilizing the energetic balance equation to compute the power consumed by the compressor:

$$\dot{W}_{in,AC} = \dot{Q}_{L,AC} + \dot{m}_2 h_2 - \dot{m}_1 h_1$$

where the properties of the air entering and exiting the compressor are taken from the air properties tables or calculated by using the EES software program, which contains a database of properties of most working fluids through a large range of temperatures and pressures. In this example, the state properties of air are calculated by using EES and given in Table 1.2. In this table, 0 is the

TABLE 1.2 Properties of the input and the output flows of air.

State point	$\dot{m}$ (kg/s)	T (°C)	P (kPa)	h (kJ/kg)	s (kJ/kg K)	ex (kJ/kg)	$\dot{Ex}$ (kW)
0	—	25	101.3	298.6	5.695	—	—
1	0.5	15	100	288.5	5.665	−0.9331	−0.4665
2	0.5	112	500	386.2	5.495	147.4	73.68

reference state, $\dot{m}$ is the mass flow rate in kg/s, T is the temperature in °C, P is the pressure in kPa, h is the enthalpy in kJ/kg, s is the entropy in kJ/kg K, ex is the specific exergy in kJ/kg, and $\dot{E}x$ is the exergy in kW.

Based on the MBE for the compressor ($\dot{m}_1 = \dot{m}_2 = \dot{m}_{air}$) and $\dot{Q}_{L,Cmp} = 0.45\dot{W}_{in,Cmp}$, the power consumed rate equation of compressor can be rewritten:

$$\dot{W}_{in,AC} = 0.45\dot{W}_{in,AC} + \dot{m}_{air}(h_2 - h_1)$$

$$\dot{W}_{in,AC} = 0.45\dot{W}_{in,AC} + 0.5(386.2 - 288.5)$$

$$\dot{W}_{in,AC} = 88.82 \text{ kW}$$

2. The energy efficiency of the compressor is computed:

$$\eta_{AC} = \frac{(\dot{m}_2 h_2 - \dot{m}_1 h_1)}{\dot{W}_{in,AC}} = \frac{\dot{m}_{air}(h_2 - h_1)}{\dot{W}_{in,AC}} = \frac{0.5x(386.2 - 288.5)}{88.82} = 0.55 = 55.0\%$$

The exergy efficiency of the air compressor is then computed:

$$\psi_{AC} = \frac{(\dot{m}_2 ex_2 - \dot{m}_1 ex_1)}{\dot{W}_{in,AC}} = \frac{\dot{m}_{air}(ex_2 - ex_1)}{\dot{W}_{in,AC}} = \frac{0.5x(147.4 - (-93.31))}{88.82} = 0.8348$$

$$= 83.48\%$$

Note that the specific exergy of the airflow is computed:

$$ex_i = (h_i - h_o) + T_o(s_i - s_o)$$

where i shows the state point, and o shows the properties at the dead-state environment conditions.

3. The effect of reference temperature on the energy and exergy efficiencies of the air compressor is shown in Fig. 1.11. As shown in this figure, the energetic efficiency of the component does not change with the increasing reference temperature from 0°C to 40°C, whereas the exergy efficiency is increased from 81.09% to 84.91% in the examined reference temperature change.

Example 1.2: Geothermal water at 900 kPa pressure and 82°C temperature enters a pump, and the exit pressure and temperature of the pumped geothermal water are increased to 1300 kPa and 86°C, respectively, as shown in Fig. 1.12. The mass flow rate of geothermal water is 32.4 kg/s. Also, the heat loss during the pumped process is assumed to equal an amount of 15% of the total work rate running the pump. Calculate:

1. the work rate consumed by the pump,
2. both energy and exergy efficiencies of the pump,
3. the variation of the power consumption rate of the pump when the mass flow rate of the geothermal water increases from 25 to 50 kg/s.

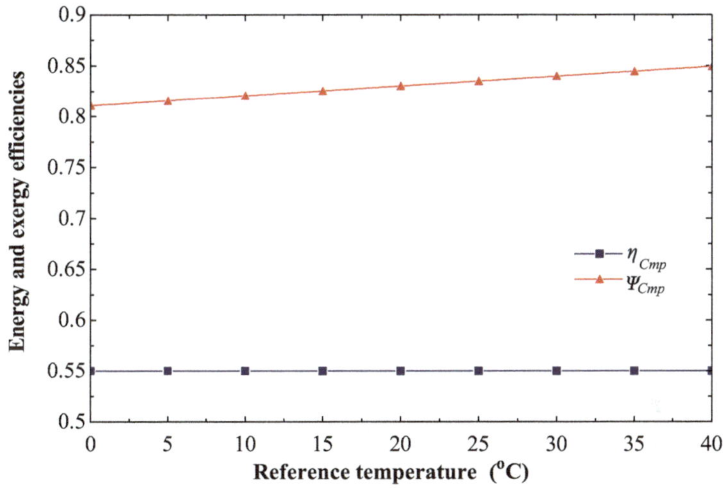

FIGURE 1.11 Effect of reference temperature on the energy and exergy efficiencies of the air compressor.

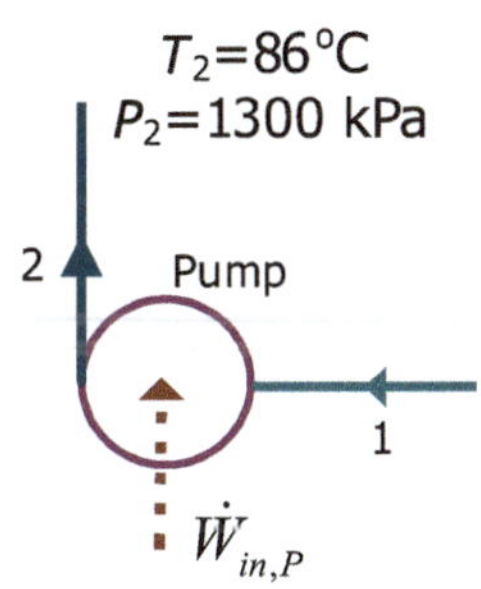

FIGURE 1.12 Schematic diagram of Example 1.2 pump.

Solution: The pumped process is steady, and power input for the pump is required.

Assumptions:

- The reference temperature and pressure are taken as 101.3 kPa and 25°C, respectively.
- The changes in the kinetic and potential energies and exergies are neglected.
- A thermal and physical property of the geothermal water is considered to be water in the thermodynamic analysis. Also, the EES software program is utilized for the determination of the thermodynamic properties of the geothermal water.

Analysis:

It is first necessary to write the mass, energy, entropy, and ExBEs for the pump:

$$\text{Mass balance: } \dot{m}_1 = \dot{m}_2$$

$$\text{Energy balance: } \dot{m}_1 h_1 + \dot{W}_{in,P} = \dot{m}_2 h_2 + \dot{Q}_{L,P}$$

$$\text{Entropy balance: } \dot{m}_1 s_1 + \dot{S}_{g,P} = \dot{m}_2 s_2 + \frac{\dot{Q}_{L,P}}{T_P}$$

$$\text{Exergy balance: } \dot{m}_1 ex_1 + \dot{W}_{in,P} = \dot{m}_2 ex_2 + \dot{E}x^{Q}_{L,P} + \dot{E}x_{D,P}$$

1. The power consumption rate of the pump component needs to be calculated in this section. Use the energetic balance equation to calculate the power consumption rate of the pump:

$$\dot{W}_{in,P} = \dot{Q}_{L,P} + \dot{m}_2 h_2 - \dot{m}_1 h_1$$

where the properties of the geothermal water entering and exiting the pump are computed by utilizing the EES software program, and the state properties of the geothermal water are written in Table 1.3.

Based on the MBE for the pump ($\dot{m}_1 = \dot{m}_2 = \dot{m}_{gw}$) and $\dot{Q}_{L,P} = 0.15\dot{W}_{in,P}$, the power consumed rate equation of the pump is calculated:

$$\dot{W}_{in,P} = 0.15\dot{W}_{in,P} + \dot{m}_{gw}(h_2 - h_1)$$

$$\dot{W}_{in,P} = 0.15\dot{W}_{in,P} + 32.84x(361.1 - 344)$$

$$\dot{W}_{in,P} = 651.7 \text{ kW}$$

TABLE 1.3 Properties of the input and the output flows of geothermal water for the pump component.

State point	$\dot{m}$ (kg/s)	T (°C)	P (kPa)	h (kJ/kg)	s (kJ/kg K)	ex (kJ/kg)	$\dot{E}x$ (kW)
0	—	25	101.3	104.8	0.3669	—	—
1	32.84	82	900	344	1.098	21.05	682.2
2	32.84	86	1300	361.1	1.145	24.22	784.8

2. The energy and exergy efficiencies of the pump are calculated:

$$\eta_P = \frac{(\dot{m}_2 h_2 - \dot{m}_1 h_1)}{\dot{W}_{in,P}} = \frac{\dot{m}_{gw}(h_2 - h_1)}{\dot{W}_{in,P}} = \frac{32.84x(361.1 - 344)}{651.7} = 0.85 = 85.0\%$$

and

$$\psi_P = \frac{(\dot{m}_2 ex_2 - \dot{m}_1 ex_1)}{\dot{W}_{in,P}} = \frac{\dot{m}_{gw}(ex_2 - ex_1)}{\dot{W}_{in,P}} = \frac{32.84x(24.22 - 21.05)}{651.7} = 0.1575$$

$$= 15.75\%$$

3. The effect of the geothermal water mass flow rate on the power consumption rate of the pump is shown in Fig. 1.13. As shown in the figure, the power consumption rate of the pump is increased from 502.9 to 1006 kW with the increasing geothermal water mass flow rate from 25 to 50 kg/s.

Example 1.3: An adiabatic steam turbine, as illustrated in Fig. 1.14, receives 4.8 kg/s of steam at a temperature of 152°C and a pressure of 500 kPa. This adiabatic steam turbine converts part of the energy in the steam to shaft work. The working fluid leaves from the turbine at a pressure of 50 kPa with 90% quality of steam.

1. Calculate the work production rate of the steam turbine.
2. Calculate the exergy destruction rate of the steam turbine.
3. Calculate the energy and exergy efficiencies of the steam turbine.

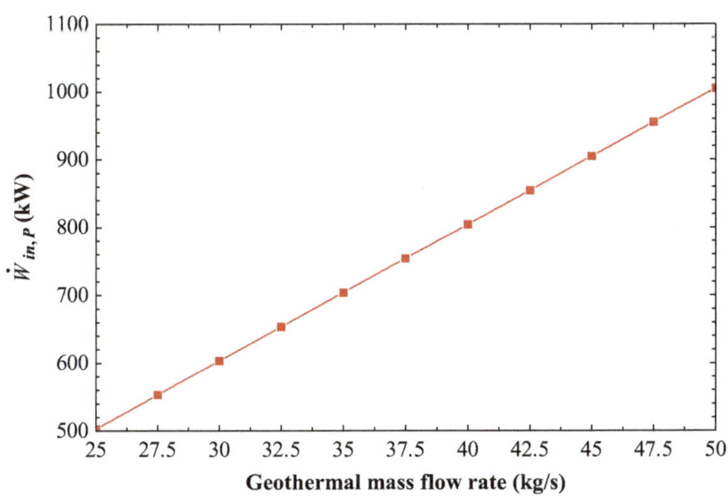

FIGURE 1.13 The effect of geothermal water mass flow rate on the power consumption rate of the pump.

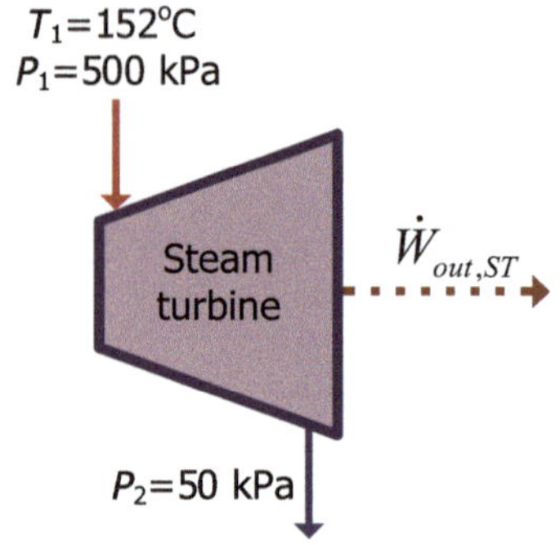

FIGURE 1.14 Schematic diagram of Example 1.3 turbine.

4. Investigate the effect of increasing the mass flow rate of steam from 2 to 10 kg/s on the power production rate and exergy destruction rate of the steam turbine.

Solution: The work production process is steady, and the work production rate for the adiabatic steam turbine is required.

Assumptions:

- The reference temperature and pressure are chosen as 101.3 kPa and 25°C, respectively.
- The real fluid properties of steam are taken from EES.
- The steam turbine is adiabatic.

Analysis:

The mass, energy, entropy, and ExBEs for the steam turbine are defined:

Mass balance: $\dot{m}_1 = \dot{m}_2$

Energy balance: $\dot{m}_1 h_1 = \dot{m}_2 h_2 + \dot{W}_{out,ST}$

Entropy balance: $\dot{m}_1 s_1 + \dot{S}_{g,ST} = \dot{m}_2 s_2$

Exergy balance: $\dot{m}_1 ex_1 = \dot{m}_2 ex_2 + \dot{W}_{out,ST} + \dot{E}x_{D,ST}$

1. The work production rate of the adiabatic steam turbine needs to be calculated in this section.

Use the energetic balance equation to calculate the work generation rate of the steam turbine:

$$\dot{W}_{out,ST} = \dot{m}_1 h_1 - \dot{m}_2 h_2$$

where the properties of the steam entering and exiting the steam turbine are computed by using the EES software program, and the state properties of the steam are given in Table 1.4. Here x shows the quality of steam.

TABLE 1.4 Properties of the input and the output flows of steam for the turbine component.

State point	$\dot{m}$ (kg/s)	T (°C)	P (kPa)	h (kJ/kg)	s (kJ/kg K)	x	ex (kJ/kg)	$\dot{E}x$ (kW)
0	—	25	101.3	104.8	0.3669	—	—	—
1	4.8	152	500	2749	6.822	—	719.5	3453
2	4.8	81.34	50	2415	6.943	0.90	349.4	1677

Based on the MBE for the steam turbine ($\dot{m}_1 = \dot{m}_2 = \dot{m}_s$), the power generation rate equation of this turbine is computed:

$$\dot{W}_{out,ST} = \dot{m}_s(h_1 - h_2)$$

$$\dot{W}_{out,ST} = 4.8x(2749 - 2415)$$

$$\dot{W}_{out,ST} = 1604 \text{ kW}$$

2. Utilize the ExBE to calculate the exergy destruction rate of the steam turbine:

$$\dot{E}x_{D,ST} = \dot{m}_1 ex_1 - \dot{m}_2 ex_2 - \dot{W}_{out,ST}$$

$$\dot{E}x_{D,ST} = 4.8x(719.5 - 349.4) - 1604$$

$$\dot{E}x_{D,ST} = 172.5 \text{ kW}$$

3. The energy and exergy efficiencies of the steam turbine are computed:

$$\eta_{ST} = \frac{\dot{W}_{out,ST}}{\dot{E}n_{in,ST}} = 0.8824 = 88.24\%$$

and

$$\psi_{ST} = \frac{\dot{W}_{out,ST}}{\dot{E}x_{in,ST}} = 0.7967 = 79.67\%$$

4. The effect of steam mass flow rate on the power production rate and exergy destruction rate of the steam turbine is illustrated in Fig. 1.15. As given in this figure, the power production rate of the steam turbine is increased from 668.2 to 3341 kW, and the exergy destruction rate of the steam turbine is increased from 71.88 to 359.4 kW with increasing the steam mass flow rate from 2 to 10 kg/s.

Example 1.4: The geothermal water leaving a production process at 146°C and 162 kPa at a rate of 43 kg/s is to be used as hot water at 87°C and

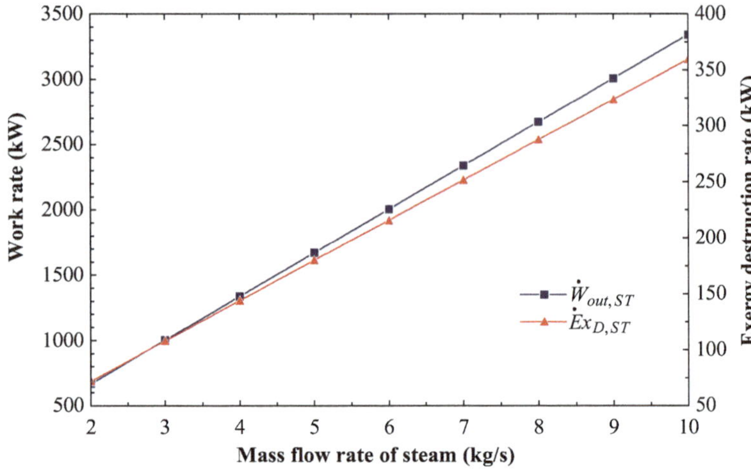

FIGURE 1.15 Effect of steam mass flow rate on the power production rate and exergy destruction rate of the steam turbine.

ambient pressure in the heat exchanger (HEX), as illustrated in Fig. 1.16. The heat loss during the heat transfer process in the HEX is assumed to equal an amount of 15% of the total thermal energy input. Water at an ambient temperature and pressure of 25°C and 101.3 kPa enters the HEX, and the geothermal water leaves the HEX at 98°C and 162 kPa.

1. Calculate the hot water production rate.
2. Calculate the exergy destruction rate of the HEX.
3. Calculate the energy and exergy efficiencies of the HEX.
4. Investigate the effect of increasing the mass flow rate of steam from 2 to 10 kg/s on the power production and exergy destruction rate of the steam turbine.

Solution: The heat transfer process is steady, and the energy and ExBEs are required.

Assumptions:

- The reference temperature and pressure are chosen as 101.3 kPa and 25°C, respectively.
- The working fluid properties of steam and water are taken from EES.

Analysis: The mass, energy, entropy and ExBEs for the HEX are written:

$$\text{Mass balance: } \dot{m}_1 = \dot{m}_2 \text{ and } \dot{m}_3 = \dot{m}_4$$

$$\text{Energy balance: } \dot{m}_1 h_1 + \dot{m}_3 h_3 - \dot{Q}_{L,HEX} = \dot{m}_2 h_2 + \dot{m}_4 h_4$$

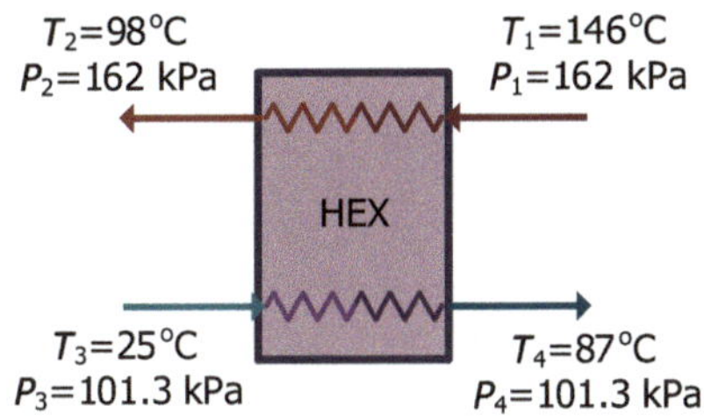

FIGURE 1.16 Schematic diagram of Example 1.4 heat exchanger.

TABLE 1.5 Properties of the input and the output flows of the HEX working fluids.

State point	$\dot{m}$ (kg/s)	T (°C)	P (kPa)	h (kJ/kg)	s (kJ/kg K)	ex (kJ/kg)	$\dot{E}x$ (kW)
0	—	25	101.3	104.9	0.3672	—	—
1	43	146	162	2764	7.364	572.8	24628
2	43	98	162	410.8	1.285	32.37	1392
3	331.4	25	101.3	104.9	0.3672	—	—
4	331.4	87	101.3	364.5	1.158	23.76	7875

Entropy balance: $\dot{m}_1 s_1 + \dot{m}_3 s_3 - \dot{Q}_{L,HEX}/T_{HEX} + \dot{S}_{g,HEX} = \dot{m}_2 h_2 + \dot{m}_4 h_4$
Exergy balance: $\dot{m}_1 ex_1 + \dot{m}_3 ex_3 - \dot{E}x^{Q}_{L,HEX} = \dot{m}_2 ex_2 + \dot{m}_4 ex_4 + \dot{E}x_{D,HEX}$

1. Firstly, the hot water production rate from the HEX component needs to be calculated in this section.

Use the energetic balance equation to compute the hot water production rate of the HEX subcomponent:

$$\dot{m}_1 h_1 + \dot{m}_3 h_3 - \dot{Q}_{L,HEX} = \dot{m}_2 h_2 + \dot{m}_4 h_4$$

where the properties of the working fluid entering and exiting the HEX are calculated by using the EES software program, and the state properties of the working water are in Table 1.5.

Based on the MBE for the HEX ($\dot{m}_1 = \dot{m}_2 = \dot{m}_{gw}$), ($\dot{m}_3 = \dot{m}_4 = \dot{m}_w$), and $\dot{Q}_{L,HEX} = 0.15x(\dot{m}_1 h_1 - \dot{m}_2 h_2)$, the power consumed rate equation of this component is:

$$\dot{m}_{gw} h_1 + \dot{m}_w h_3 - 0.15\dot{m}_{gw}(h_1 - h_2) = \dot{m}_{gw} h_2 + \dot{m}_w h_4$$

$$\dot{m}_w = 331.4\ \mathrm{kg}/s$$

2. Use the ExBE to compute the exergy destruction rate of the HEX:

$$\dot{m}_1 ex_1 + \dot{m}_3 ex_3 - \dot{E}x^Q_{L,HEX} = \dot{m}_2 ex_2 + \dot{m}_4 ex_4 + \dot{E}x_{D,HEX}$$

$$\dot{E}x_{D,HEX} = \dot{m}_{gw}(ex_1 - ex_2) + \dot{m}_w(ex_3 - ex_4) - \dot{E}x^Q_{L,HEX}$$

Here:

$$\dot{E}x^Q_{L,HEX} = \dot{Q}_{L,HEX}\left(1 - \frac{T_o}{T_{HEX}}\right)$$

Based on these equations:

$$\dot{E}x_{D,HEX} = 11640 \text{ kW}$$

3. The energy and exergy efficiencies of the HEX are calculated:

$$\eta_{HEX} = \frac{(\dot{m}_4 h_4 - \dot{m}_3 h_3)}{(\dot{m}_1 h_1 - \dot{m}_2 h_2)} = \frac{\dot{m}_w (h_4 - h_3)}{\dot{m}_{gw}(h_1 - h_2)} = \frac{331.4x(364.5 - 104.9)}{43x(2764 - 410.8)} = 0.8502 = 85.02\%$$

and

$$\psi_{HEX} = \frac{(\dot{m}_4 ex_4 - \dot{m}_3 ex_3)}{(\dot{m}_1 ex_1 - \dot{m}_2 ex_2)} = \frac{\dot{m}_w (ex_4 - ex_3)}{\dot{m}_{gw}(ex_1 - ex_2)} = \frac{331.4x(23.76 - 0)}{43x(572.8 - 32.37)} = 0.338 = 33.89\%$$

4. The effect of the geothermal water mass flow rate on the exergy destruction rate and exergy efficiency of the HEX is illustrated in Fig. 1.17. As given in the figure, the exergy destruction rate of the HEX is increased

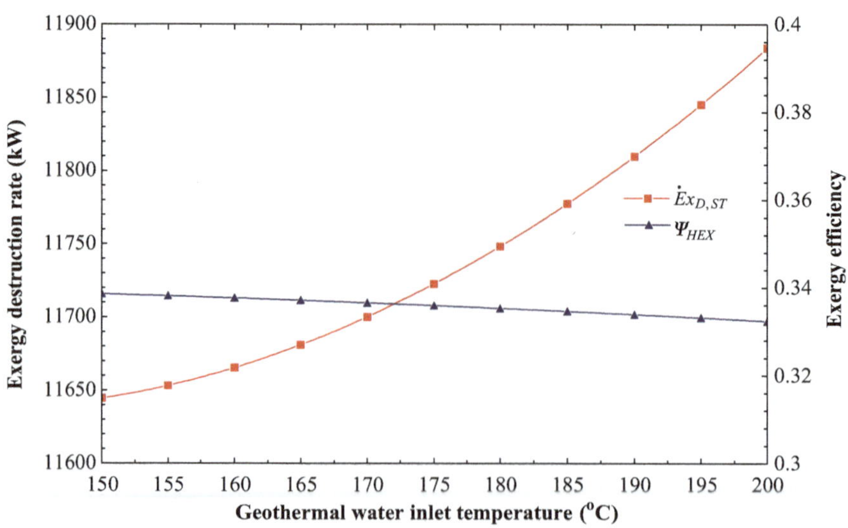

FIGURE 1.17 Effect of geothermal water mass flow rate on the exergy destruction rate and exergy efficiency of the HEX component. *HEX*, Heat exchanger.

from 11,645 to 11,884 kW, and the exergy efficiency of the HEX is decreased from 33.86% to 33.24% with increasing the geothermal working fluid temperature from 150°C to 200°C.

1.4 Closing remarks

In the design, analysis, assessment, and improvement of geothermal energy systems, it is essential to use thermodynamic tools correctly and consistently, and such thermodynamic tools are the first and second laws of thermodynamics. In dealing with these laws, we need to write the balance equations for mass, energy, entropy, and exergy correctly and accordingly for both types of thermodynamic systems, namely closed and open types. This chapter has aimed to provide the fundamental background information, basic definitions, key concepts, and true statements to better guide students, researchers, scientists, and engineers for better system design, analysis, and assessment. True efficiency assessment and evaluation are achieved by using both energy and exergy efficiencies. Also, some illustrative examples were presented to better illustrate how to analyze systems for practical applications.

Nomenclature

E	Energy (kJ)
e	Specific energy (kJ/kg)
$\dot{E}$	Energy rate (kW)
ex	Specific exergy (kJ/kg)
$\dot{E}x$	Exergy rate (kW)
$\dot{E}x_d$	Exergy destruction rate (kW)
$\dot{E}x^Q$	Exergy transfer rate associated with heat transfer (kW)
g	Gravitational acceleration (m^2/s)
G	Gibbs free energy (kJ)
h	Specific enthalpy (kJ/kg)
H	Enthalpy (kJ)
m	Mass (kg)
$\dot{m}$	Mass flow rate (kg/s)
P	Pressure (kPa)
q	Specific heat transfer (kJ/kg)
Q	Heat (kJ)
$\dot{q}$	Specific heat transfer rate (kW/kg)
$\dot{Q}$	Heat rate (kW)
s	Specific entropy (kJ/kg K)
S	Entropy (kJ/K)
$\dot{S}$	Entropy rate (kW/K)
t	Time (s)
T	Temperature (°C, K)
u	Internal energy (kJ/kg)
v	Velocity (m/s)

W	Work (kJ)
$\dot{w}$	Specific work rate (kW/kg)
$\dot{W}$	Work rate (kW)
z	Elevation (m)

Greek letters

Δ	Change in variable
θ	Total specific energy of flowing materials (kJ/kg)
η	Energy efficiency
ψ	Exergy efficiency

Subscript

a	Air
AC	Air compressor
ch	Chemical
Cmp	Compressor
Con	Condenser
D	Destruction
dh	Dehumidifier
df	Diffuser
e	Exit condition
gen	Generation
kn	Kinetic
i	Inlet condition
nz	Nozzle
ofw	Open feedwater heater
P	Pump
ph	Physical
pt	Potential
RO	Reserve osmosis
Sp	Separator
ST	Steam turbine
sys	System
tot	Total
T	Turbine
Val	Valve
wf	Working fluid
$3WV$	3-way valve
1...74	State numbers
0	Ambient or reference condition

Superscripts

.	Rate
Ch	Chemical

Acronyms

COP	Coefficient of performance
CV	Control volume
EBS	Energy balance equation
EnBS	Entropy balance equation
ExBS	Exergy balance equation
EES	Engineering equation solver
HEX	Heat exchanger
MBE	Mass balance equation
OS	Open system
SB	System boundary

References

[1] I. Dincer, Thermodynamics: A Smart Approach, Wiley, London, 2020.
[2] Y.A. Cengel, M.A. Boles, Thermodynamics: An Engineering Approach, ninth ed., McGraw Hill, NY, 2019.
[3] I. Dincer, M.A. Rosen, Exergy: Energy, Environment and Sustainable Development, second ed., Elsevier, Oxford, 2013.
[4] I. Dincer, C. Zamfirescu, Sustainable Hydrogen Production, Elsevier, NY, 2016.
[5] I. Dincer, C. Zamfirescu, Sustainable Energy Systems and Applications, Springer, NY, 2011.
[6] I. Dincer, C. Zamfirescu, Advanced Power Generation Systems, Elsevier, NY, 2014.
[7] I. Dincer, Refrigeration Systems and Applications, third ed., Wiley, London, 2017.
[8] I. Dincer, M.A. Rosen, Thermal Energy Storage: Systems and Applications, second ed., Wiley, London, 2011.
[9] I. Dincer, M.A. Rosen, Exergy Analysis of Heating, Refrigerating, and Air: Conditioning Methods and Applications, Elsevier, NY, 2015.

Study questions and problems

1.1. Is it possible that a CS exchanges mass with the surroundings?
1.2. Please explain the difference between the closed thermodynamic system and open thermodynamic system.
1.3. Please explain the difference between the adiabatic thermodynamic system and isolated thermodynamic system.
1.4. Can the entropy of a system decrease during a process? If so, does this violate the increase of entropy principle?
1.5. Explain the difference between internal energy and enthalpy.
1.6. What is the relationship between the entropy generation and irreversibility?
1.7. Which is a more effective way of increasing the performance of a plant: decreasing entropy generation or decreasing exergy destruction due to irreversibilities?
1.8. Does an exergy analysis replace an energy analysis? Describe any advantages of exergy analysis over energy analysis.

1.9. Can you perform an exergy analysis without an energy analysis? Explain.

1.10. What is the relationship between entropy change of a plant, entropy change of its surroundings, and generated entropy?

1.11. Does the reference environment have any effect on the results of an exergy analysis?

1.12. Do you recommend using standard atmospheric conditions (101.3 kPa, 25°C) or actual atmospheric conditions as a reference environment in an exergy analysis? Explain.

1.13. Please describe energy and exergy efficiencies. Define both for an adiabatic steam turbine.

1.14. The most common application of geothermal energy is power generation. Some other uses are process heating, district heating, district cooling, greenhouse heating, and heating for fish farming. From a thermodynamic point of view and considering the quality of energy, explain which of these uses you recommend most.

1.15. Noting that heat transfer does not occur without a temperature difference and that heat transfer across a finite temperature difference is irreversible, is there such a thing as reversible heat transfer? Explain.

1.16. What is the effect of ambient air temperature on the exergetic performance of an adiabatic steam turbine?

1.17. Write the mass, energy, entropy and exergy balance equations for the following devices:

a. an adiabatic steam turbine,

b. an air compressor with heat loss from the air to the surroundings,

c. a pump,

d. a heat exchanger with heat loss to the surroundings.

1.18. Using thermodynamic tables or EES software, calculate the thermomechanical exergy of saturated steam at 10,000 kPa. Assume standard values for temperature and pressure of the environment.

1.19. The intake of an air compressor receivers air at 128 kPa and 36°C and compresses air to 826 kPa and 207°C. The rate of airflow is 2.3 kg/s. The heat loss during this process is calculated as 48 kJ/kg. Calculate the power required to work the compressor, exergy destruction rate, energy and exergy efficiencies of the compressor, assuming that the net change in kinetic and potential energies and exergies are zero.

1.20. Steam enters a steam turbine at 3000 kPa and 500°C and leaves at 18 kPa with 90.2% quality of steam. This steam turbine produces 6000 kW.

a. Calculate the change in enthalpy.

b. Calculate the work done per kilogram of steam.

c. Determine the mass flow rate of steam.

d. Calculate the exergy destruction rate of the steam turbine.

e. Calculate the energy and exergy efficiencies of the steam turbine.

Chapter 2

Energy, environment, and sustainable development

2.1 Introduction

Energy, environment, and sustainable development concepts are applicable to all areas of science and engineering. This chapter primarily provides the necessary background for understanding these concepts, as well as basic principles, general definitions, practical applications, and implications. The scope of this chapter is partly illustrated in Fig. 2.1, where the domains of energy, environment, and sustainable development are given for meeting the needs of the world. The relationship between sustainable development and the utilization of sources, particularly energy sources, is of great significance to countries. Attaining sustainable development requires that sustainable energy sources are utilized effectively and efficiently without causing any negative environmental and societal problems. Thermodynamic modeling

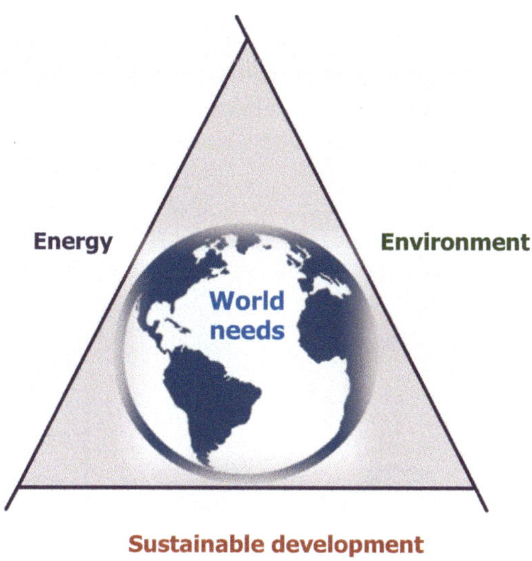

FIGURE 2.1 Interactions among the domains of energy, environment, and sustainable development for a cleaner environment.

Geothermal Energy Systems. DOI: https://doi.org/10.1016/B978-0-12-820775-8.00005-2

studies are important since they are useful for improving the performance of energy plants. The relationships between the world's needs and both energy and the environment make it clear that clean energy production options are directly related to sustainable development.

Energy as a basic physics term is defined as the ability to do work. Besides this physics definition, energy connotes many things, such as an indicator for development, responsibility for global warming, the supply source for transportation and heating, etc. People have learned how to make fire and use it then this small incident caused to begin the civilization [1]. At first, human beings burned wood to cook or heat spaces. Burning wood is a relatively cleaner method than burning fossil fuels because of its CO_2 content, which plants use for photosynthesis. At some point, wood started to be insufficient for human needs, and we started using coal and oil, which were formed over millions of years. Then people learned to benefit from fossil fuels in many ways.

Energy, as mentioned previously, is defined in physics as the ability to do work. On the other hand, energy is not an issue limited to physics; it is critical in every aspect of life. Energy is so important that many countries form their environments according to their energy resources and energy uses. To prevent any misconceptions, it is highly crucial to define a few terms related to energy, such as energy form, energy sources, and energy carriers, before pointing out any energy issue [2]. It is important to come up with a common triangle to illustrate the close relationships of the three aspects of generational drivers—energy, environment, and sustainable development.

Energy forms, sources, and carriers are illustrated in Fig. 2.2. As shown in the figure, electricity and hydrogen can be generated from available energy resources and utilized in applications where fossil fuels are being utilized today. In such a plant, electricity and hydrogen can be generated in large industrial systems as well as in small, decentralized processes, wherever the primary energy resource (solar, nuclear, and even fossil) is available. Electricity can be

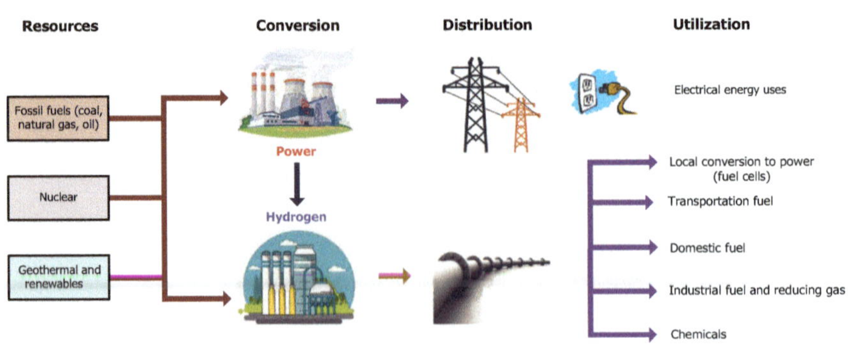

FIGURE 2.2 Energy forms, sources, and carriers. *Modified from Ref. [3].*

utilized directly or transformed into hydrogen. For large-scale storage applications, hydrogen can be stored underground in ex-mines, caverns, and/or aquifers. Power transport to end users, depending on distance and overall economics, takes place either in the form of electricity or in the form of hydrogen. Hydrogen is transported by means of pipelines or supertankers and is then utilized in the transportation, industrial, residential, and commercial sectors as a fuel [4]. Some of the produced and stored hydrogen may be utilized to produce power (via fuel cells), depending on demand, geographical location, or time of day. Nowadays, fuel cell plants are available in megawatt power plant size or as individual devices (several kilowatts) suitable for distributed electricity production.

- *Energy forms*: Energy can be found in various forms such as fossil fuels or their products (gasoline, diesel, etc.), electricity, work, heat, heated substances, light, etc.
- *Energy sources*: This term states where energy can be obtained. These sources can be classified into two groups: having finite quantities and renewables.
- *Energy carriers*: Energy carriers mean the energy forms used both as resources and as processed energy forms (gasoline, heat, hydrogen, etc.). Processed energy forms are the forms created by treating the sources in some processes.

At this point, the difference between energy sources and energy carriers should not be neglected. Many types of energy carriers may be converted from one to another. However, an energy source is the original resource from which energy carriers can be obtained. Indeed, some energy sources are also energy carriers. For example, hydrogen seems to be a future energy source for many people. However, this is a misconception. Hydrogen is an energy carrier but not an energy source. The role of hydrogen is enormous because it is an ultimately clean energy carrier, fuel, and main element for other chemicals and fuels.

Because the Earth is fed by solar radiation, more than millions of years ago, plants on the planet were converting sunlight to chemical energy by photosynthesis. Some of the produced chemical energy was consumed by plants or animals, and some of that energy was hiding under the soil and starting to decay to yield coal, oil, and natural gas. About a hundred years ago, people started using that chemical energy in different machines to produce heat or mechanical energy or to manufacture plastic materials. The quantity of coal, oil, and natural gas (fossil fuels) that had formed over millions of years started to decrease in the last few decades. The shares of energy sources of the world between 1990 and 2018 are shown in Fig. 2.3. Due to their decreasing quantities, fossil fuels are called an "energy resource," but renewables are also called "energy sources" because they cannot be depleted [6]. Coal, natural gas, and oil usage caused much

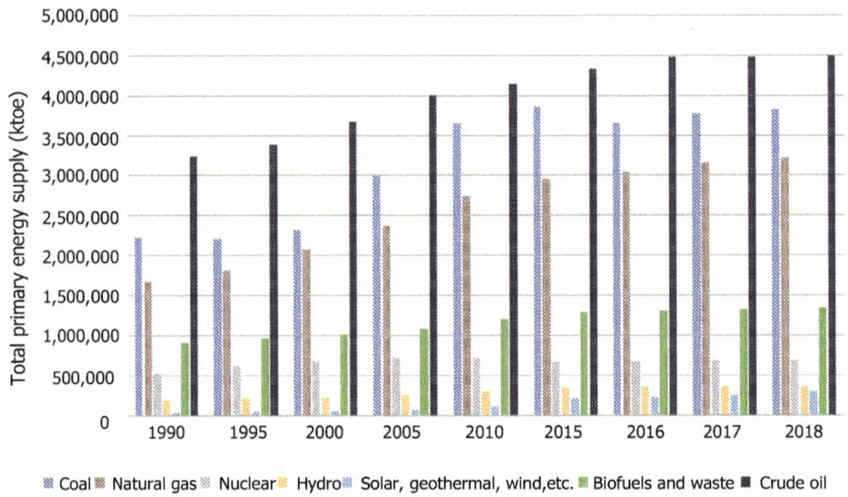

FIGURE 2.3 Total primary energy supply by source for the world between 1990 and 2017. *Data from IEA [5].*

environmental damage to the environment, such as atmospheric pollution, impacts on global warming, and impacts on water quality where mining is performed. In the same context, atmospheric pollution affects human health, crops, forests, freshwater fisheries, and unmanaged ecosystems.

Even though the Earth has massive renewable energy sources, world energy demand is met mostly by fossil fuels. After the Industrial Revolution, coal, oil, and gas have been used in many industrial processes, producing harmful gases and causing acid rains, global warming, global climate change, etc. Since the Industrial Revolution, the scenario has only gotten worse. When the reasons for global climate change and global warming are analyzed, human activities have been identified as having the main responsibility.

The environment generally covers all living and nonliving things that occur naturally on Earth. The "environment," as a term, can also be used in some disciplines such as thermodynamics to indicate "surrounding." "Environmental impact," as a term, is used for any effect that any process has on the natural environment. Environmental impacts can be observed at a certain time or in a period, such as over the life cycle of a system, product, or process [7].

With improving technologies that we use in daily life, the habits of people change. We use more technological devices, and we have many electrical devices; therefore we need more energy than we did the day before. To supply this demand, we consume energy sources and harm the environment [8]. Increasing the amount of CO_2 and other greenhouse gases will continue to cause global warming, the melting of ice caps, and unpredictable environmental hazards. Each year, the number of deaths caused by air pollution rises. Hence, precautions must be taken to prevent this scenario and to save the environment for future generations.

Oil is still the most used fuel in 2018, when global primary energy consumption by fuel is analyzed [9]. Coal comes in second with a share of 27%, which is the lowest level for the last 15 years. The share of natural gas reached 24%; as a consequence, the share of fossil fuels reaches about 85%. The share of renewables except hydroelectricity increased by 4% in 2018. A large amount of fossil fuel share of total energy consumption makes it difficult to decrease greenhouse gases and thus to save the environment. Every thinking human being will agree that the share of renewables must be increased, but this is a long-term precaution. Before renewables' share increases, other short-term, practical energy efficiency measures can be taken:

- More energy saving practices should be implemented.
- More efficient devices should be utilized.
- Comfort conditions should be adjusted to be more effective and energy saving modes without compromising comfort and quality.
- Public transportation, cycling, or walking should be encouraged.
- More efficient systems and applications should be designed and operated.
- Fossil fuel–based plants should be utilized in more effective and environmentally benign systems.
- Exergy-based system design, analysis, and assessments are made daily practices.

All these measures are necessary for all economic sectors to achieve energy conservation. In energy-generation plants, single-generation options should be given up, and at least combined heat and power systems should be replaced. Environmentally benign electricity options, such as electricity by renewables, can be used more. To support these changes, some changes in politics about renewable energy usage and its economies need to be made. So policy makers have also the responsibility in this transition time from fossils to renewables. Some energy sectors are facing difficulties in meeting energy demands, and resultantly, greenhouse gases continue to be emitted, and global climate change can be expected. At the same time, energy prices should be at an affordable level. Renewable energy systems offer a clean and continuous energy supply; however, at this time they cannot challenge fossil fuels in terms of cost. With improving technologies, the efficiency of renewable energy systems will be higher, and consequently the price of such energy will be decreased. However, to completely rely on some renewables has disadvantages because of the intermittency of some renewable energy sources such as solar and wind. To cope with intermittency problems and the other disadvantages of renewables, integrated multigeneration energy systems can be used. Integrated energy systems offer multiple products like power, hydrogen, chemicals, heating and cooling, hot water, and freshwater by using renewable energy sources and even fossil fuels. For example, to deal with the intermittency of solar energy, the system is integrated with another source such as geothermal energy, biomass, or a coal gasification unit according to the regional energy source. Such options also offer higher efficiency than other energy-generation plants (i.e., single-generation, cogeneration, or trigeneration systems) due to the utilization of waste heat again and again. In these types of integrated multigeneration systems, even coals can be used in more environmentally ways by gasification and keeping hazardous wastes away from the environment.

2.2 The relation of energy and population

The urban population compared to the rural population is increasing every passing day. In 2017, about 4.1 billion people were living in urban areas, an amount equaling over half of the world population, nearly 55% [10]. Although many developed country populations suggest a trend of moving to rural areas instead of urban areas, people tend to shift to urban areas. Thus the population in urban areas is increasing with an annual growth rate of 1.8% [11]. This increase in urban population causes, the growing population, and complexity in cities. Interestingly, even though cities cover only 2% of the land surfaces of the world, 75% of resource consumption by humans is in cities [12]. Also, it can be said that the ecological footprints of cities are larger than that of the lands they occupy. The world population rises, especially in cities. This incremental rate requires supplying more energy even if relatively cheap oil and coal reserves are being depleted. The energy needs of people go up due not only to population growth but also to increasing living standards.

2.3 The relation of energy and environment

An increase in population and in living standards has caused increased energy consumption and demand, and more and more energy resources are being used. As a result, oil and coal reserves have started to be depleted, greenhouse gases are increasingly emitted into the atmosphere, and the global temperature of the Earth is increasing. During the past three decades, the risk of environmental problems has become more apparent, and people are more aware than ever. Human activities, especially energy-generation plants, have affected the environment, and the resultant environmental problems have been clearly observed.

Increasing industrial developments have caused greater use of road transport, and people have preferred to use their private cars for daily transportation. Therefore emissions such as NO_x and volatile organic compounds (VOCs) caused by vehicles have risen. The main consequent environmental problems are major environmental accidents, water pollution, maritime pollution, land use and siting impact, radiation and radioactivity, solid waste disposal, hazardous air pollutants, ambient air quality, acid rain, stratospheric ozone depletion, and global warming [13].

2.4 Sustainable development

Today, energy is used in many areas and is one of the indispensable requirements of humans. We use energy for manufacturing, transportation, space heating and cooling, lighting, and running electrical devices. The critical question, "How do we meet our energy needs?" *Sustainable development* was defined in 1987 as "development that meets the needs of the present without compromising the ability of future generations to meet their own needs" [14]. In other words, the present generation should not threaten the next generations' rights by meeting their own needs. If we adopt this concept to energy-related issues, then it can be said that fossil fuels are not sustainable due to their depletion, that renewable energy sources can be called sustainable, and that nuclear is more sustainable than fossil. Many factors contribute to achieving sustainable development. According to Dincer and Rosen [15], for development to be sustainable the desires and requirements of society must be satisfied, development must be environmentally and ecologically benign, and resources must be adequate.

The term "sustainable development" was used in 1980 and then popularized in the Brundtland Commission of 1987 [14]. Five years later, at the UN Conference on Environment and Development in Rio de Janeiro, sustainable development took on a global mission. The Brundtland Commission report was not only about the sustainability concept but also about environmental, social, and economic problems. Another definition of sustainable development is from the *Encyclopedia of Life Support Systems* [16]: "the wise use of resources through critical attention to policy, social, economic, technological, and ecological management of natural and human-engineered capital so as to promote innovations that assure a higher degree of human needs fulfillment, or life support, across all regions of the world, while at the same time ensuring intergenerational equity."

2.4.1 Background and goals of sustainable development

As mentioned, sustainable development not only means sustainable energy but also covers many other subjects such as no poverty, quality education, reduced inequality, etc., as seen in Fig. 2.4 and outlined by the United Nations and their global activities for better life and future. Clean water and sanitation, affordable and clean energy, sustainable cities and commodities, responsible consumption and production, and climate action are directly related goals to sustainable energy infrastructure.

If a society targets development at the same time aiming to preserve the environment, then energy, the environment, and sustainable development

FIGURE 2.4 An illustration of sustainable development goals of the United Nations. *Modified from [17].*

issues should be underlined. A society trying to reach sustainable development goals should change its energy habits to a renewable energy system that does not emit harmful emissions to the environment. Because all energy sources have some negative impacts on the environment, green and smart energy infrastructure can be set to overcome those problems.

2.4.2 Sustainable development indicators

Measuring sustainability is a crucial subject in achieving sustainable development. Therefore different tools are necessary to measure the sustainability index. The development and selection of sustainable development indicators call for parameters for reliability, appropriateness, practicality, and limitations. Fig. 2.5 shows sustainable development indicators under four subheadings: (1) society, (2) environment, (3) energy, and (4) economy, which are recognized as the four main domains in achieving sustainable development locally and globally.

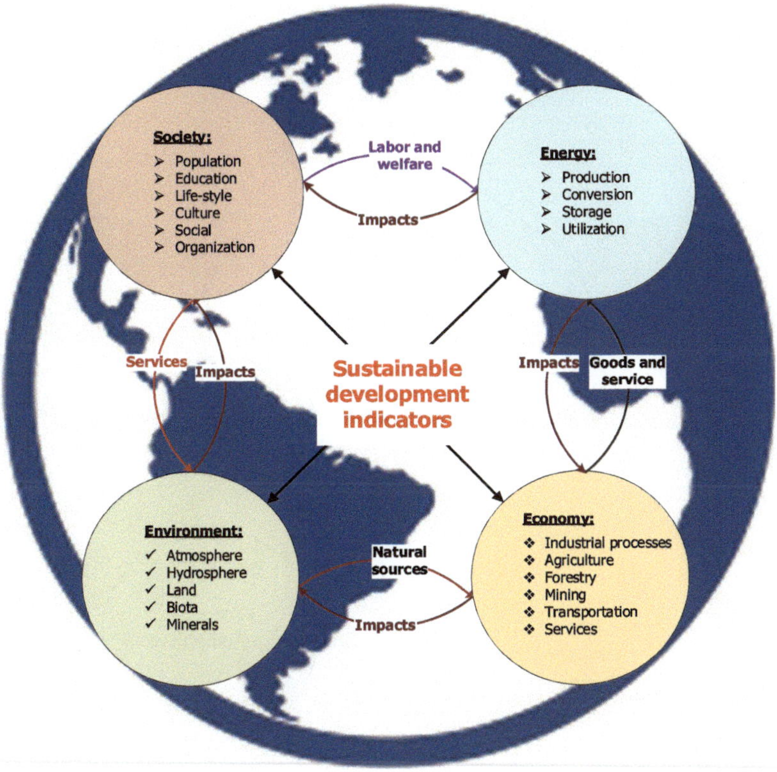

FIGURE 2.5 Illustration of some sustainable development indicators. *Modified from [18].*

For instance, the population, lifestyle, or culture of human beings affects surroundings such as the atmosphere, hydrosphere, and land, and this also affects sustainable development goals in terms of how these services are achieved and which natural sources are used. To achieve sustainable development, one can achieve five factors of sustainability: (1) social sustainability, (2) environmental sustainability, (3) economic sustainability, (4) resource sustainability, and (5) energy sustainability. These five factors are presented as the factors impacting sustainable development in Fig. 2.6.

2.4.3 Sustainable energy

Achieving sustainable energy is a relatively difficult task due to the increasing energy consumption. Sustainable energy aims to supply our energy demands without threatening the energy demands of the next generation and without damaging the environment. Fig. 2.7 demonstrates the renewable

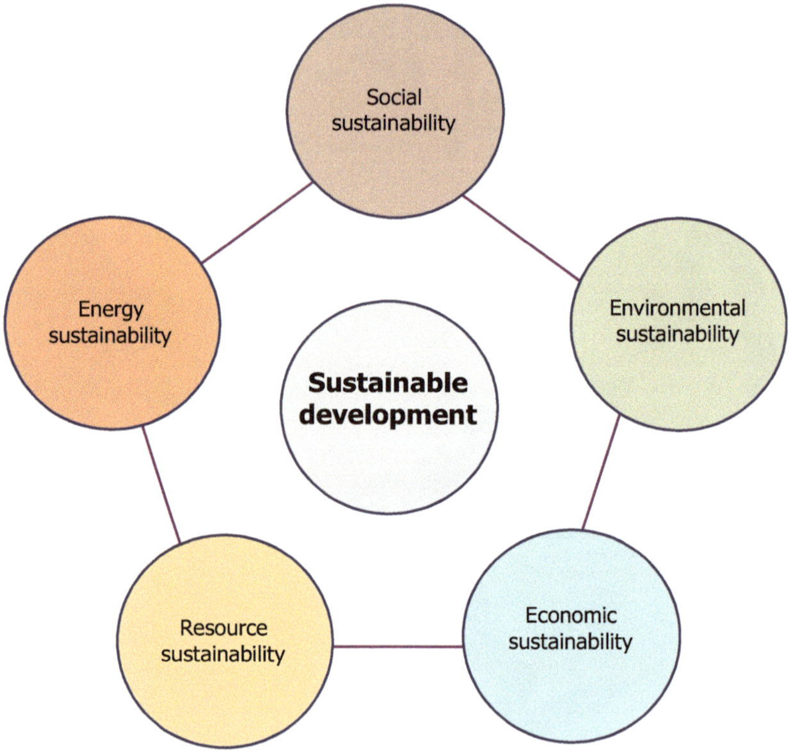

FIGURE 2.6 Factors impacting sustainable development. *Modified from [19].*

energy sources that are infinite and nonrenewable energy sources that are in limited quantities. Fossil fuels, biomass, nuclear, and wastes are nonrenewable energy sources, and, by the way, nuclear fuel lifetime depends mainly on advanced technologies [20,21].

Renewable energies are solar radiation coming to the Earth's surface, energy forms created by solar radiation, and energy caused by the core temperature, rotation, and gravitation of the Earth. Solar radiation, which reaches the Earth's surface with energy about 20,000 times greater than global energy demand, can be used as heat or electric energy by using photovoltaic panels. As seen from Fig. 2.7, hydraulic, wave, wind, ocean, and biomass energy are derived from solar energy. Hydraulic energy can be used by converting the mechanical energy of water to electricity or directly to mechanical energy. Ocean energy is based on the temperature difference between the surface and the deep. Geothermal energy can also be used as heat energy or as electricity by employing some technologies. Tidal energy is caused by the gravitational forces of the sun and moon and the rotational movement of the Earth. Biomass energy, including wood, plants, and organic wastes,

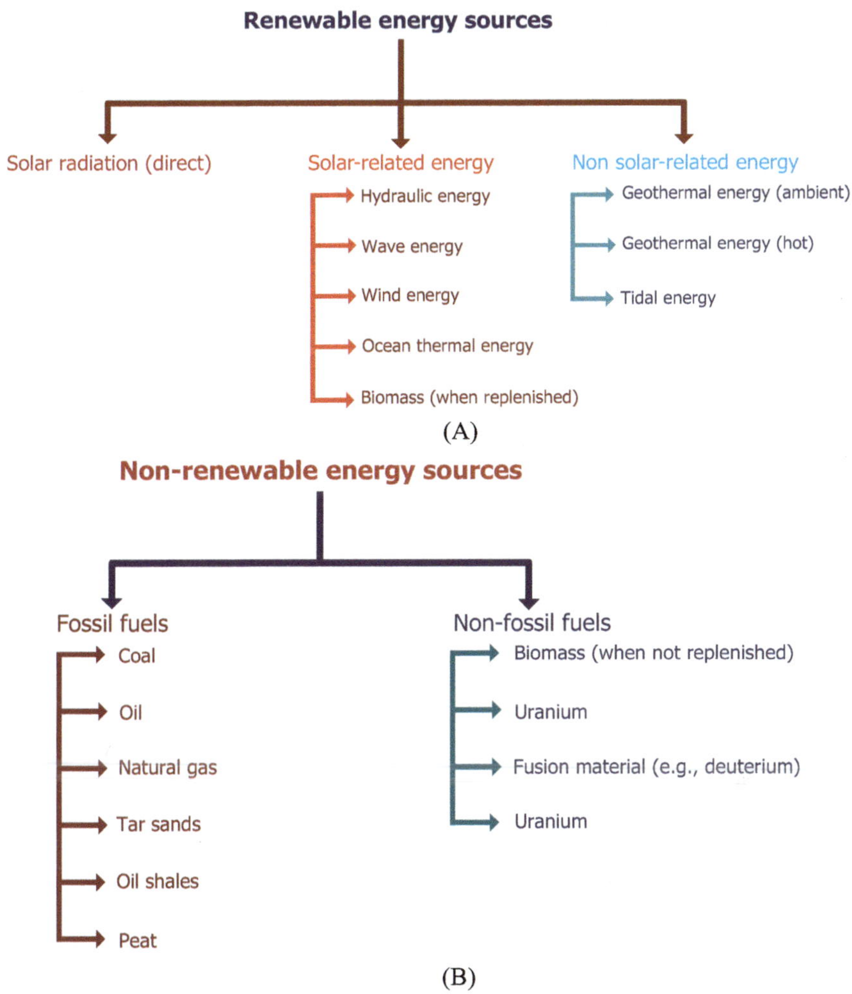

FIGURE 2.7 Energy sources: (A) renewable and (B) nonrenewable.

can be burned or converted to other energy forms such as electricity or chemicals [22].

As mentioned, energy terminology falls into three types: (1) energy forms, (2) energy sources, and (3) energy carriers. To reach sustainable development goals, after selecting and using a sustainable energy source, sustainable energy carriers should be selected. Energy carriers can be grouped into two categories: material energy carriers and nonmaterial energy carriers, as seen in Fig. 2.8. Material energy carriers can be gasoline, diesel fuel, naphtha, coal products, or other chemical fuels like hydrogen, methanol, or ammonia. Thermal energy can be transported to district energy systems

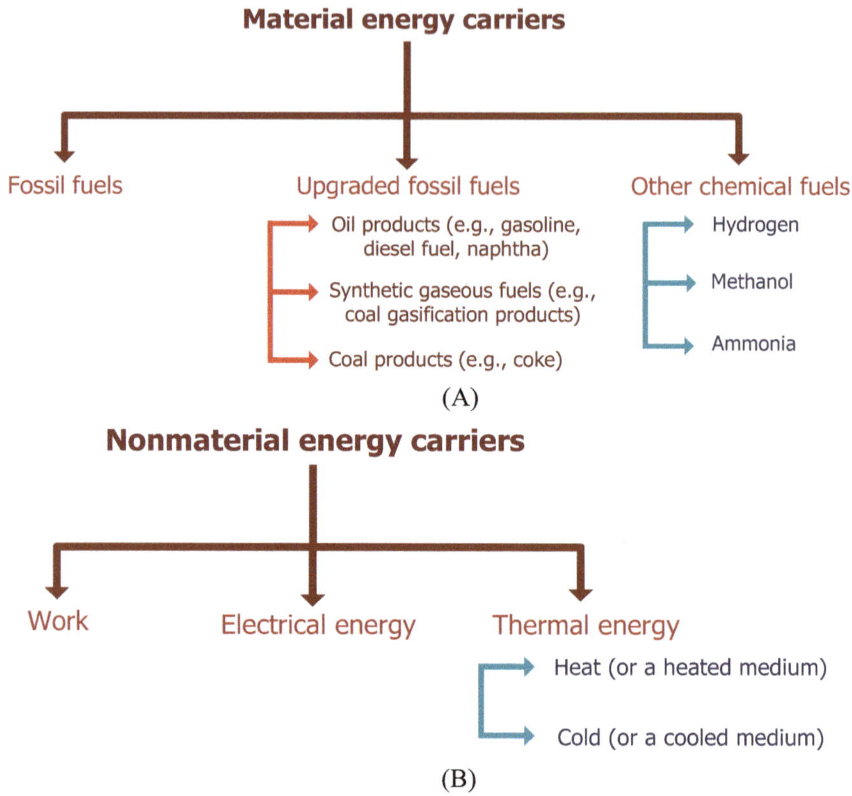

FIGURE 2.8 Energy carriers with (A) material and (B) nonmaterial.

via either hot or cold media. For example, space heating from geothermal energy can be achieved by connecting hot water pipes. Energy carriers are important to reach sustainability. Electricity as an energy carrier is used widely, yet hydrogen is seen as a future energy carrier. This does not mean that hydrogen will be replaced by electricity, but hydrogen can be converted to electricity or vice versa. If sustainable energy sources are used, then electricity and hydrogen can be used as sustainable energy carriers. Hydrogen economy will play a crucial role in the coming years to supply hydrogen to power systems, transportations, or fuel cells [23].

Some countries are developing their infrastructure to transition to the hydrogen economy [24–26]. The most important part of the hydrogen economy is how and where hydrogen is produced. Hydrogen can be produced from fossil fuels or biomass, as is mostly done today, or it can be generated from renewables. Hydrogen is found mostly in water; energy gained from renewables can be used to split water into hydrogen and oxygen with either thermal energy or electric energy [27].

Another option in reaching sustainable energy is to increase efficiency, an option that has many positive aspects such as a decrease in environmental impacts, a decrease in the unit cost of energy, and improved energy security. Increasing energy efficiency also covers energy conservation, fuel substitution, better energy management, better selection of energy carriers, and more efficient use of energy quality. Of course, the effect of increasing energy consumption due to efficient energy, known as the rebound effect, should be considered. Due to the rebound effect, optimum efficiency, which is beneficial for environmental sustainability, should be considered.

There are two types of efficiency: energy efficiency based on the first law of thermodynamics and exergy efficiency based on the second law of thermodynamics. It should be noted that energy efficiency cannot show the whole process occurring in the energy system. Therefore, when energy efficiency is considered, exergy efficiency should be taken into account for a better evaluation of any system. Exergy differs from energy in terms of quality, not quantity. Exergy is defined as the maximum work that can be produced by a flow of matter or energy as it comes to equilibrium with a reference environment. Exergy analyses show meaningful results about an energy system and indicate where and how thermodynamic losses occur in the process. Therefore a decision maker can know how to build an efficient system by decreasing inefficiencies.

Another step that can be performed to achieve sustainable energy is to reduce environmental impact. The environmental impacts of current energy systems are very well-known [15]:

- acidification (the effect on soil and water due to acidic emissions)
- depletion of the ozone layer [an increase in ultraviolet (UV) radiation coming to the Earth's surface because of the destruction of the atmospheric ozone layer]
- depletion of abiotic resources (due to usage of nonrenewable raw materials)
- radiological effects (for instance, radiogenic cancer mortality)
- water pollution
- soil pollution
- ecotoxicity (health problems due to toxic wastes)
- global warming (the rise of global temperatures due to greenhouse gas emissions and therefore global climate change)

2.4.3.1 *Atmosphere*

The atmosphere contains a layer of gases surrounding the Earth and is held by gravity. The atmosphere causes pressure on the Earth, allowing liquid water to exist on the surface, and acts as an absorbent for UV radiation coming from the Sun. Mostly known as air, the atmosphere includes about 78% nitrogen, 21% oxygen, and 1% other gases. The troposphere, stratosphere,

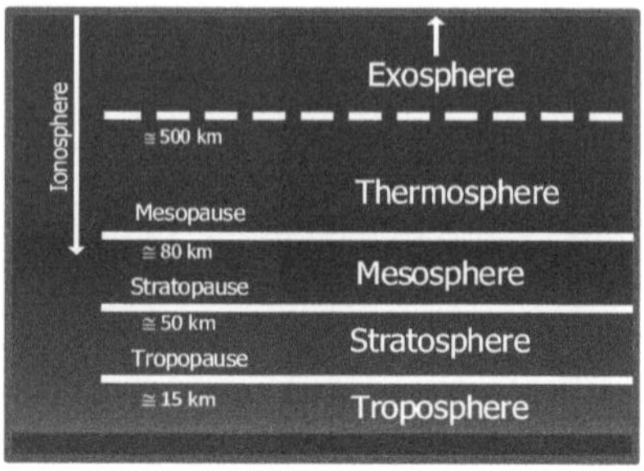

FIGURE 2.9 Layers of the atmosphere.

mesosphere, thermosphere, and exosphere are the layers of the atmosphere, respectively, from the Earth, as seen in Fig. 2.9.

2.4.3.2 Surface temperature of planets

The existence and content of an atmosphere directly affect the average temperature of the planet. The angle of incoming solar radiation on a planet determines how much energy reaches the surface; however, the existence and content of the atmosphere determine how much energy is kept on the planet. For example, the surface temperature of Mercury, which does not have any atmospheric layer, changes between −170°C and +449°C, whereas the surface temperature of Venus, which has an atmosphere consisting mostly of CO_2, is fixed about +465°C. On the other hand, the Earth, having an atmosphere consisting of about 78% N_2 and 21% O_2, has a surface temperature varying between −89°C and +58°C.

2.4.3.3 Influence of SO_x and NO_x on the atmosphere

Combusting fossil fuels produces acids, and acid can travel for kilometers through the atmosphere. Precipitation becomes acidic mainly due to SO_2 (sulfur dioxide) and NO_x (nitrogen oxide) emissions. The scheme for acid precipitation from combusting fossil fuels is presented in Fig. 2.10. Now that the awareness of environmental issues among people has increased, other harmful materials, such as VOCs, chlorides, ozone, and trace materials, have been identified as responsible for acid precipitation. SO_2 emissions are caused mainly by electric power stations, residential heating, and industrial energy use. Road transport is a fundamental source of NO_x emissions accounting for 48% of the total emissions in OECD countries [29].

2.4.3.4 Ozone layer depletion and holes

The ozone layer is named after the chemical molecule O_3 (ozone) and is found between the altitudes of 12 and 25 km. This layer plays a crucial role in absorbing the UV and infrared radiation coming from the Sun [30]. Ozone

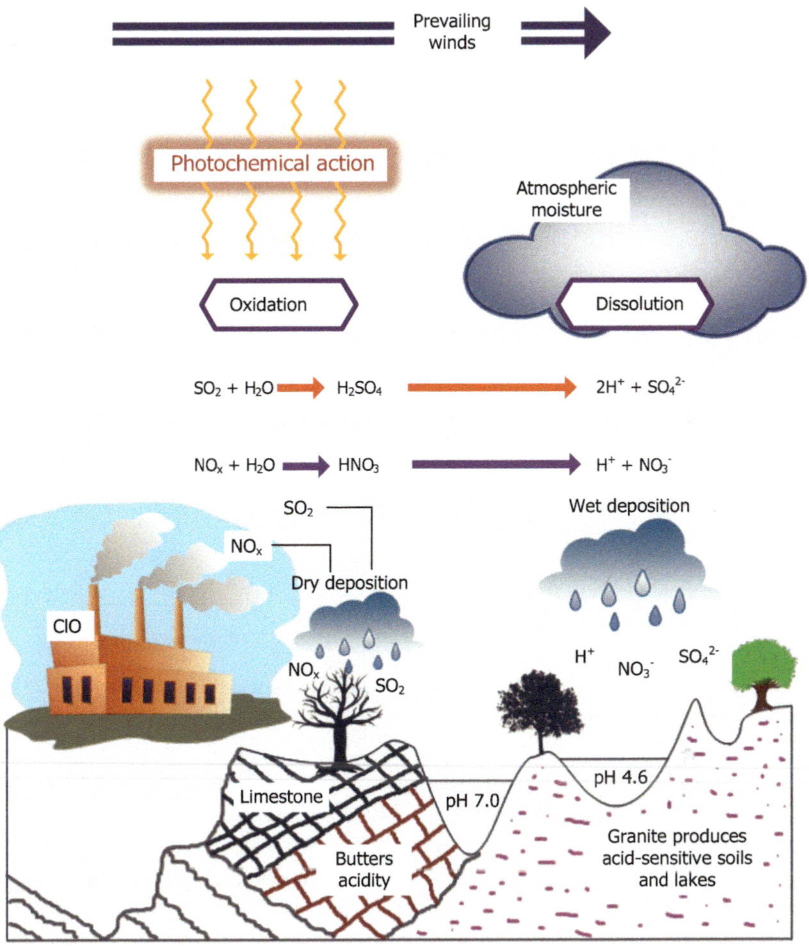

FIGURE 2.10 A schematic demonstration of acid precipitation. *Modified from [28].*

layer depletion is a global environmental problem caused by chlorofluorocarbons (CFCs), halons (chlorinated and brominated organic compounds), and NO_2 emissions. Due to ozone layer depletion in the stratosphere, an increased amount of UV radiation can reach the Earth's surface, causing skin cancer, eye damage, or harm to biologic species. Fig. 2.11 presents the representation of ozone-depleting sources.

2.4.3.5 The greenhouse effect

The most crucial environmental problem caused by energy generation may be global climate change, which started to be "real" after global warming

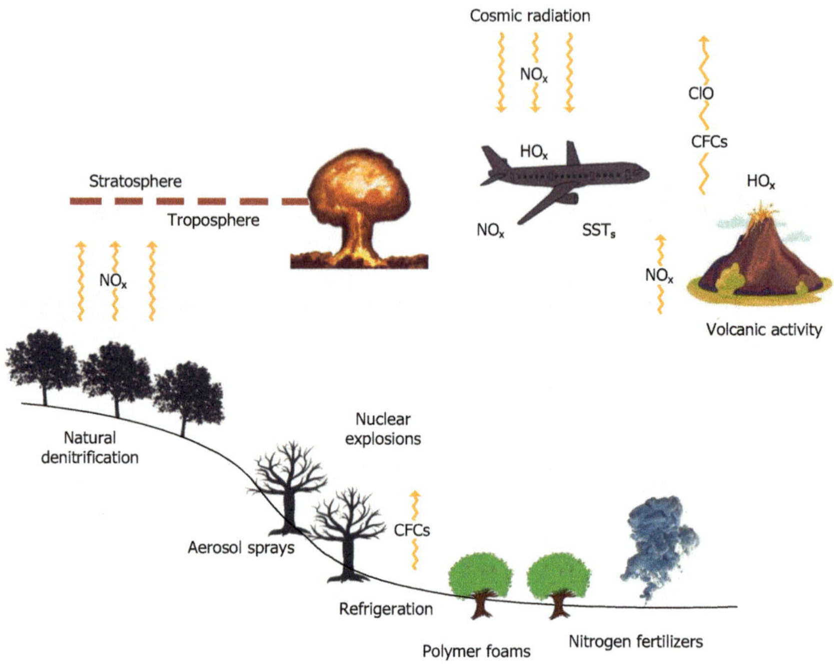

FIGURE 2.11 A schematic representation of ozone-depleting sources. *Modified from [28].*

and the greenhouse effect became apparent. Every day, atmospheric concentrations of gases, such as CO_2, CH_4, CFCs, halons, NO_2, ozone, and peroxyacetylnitrate, are increasing, causing global warming. This increase creates a trap for heat energy coming from the Sun via solar radiation, as shown in Fig. 2.12. The Earth's surface absorbs a certain amount of solar radiation and re-radiates some of it. However, this re-radiation has a longer wavelength than the original radiation coming from the Sun. Therefore, due to the existence of the atmosphere, not all of the re-radiation can escape from the Earth and comes back again to the surface. As a result, the average temperature of the Earth's surface will increase [31]. In fact, the average temperature of the Earth's surface has increased about 0.6°C over the last century, and hence the sea level is expected to rise about 20 cm. Such changes in the natural environment may have huge consequences for humans and other living things on Earth. Many scientists calculate that, if current energy usage based on mainly fossil fuels keeps up, by the next century the average surface temperature of the Earth will have increased by 2°C or 4°C. If this occurs, then the sea level will increase by 30 or 60 cm. Currently, among greenhouse gases, it is known that CO_2 is responsible for about half of global warming. The CO_2 level increases due to the burning of fossil fuels and acts as another layer in the atmosphere.

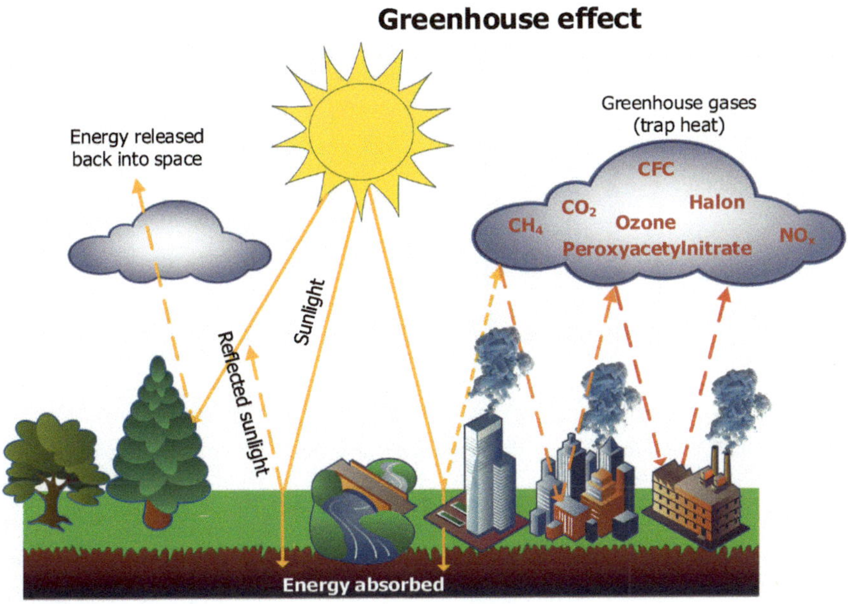

FIGURE 2.12 A schematic representation of the greenhouse effect. *Modified from [15].*

Today many countries are searching for options to assure meeting their energy demands and to continue their economic growth. To decrease the dependence on foreign sources, countries need to find local alternative energy systems of their own that are environmentally benign using. As the human population increases and human lifestyles become more consuming, energy demands will be increased. It is known that about 75% of the world's energy supply is consumed by about 25% of the total population. In 2018 the world's energy consumption per year was calculated as 14,302 Mtoe (million tons of oil equivalent), as seen in Fig. 2.13. To meet this energy demand, 45 billion tons of CO_2 are released into the atmosphere. This high amount of CO_2 is due to the use of fossil fuels for energy supply. If this rate of energy usage continues, not to mention increasing, there will shortly be no coal and oil reserves.

2.4.3.6 Sustainable energy options

Studies on decreasing CO_2 and other greenhouse gas emissions are being done to overcome the resultant environmental problems. However, to successfully reach sustainable development goals in terms of energy, long-term actions have to be taken [32]. One of the possible and highly important solutions is a reduction of fossil fuel usage and changing to renewable sources. Another solution is to improve energy efficiency. To capture CO_2 emissions,

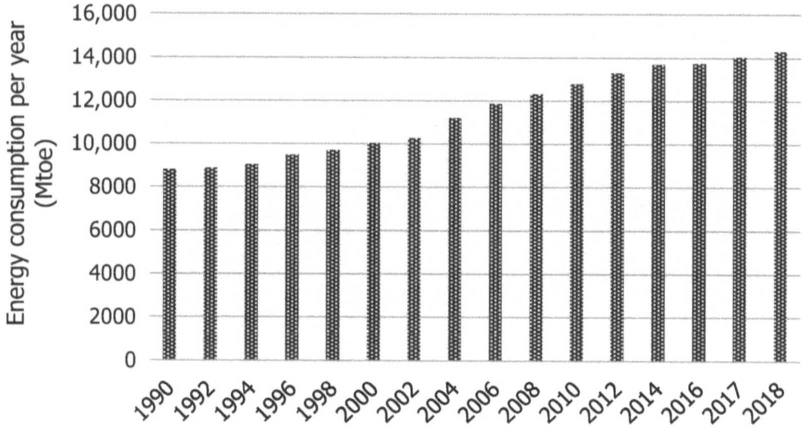

FIGURE 2.13 World total energy consumption per year from 1990 to 2018. *Data taken from IEA [5].*

forestation may be a solution as well. Lastly, changing lifestyles and increasing public awareness will help to save the environment by emitting less of these harmful gases.

One action people can take to integrate renewable energy sources into their lives and use those sources to meet our daily needs, such as power, transportation, space heating and cooling, hot water, etc. Unfortunately, renewable energy sources have some disadvantages; hence their usage and ability to meet our energy needs rely on two conditions: (1) The capture area of renewable energy sources should be greater than the area of end users, or (2) the needs of people should be decreased to the capacity of renewable energy sources. Because of limited conversion efficiencies, the only solution is to change our lifestyles so as to lower our energy demands to levels that can be met by renewables.

2.4.3.7 *Life cycle assessment*

To address energy-related environmental impacts, a novel viewpoint is necessary. For example, some car manufacturers produce electric cars and sell them as zero-emission cars. Let's consider this assertion with "from cradle to grave" viewpoint. How much material and energy are used in the manufacturing of the car? Where and how is the electricity running the car supplied? What is the lifetime of the car? These questions should be answered before we decide whether a car can be called zero-emission. Such an approach is called a life cycle assessment (LCA), which is a very beneficial tool for the evaluation of environmental issues. As presented in Fig. 2.14, any process can be evaluated from production and even from the raw material resources to the end of life. After gaining attention due

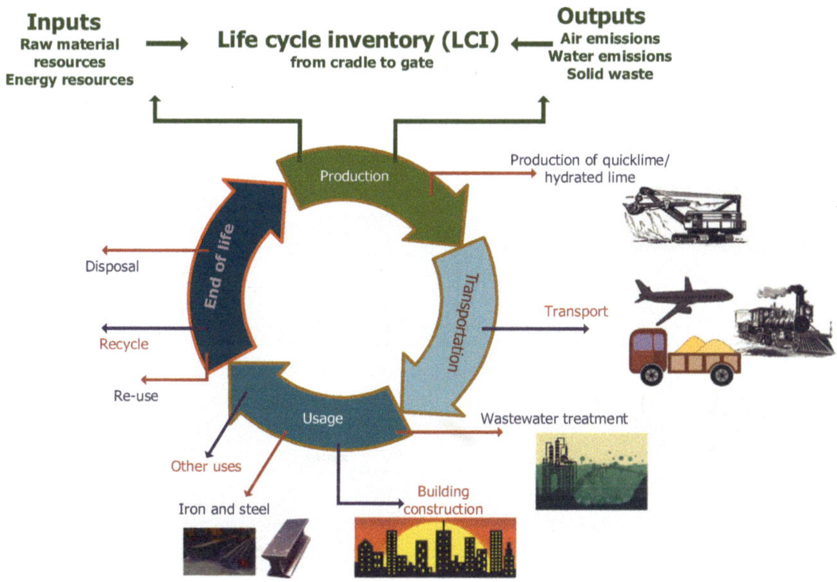

FIGURE 2.14 Schematic representation of life cycle inventory.

to its effectiveness, LCA has been standardized by the International Organization of Standardization (ISO 14040 Life Cycle Assessment—Principles and Guidelines).

There are three main steps in LCA: (1) inventory assessment, (2) impact assessment, and (3) improvement assessment. From the first step to the last, the environmental burdens associated with the product or process are classified, and the amount of resource and waste to the environment is calculated. Then in the inventory assessment, environmental stresses are quantified, and, lastly, environmental improvements are determined for the process [33]. Fig. 2.15 shows the scope and steps of the LCA approach.

2.4.3.8 Thermal energy storage

Thermal energy storage (TES) is a device that stores thermal energy to be used when necessary. The TES method can be divided into two types: (1) sensible heat storage and (2) latent heat storage. TES applications, also known as heat or chilled water storage processes, have been utilized intensely in domestic heating-cooling plants since the beginning of the 1970s [34]. TES is a very useful system when demand and supply times do not coincide. For instance, some renewable energy resources are intermittently available and can peak during daytime or nighttime. Therefore the increased integration of intermittent renewable energy sources, such as wind and solar power, with geothermal energy is a growing challenge to the new flexible model of power generation. For example, a high-temperature working fluid

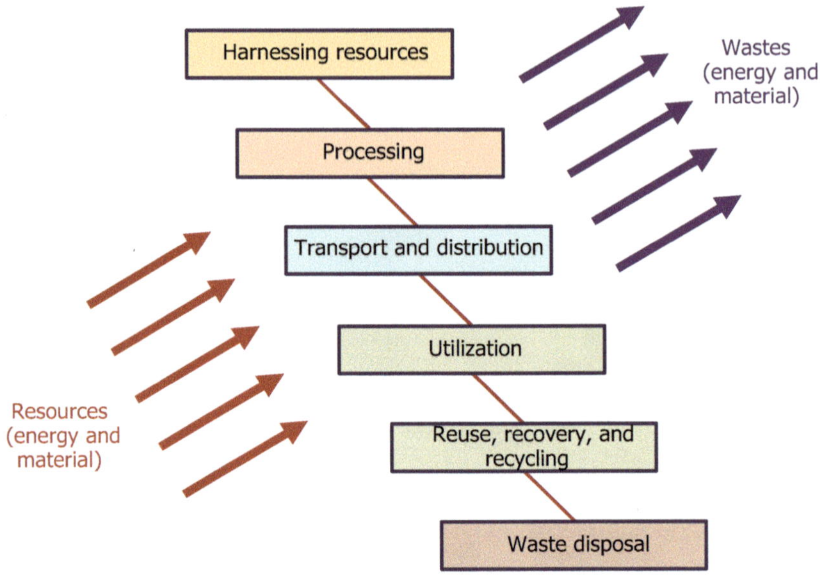

FIGURE 2.15 Scope of life cycle assessment of a product or process.

can be produced by means of a geothermal power plant with concentrating solar collectors for integrated power production systems. TES can then store the heat energy gained at noontime and serve as a heat source when demand is on. TES has three main stages—charging, storing, and discharging—as seen in Fig. 2.16.

The use of TES can reduce energy consumption and conserve fossil fuel use by increasing the efficiency of energy systems and therefore decreasing emissions. Besides increasing efficiency and being environmentally benign, TES also helps to reduce the cost of energy. TESs are also very helpful in tandem with intermittent energy sources, whose advantages are presented in Fig. 2.17. Nowadays, long-term energy storage applications, also known as seasonal energy storage technologies, have gained a place in the district power market, allowing thermal power to be stored for days or even weeks without significant heat losses. Long-term energy storage can be created as great ponds or pools of water, heated by renewable energy sources, with a depth of around 12–17 m, and covered by useful insulation materials. It is possible that TES systems will play an important role in future energy systems, due to the rising share of renewable energy technologies, whose plants must be connected with the main distribution plant.

2.4.3.9 Heat pumps

Heat pumps are used to extract heat from a low-temperature region and move it to a higher-temperature region by means of electricity. Heat pumps

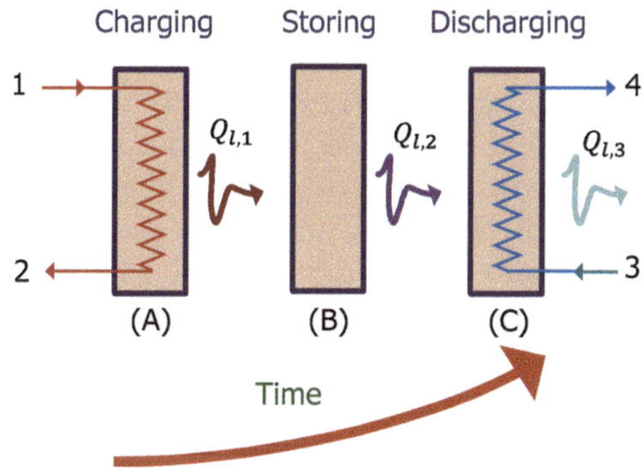

FIGURE 2.16 Three stages of TES: (A) charging period, (B) storing period, and (C) discharging period. *TES*, Thermal energy storage.

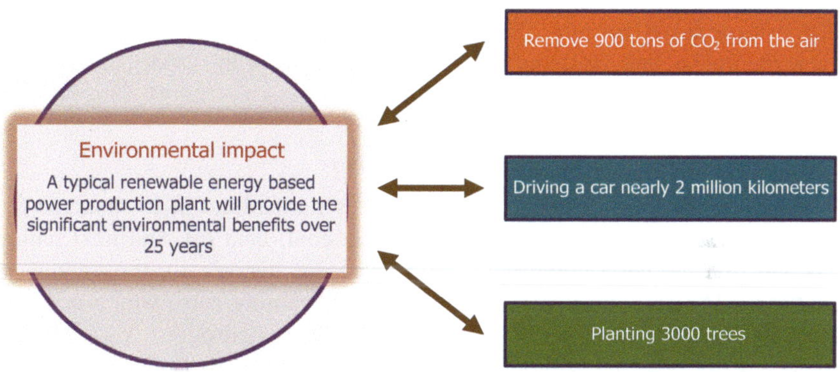

FIGURE 2.17 The simple advantage of the use of renewable energy.

contain an evaporator, compressor, condenser, and expansion valve. Heat pumps are very efficient devices; compared to electric heaters, they are four times more efficient. They are seen as a key factor in decreasing energy consumption rates and increasing efficiency in buildings [35].

2.4.3.10 Cogeneration, trigeneration, and multigeneration

Due to the low efficiency of single-generation systems and their release of a high amount of waste heat to the environment, engineers have tried to find alternatives to utilize that waste heat and increase efficiency. First, a heating system was integrated into the power generation system, a technique called cogeneration. Cogeneration is the simultaneous production of thermal and electrical energy using fewer energy sources than a single-generation system. Having gained positive results and higher efficiencies, trigeneration systems, integrating heating, cooling, and power systems, have come into use. Single-generation energy systems are thermal power plants in which fuel is converted to heat energy as steam or hot gas, which is then converted to mechanical energy and lastly to electric energy. In that type of system, waste heat is rejected to the environment. Hence their conversion efficiency is between 25% and 50%. Cogeneration systems offer the advantage of utilizing that waste heat. They produce electricity with the same methods as a single generation but then use the waste heat as a product of the system. Thus they produce heat and power simultaneously. Cogeneration systems can convert 80% of input energy to power and heat energy. Trigeneration systems are a little more complex than cogeneration systems. They offer power, heating, and cooling simultaneously. Multigeneration systems produce multiple products to meet the demands of users, such as power, chemicals, heating and cooling, hot water, etc. A schematic view of the progress made in geothermal power–based energy conversion plants is shown in Fig. 2.18. Due to the utilization of waste heat to produce multiple products, multigeneration systems have higher energy and exergy efficiencies than trigeneration systems. Using cogeneration or trigeneration systems or even multigeneration systems has many advantages with respect to achieving sustainable energy pathways [36].

2.4.3.11 Hydrogen as a magic solution

If one overviews the history of energy use by humankind, it is observed that the energy spectrum started with wood, next coal, next oil, and next natural gas and that the carbon/hydrogen ratio decreases historically throughout the centuries. The carbon/hydrogen ratio is now becoming zero, which critically addresses the need for a carbon-free fuel such as hydrogen. A carbon-free society is not possible without hydrogen. Furthermore, hydrogen is now recognized as a true energy carrier, a clean fuel, and a key commodity for other chemicals and fuels. See Fig. 2.19.

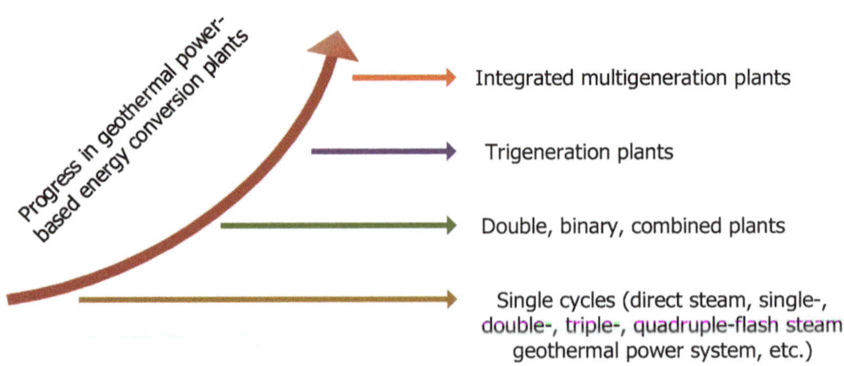

FIGURE 2.18 Schematic view of progress in geothermal power – based energy conversion plants.

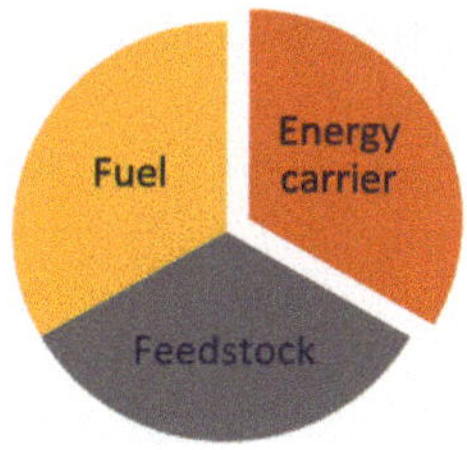

FIGURE 2.19 Hydrogen as energy carrier, fuel, and feedstock for other chemicals and fuels.

With its many advantages, hydrogen is seen as a future energy carrier alternative to electricity or other chemical fuels. Hydrogen is bountiful in the universe as it forms molecules with other elements. Hydrogen is essentially known as a clean energy carrier; when it is combusted, the only waste of combustion is pure water. Hydrogen is also a nontoxic gas. When compared to other energy sources, hydrogen is far more efficient than other proposed solutions. Of course, the infrastructural requirements, safety, standards, codes, and guidelines are recognized as important subjects if one wants to deal with and implement such energy solutions. The most important issue about hydrogen is how it is produced. For instance, hydrogen can be generated from fossil fuels by combusting them and using the heat energy that is required to split water into hydrogen and oxygen. Hydrogen can also be generated by splitting water using solar energy.

2.5 Closing remarks

This chapter has presented a comprehensive description of energy, environment, and sustainability. The steps in achieving sustainability and reaching sustainable development goals have been presented. Without threatening the next generations' rights in every respect, we need to develop. We therefore need to save the environment by reducing the emissions causing the greenhouse effect and the wastes that are hazardous for both the land and water. Some key points should not be forgotten in reaching sustainable energy pathways: using sustainable energy sources, utilization of sustainable energy carriers, increasing the efficiency of both electrical devices and energy-generation systems, reducing environmental impact, improving socioeconomic reachability, and improving public awareness. Moreover, some important tools are to be used in energy-related issues like exergy analysis and LCA, which show better analytical results compared to conventional analyses. The use of those types of analyses helps to enlighten people. If we want to keep the world as a livable place for the coming generations, we should follow and be faithful to sustainable development goals, change our behaviors and habits, and save the environment.

Nomenclature

A	Area (m^2)
E	Energy (kJ)
P	Pressure (kPa)
t	Time (s)
T	Temperature (K)
V	Volume (m^3)
W	Work (kJ)

Greek letters

Δ	Change in variable
η	Energy efficiency
ψ	Exergy efficiency

Subscript

a	Air
cooling	Cooling load
f	Fuel
heating	Heating load
Tur	Turbine

Acronyms

CFCs	Chlorofluorocarbons
LCA	Life cycle assessment
TES	Thermal energy storage
UV	Ultraviolet
VOCs	Volatile organic compounds

References

[1] I. Dincer, M.A. Rosen, Energy, environment and sustainable development, Appl. Energy 64 (1999) 427–440.

[2] M.A. Rosen, Natural energy versus additional energy, Encyclopedia of Energy Engineering and Technology, Taylor & Francis, New York, 2007, pp. 1088–1095.

[3] S.A. Sherif, F. Barbir, T.N. Veziroglu, Hydrogen energy solutions, in N.L. Nemerow, in: F. J. Agardy (Ed.), Environmental Solutions, Elsevier/Academic Press, Burlington, MA, 2005, pp. 143–180.

[4] M. Ozturk, I. Dincer, Thermodynamic analysis of a solar-based multi-generation system with hydrogen production, Appl. Therm. Eng. 51 (1–2) (2013) 1235–1244.

[5] IEA, Sustainable Recovery: World Energy Outlook Special Report, International Energy Agency, Paris, 2020.

[6] I. Dincer, C. Zamfirescu, Environmental dimensions of energy, Comprehensive Energy Systems, Elsevier, Cambridge, MA, 2018, pp. 49–100.

[7] M.A. Rosen, D. Mohsen, Environment, Ecology and Exergy, Nova Science Publishers, New York, 2016.

[8] I. Dincer, C. Zamfirescu, Advanced Power Generation Systems, Elsevier, New York, 2014.

[9] BP Company, BP Statistical Review of World Energy, British Petroleum, London, 2019.
[10] R. Hannah, R. Max, Urbanization, 2019. Available from: https://ourworldindata.org/utbanization. (Accessed 15 March 2020).
[11] UNIDO, How Industrial Development Matters to the Well-Being of the Population: Some Statistical Evidence, United Nations Industrial Development Organization, Vienna, 2020.
[12] WRI, Estimating Power Plant Generation in the Global Power Plant Database, World Resource Institute, Washington, DC, 2020.
[13] Y.E. Yuksel, M. Ozturk, I. Dincer, Thermodynamic performance assessment of a novel environmentally-benign solar energy based integrated system, Energy Convers. Manage. 119 (2016) 109–120.
[14] UNGA, Report of the World Commission on Environment and Development: Our Common Future, United Nations General Assembly, Oslo, 1987.
[15] I. Dincer, M.A. Rosen, Exergy: Energy, Environment and Sustainable Development, Elsevier, Oxford, 2012.
[16] EOLSS, Encyclopedia of Life Support Systems: Conceptual Framework, EOLSS Publishers, Oxford, 1998.
[17] UN, The Millennium Development Goals Report 2012, United Nations, 2012.
[18] J. Elkington, Towards the sustainable corporation: win-win-win business strategies for sustainable development, Calif. Manage. Rev. 36 (2) (1994) 90–100.
[19] I. Dincer, M.A. Rosen, M. Al-Zareer, Exergy management, Comprehensive Energy Systems, Vol. 5, Elsevier, 2018, pp. 166–201.
[20] J. Cleveland, Advanced plants to meet rising expectations for nuclear power", Int. J. Glob. Energy Issues 30 (2008) 393–412.
[21] H.H. Rogner, A. McDonald, K. Riahi, Long-term performance targets for nuclear energy. Part 2: Markets and learning rates, Int. J. Glob. Energy Issues 30 (1–4) (2008) 77–101.
[22] I. Dincer, C. Acar, A review of clean energy solutions for better sustainability, Int. J. Energy Res. 39 (2015) 585–606.
[23] M. Ozturk, Y.E. Yuksel, Energy structure of Turkey for sustainable development, Renew. Sustain. Energy Rev. 53 (2016) 1259–1272.
[24] Dalcor Consultants; Intuit Strategy, Canadian Hydrogen: Current Status and Future Prospects, Natural Resources Canada, Ottawa, ON, 2004.
[25] W.C. Lattin, V.P. Utkigar, Transition to hydrogen economy in the United States: a 2006 status report, Int. J. Hydrogen Energy 32 (2007) 3230–3237.
[26] B. Arnason, T.I. Sigfusson, Iceland—a future hydrogen economy, Int. J. Hydrogen Energy 25 (5) (2000) 389–394.
[27] I. Dincer, C. Zamfirescu, Sustainable Hydrogen Production, Elsevier, New York, 2016.
[28] R. Perman, Y. Ma, J. McGilvray, Natural Resource and Environmental Economics, Longman, London, 1996.
[29] IEA, Data and Statistics, International Energy Agency, 2020. Available from: https://www.iea.org/data-andstatistics/datatables?country = WORLD&energy = Balances&year = 1990. (Accessed 15 March 2020).
[30] I. Dincer, Energy and environmental impacts: present and future perspectives, Energy Sources 20 (4–5) (1998) 427–453.
[31] I. Dincer, C. Acar, Smart energy systems for a sustainable future, Appl. Energy 194 (2017) 225–235.
[32] S. Chu, A. Majumdar, Opportunities and challenges for a sustainable energy future, Nature 488 (2012) 294–303.

[33] M. Granovskii, I. Dincer, M.A. Rosen, Exergetic life cycle assessment of hydrogen production from renewables, J. Power Sources 167 (2007) 461–471.
[34] I. Dincer, M.A. Rosen, Thermal Energy, Storage Systems and Applications, second ed., John Wiley & Sons, London, 2011.
[35] M. Ozturk, Energy and exergy analysis of a combined ground source heat pump system, Appl. Therm. Eng. 73 (1) (2014) 362–370.
[36] Y.E. Yuksel, M. Ozturk, I. Dincer, Development of a geothermal based integrated plant for generating clean hydrogen and other useful commodities, ASME-J. Energy Resour. Technol. 142 (9) (2020) 1–13.

Study questions and problems

2.1 Define the following terms and explain: energy, environment, and sustainable development.
2.2 Define the following terms and explain, where appropriate, their differences: energy sources, energy conversion, energy distribution, and energy utilization.
2.3 Describe the terms "system," "plant," "process," and "cycle."
2.4 Name different important environmental concerns, and describe how the exergy concept can help reduce or mitigate them.
2.5 List the major environmental problems faced by people and societies.
2.6 Describe the process of global warming.
2.7 List the main four greenhouse gases.
2.8 Describe the concept of "greenhouse gas emissions trading" and discuss its implications.
2.9 Why is geothermal energy environmentally friendly?
2.10 What are the environmental impacts of using geothermal energy?
2.11 What are the sustainable development indicators? Why are these very important for society and the environment?
2.12 What is the general index of sustainability?
2.13 What are the key indicators in achieving sustainable development? Rank the various indicators important to achieving sustainable development in society from most to least important, justifying your rankings as much as possible.
2.14 What are the differences between renewable and nonrenewable energy resources?
2.15 What benefits do energy storage systems offer?
2.16 What are the benefits of TES systems?
2.17 What criteria should be considered in TES evaluation?
2.18 What is seasonal TES? Describe how it works for cooling and heating options.
2.19 What are the potential utilizations of hydrogen and its connection to energy storage systems?
2.20 What are the potential uses of hydrogen and its connection to energy storage options?

Chapter 3

Geothermal energy sources

3.1 Brief geothermal history

Geothermal energy has been used as a renewable and important energy source throughout human history. Its use has gone through certain stages over time. In this section, the specific phases of the use of geothermal energy are presented.

Geothermal energy is accepted to have existed since the formation of the world billions of years ago. Therefore, it is considered to be one of the oldest energy production sources in human history. At the same time, the fact that this energy source is renewable reveals the potential of geothermal energy. Archeological research indicates that the first use of geothermal energy coincides with the Paleolithic age. During this period, people used geothermal energy to meet their daily needs such as eating and bathing. The use of geothermal energy in this period had the aim of meeting personal needs in general [1].

When we look at the use of geothermal energy before the 20th century, personal uses such as bathing predominate. Electricity generation using geothermal energy begins in the 20th century. With respect to the production of electricity from geothermal energy, the first geothermal system to produce electricity was established in the Larderello region of Italy. Installed in 1904, this system took its place in history as the first that could generate electrical energy from geothermal energy. This geothermal power system started to produce electricity with a capacity of 250 kW in 1913 [2]. It is generally accepted that electricity production by means of a geothermal power system has made a great contribution not only to the industrial use of geothermal energy but also to the development of systems producing electricity from geothermal energy in general. After this turning point, steps were taken in other parts of the world, and the usage area of geothermal energy expanded. As of 1942, electricity generation capacity from geothermal energy was 127,650 kW. Later, in many countries, valuable progress has been made toward the greater use of geothermal energy. The first geothermal wells were drilled in the Beppu region of Japan in 1919 and in the U.S. State of California in 1921. Later, small-scale geothermal power systems in various countries started to generate electricity, and the areas of use of geothermal energy began to expand. Especially between 1950 and 1960, developments related to geothermal energy systems were recorded in many countries. These countries, which have firsts in generating electricity from geothermal energy, are New Zealand, the United States, and Mexico [3].

Geothermal Energy Systems. DOI: https://doi.org/10.1016/B978-0-12-820775-8.00004-0

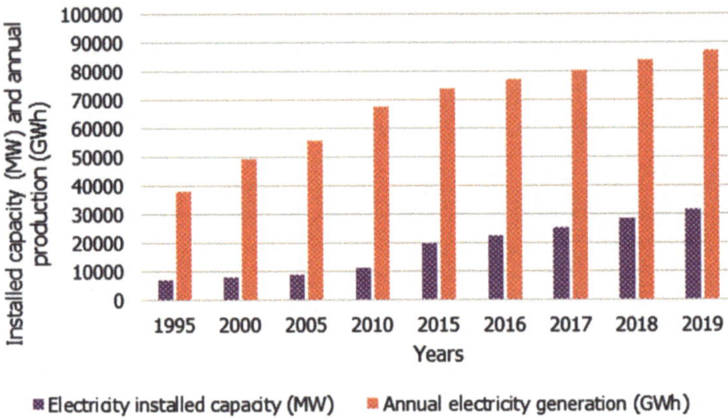

FIGURE 3.1 Geothermal installed capacity and annual electricity generation in the world. *Data from IRENA [4].*

By the 21st century, studies have shown that geothermal power systems exhibit a successful behavior in electricity generation. In a study conducted by the International Renewable Energy Agency (IRENA) for the time interval between 1995 and 2019, the capacity of the installed geothermal energy systems in the world is [4] summarized in Fig. 3.1. It can be clearly seen that the capacities of installed geothermal power systems have increased to 13,277 MW, as evidenced in the IRENA study. The 1995–2019 study carried out by the agency showed that electricity production has accelerated in the last three years (Fig. 3.1). This study is important for systems where electrical output is used as input because electricity is accepted as the first source of energy. IRENA [4] also determined that the worldwide generation of electricity from geothermal energy in 2019 was 85,978 GWh. This agency also conducted research on electricity output produced by geothermal energy. The production of electrical output, which has great potential in terms of efficiency, makes this study valuable [5].

3.2 Nature of geothermal resources

For human beings, geothermal energy is one of the most important energy resources that the Earth's crust offers. When looking at the nature of geothermal energy sources, it is seen that this energy is provided by underground hot fluid sources, and utilization of this type of energy varies according to its temperature value. Given the nature of the geothermal energy source, if its temperature is high and technology enables utilization of it, then utilization of the source will be greater. As the technology develops, the level of utilization of geothermal energy will be associated with technology as it provides access to geothermal energy sources that are deeper in the Earth's crust.

Today's technology reaches a depth of 10,000 m into the Earth's crust, and technologies utilizing geothermal energy at this depth are being developed. The term *geothermal gradient* defines temperature behavior related to the increase in depth in the Earth's crust. This term is used in studies on the Earth's crust and geothermal fluids. The increase in geothermal gradient every 100 m from the Earth's surface is approximately 3°C. According to the research, the temperature on the surface of the Earth's crust is measured as approximately 14°C, while the temperature within the Earth's crust changes from 1000°C to 3500°C. The reason for the temperature increase with depth is the heat source in the center of the world. Although the amount of energy utilized from this energy source of the Earth is less than the amount utilized from that of the Sun, one of the largest energy sources in our world, the heat source at the center of the world offers temperatures above 5000°C. Although it is not possible to see this energy source in the center of the world directly, many models related to it have been suggested. This energy source in the center of the Earth is an important heat source because it is an indispensable and utilizable energy source for the world. Fig. 3.2 depicts a huge, utilizable geothermal energy source in the center of the world at a depth of 6370 km from the Earth's surface. This enormous energy source provides high temperature levels to the regions around the core. Geothermal fluids with high temperature levels are used in our world by transferring the heat energy contained in the Earth's core to water bodies, which are made into geothermal energy sources located in various parts of this Earth's crust.

The Earth globe model in Fig. 3.2 is a simplification. However, the general geological structure of the world is not so regular. Temperature increases depending on depth occur curvilinearly rather than linearly; that is, the characteristic of the temperature increase depending on the depth varies across the Earth sphere. For this reason, the increase in the geothermal gradient is

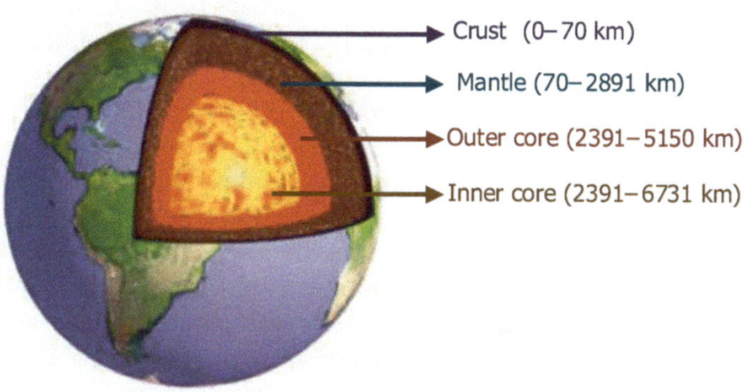

FIGURE 3.2 3-D representation of the energy source in the Earth's crust. *Modified from [6].*

approximately 2°C per 100 m, but it can be seen as 1.5°C in different parts of the world. This variability causes differences in the amount of energy stored by geothermal energy resources at the same depth in the Earth. These temperature differences in geothermal energy sources also offer varying energy potentials to the countries in the region.

3.3 Geothermal sources potential

Since geothermal energy sources support both renewable and environmental policies, they make many countries privileged. Many countries in the world, especially countries rich in geothermal energy, have advantages in both environmental policies and energy. These countries, by obtaining various useful products from this energy source, both increase the welfare level of the people in the country and prevent vulnerabilities in their economies. The strategic positions of countries in the world are of great importance in terms of the level of utilization of geothermal energy because, as previously mentioned, changes in the geothermal gradient occur curvilinearly, that is nonlinear, due to the irregular structure of the Earth's crust.

The utilization level of the country of geothermal energy is directly proportional to its technological level because, in order to make use of geothermal energy, the technological potential of the country must be suitable to use this energy. For this reason, countries whose technological facilities are at the highest level and whose location in the world is suitable for making use of geothermal energy resources have the opportunity to make maximum use of geothermal energy. Many studies on the level of geothermal energy utilization of countries around the world have been and are still being carried out.

Generally, the level of utilization of geothermal energy is gradually increasing. The first reason for this increase can be shown as the increasing rate of global warming. The use of geothermal energy, which is one of the renewable energy sources, is important in order to provide a more livable environment and sustainability in the world, which is the living environment for all living things. Many national and international organizations are investigating the level of geothermal energy utilization of countries in the world and are presenting this level of utilization to inform people through analysis. In addition to these studies, many studies have been carried out and continue to be carried out. These studies are extremely valuable for our world in order to see the applications of geothermal energy in the world.

It can be clearly seen in the studies that the use of geothermal energy in the world is gradually increasing. The shares in this increase in these countries vary. Two factors are at play in this variability: advanced technology and strategic position in the world. These two issues are addressed in the first part of this section.

According to the research, the United States comes first among the countries that benefit from geothermal energy. As seen in Fig. 3.3, the

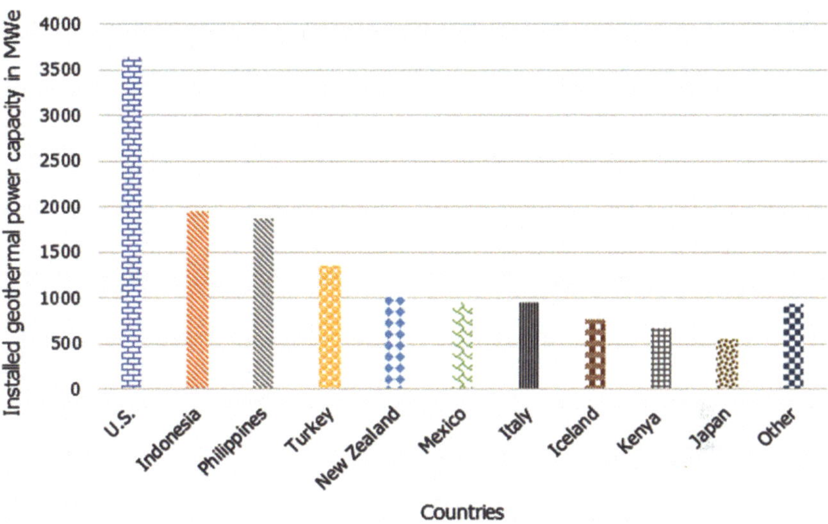

FIGURE 3.3 Illustration of top 10 geothermal countries. *Data from IRENA [7].*

United States leads this race. Undoubtedly, one of the most important features of the United States' lead in this race is the importance it attaches to advanced technology and its investments in research and development activities. In this study carried out for 2018, the top 10 countries in the world are listed according to the level of utilization of geothermal energy. As seen in the figure, Indonesia has the second largest geothermal facility installation after the United States. Although the technology of this country is not as great as that of the United States, its strategic position in the world enables it to benefit from geothermal energy. In the ranking in the figure, Indonesia and the Philippines are similar. Other countries in the ranking are Turkey, New Zealand, Mexico, Italy, Iceland, Kenya, and Japan.

Although research is conducted on countries that make use of geothermal energy internationally, each country publishes statistics to analyze its own geothermal energy use as given in Table 3.1. In this chapter, geothermal energy utilization in the world is presented briefly. Information about the 10 largest geothermal energy utilizer countries is discussed: the United States, Indonesia, Philippines, Turkey, New Zealand, Mexico, Italy, Iceland, Kenya, and Japan. In 2018, about 0.5 GW worth of new geothermal power–generating systems were reported, resulting in the total capacity in the world becoming 13.3 GW. About two-thirds of this increase belonged to Turkey and Indonesia. Croatia founded its first geothermal power plant that year. A short overview of the most active geothermal energy user countries is given in the following subchapters.

TABLE 3.1 Top countries in terms of increase rate on geothermal installed capacity and total amounts.

Country	Additional (MWe)	Total (MWe)
Turkey	219	1283
Indonesia	140	1946
Unites States	58	2541
Iceland	45	753
New Zealand	25	966
Croatia	18	20

Source: Data from [8].

3.3.1 United States

The United States is the first country with a ranked installed capacity of geothermal energy of 2541 MW in 2018, as seen in Fig. 3.4. Midwestern and Eastern states have the highest growth rate compared to other states, having more than 1.4 million units of installed capacity of 12 kW [9]. Some states offer more advantages than others in terms of tax breaks or refunds or encouragements for people to use geothermal energy as an energy source. Moreover, the U.S. Department of Energy (DOE) bestows special importance on investments in enhanced geothermal systems (EGS). According to the US-DOE report, by using EGS, tens of millions of American homes and businesses can benefit [10]. For this purpose, the geothermal technologies office is working on new technologies and methods to reduce costs and improve performance. The advantages and disadvantages with the limitations of using geothermal energy are discussed. In the near future, it is expected to be economically viable for EGSs to compete in the energy markets. Fig. 3.4 demonstrates the increase in installed capacity of geothermal energy from 2005 to 2018. As seen in the figure, a sharp increase occurred between 2011 and 2012 with about 200 MW. An installed capacity of 2541 MW has been reported in 2018 for the United States. The red line in the figure shows the amount of produced energy from geothermal energy systems in gigawatt-hours. Between 2005 and 2018, intense development can be seen reaching 18,710 GWh. Although the United States is the highest geothermal energy user in the world, the population is about 327 million people in 2018. So if geothermal energy utilization per capita is considered, the United States is still behind some other countries. In the last five years from 2013 to 2018, a 350 MWe-capacity new geothermal plant was installed in the State of California. Alaska has the lowest-temperature geothermal power plant with the only 74°C. This value can be a world record for power generation from such a low source [5]. The

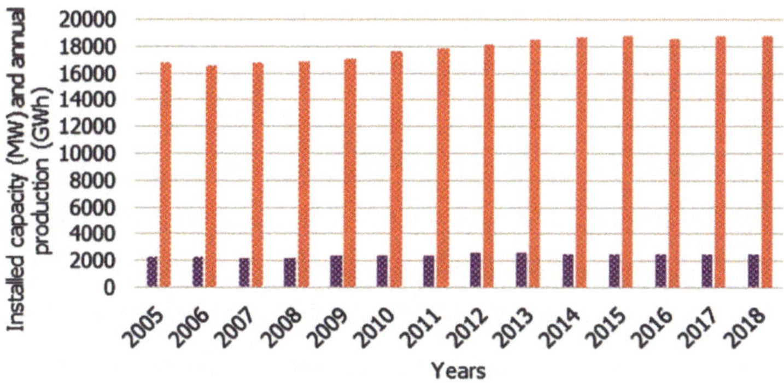

FIGURE 3.4 Geothermal energy–based electricity installed capacity and annual electricity generation for the United States. *Data from IRENA [7].*

United States has a total of 2 million km^2 area of geothermal sources that are able to compete economically with other energy sources by improving the technologies.

3.3.2 Indonesia

Indonesia is one of the fastest growing countries in terms of geothermal energy investments. Indonesia has about 28 GW expected capacity of geothermal energy, making the country the largest geothermal energy utilizer in the world if the source is used efficiently and if utilization is economically viable. Geothermal sources in the country are associated with volcanoes in Sumatra, Java, Bali, and islands in the eastern part of the country. Fig. 3.5 summarizes the Indonesia geothermal energy outlook between 2005 and 2018. In 2005, the installed capacity was about 850 MW, and it became 1946 MW in 2018. As a result of this installed capacity, the country was able to produce 13,296 GWh of energy from geothermal sources in 2018. In addition, Indonesia has another advantage, that is, geothermal energy sources are spreading over the country.

The 10 locations utilizing geothermal energy are Darajat (260 MW), Dieng (60 MW), Kamojang (200 MW), Gunug Salak (377 MW), Sibayak (11 MW), Lahendong (87 MW), Wayang Windu (227 MW), Ulu Belu-South Sumatra (110 MW), Ulumbu-Flores (5 MW), and Mataloko (2.5 MW) [11]. About 300 explored areas are suitable for power or heat applications with high-enthalpy sources. Indonesia is also a highly populated country with about 260 million people in 2018. Hence produced energy per capita from geothermal sources equals about 49 kWh, which is still a low number.

3.3.3 Philippines

After the government promoted new laws, incentives, and promotions, the installed capacity of geothermal energy sources increased twice in 2014 and

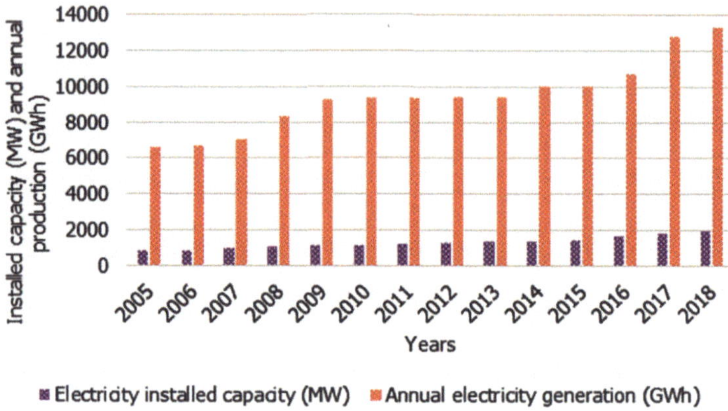

FIGURE 3.5 Geothermal energy–based electricity installed capacity and annual electricity generation for Indonesia. *Data from IRENA [7].*

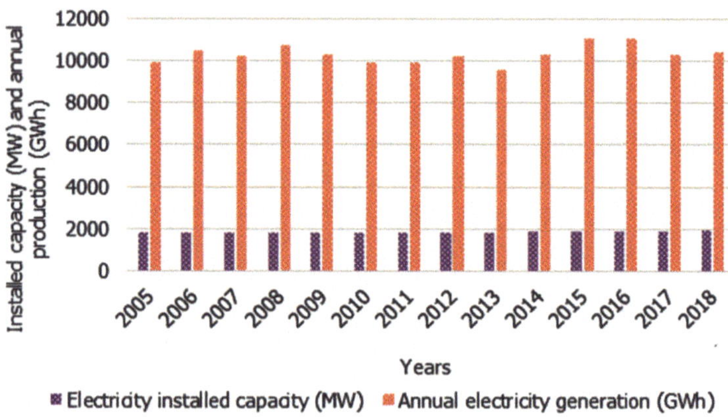

FIGURE 3.6 Geothermal energy–based electricity installed capacity and annual electricity generation for the Philippines. *Data from IRENA [7].*

2018. Fig. 3.6 presents the installed capacity and amount of annual produced energy from geothermal energy between 2005 and 2018. In the Philippines, the geothermal energy structure is relatively stable, as seen in the figure. From 2005 to 2013 and from 2014 to 2017, installed capacity did not change and was fixed at 1847 MWe and 1916 MWe, respectively. Then it increased one more time to 1942 MW in 2018. This amount of capacity was able to meet about 15% of the country's requirement. As a result of this installed capacity for the Philippines, the country was able to produce 10,465 GWh of electrical energy from geothermal resources in 2018.

3.3.4 Turkey

Turkey is one of the highest in the world in terms of geothermal capacity. Turkey is settled on a relatively young land, called Anatolia, which has many small earthquake activities. Due to its fault lines, the geothermal energy capacity of Turkey is high. Geothermal energy has been utilized in the country for many years for district heating or thermal tourism [12]. Also, there are many agricultural lands in the country. Farmers use greenhouses to keep the temperature stable due to the changing weather in the country, and so geothermal energy is utilized to heat greenhouses [13]. After 2010, with new laws and regulations, Turkey has become the first country to increase its capacity. The total installed capacity of Turkey was only 15 MW in 2005. Then it started to increase slightly until 2012, after which the installed capacity nearly doubled in one year and reached 311 MW in 2013. In 2018, the total installed capacity reached 1283 MW, and in 2018 the amount of produced energy was 7431 GWh, as presented in Fig. 3.7.

Turkey has made many investments after 2002 in the geothermal energy area, the number of geothermal fields has been increased, geothermal heating for greenhouses has been increased some 686%, geothermal space heating has increased by 281%, geothermal electricity generation has increased sharply by 3925% [9]. The reason behind these increases is that the Turkish government has provided purchase guarantees for the electricity generated by geothermal sources with a cost of 0.20 \$/kWh [14]. Thus private sectors have made their investments in geothermal energy–based electricity generation.

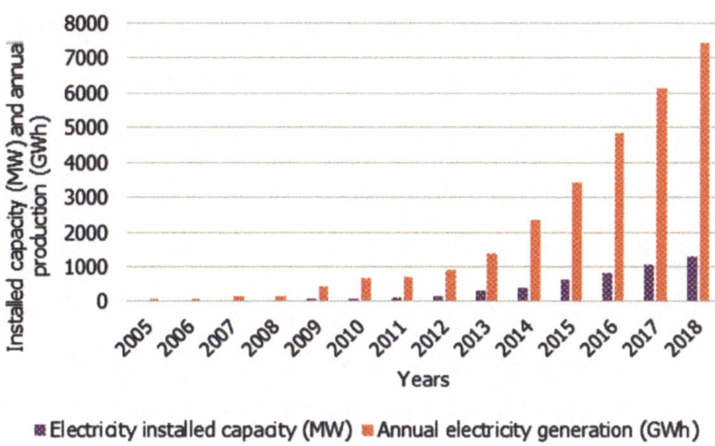

FIGURE 3.7 Geothermal energy–based electricity installed capacity and annual electricity generation for Turkey. *Data from IRENA [7].*

3.3.5 New Zealand

New Zealand is one of the countries that raised its geothermal capacity. The advantage in New Zealand is the high-temperature geothermal sources that are suitable for electricity generation at a low cost. Even the cost of electricity production from geothermal energy can compete with fossil fuel–based electricity production. New Zealand, with a total installed capacity of 966 MW at the end of 2018, is the fifth most productive country in the world. They were able to produce 7815 GWh energy from geothermal sources in 2018, as seen in Fig. 3.8. Geothermal–based electricity production is about 17% of total electricity production in New Zealand in 2017 [15].

3.3.6 Mexico

Mexico has four main regions for geothermal energy production: Cerro Prieto, Los Humeros, Los Azufres, and Tres Virgenes. The total installed capacity of geothermal energy in the country was 951 MW in 2018 [5]. Actually, Mexico has geothermal plants that would add to this value, but some of them are not working. Therefore, a decline can be seen after 2010 in Fig. 3.9. In 2010, the installed capacity of geothermal energy was 965 MW. It started to decline at that time, and, in 2014, installed capacity dropped to 813 MW. In 2018, this value reached 951 MW, with 2% of total electricity production met by geothermal in Mexico. However, this is a very small portion of the country. Mexico still depends mainly on fossil fuels to meet the country's needs. Fossil fuels provide 80% of the total energy consumption in the country [16]. According to the Paris agreement and Mexican Law of Climatic Change, Mexico has to decrease greenhouse gaseous emissions by 30% by 2020 and 50% by 2050. In addition, clean electricity production must be at least 35% in 2024 [17].

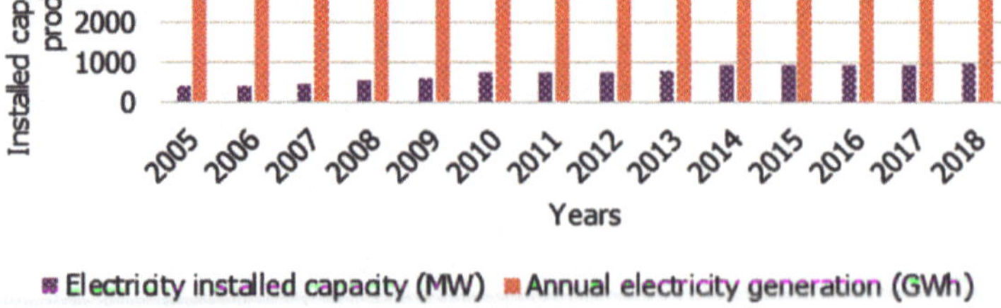

FIGURE 3.8 Geothermal energy–based electricity: installed capacity and annual electricity generation for New Zealand. *Data from IRENA [7].*

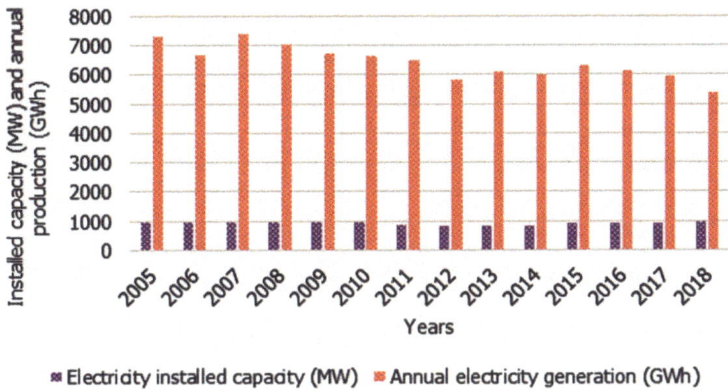

FIGURE 3.9 Geothermal energy–based electricity: installed capacity and annual electricity generation for Mexico. *Data from IRENA [7].*

There are more than 225 geothermal production wells in the country, producing 5946 GWh of energy in 2017 and 5375 GWh in 2018.

3.3.7 Italy

Italy has had geothermal plants using its abundant geothermal sources for more than a century. Those geothermal sources range widely from low temperature, medium temperature (above 90°C), and high temperature (above 150°C). Especially high-temperature sources are situated on tectonically active fault lines [18]. Today, many of the geothermal sources are utilized for power production or district heating systems in the country. District heating systems are generally located in the Tuscany region. Geothermal energy is one of the most used renewable energy sources in Italy. The first one is bioenergy, and the second is air-source heat pump systems. Geothermal energy comes after them, contributing a 2% portion to the total renewable heat consumption [19]. Fig. 3.10 presents the installed capacity of geothermal energy and the amount of produced energy from geothermal energy. In 2010, Italy's government implemented a law that liberalizes geothermal energy utilization as a result of increasing demand for energy, especially renewable–based energy. In 2005, the installed capacity of geothermal sources was 671 MW, and it became 767 MW in 2018 after the installation of new 39 MWe plants. By utilizing geothermal sources with those installed plants, Italy was able to produce about 6105 GWh of energy in 2018.

3.3.8 Iceland

Due to its geologic characteristic, Iceland (located on the Mid-Atlantic Ridge) is one of the countries using geothermal energy the most effectively,

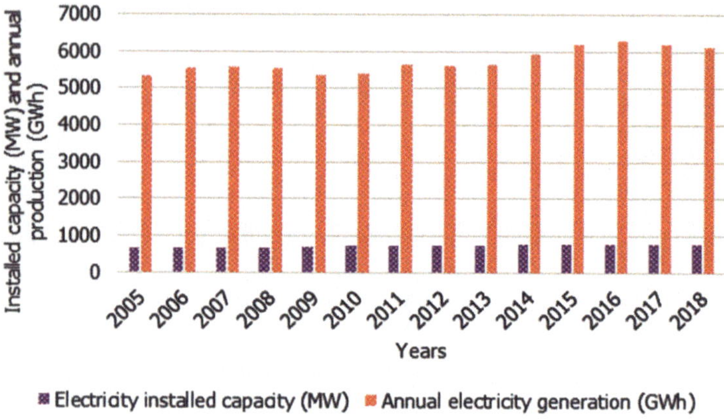

FIGURE 3.10 Geothermal energy–based electricity: installed capacity and annual electricity generation for Italy. *Data from IRENA [7].*

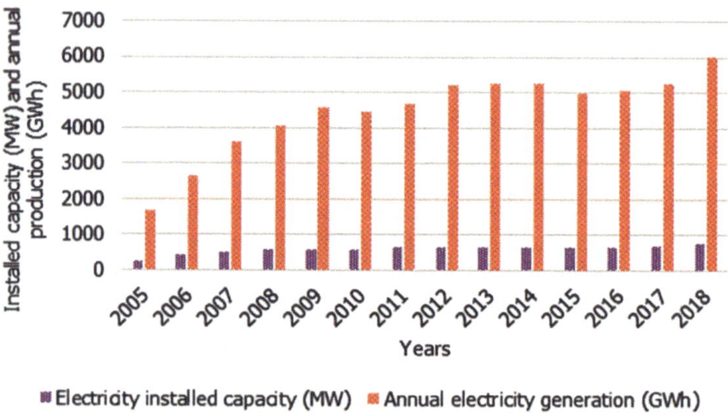

FIGURE 3.11 Geothermal energy–based electricity: installed capacity and annual electricity generation for Iceland. *Data from IRENA [7].*

with 69% of the primary energy supply of Iceland met by geothermal energy and 90% of house heating provided by geothermal sources [11]. Electricity generation from geothermal energy started about 50 years ago in the country and supplied more than 30% of total needs. In 2017, Iceland met 98% of electricity production from renewable energy sources, namely hydro, and geothermal [20]. As indicated in Fig. 3.11, the installed capacity of geothermal energy was 232 MW in 2005 and reached 753 MW in 2018. Iceland was also able to produce about 6010 GWh of electrical energy from geothermal in 2018. Other than power production and house heating, heat energy for swimming and bathing is also supplied by geothermal energy. Due to the long winters, outdoor swimming pools or recreational bathing are popular in the country. Having provided heat energy from geothermal sources has reduced dependence on fossil fuels as well. Another low-temperature geothermal application is snow melting in Iceland. Because of its climate, the country is snowy in winters and at least walkways and roads need to be melted. About 1900 TJ of energy is spent on ice melting every year in Iceland. Besides these applications, other utilizations of geothermal energy are fish drying, salt production, greenhouse heating, aquaculture heating, liquid CO_2 production, and methanol production. Iceland is the world leader in the utilization of geothermal energy per capita. The various production options and low population of the country make them advantageous.

3.3.9 Kenya

In terms of installed capacity of geothermal energy, Kenya is ninth in the world after increasing its geothermal capacity three times from 2005 to 2018. Kenya has a large amount of total capacity, and the country tries to generate electricity from geothermal energy resources. Due to its climate, space heating is not very necessary, but it might be needed only between 02:00 a.m. and 07:00 a.m. High-temperature geothermal sources are located within the Kenya Rift Valley. Fig. 3.12 demonstrates the improvement of the installed capacity of geothermal energy in Kenya. While in 2005 installed capacity was 128 MW, it reached 663 MW in 2018. The country was able to generate 5005 GWh of energy in 2018 from geothermal resources.

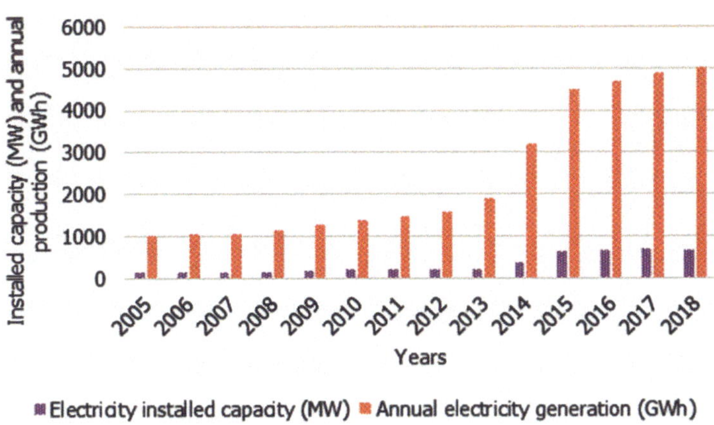

FIGURE 3.12 Geothermal energy–based electricity: installed capacity and annual electricity generation for Kenya. *Data from IRENA [7].*

3.3.10 Japan

Tenth in the world in geothermal capacity is Japan. The potential of geothermal in Japan is huge (20 GW); however, the total installed capacity was still near 490 MW in 2018. After the nuclear accident in Fukushima in 2011, the government encouraged the private sector to build up small binary systems. Akita, Fukushima, Hachijojima, Hokkaido, Iwate, Kagoshima, Kumamoto, Miyagi, Oita, and Tokamachi are active fields in Japan [21]. In Fig. 3.13, it is seen that the installed capacity of geothermal energy in Japan was 535 MW in 2005 and 486 MW in 2018. In the last 10 years, there was not a big change in Japan's installed capacity. In 2017 and 2018, the country produced 2457 GWh of energy from geothermal energy resources.

3.4 Classification of geothermal resources

The definition and classification of geothermal energy as a renewable energy source can differ in geothermal power systems studies. In this section, geothermal energy resources will be classified according to geothermal energy source enthalpy (or temperatures). This method is the most common method for classifying geothermal energy sources because it is clear, simple, and understandable in studies on geothermal energy. Whichever temperature or enthalpy is chosen here, this phenomenon is all about the thermal energy of the geothermal fluid. Since enthalpy is a term used to express the heat energy of geothermal fluid, it is an important parameter for the classification of geothermal energy sources. Although different temperature values are used to classify the geothermal energy source, generally geothermal energy sources are examined in three groups: low level, medium level, and high

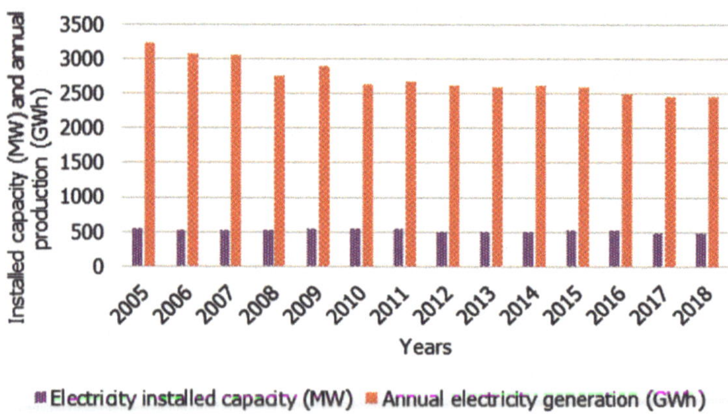

FIGURE 3.13 Geothermal energy–based electricity: installed capacity and annual electricity generation for Japan. *Data from IRENA [7].*

TABLE 3.2 Classification of geothermal energy resources based on the well temperature range.

Enthalpy classification of geothermal sources	Temperature range (°C)	Reference
Low-enthalpy resources	<90	[22]
	<125	[23]
	<100	[24]
	≤150	[25]
	≤190	[26]
Intermediate-enthalpy resources	90–150	[22]
	125–225	[23]
	100–200	[24]
High-enthalpy resources	>150	[22]
	>225	[23]
	>200	[24]
	>150	[25]
	>190	[26]

level of enthalpy or temperature. This classification is determined for different temperature or enthalpy values in different studies. Which temperature or enthalpy values of geothermal energy sources are classified in different studies can be clearly seen in Table 3.2.

Muffler and Cataldi [22] consider a geothermal energy source as a low-enthalpy class if the temperature of the geothermal energy source is below 90°C. If the source temperature is between 90°C and 150°C, they define the source as medium and reserve the high-enthalpy class for 150°C and above. Hochstein [23] utilizes a classification with different temperature values: low-enthalpy geothermal resources for temperatures below 125°C, medium-enthalpy geothermal resources for temperatures between 125°C and 225°C, and high-enthalpy geothermal resources for temperatures above 225°C. Benderitter and Cormy [24], on the other hand, use the classification of 100°C and 200°C in their study. They classify a low-enthalpy geothermal resource for temperatures below 100°C, a medium-enthalpy geothermal resource for temperatures between 100°C and 200°C, and a high-enthalpy geothermal resource for temperatures above 200°C. In these three studies, geothermal energy sources are divided into three classes. In other studies, geothermal resources are classified by taking the same parameter, enthalpy, as the reference. Nicholson [25] examines geothermal energy sources in two groups, and defines the low-enthalpy geothermal resource classification for temperatures below 150°C and the high-enthalpy classification for temperatures above 150°C. Another such definition is made by Axelsson and Gunnlaugsson [26]. They classify a low-enthalpy geothermal resource for temperatures below 190°C and a high-enthalpy geothermal resource for temperatures above 190°C.

There are many studies in the literature on the classification of geothermal energy sources. In performing this classification, the best accepted method is to avoid uncertainty and complexity by using the temperature or enthalpy parameter [27]. For this reason, the just papers mentioned in this section for the classification of geothermal energy sources form the basis of many studies. A more comprehensive version of classification studies can be seen in Fig. 3.14. The number of these classifications can be increased by more comprehensive studies.

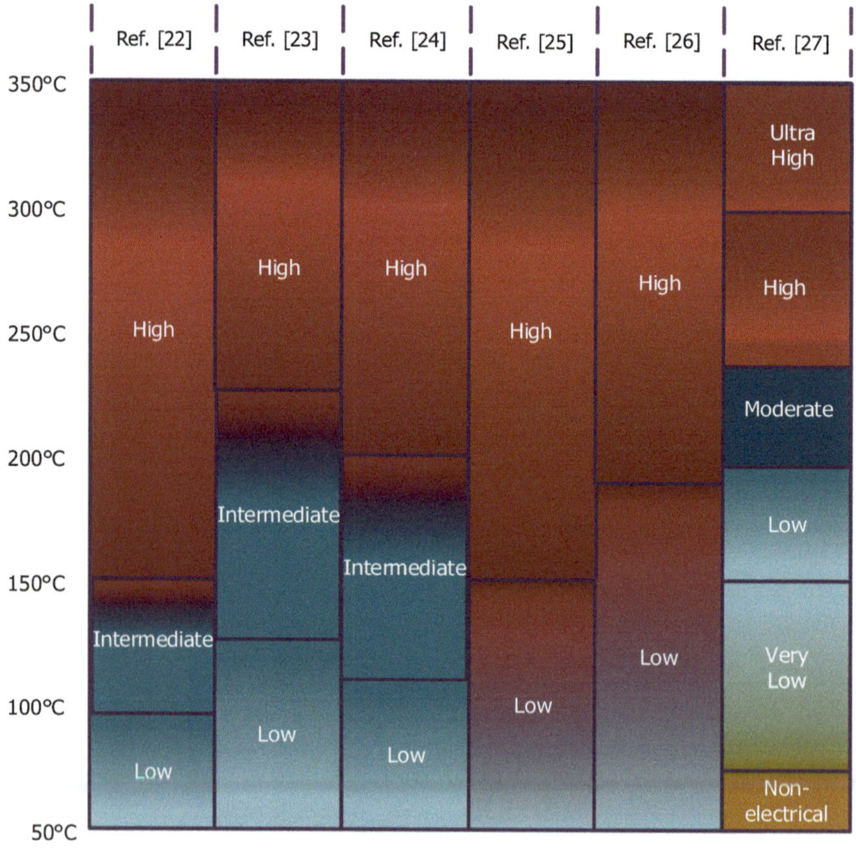

FIGURE 3.14 Classification of geothermal sources based on production well temperature.

3.5 Benefits of geothermal energy for sustainable development

Geothermal energy sources are among the renewable energy sources that can be used effectively in energy generation systems. Since geothermal energy is a renewable energy source that supports both environmental impact and environmental policies, it has a potential place in energy resources. Geothermal energy sources have many benefits for both humans and other living creatures. It can be accepted that its most important benefit for our world is that it is an environmentally friendly energy source. A brief view of the pros for geothermal energy sources is shown in Fig. 3.15A. Beneficial output production can be realized directly with this geothermal energy source without any combustion cycle. The emission values emitted by the energy generation systems that do not perform combustion cycles are very low. Another benefit of geothermal energy is the reliability it provides to the country where it is used because there is continuity in the type of energy. In other words, geothermal energy sources are not affected by factors such as weather conditions; it will be able to provide continuous electricity or useful output to the country where it is located. In addition, geothermal energy sources can be integrated with different renewable energy sources in an energy generation system. The geothermal energy source can also contribute to employment in the countries where it is available. Because, in order to benefit from geothermal energy, at least one energy generation system must be established and operated, trained, knowledgeable, and skilled engineers and employees are needed.

Since geothermal energy sources are located within the Earth, there are large regions to be used. The factor that determines the level of utilization of this energy is the degree of development of the country in terms of advanced technology. The more advanced the technology is in the country with a

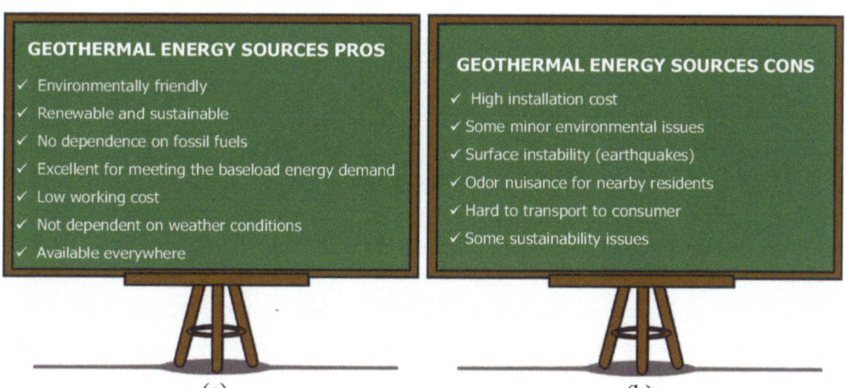

FIGURE 3.15 (A) Pros and (B) cons for geothermal energy sources.

geothermal energy source, the more electricity that can be obtained from geothermal energy. In addition, countries that benefit from geothermal energy also help their national financial policies, because once the system's initial investment costs are recovered, the plant starts to turn a profit and contributes to the national economy.

Geothermal energy sources also reduce dependence on fossil resources. Increasing the utilization level of geothermal energy resources contributes to the replacement of high-emission energy sources with geothermal energy resources. For this reason, geothermal energy sources also provide environmental security to countries. Since geothermal energy systems do not cause sound pollution, it is thought that energy production can be realized at acceptable sound levels. Another benefit of geothermal energy sources is that hot water production is carried out more efficiently than with traditional energy production systems. An installed geothermal system can produce hot water at a low cost.

Since geothermal power systems are designed and installed with very few sections, they can be used for energy production in different locations. This ensures that these systems have high uptime and useful output. The maintenance costs of geothermal power systems are low as they are user-friendly energy systems.

Of course, there are different renewable energy sources in our environment. The low cost of utilizing geothermal energy resources and the high benefits gained at low cost put geothermal energy resources in an advantageous position among other energy resources. Countries also support the use of these energy resources and show how important these systems are at the national level.

3.6 Disadvantages of geothermal energy resources

Although geothermal energy sources have many advantages, they also have some disadvantages. But, given their advantages, the downside is not great. A brief view of the cons for geothermal energy sources is illustrated in Fig. 3.15B. In the previous section, it was mentioned that the emission values of the energy generation systems using geothermal energy sources are lower than those of conventional systems. Although they emit low emissions, the systems using underground geothermal energy sources cause the release into the atmosphere of some gases harmful to the environment. The amount of these emitted gases is usually at the maximum near the installed geothermal power systems. Even if these emission gases have an environmentally harmful effect, however, the harmful effect of geothermal power systems on the environment does not exceed the harmful effect of traditional fossil resource systems. Another disadvantage of a geothermal power system for our environment is the possibility of triggering an earthquake, a natural disaster because, during the installation of the system, the shape of the

Earth's crust may be distorted. Specifically, the installation of developed geothermal energy systems can further disrupt the natural structure of the Earth's crust. This possibility should not be ignored when installing these systems.

As mentioned, the level of development in advanced technology plays an important role in being able to benefit from a geothermal energy source. The initial installation cost of geothermal energy systems can be high. Since the first investment cost includes accessing the geothermal energy, the energy access item is added to the investment cost. Understandably, the initial investment cost is higher for obtaining deeper geothermal energy resources. Generally, geothermal energy systems are long-term energy generation systems. In other words, they become self-paying systems after many years of use. The first investment cost of geothermal energy is mostly determined by the system's level of development.

Since geothermal energy sources bear a close relationship with a country's location, countries that do not have geothermal energy resources are not able to benefit from these resources efficiently. If countries that do not have this energy source want to transport the generated energy over some distance, this is not a very effective use since there will be high temperature losses. Another problem for geothermal energy sources is sustainability. When we look at the formation process of geothermal energy resources, we can see that many very long years are required. Geothermal energy sources are formed when rainwater seeps down through the gaps between the crust of the Earth over years. If the output in the geothermal energy source is more than the input, we face a sustainability problem of this energy source. To prevent this from happening in geothermal energy generation systems, the used geothermal fluid is generally sent underground again. But since all systems that utilize geothermal energy resources do not follow such a method, the sustainability problem of geothermal energy sources remains a daily problem.

In this section, the general disadvantages of geothermal energy sources have been given. These disadvantages should be considered together with the advantages of geothermal energy sources, on which studies should be carried out. In general, the following example can be given. Sometimes the long-term potential of geothermal energy sources can be ignored in order to save on the initial investment cost. Some steps may be therefore taken to save the initial cost of the energy plant, forgoing the benefits in the long term. However, considering the potential of the system, the savings made in the initial investment cost, after a certain time, are not actually savings. As can be seen from this example, the advantages of geothermal power generation systems are greater and their disadvantages lessened if the long term is kept in mind. Despite their disadvantages, in general, considering the benefits obtained from geothermal energy systems, geothermal energy sources are a potential energy source for humanity and other living things.

3.7 Future perspective of geothermal energy

As long as geothermal energy resources exist, countries around the world will continue to use these energy resources to both strengthen their economies and to minimize harmful effects on the environment. The level of development of countries in advanced technology will determine the level of utilization of this energy [28].

People need to be informed for countries to benefit socially from this energy. If society adopts geothermal energy, the usage areas of geothermal energy will expand. Here, the biggest responsibility rests with universities and other educational institutions. The more effectively topics, such as environmental impact and cost, are explained, the more common the use of geothermal energy resources will be in society.

In order to reduce the costs of geothermal energy systems, the drilling costs included in the initial investment cost should be reduced. Advances in enhanced technology play an important role in reducing these costs. Another cost reduction work that needs to be done is to improve the performance of the systems. In order to increase the efficiency of the systems, environmental impact and performance analyses should be performed using the data of existing systems [29,30]. In the pursuit of these studies, support and investment for research and development activities for geothermal energy use should be increased. These research and development activities will also bring technological advances.

Although the harmful effect of geothermal energy systems on the environment is low, this effect should not be ignored. Therefore, in order to reduce such harmful effects, it is necessary to carry out studies that will minimize this effect during the initial installation of geothermal energy systems.

Studies that contribute most to the future of geothermal energy systems are modeling studies. Again, at this point, a great role must be played by research and development activities. In general, geothermal energy sources have a potential place in the energy economy because geothermal energy sources can increase the performance of the systems by being integrated with existing energy systems. Usage areas of geothermal energy resources should be increased in order to meet the increasing energy demand and to contribute to the global and national economy. At the same time, the integration efforts of this energy resource into green and smart cities and countries for energy production systems should be accelerated.

Considering renewable energy sources in general, solar energy, wind energy, and geothermal energy sources are expected to become more popular for power generation. Although the main reasons for this have already been mentioned, it will be useful to indicate what is important in this chapter as well. Global warming and environmental impact policies have a great impact on the utilization of the great potential of renewable energy sources. For this reason, many companies that generate income from energy source production are interested in the increasing popularity of renewable energy sources. Especially in academic

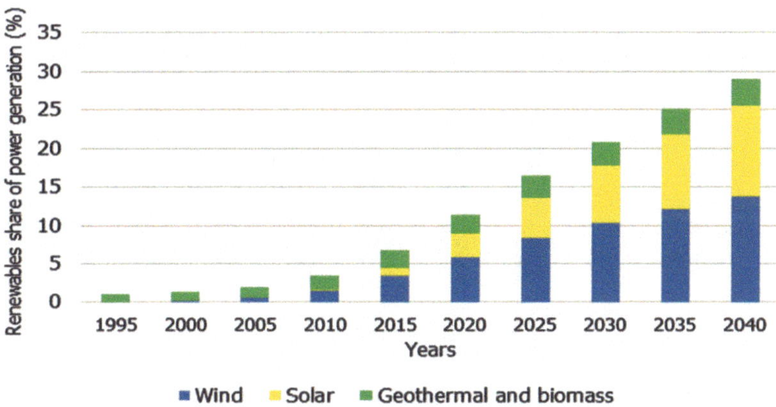

FIGURE 3.16 Renewables share of power generation by source up to 2040. *Data from BP [31].*

studies, energy generation systems using renewable energy sources are designed and performance analyses are carried out. Studies on the potential of future renewable energy sources are carried out on both platforms by providing various statistical data. As seen in Fig. 3.16, BP company carries out a study on the use of renewable energy sources in power generation [31]. As seen in the figure, in the breakout of renewable energy sources over the years, geothermal energy sources have been used for power generation in almost every period. It is also seen from this graph that it was the only renewable energy source used for power generation until 2005. Over the years and with the development of technology, it is seen that other renewable energy sources have had a share in power generation. The chart also includes forecasts for the post-2020 period to show the potential of the share of renewable energy sources in power generation. It is predicted that the share of geothermal energy resources in power generation will increase by approximately 1% every five years. Statistical data of other renewable energy sources are also available in this chart, and it is useful to briefly touch on them to explain the shares of renewable energy sources in power generation.

As seen in Fig. 3.16, the share of wind energy in power generation is gradually increasing. Compared to other renewable energy sources, it is seen that the trend in wind energy is higher. Since future estimates are based on this increasing trend, it is predicted that wind will have a higher share in power generation compared to other renewable energy sources in the coming years. The share of solar energy in power generation also shows similar behavior to wind energy. Although the use of solar energy in power generation began later than others, it is competing with wind energy especially with advances in advanced technology.

In general, in Fig. 3.16, the share of renewable energy sources as of 2020 is expected to be approximately 11.5%. Considering advances in

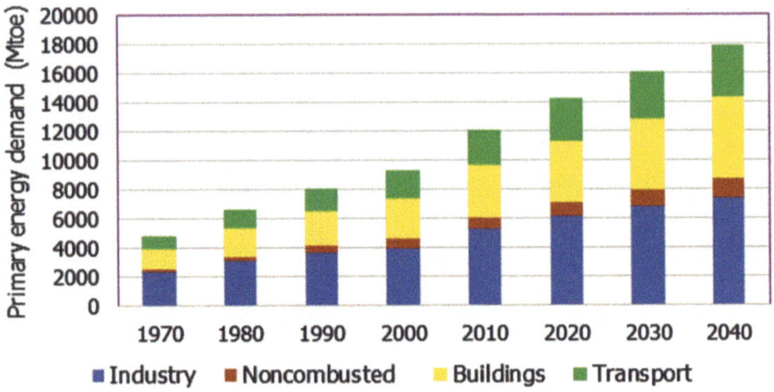

FIGURE 3.17 Primary energy demand for end use sector up to 2040. *Data from BP [31].*

technology, this share in power generation is expected to be approximately 29% by 2040, showing us the potential of renewable energy sources in power generation.

The amount of energy demand naturally also varies because the sectors in which energy is used differs. Energy demand varies for different areas, such as industrial applications, and needs to be met continuously. So natural energy needs are higher in cases where the energy need is to be met continuously. But when energy demand such as transportation is only partially needed in different periods, the amount of energy per unit of time that needs to be met may be lower than the other. A comparative analysis of the energy demand in various fields as million tonnes of oil equivalent (Mtoe) over the years realized by BP company is given in the graphic in Fig. 3.17. As can be clearly seen from the graph, the energy demand in industrial applications is higher in each period. The difference between energy demands in industrial applications, especially between 2000 and 2010, is higher than in other periods. Among the reasons for this is the fact that industrialization gained speed in the world between these periods. In this graph, it is seen that the energy demand in industrial applications was 6195 Mtoe by 2020, and accordingly, this demand will be 7443 Mtoe by 2040. As can be seen from Fig. 3.17, energy demand in the transportation sector shows a marked increase trend between periods. The reason for this is the alternative transportation channels that come with developing technology in the transportation sector. Another reason may be the increasing world population and the accompanying need for transportation. Looking at the graph in general, another area that is similar to the transportation sector in that it needs more energy is that of residential areas whose energy demand is increasing gradually. Energy demand between periods in residential areas varies between 400 Mtoe and 800 Mtoe. From this graph, it can be seen approximately how much energy demand corresponds to oil in

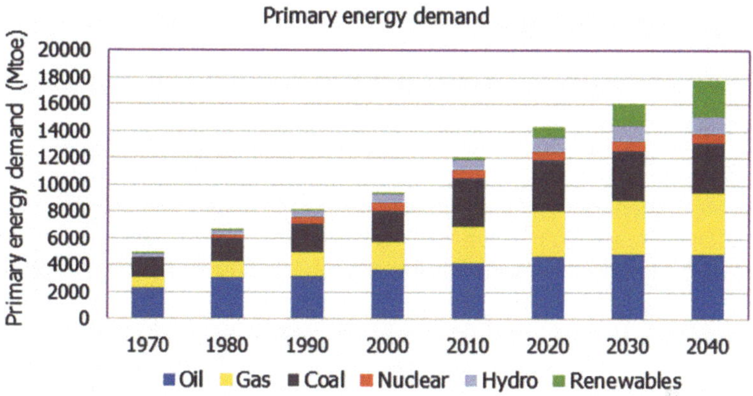

FIGURE 3.18 The primary energy demand of the world up to 2040. *Data from BP [31].*

various fields. Therefore, the potential of renewable energy sources should be used to meet energy demand, both to reduce the global warming rate and to support environmental policies.

Various energy sources are used to meet the increasing energy demand. The quality of these energy sources varies naturally. In addition to the quality of the energy source, many parameters affect the usage potential of this energy source. One of the most important of these parameters is the environmental impact. Increasing global warming and environmental impact policies encourage energy systems designed to use cleaner energy sources because sustainability is an important parameter for the Earth. Comparative analysis of the shares of different energy sources in energy demand by BP company over the years is given in Fig. 3.18. As seen from the graph, the largest share in energy demand is clearly oil, and the share of gas in energy demand has increased over the years. Accordingly, the share of energy demand is predicted to increase in the future. The share of renewable energy sources in energy demand can be clearly seen in 2010 and after. Studies on renewable energy sources were carried out in 2010 and before then, but the share of other resources in energy demand is greater. In general, however, the share of renewable energy sources in energy demand has increased exponentially. In the period between 2000 and 2020, the data in the graph show that great progress has been made in renewable energy sources. During these years, the share of renewable energy resources in energy demand in 2000 was 59 Mtoe, its share in 2010 was 234 Mtoe, and its share in 2020 is given as 802 Mtoe. As of 2010, the share of renewable energy resources in energy demand increased to nearly four times that of 2000. As of 2020, this share increased to approximately three times that of 2010. By 2040, the share of renewable energy resources in energy demand is expected to be 2748 Mtoe. This shows that the share and potential of renewable energy resources in energy demand will increase exponentially.

3.8 Closing remarks

According to recent reports, geothermal energy plants can supply a strong, long-lasting alternative with attributes that complement other significant power generation systems from clean coal gasification, nuclear, solar, wind, hydropower, and biomass. Geothermal power sources show numerous benefits over alternative energy resources. As an example, unlike biomass, fossil fuel, or nuclear energy sources, geothermal energy is location specific and requires no transportation of raw material from the source of extraction to the power plant.

The primary utilization of geothermal energy covers a wide range of applications, such as residential heating and domestic hot water supply, aquaculture, greenhouse heating, swimming pool and balneology, industrial heating processes, heat pumps and power production. But for the higher geothermal system efficiency, decreased thermal losses and wastes, decreased operating costs, decreased harmful gaseous emissions, better use of geofluid sources, multiple generation options and increased reliability, the geothermal energy—based integrated systems depending on the local availability of resources should be designed and operated. Geothermal energy should be incorporated with different alternative energy sources, such as solar, biomass, wind, etc. in the actual processes, depending on the local availability of resources for generating useful outputs.

This chapter provides detailed information on the history and nature of geothermal energy resources and presents comprehensive data retrieved from several countries around the world to investigate geothermal source potential. The presented data show that electricity installed capacity and annual electricity generation in the world reached 31462 MW and 87120 GWh, respectively, in 2019. Also, the highest installed geothermal power capacity and the highest number of direct and indirect geothermal energy varieties are applied in the United States, followed by Indonesia, the Philippines, Turkey, New Zealand, Mexico, Italy, Iceland, Kenya, and Japan.

Iceland, with its population of 364,000, makes up 85% of its overall space heating demands with its abundant geothermal energy resources, thus making the country as a whole almost zero-emission in terms of space heating. On the other hand, considering the abundance of geothermal energy on Earth, Iceland might be seen as visionary for the rest of the world. In addition, Italy and Turkey also illustrate high performance, comparing their annual energy generation rates to their populations. As to continents, Africa, with its high potential, illustrates the minimum utilization rate, which indicates that the potential should be utilized for energy generation from the geothermal energy resources for communities and even for remote areas.

As a final comment, it can be noted that geothermal energy is one of the most abundant renewable energy resources, without showing any intermittence as in solar, wind, and biomass applications. Considering 90% of the Earth is hotter than 100°C, the utilization of and further investigations into the geothermal power usage for many applications, ranging from direct utilization to power generation, are the sustainable solution to the world's challenges on energy shortcomings and a logical alternative to carbon-based fuel utilization.

Nomenclature

h	Specific enthalpy (kJ/kg)
T	Temperature (°C; K)

Acronyms

BP	British Petroleum
IRENA	International Renewable Energy Agency
toe	Tonne of oil equivalent
US	United States
USA	United States of America
US-DOE	United States Department of Energy

References

[1] A. Toth, E. Bobok, Flow and Heat Transfer in Geothermal Systems: Basic Equations for Describing and Modeling Geothermal Phenomena and Technologies, Elsevier, Amsterdam, Netherlands, 2017.

[2] Kutscher, C. "The status and future of geothermal electric power", No. NREL/CP-550-28204, National Renewable Energy Laboratory, Golden, Colorado, US, 2000.

[3] M.H. Dickson, M. Fanelli, Geothermal Energy: Utilization and Technology, Renewable Energy Series, UNESCO Publishing, John Wiley & Sons, 1995.

[4] IRENA, "Renewable capacity statistics 2020" International Renewable Energy Agency (IRENA), Abu Dhabi, 2020.

[5] R. Bertani, Geothermal power generation in the world 2010–2014 update report, Geothermics 60 (2016) 31–43.

[6] I. Stober, K. Bucher, Geothermal Energy: From Theoretical Models to Exploration and Development, Springer, New York, 2013.

[7] IRENA, "Geothermal energy data", International Renewable Energy Agency, 2020. Available: https://www.irena.org/geothermal. [Accessed 21 March 2020].

[8] TGE, Geothermal energy data in Think geo-energy 2018, https://www.thinkgeoenergy.com/the-top-10-geothermal-countries-2018-based-on-installed-generation-capacity-mwe, [Accessed 14 February 2020].

[9] I. Dincer, H. Ozcan, Geothermal energy, Comprehensive Energy Systems, Elsevier, 2018, pp. 702–732.

[10] US-DOE, 2020, US Department of Energy [Online]. Available: https://www.energy.gov/eere/geothermal/geothermal-energy-us-department-energy [Accessed 30 March 2020].

[11] Mertoglu, O. Simsek, S., Basarir, N. Geothermal country update report of Turkey (2010–2015), Proceedings World Geothermal Congress 2015 Melbourne, Australia, 19–25 April 2015.

[12] M. Ozturk, N.C. Bezir, N. Ozek, Energy market structure of Turkey, Energy Sources, Part. B 3 (4) (2008) 384–395.
[13] REN21, "Renewables 2019: Global status report" REN21 Secretariat, 2019.
[14] M. Ozturk, Y.E. Yuksel, Energy structure of Turkey for sustainable development, Renew. Sustain. Energy Rev. 53 (2016) 1259–1272.
[15] E. Pérez-Denicia, F. Fernández-Luqueño, D. Vilariño-Ayala, L. Montaño-Zetina, L. Maldonado-López, Renewable energy sources for electricity generation in Mexico: a review, Renew. Sustain. Energy Rev. 78 (2017) 597–613.
[16] R.M. Prol-Ledesma, D.J. Morán-Zenteno, Heat flow and geothermal provinces in Mexico, Geothermics 78 (2019) 183–200.
[17] A. Santilano, A. Manzella, G. Gianelli, A. Donato, G. Gola, I. Nardini, et al., Convective, intrusive geothermal plays: what about tectonics? Geotherm. Energy Sci. 3 (2015) 51–59.
[18] A. Manzella, R. Bonciani, A. Allansdottir, S. Botteghi, A. Donato, S. Giamberini, et al., Environmental and social aspects of geothermal energy in Italy, Geothermics 72 (2018) 232–248.
[19] Orkustofnun "Primary Energy Use in Iceland 1940-2017, OS-2018-T009-01" Orkustofnun - National Energy Authority, [Online]. Available: https://rafhladan.is/bitstream/handle/10802/15253/OS-2018-T009-01.pdf?sequence = 1 [Accessed 11 February 2020].
[20] Yasukawa K., Sasada, M. "Country update of Japan: renewed opportunities" Proceedings World Geothermal Congress 2015 Melbourne, Australia, 19–25 April 2015.
[21] B. Ruggero, Geothermal power generation in the world 2010–2014 update report, Geothermics 60 (2016) 31–43.
[22] P. Muffler, R. Cataldi, Methods for regional assessment of geothermal resources, Geothermics 7 (2-4) (1978) 53–89.
[23] M.P. Hochstein, Classification and assessment of geothermal resources, in: M.H. Dickson, M. Fanelli (Eds.), Small Geothermal Resources: A Guide to Development and Utilization, UNITAR, New York, 1990, pp. 31–57.
[24] Y. Benderitter, G. Cormy, Possible approach to geothermal research and relative costs, in: M.H. Dickson, M. Fanelli (Eds.), Small Geothermal Resources: A Guide to Development and Utilization, UNITAR, New York, 1990, pp. 59–69.
[25] K. Nicholson, Geothermal Fluids: Chemistry and Exploration Techniques, Springer Verlag, Berlin, 1993.
[26] G. Axelsson, E. Gunnlaugsson, Geothermal utilization, management and monitoring, in long-term monitoring of high and low-enthalpy fields under exploitation, World Geotherm. Congress: Short. Course, Mori oka (2000) 3–10.
[27] Sanyal, S.K. "Classification of Geothermal systems–a possible scheme" Proceedings 30th workshop on geothermal reservoir engineering. Stanford, CA: Stanford University; January 31–February 2, 2005.
[28] A.A. Kocer, M. Ozturk, Thermodynamic analysis of power and hydrogen production from renewable energy-based integrated system, Int. J. Exergy 19 (4) (2016) 519–543.
[29] Y.E. Yuksel, M. Ozturk, Thermodynamic and thermoeconomic analyses of a geothermal energy based integrated system for hydrogen production, Int. J. Hydrog. Energy 42 (4) (2017) 2530–2546.
[30] Y.E. Yuksel, M. Ozturk, I. Dincer, Development of a geothermal based integrated plant for generating clean hydrogen and other useful commodities, J. Energy Resour. Technol. 142 (9) (2020) 1–13.
[31] BP Energy, BP Energy Outlook 2019 edition https://www.bp.com/content/dam/bp/business-sites/en/global/corporate/pdfs/energy-economics/energy-outlook/bp-energy-outlook-2019.pdf [Accessed 14 February 2020].

Study questions and problems

3.1. From an ecological viewpoint, how does geothermal power utilization differ from that of carbon-based-fueled plants? What are the advantages? What are the disadvantages?
3.2. What geological fields are most likely to have high heat flows?
3.3. How can you express the nature of geothermal energy resources?
3.4. What is the relationship among energy efficiency, environment, and sustainability?
3.5. What regions in the world are best suited for geothermal energy plants? Why? Which regions are the least well suited?
3.6. What are the key points that determine the sustainability of a geothermal project?
3.7. Explain the difference between geothermal energy—based electricity installed capacity and annual electricity generation.
3.8. What indicators determine the sustainability of the geothermal energy resources? What operational programs can be utilized to improve the sustainability?
3.9. Describe the nature of geothermal resources.
3.10. What are the main environmental effects of geothermal energy—based power generation systems?
3.11. Categorize the methods for geothermal source temperature classification.
3.12. What are the benefits of geothermal energy resources?
3.13. What are the disadvantages of geothermal energy resources?
3.14. What are the significant indicators that are generally required to successfully develop a geothermal energy—based design project?
3.15. Describe the global warming effect.
3.16. Describe acid precipitation formation and its environmental effect.
3.17. Comment on the energy efficiency impact on the environment.
3.18. Identify and describe several methods for increasing the performance and decreasing the environmental effect of the geothermal energy—based plants.
3.19. The most common application of geothermal energy is electricity production. Some other utilizations are process heating and cooling, district heating and cooling, greenhouse heating, and heating for fish farming. From the economic, thermodynamic, and environmental points of view and considering the quality of energy, which of these uses do you recommend most? Explain.
3.20. Geothermal resources should be classified based on resource temperature or resource exergy. Which classification is more suitable if geothermal energy is to be used for:
(a) district heating,
(b) cooling, and
(c) power generation? Explain.

Chapter 4

Geothermal energy utilization

4.1 Introduction

The word "geothermal" comes from the Greek words *geo* (earth) and *therme* (heat), or "Earth heat." In other words, geothermal heat is the thermal heat power within the Earth's crust, that is, the warm rock and fluid (steam or geothermal water containing large amounts of dissolved solids) that fills the pores and fractures within the rock and flows within sand and gravel. Geological investigations illustrate that the Earth, originating from a completely molten situation, would have cooled and become completely solid many thousands of years ago without thermal power input, in addition to that of solar energy. Also, it is accepted that the main resource of geothermal heat energy is radioactive decay within the Earth. The origin of this heat is linked with the internal structure of the planet and the physical processes occurring within it.

Geothermal energy resources have been utilized commercially for over 85 years and four decades on the scale of hundreds of megawatts for power production and direct utilization. Electricity, considered as an energy source, is the energy source that provides a more comfortable relationship between input given to the system and output from the system. In other words, it gives satisfactory results in terms of system performance. With geothermal energy sources, many useful outputs can be obtained with or without electricity. The parameter to be considered in the production of these useful outputs is the temperature of the geothermal energy source [1]. The temperatures at which electricity generation occurs are generally high. Table 4.1 lists the usages of these geothermal energy sources at different temperatures.

As can be seen from Table 4.1, a wide range of geothermal energy sources are used. From this table, it is clearly seen that they are used in applications with higher energy requirements as the geothermal source temperature increases. In this section, general information about the ways of utilizing geothermal energy and the useful outputs obtained are given.

4.2 Heating applications

One of the important useful outputs obtained from geothermal energy systems is heat that can be used for heating purposes. When the geothermal energy resources were first used, personal needs, such as bathing and cleaning were

Geothermal Energy Systems. DOI: https://doi.org/10.1016/B978-0-12-820775-8.00003-9

TABLE 4.1 Geothermal energy classification by resource temperature and application areas.

Classification of sources	Source temperature range (°C)	Geothermal fluid phase in reservoir	Geothermal fluid state at utilization	Power transformation technology	Applications	Operational problems
Nonelectrical	20 to 50	Liquid water	Liquid water	Direct use	– fish farming – swimming pool – thermal bath – fermentation – aquaculture – soil warming – mushroom – growing – heat pumps	The cleaning, filtration, control of outlet temperature, material selection, installation location selection, scaling and corrosion, design and construction are the main problems.
	50 to <100				– greenhouse/space heating – dry air – grains/fruits/vegetable drying – freshwater – process heating	
Very low temperature	100 to <150	Liquid water	Liquid water Steam-water mixture	Binary	– food drying – drying of stock fish – leather and fur treatment – washing and dying of textiles – pulp and paper processing – space cooling – direct steam – evaporation in sugar refining – power production	The cleaning, filtration, control of outlet temperature, material selection, installation location selection, scaling and corrosion, design and construction are the main problems.

Low temperature	150 to <190	Liquid water	Liquid waterSteam-water mixture	BinaryTwo-stageFlashHybridKalina cycle	– drying farm products at high rates – canning of food – refrigeration by ammonia absorption – digestion in paper pulp – chemical production – power production	Calcite scaling in production wells and stibnite scaling in binary plant are occasional problems.
Moderate temperature	190 to <230	Liquid water	Steam-water mixture	Single-stageFlashTwo-stageFlashHybrid	– chemical production – power production	Calcite scaling in production wells occasional problem, aluminosilicate scale in injection system a rare problem
High temperature	230 to <300	Liquid water; Liquid-dominated two-phase	Steam-water mixtureSaturated steam	Single-stageFlashHybrid	– hydrogen production – alternative fuel production – conventional power production – multigeneration	Silica scaling in the injection system; occasionally corrosion; occasionally high NCG content
Ultra high temperature	>300	Liquid-dominated two-phase	Steam-water mixtureSaturated steamSuperheated steam	Single-stageFlash	– hydrogen production – alternative fuel production – conventional power production – multigeneration	High NCG content; silica scaling in injection system; occasionally corrosion; silica scaling potential in production wells at lower wellhead pressures
Steam field	240°C (33.5 bar – a pressure; 2800 kJ/kg enthalpy)	Steam	Saturated or superheated steam	Direct steam	– hydrogen production – alternative fuel production – conventional power production – multigeneration	High NCG content or corrosion or silica particulate deposition on turbine blades

Source: Modified from [2].

provided by these resources. Historical structures related to meeting these personal needs are still available in some countries [3]. As time goes on and with the advancement of technology, the ways of utilizing geothermal energy have also become diversified. The heat output from the geothermal energy source may take different forms. With the design of the geothermal energy system, the heating output can be obtained in various parts of the system. Therefore, where this heating output is used makes the geothermal energy system valuable. The use of heating output obtained by geothermal energy systems is generally carried out in two sectors: space heating, which is specific to a place in a house or in a building, and district heating, which is an enhanced version of space heating for a district where multiple residential buildings benefit for heating purposes.

The first use of geothermal energy sources is generally for space heating, which is widely used today with the developing technology [4]. The geothermal energy source to be used for space heating must have a certain temperature value because the performance of the system in the heating process must cover the cost of the system. In general, when the temperatures of geothermal energy sources are examined, temperatures between 30°C and 125°C are suitable for space heating. However, high temperatures can be a problem in terms of system components because geothermal fluid may need to be conditioned in order to use the component in a sustainable way.

4.2.1 Ground source heat pumps

In the world, the residential sector is considered one of the most energy consuming and hence most polluting. Therefore, there has been increasing interest in using renewable energy–based solutions for heating applications, and geothermal energy options are in this regard recognized as one of the most suitable possibilities.

Effective use of geothermal energy for heating applications can decrease carbon-based source consumption. When geothermal energy is desired for space heating, some requirements should be considered and studied [5,6]. These can be considered mainly as climate, population, building type, the technological aspect, and the economic aspect. Ground source heat pumps (GSHPs) utilize the stable temperature level of the planet as the heat exchanging area instead of the reference air [7]. Based on these phenomena, GSHP leads to a higher temperature variation between higher and lower working point temperatures, bringing a higher Carnot effectiveness. Also, indicators affecting the running of GSHPs and their performances are investigated in this subsection. The geothermal energy–based heating and cooling systems consist of three main components: (1) heat pump, (2) underground heat exchanger, and (3) distribution system such as air ducts. Ground source heat pump processes with horizontal circulation loops and vertical circulation loops for heating and cooling applications are illustrated in Figs. 4.1 and 4.2, respectively.

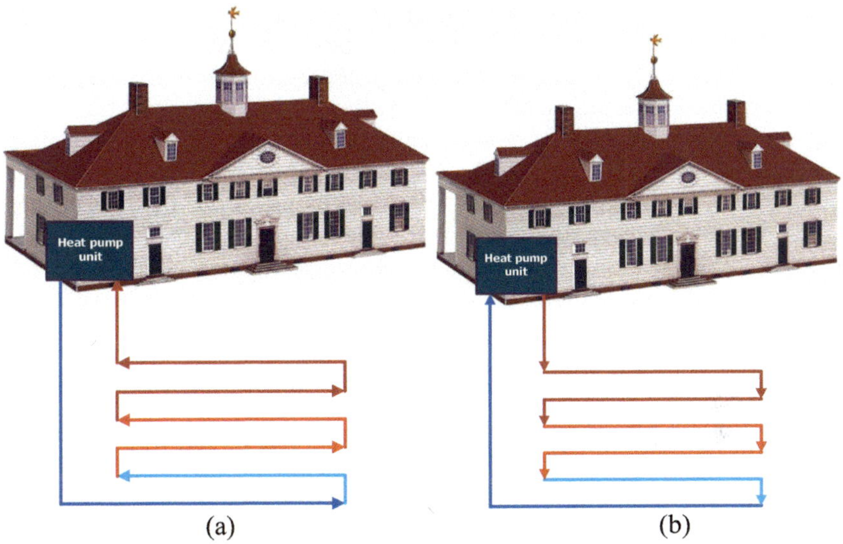

FIGURE 4.1 Ground source heat pump process with horizontal circulation loops for (A) heating and (B) cooling applications.

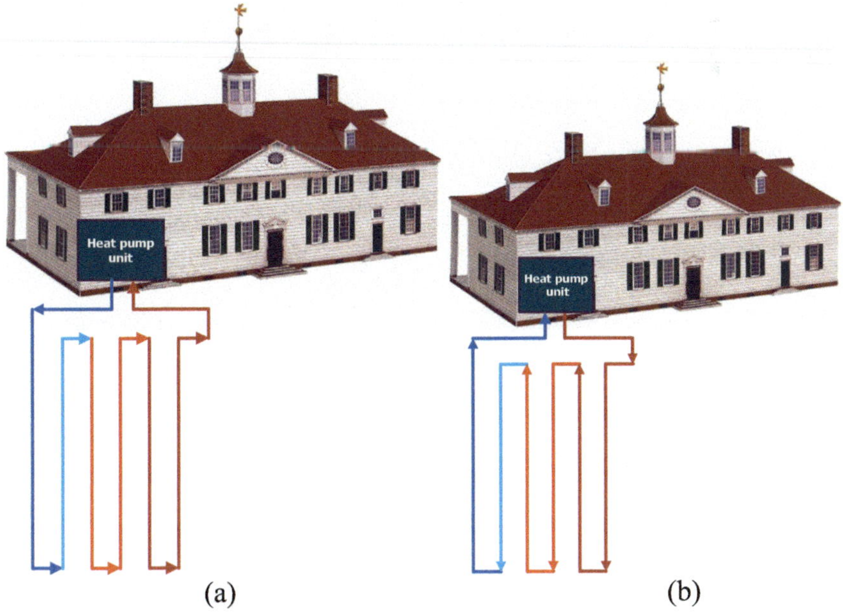

FIGURE 4.2 Ground source heat pump process with vertical circulation loops for (A) heating and (B) cooling applications.

In addition, more detailed working and design comparisons of ground source heat pumps with horizontal and vertical circulation loops are tabulated in Table 4.2. The design practices of these ground source heat pumps with horizontal circulation loops and vertical circulation loops for heating and cooling aims should be considered based on the properties comparison given in Table 4.2.

Generally, as given in Fig. 4.3, the heat pumps work between a high-temperature point (heat sink at T_H) and a low-temperature point (heat source at T_L). The GSHPs, also called ground-coupled heat pumps, geothermal heat

TABLE 4.2 Working and design comparison of ground source heat pumps with horizontal and vertical circulation loops.

Properties	Vertical ground-source heat pump	Horizontal ground-source heat pump
Performance	3	2
Design standard		
Practicality	1	1
Fabrication difficulty	3	2
Life cycle cost		
Installation	3	2
Operation	1	1
Maintenance	1	1
Total	1	1
Environmental		
Carbon emission	1	1
Land disturbance	2	1
Water contamination	0	0
Durability	2	2
Practical issues		
Operating restrictions	1	1
Aesthetics	2	2
Quietness	3	3
Vandalism	0	0
Safety	2	2

Note: Numbers given in this table show the significance of the characteristic: 3 = high significance, 2 = moderate significance, 1 = minor significance, and 0 = no significance.
Source: Data compiled from [8,9].

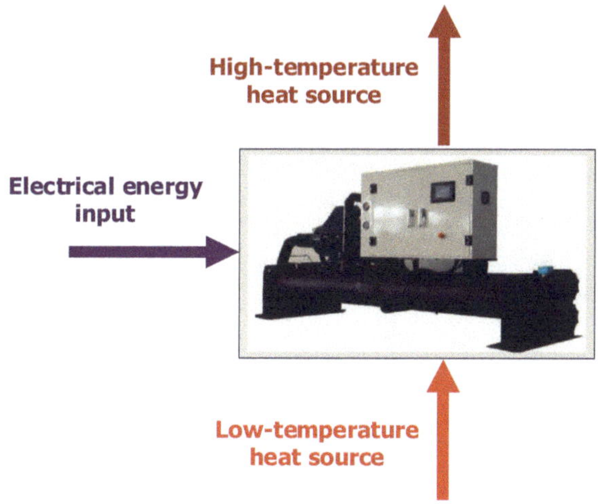

FIGURE 4.3 Schematic diagram of an electric heat pump.

pumps, or ground heat pumps, are the heat pumps that utilize the ground as a heat sink. The heat energy exchange between the ground and the heat pump process happens by using the ground heat exchangers (GHEs). In order to heat a space for long periods, the heat exchange between the heat pump cycle and space is done by using the heating coils.

As shown in Fig. 4.4, a heat pump system primarily has four subcomponents: (1) expansion valve, (2) evaporator, (3) compressor, and (4) condenser. These systems are commonly utilized heat pumps for heating applications in the winter season and in a reversed mode for cooling applications in the summer season, as defined in Chapter 2. The working conditions are the same as for refrigeration plants.

In the heating time (ground heat energy removal), the working temperature of the heating coil can be taken into account as the high-temperature source, while the working fluid temperature of the GHEs can be taken into account as the low-temperature source. For the steady-state condition, the coefficient of performance of GSHPs for heating applications is defined as:

$$COP_{\text{heating}} = \frac{1}{1 - \frac{T_{\text{wf}}}{T_{\text{hc}}}} \tag{4.1}$$

Also, the energetic COP and exergetic COP terms are used to describe the heating effects for heat pump systems for heating applications, and these terms are given as follows:

$$COP_{\text{en,heating}} = \frac{\dot{Q}_{\text{heating}}}{\dot{W}_{\text{Comp}}} \tag{4.2}$$

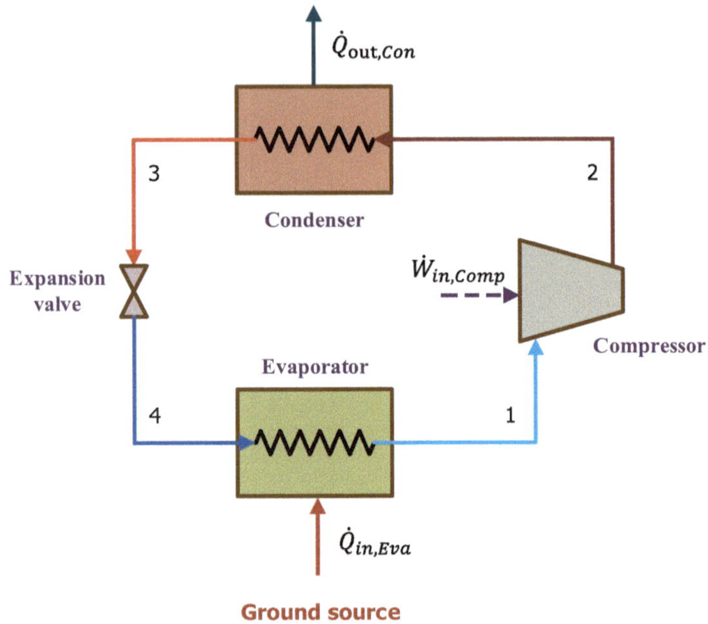

FIGURE 4.4 Schematic diagram of a simple heat pump for heating mode.

and

$$COP_{\text{ex,heating}} = \frac{\dot{E}x^{Q}_{\text{heating}}}{\dot{E}x^{W}_{\text{Comp}}} \tag{4.3}$$

Here, T_{hc} is the heating coil temperature and T_{wf} is the working fluid temperature. It can be noted that the coefficient of performance equations which given above can only be utilized for reversible heat pump systems.

The design of heat exchangers (HEXs) is essential for ground source heat pumps. Generally, HEX subsystems of geothermal heat pumps can be divided into main two types: low-temperature and high-temperature heat exchangers. Also, as seen from Fig. 4.5, the low-temperature geothermal heat exchangers have mainly three categories: (1) direct exchange, (2) open loop, and (3) closed loop (horizontal, vertical, Slinky, and pond/lake).

4.2.1.1 Direct exchange

The direct exchange connection, called the geothermal heat exchanger, is in direct connection with the ground source heat pump by the flowing working fluid in the ground. The thermal energy is exchanged directly between the working fluid and ground by using copper pipes. This type of thermal energy transfer is more effective than an extra heat exchanger

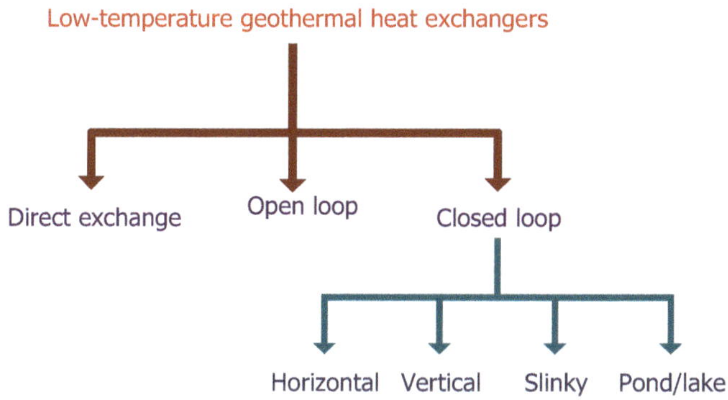

FIGURE 4.5 Low-temperature geothermal heat exchanger types.

integrated with the condenser or evaporator of the heat pump plant. Moreover, the utilization of a cooling fluid gives higher thermal energy transfer rates through the phase change of the cooling fluid. Hence, the direct exchange ground connections are frequently smaller in dimensions and can create lower installation and maintenance costs for a desired thermal power injection or extraction load.

4.2.1.2 Open loop

Geothermal energy sources, which have potential among renewable energy sources, are an effective source for heat pump cycles based on the potential temperature that they consistently contain. So the fluid temperature at the source is almost constant. When the use of this geothermal fluid in open cycle geothermal systems is examined, it is seen that the fluid is transferred from a geothermal well to the surface or to another well. In the region between these two transfer points, useful products are obtained directly from the geothermal fluid. One of the biggest contributors to this geothermal fluid utilization process is the heat pump. In general, considering the operation of the system, the inlet and outlet points of the geothermal fluid are not interconnected. Systems that do not have a connection between the inlet of the geothermal fluid into the system and its exit in the system are defined as open loop [10]. Open loop geothermal energy systems have the potential to be used in many applications. Heating or cooling of the ambient air flowing in the pipes can be realized with geothermal energy sources. Air entering any residential area can be heated and cooled if needed. Many situations affect the performance of open loop geothermal energy systems. The efficient operation of system components affects the performance of these systems. At the same time, the quality of the geothermal fluid is an important parameter for efficiency. These parameters should be taken into account in system setup and operation.

4.2.1.3 Closed loop

Closed loop geothermal systems are systems where useful products are obtained that utilize the potential of geothermal energy sources. Closed loop geothermal systems have continuous pipe loops to carry thermal energy. In these systems, the ends of these pipe loops are interconnected, and their arrangement is formed vertically or horizontally. In fact, the systems are named according to the arrangement of these pipes. Some systems with this layout are shown in Fig. 4.6, which shows horizontal ground heat exchangers (HGHE). These types of heat exchanger pipes are arranged in two different ways, and their ends are connected at one point. The plastic pipes of this type of heat exchanger are arranged either in series or in parallel to each other in a horizontal trench. These two types of heat exchangers are shown in Fig. 4.6A and Fig. 4.6B. The number of pipes and trenches in these heat exchangers naturally varies according to the size of the useful product output obtained from the system and the thermal properties of the area where the heat exchanger is located. When a cost analysis of this type of heat exchanger is done, it is seen that the exchanger is economical if sufficient space is available. Another model has been designed to reduce this needed space. This is defined as the Slinky type heat exchanger where the pipes are crimped between themselves as seen in Fig. 4.6C. Horizontal ground heat exchangers are installed in trenches at a depth of 1–2 m. The beneficial output of heating or cooling from these systems requires a horizontal ground heat exchanger with a length of 35–60 m/kW [10].

Some settlements are located near areas such as lakes or ponds. For these settlements, lake-/pond-type heat exchangers shown in Fig. 4.6D are recommended. This type of heat exchanger is similar to a Slinky-type heat

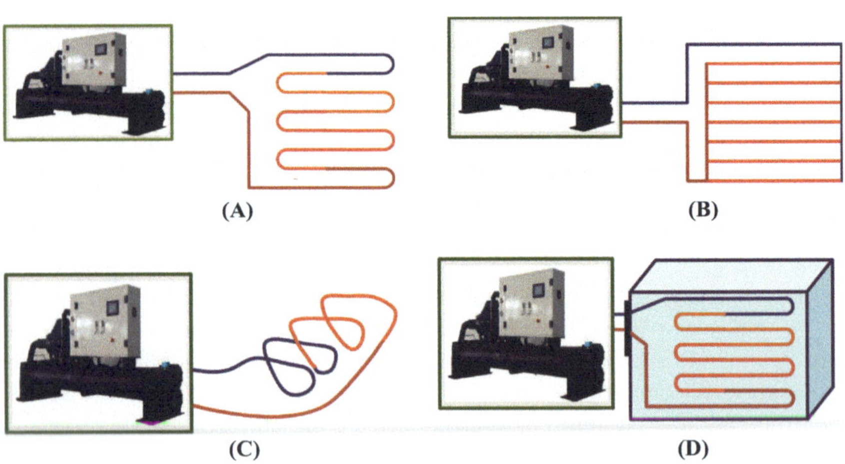

(A) (B) (C) (D)

FIGURE 4.6 Schematic diagrams of the closed-loop ground heat exchangers: (A) series connections, (B) parallel connection, (C) Slinky, and (D) pond/lake.

exchanger. Coils are used for this type of heat exchangers and, in Slinky-type heat exchangers, the coils are twisted like pipes.

Another closed-loop geothermal energy system is the vertical ground heat exchanger. These heat exchangers are called borehole heat exchangers (BHE) due to their design, and they are more costly to install than HGHE. But such heat exchangers require less piping for the desired useful product (heating or cooling) because deep underground is colder than the surface in hot seasons and hotter in cold seasons [11].

Several parameters affect the size of the geothermal heat exchanger and the number of pipes [12]. The main parameters are the geometry of the heat exchanger component, the capacity of the heat pump, and the temperature and thermal properties of the ground. The different types of heat exchangers just described have their advantages and disadvantages. The closed-loop connection option can be divided into three configurations: (1) closed-loop horizontal, (2) closed-loop Slinky, (3) closed-loop vertical, and (4) closed-loop pond.

The advantages and disadvantages of closed-loop and open-loop geothermal energy systems should be analyzed to effectively design applications. The advantages of closed-loop horizontal are that the cost of creating an underground space of closed loops with horizontal architecture is lower than the cost of closed-loop architectures with vertical architecture. More equipment, such as trenchers, is needed to carry out the excavation work for this operation and for the creation of the underground area of systems with this architecture. Different piping models can be tried due to the low depth of the system. The disadvantages of closed-loop horizontal are the large space requirement for its installation; system efficiency affected by seasonal conditions and the depth at which the system is placed; the piping that might be needed if the soil structure and the geothermal fluid level in the areas where the system is intended to be installed are not in the desired state; the possibility of damaging underground pipes during filling; requiring more piping than systems with vertical architecture performing equivalent production; the need for antifreeze against the risk of freezing in the cold season.

Closed-loop Slinky models have many advantages. One of them is the loop with Slinky architecture requires less trenching space compared to other horizontal architectures. The other advantage of this model is that the installation costs of such systems are naturally lower. The disadvantages of the closed-loop Slinky are that the pumping energy required by systems with this architecture is higher than that required by horizontal architectures. Further, due to the curved structure of the coils placed in the opened area for this system, filling the coils may cause problems depending on some soil types; in particular, filling the curved areas of the vertical coils in narrow spaces creates more difficulty than the coils placed flat in a wide area.

The other geothermal energy system configuration is closed-loop vertical, which has many advantages. This configuration requires fewer pipe length requirements than almost all other closed-loop architectures and requires less space to install the system. Also, this configuration is less affected by the temperature levels that may occur depending on the seasonal conditions. But the configuration has some disadvantages. One of them is that the drilling cost required for the installation of the system is higher than for systems with horizontal architecture. These systems, which generally have vertical architecture, tend to be in the category of geothermal heat pumps that require the highest cost. Also, soil temperature changes if there is not enough distance between the drill holes opened in order to install the system.

The closed-loop pond configuration has some advantages. Among all the architectures, this configuration may be the one that requires the least pipe length, and, if there is an existing body of water for the installation of the system, it may be the least expensive of closed loop architectures. At the same time, this architecture has some disadvantages. This architecture is known to require more permits than other embedded closed-loop systems to place the coils desired in the existing water body. Also, if the locations of the coils are not clearly shown, there is a possibility of damage from vessels anchoring in and moving over the water.

One of the geothermal energy system configurations is the open-loop pond. This configuration has many advantages. These are easy to design and install. Compared to the drilling cost of cycles with vertical architecture their cost is low. In order not to freeze the working fluid in the pipe, antifreeze should be used in cold weather conditions. Systems with this architecture may require less initial setup costs if existing wells can be used for water or irrigation applications used in living areas and if the wells have sufficient production capacity for the geothermal heat pump plant.

This configuration has some disadvantages. The configuration requires permitting from relevant authorities to use the groundwater and surface water subject to these authorities. Also, the architecture needs large amounts of water flow that may exceed the amount of water available. In addition, to provide a sufficient amount of water to the heat exchanger, harmful effects exposed to the environment will increase. Also, this configuration requires greater power requirements than pumps in closed-loop architectures to ensure the fluid movement of primary pumps in open-loop architectures. In addition, water discharge regulations may not allow the installation of single-well systems, and these regulations can also limit the design of systems with a vertical column open-loop architecture. The configuration has a higher initial cost of installation if a new well has to be created for draining fluid in the cycle.

Example 4.1: A geothermal energy–based radiator heating system, as illustrated in Fig. 4.7, where geothermal water at a temperature of 78°C, a pressure of 101.3 kPa, and a mass flow rate of 40 kg/s is utilized to maintain the temperature of a building at 22°C; the water exits the building heating system at a temperature of 50°C and a pressure of 120 kPa. Also, the working

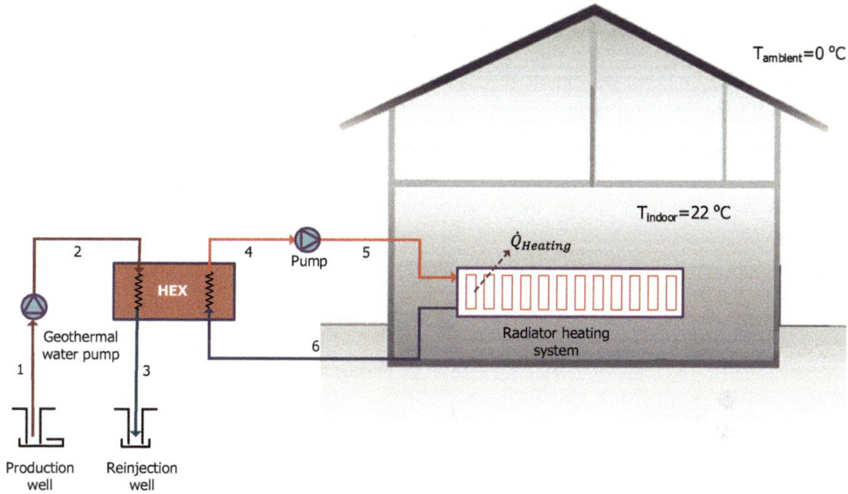

FIGURE 4.7 Schematic diagram of geothermal energy-based radiator heating system.

fluid in radiators of the building heating system has a temperature of 26°C and pressure of 110 kPa. Calculate:

1. the amount of heat extracted from hot water for the building,
2. the exergy destruction rate of heating system,
3. both the energy and exergy efficiencies of the radiator heating system,
4. the variation of exergy efficiency and exergy destruction rate of the radiator heating system, when the reference temperature increases from −20°C to +15°C.

Solution: First, some assumptions are required for geothermal energy-based radiator heating system analysis.

Assumptions:

- The system component operates under steady-state conditions.
- The reference temperature and pressure are taken as 0°C and 101.3 kPa, respectively.
- The pressure losses in the connections are negligible.
- The changes in the kinetic and potential energies and exergies are neglected.

Analysis:

The mass, energy, entropy, and exergy balance equations for the geothermal water pump are defined:

$$\dot{m}_1 = \dot{m}_2$$

$$\dot{m}_1 h_1 + \dot{W}_{GWP} = \dot{m}_2 h_2$$

$$\dot{m}_1 s_1 + \dot{S}_{g,\mathrm{GWP}} = \dot{m}_2 s_2$$

$$\dot{m}_1 ex_1 + \dot{W}_{\mathrm{GWP}} = \dot{m}_2 ex_2 + \dot{E}x_{D,\mathrm{GWP}}$$

The mass, energy, entropy, and exergy balance equations for the heat exchanger are written:

$$\dot{m}_2 = \dot{m}_3 \text{ and } \dot{m}_6 = \dot{m}_4$$

$$\dot{m}_2 h_2 + \dot{m}_6 h_6 = \dot{m}_3 h_3 + \dot{m}_4 h_4$$

$$\dot{m}_2 s_2 + \dot{m}_6 s_6 + \dot{S}_{g,\mathrm{HEX}} = \dot{m}_3 h_3 + \dot{m}_4 h_4$$

$$\dot{m}_2 ex_2 + \dot{m}_6 ex_6 = \dot{m}_3 ex_3 + \dot{m}_4 ex_4 + \dot{E}x_{D,\mathrm{HEX}}$$

The mass, energy, entropy, and exergy balance equations for the pump are given:

$$\dot{m}_4 = \dot{m}_5$$

$$\dot{m}_4 h_4 + \dot{W}_{\mathrm{P}} = \dot{m}_5 h_5$$

$$\dot{m}_4 s_4 + \dot{S}_{g,P} = \dot{m}_5 s_5$$

$$\dot{m}_4 ex_4 + \dot{W}_{\mathrm{P}} = \dot{m}_5 ex_5 + \dot{E}x_{D,P}$$

The mass, energy, entropy, and exergy balance equations for the radiator heating system are defined:

$$\dot{m}_5 = \dot{m}_6$$

$$\dot{m}_5 h_5 = \dot{m}_6 h_6 + \dot{Q}_{\mathrm{heating}}$$

$$\dot{m}_5 s_5 + \dot{S}_{g,\mathrm{RHS}} = \dot{m}_6 s_6 + \frac{\dot{Q}_{\mathrm{heating}}}{T_s}$$

$$\dot{m}_5 ex_5 = \dot{m}_6 ex_6 + \dot{E}x^{Q}_{\mathrm{heating}} + \dot{E}x_{D,\mathrm{RHS}}$$

The properties of the geothermal water and radiator water entering and exiting the radiator heating system are computed by using the EES software program, and the state properties of these working fluids are given in Table 4.3.

1. The amount of heat extracted from hot water for the building is:

$$\dot{Q}_{\mathrm{heating}} = \dot{m}_5(h_5 - h_6) = 31.69x(210.3 - 109.1) = 3205 \text{ kW}$$

2. The exergy destruction rate of the heating system is:

$$\dot{E}x_{D,\mathrm{RHS}} = \dot{m}_5(ex_5 - ex_6) - \dot{E}x^{Q}_{\mathrm{heating}} = \dot{m}_5(ex_5 - ex_6) - \dot{Q}_{\mathrm{heating}}x\left(1 - \frac{T_o}{T_s}\right)$$

$$= 31.69x(8.335 - 0.8652) - 3205x\left(1 - \frac{0 + 273}{22 + 273}\right) = 152.3 \text{ kW}$$

TABLE 4.3 Properties of the input and the output flows of working fluids for the geothermal energy–based heating system.

State point	$\dot{m}$(kg/s)	T (°C)	P (kPa)	h (kJ/kg)	s (kJ/kg K)	ex (kJ/kg)	$\dot{E}x$(kW)
0		15	101.3	63.08	0.2244		
1	18	78	101.3	326.7	1.052	25.21	453.8
2	18	78.2	110	327.5	1.054	25.37	456.7
3	18	36	110	150.9	0.5186	3.06	55.09
4	31.69	50	110	209.4	0.7038	8.235	803.9
5	31.69	50.2	120	210.3	0.7063	8.335	264.1
6	31.69	26	110	109.1	0.3812	0.8652	27.42

3. The energy efficiency of the radiator heating system is:

$$\eta_{RHS} = \frac{\dot{Q}_{heating}}{\dot{m}_5(h_5 - h_6)} = \frac{3205}{31.69x(210.3 - 109.1)} = 1 = 100\%$$

As one can see, the energetic performance is 100%, which does not represent what is actually happening thermodynamically and does not account for irreversibility or exergy destruction rates. This impractical situation makes it a requirement to go one step ahead to include exergetic assessment and exergetic performance analysis in order to reflect reality. In this regard, the energy efficiency of the radiator heating system is:

$$\psi_{RHS} = \frac{\dot{E}x^{Q}_{heating}}{\dot{m}_5(ex_5 - ex_6)} = \frac{238.9}{31.69x(8.335 - 0.8652)} = 0.6107 = 61.07\%$$

4. The effect of reference temperature on the exergy destruction rate and exergy efficiency of the geothermal energy–based radiator heating system is shown in Fig. 4.8. As shown in the figure, the exergy destruction rate is increased from 141.2 kW to 160.7 kW, and the exergy efficiency of the geothermal energy–based radiator heating system is decreased from 0.7637 to 0.3212 with the increasing reference temperature from −20°C to 15°C.

Example 4.2: A heat pump–based heating plant, as shown in Fig. 4.9, uses R-134a as the working fluid for a heating application. R-134a enters the compressor at 144.2 kPa at a flow rate of 768 l/min as a saturated vapor. The isentropic efficiency of the compressor is 85%. The refrigerant then leaves

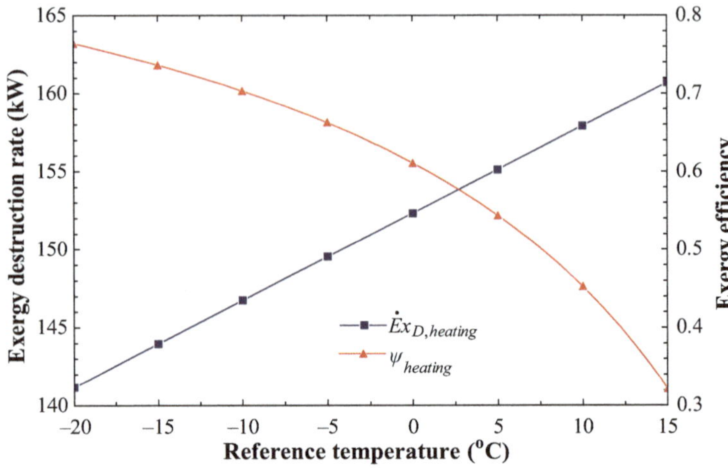

FIGURE 4.8 Effect of reference temperature on the exergy destruction rate and exergy efficiencies of the geothermal energy–based radiator heating system.

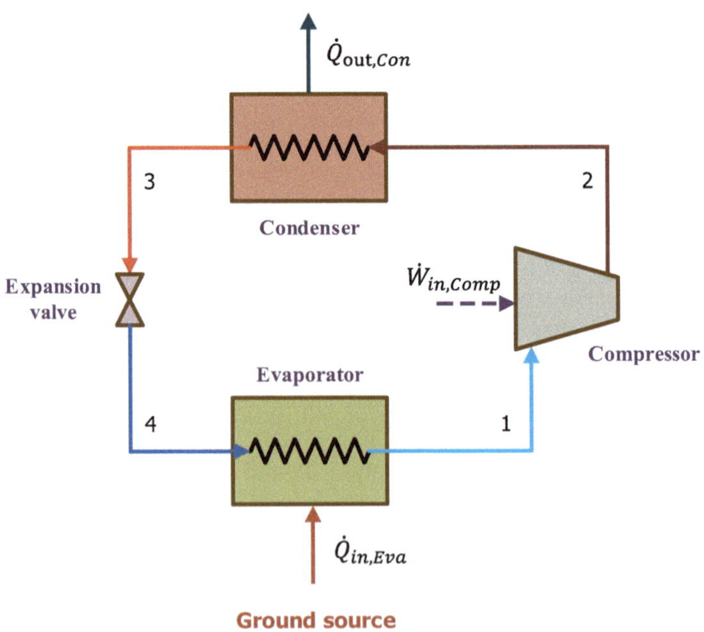

FIGURE 4.9 Schematic diagram of the heat pump–based heating plant.

the condenser at 48.17°C as a saturated liquid. The cold and hot reservoir temperatures are 0°C and 25°C, respectively. Calculate:

1. the heat input rate of the evaporator, the work consumption rate of the compressor, and the heat output rate from the condenser;

2. the energetic COP of the heating cycle;
3. the exergy destruction in each component of the cycle and the total exergy destruction in the heating plant;
4. the exergetic COP of the heating cycle;
5. the variation of energetic and exergetic COP of the heat pump–based heating plant, when the reference temperature increases from 0°C to 40°C.

Solution: First, some assumptions are required for the heating system analysis.

Assumptions:

- All components operate under steady-state conditions.
- The reference temperature and pressure are taken as 20°C and 101.3 kPa, respectively.
- The pressure losses in the connections between the components are negligible.
- The pressure heat energy losses from the components are negligible.
- The changes in the kinetic and potential energies and exergies are neglected.

Analysis:

The mass, energy, entropy, and exergy balance equations for the evaporator are written:

$$\dot{m}_4 = \dot{m}_1$$

$$\dot{m}_4 h_4 + \dot{Q}_{\text{in,Eva}} = \dot{m}_1 h_1$$

$$\dot{m}_4 s_4 + \frac{\dot{Q}_{\text{in,Eva}}}{T_{\text{Eva}}} + \dot{S}_{g,\text{Eva}} = \dot{m}_1 s_1$$

$$\dot{m}_4 ex_4 + \dot{E}x^{Q}_{\text{in,Eva}} = \dot{m}_1 ex_1 + \dot{E}x_{D,\text{Eva}}$$

$\dot{E}x^{Q}_{\text{in,Eva}}$ is defined:

$$\dot{E}x^{Q}_{\text{in,Eva}} = \left(\frac{T_o}{T_{\text{Eva}}} - 1\right)\dot{Q}_{\text{in,Eva}}$$

The mass, energy, entropy, and exergy balance equations for the compressor are expressed:

$$\dot{m}_1 = \dot{m}_2$$

$$\dot{m}_1 h_1 + \dot{W}_{\text{in,Cmp}} = \dot{m}_2 h_2$$

$$\dot{m}_1 s_1 + \dot{S}_{g,\text{Cmp}} = \dot{m}_2 s_2$$

$$\dot{m}_1 ex_1 + \dot{E}x^{W}_{\text{in,Cmp}} = \dot{m}_2 ex_2 + \dot{E}x_{D,\text{Cmp}}$$

The mass, energy, entropy, and exergy balance equations for the condenser are written:

$$\dot{m}_2 = \dot{m}_3$$

$$\dot{m}_2 h_2 = \dot{m}_3 h_3 + \dot{Q}_{\text{out,Con}}$$

$$\dot{m}_2 s_2 + \dot{S}_{g,\text{Con}} = \dot{m}_3 s_3 + \frac{\dot{Q}_{\text{out,Con}}}{T_{\text{HR}}}$$

$$\dot{m}_2 ex_2 = \dot{m}_3 ex_3 + \dot{E}x^{Q}_{\text{out,Con}} + \dot{E}x_{D,\text{Con}}$$

$\dot{E}x^{Q}_{\text{out,Con}}$ can be defined as:

$$\dot{E}x^{Q}_{\text{out,Con}} = \left(1 - \frac{T_o}{T_{\text{Con}}}\right)\dot{Q}_{\text{out,Con}}$$

The mass, energy, entropy, and exergy balance equations for the expansion valve are defined:

$$\dot{m}_3 = \dot{m}_4$$

$$\dot{m}_3 h_3 = \dot{m}_4 h_4$$

$$\dot{m}_3 s_3 + \dot{S}_{g,\text{EV}} = \dot{m}_4 s_4$$

$$\dot{m}_3 ex_3 = \dot{m}_4 ex_4 + \dot{E}x_{D,\text{EV}}$$

1. Before calculating the heat input, heat output, and work input rates, the thermodynamic variables needs to be calculated for each flow given in Fig. 4.9.

For flow 1:

As given in the question, $P_1 = 144.2$ kPa and $x_1 = 1$ (saturated vapor), and based on these data, h_1, s_1, and v_1 can be determined from the R-134a thermodynamic tables or EES as given in Table 4.4.

For flow 2:

In case there is no pressure loss in the condenser plant, $P_2 = P_3 = P_{3,\text{sat}@T_3}$, and also as given in the question, T_3 is equal to 48.17°C. Based on these data, P_2 and P_3 can be determined from the R-134a thermodynamic tables or EES as 1223 kPa, and h_{2s} can be determined as 284.9 kJ/kg. To calculate h_2, the isentropic efficiency of compressor can be used as:

$$\eta_{Cmp} = \frac{h_{2s} - h_1}{h_2 - h_1} \Rightarrow 0.85 = \frac{284.9 - 239.6}{h_2 - 239.6} \Rightarrow h_2 = 292.9 \text{ kJ/kg}$$

For flow 3:

As given in the equation, $x_3 = 0$ (saturated liquid), and P_3 is calculated as before. Based on these thermodynamical variables, h_3 and s_3 can be determined from the R-134a thermodynamic tables or EES, as given in Table 4.4.

TABLE 4.4 Chosen and calculated thermodynamic state points of the heating plant.

State point	$\dot{m}$(kg/s)	T (°C)	P (kPa)	h (kJ/kg)	s (kJ/kg K)	ex (kJ/kg)	$\dot{E}x$(kW)
0	—	20	101.3	272.1	1.091	—	—
1	0.09393	−18.08	144.2	239.6	0.944	10.46	0.9821
2	0.09393	63.94	1259	292.9	0.9679	56.74	5.33
3	0.09393	48.17	1259	120.6	0.4332	41.24	3.873
4	0.09393	−18.08	144.2	120.6	0.4777	28.2	2.649

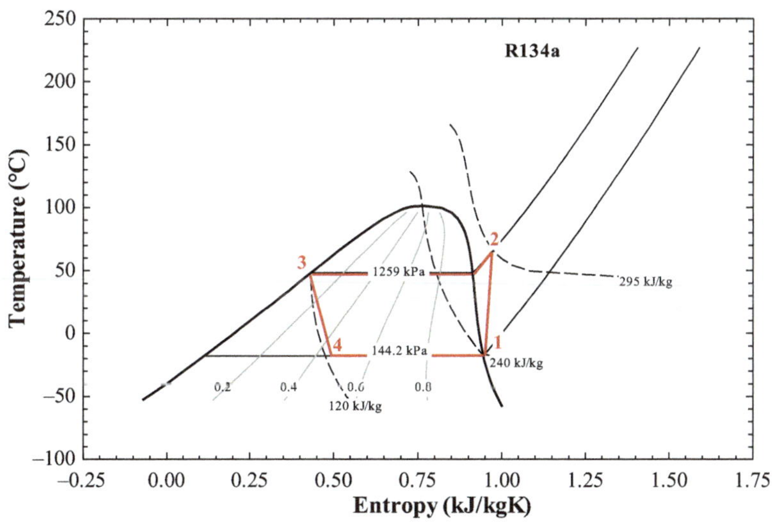

FIGURE 4.10 T-s diagram of the heat pump cycle for heating application.

For flow 4:

In the expansion valve, $h_3 = h_4$, and P_4 is equal to 144.2 kPa. Based on these data, s_4 can be determined from the R-134a thermodynamic tables or EES. Finally, the mass flow rate of R-134a can be calculated:

$$\dot{m}_{wf} = \frac{\dot{V}_1}{v_1} = \frac{(0.768/60)\text{m}^3/s}{0.1363m^3/\text{kg}} = 0.09393 \text{ kg}/s$$

The temperature−entropy diagram of the heat pump cycle−based heating plant is given in Fig. 4.10.

The heat input rate of the evaporator is:

$$\dot{Q}_{\text{in,Eva}} = \dot{m}_{\text{wf}}(h_1 - h_4) = \left(0.09393\frac{\text{kg}}{\text{s}}\right) \times (239.6 - 120.6)\frac{\text{kJ}}{\text{kg}} = 11.17\ \text{kW}$$

The work consumption rate of the compressor is:

$$\dot{W}_{\text{in,Cmp}} = \dot{m}_{\text{wf}}(h_2 - h_1) = \left(0.09393\frac{\text{kg}}{\text{s}}\right) \times (292.9 - 239.6)\frac{\text{kJ}}{\text{kg}} = 5.008\ \text{kW}$$

The heat output rate from the condenser is:

$$\dot{Q}_{\text{out,Con}} = \dot{m}_{\text{wf}}(h_2 - h_3) = \left(0.09393\frac{\text{kg}}{\text{s}}\right) \times (292.9 - 120.6)\frac{\text{kJ}}{\text{kg}} = 16.18\ \text{kW}$$

2. The energetic COP of the heating cycle is:

$$COP_{\text{en}} = \frac{\dot{Q}_{\text{in,Eva}}}{\dot{W}_{\text{in,Cmp}}} = \frac{11.17\ \text{kW}}{5.008\ \text{kW}} = 2.231$$

3. The exergy destruction rate of the evaporator is:

$$\dot{E}x_{D,\text{Eva}} = \dot{m}_{\text{wf}}(ex_4 - ex_1) + \dot{E}x^{Q}_{\text{in,Eva}} = \left(0.09393\frac{\text{kg}}{\text{s}}\right) \times (28.2 - 10.46)\frac{\text{kJ}}{\text{kg}} + \left(\frac{293}{273} - 1\right)11.17\ \text{kW} = 2.481\ \text{kW}$$

The exergy destruction of the compressor is:

$$\dot{E}x_{D,\text{Cmp}} = \dot{m}_{\text{wf}}(ex_1 - ex_2) + \dot{E}x^{W}_{\text{in,Cmp}} = \left(0.09393\frac{\text{kg}}{\text{s}}\right) \times (10.46 - 56.74)\frac{\text{kJ}}{\text{kg}} + 5.008\ \text{kW} = 0.6602\ \text{kW}$$

The exergy destruction rate of the condenser is:

$$\dot{E}x_{D,\text{Con}} = \dot{m}_{\text{wf}}(ex_2 - ex_3) - \dot{E}x^{Q}_{\text{out,Con}} = \left(0.09393\frac{\text{kg}}{\text{s}}\right) \times (56.74 - 41.24)\frac{\text{kJ}}{\text{kg}} - 16.18\ \text{kW}\left(1 - \frac{293}{298}\right) = 1.185\ \text{kW}$$

The exergy destruction rate of the expansion valve is:

$$\dot{E}x_{D,EV} = \dot{m}_{wf}(ex_3 - ex_4) = \left(0.09393\frac{\text{kg}}{\text{s}}\right) \times (41.24 - 28.2)\frac{\text{kJ}}{\text{kg}} = 1.224\ \text{kW}$$

The total exergy destruction in the heating system is:

$$\dot{E}x_{D,\text{Total}} = \dot{E}x_{D,\text{Eva}} + \dot{E}x_{D,\text{Cmp}} + \dot{E}x_{D,\text{Con}} + \dot{E}x_{D,\text{EV}} = 2.481 + 0.6602 + 1.185 + 1.224 = 5.5502\ \text{kW}$$

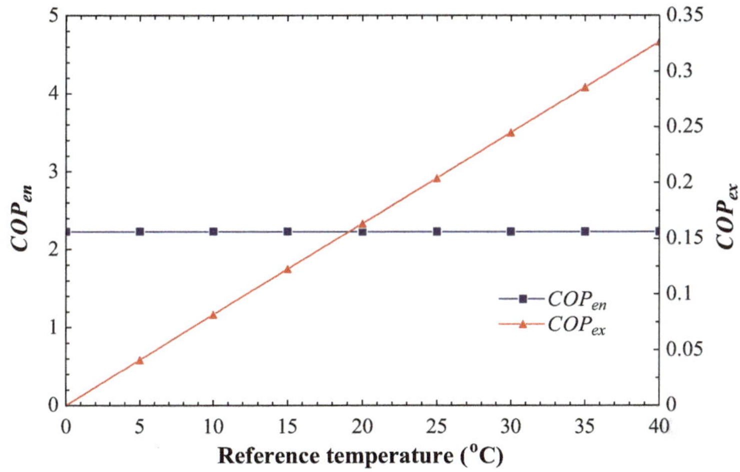

FIGURE 4.11 Effect of reference temperature on the energetic and exergetic COP of the heating plant.

4. The exergetic COP of the heating cycle is:

$$COP_{ex} = \frac{\dot{E}x^{Q}_{\text{in,Eva}}}{\dot{E}x^{W}_{\text{in,Cmp}}} = \frac{0.818 \text{ kW}}{5.008 \text{ kW}} = 0.1634$$

5. The effect of reference temperature on the energetic and exergetic COP of the heat pump–based heating plant is shown in Fig. 4.11. As shown in this figure, the energetic COP of the heating plant does not change with the increasing reference temperature from 0°C to 40°C, whereas the exergetic COP of the heating plant is increased from 0.0114 to 0.3267 in the examined reference temperature change.

4.3 Cooling production

One of the useful outputs obtained from geothermal energy sources is cooling output. This cooling output is realized by taking the geothermal energy source as an input to an energy production plant. Two fluids are generally used in geothermal energy generation systems to obtain cooling output as a useful product. Cycles using a single fluid are known as the vapor compression cycles. Cycles using two fluids to achieve cooling output are defined as absorption refrigeration cycles. Therefore, it is possible to obtain cooling output from the system by using a geothermal energy source with two fluids and system input. The difference that these two cycles make while obtaining the cooling product is the difference in the pressure parameter. Another difference in these two cycles is the cycle followed by the fluid used to perform the cooling operation in the cycle. In fact, these two differences occur interconnectedly in cycles. In some studies, these two differences are also given as a single difference. Explaining these differences in more detail helps to understand the difference between the cooling cycles [13]. In order to move the working fluid in the vapor compression cycle, it is necessary to create a pressure difference. This pressure difference is realized with the help of a compressor included in the system. In absorption refrigeration cycles, another working fluid is included in the system so that the working fluid can move in the cycle. These two cycles are summarized and depicted in Figs. 4.12 and 4.13, respectively. The typical single-stage vapor compression refrigeration plant consists of mainly four subcomponents: (1) compressor, (2) condenser, (3) expansion valve, and (4) evaporator. Also, a fan should be used to rapidly remove the warm indoor air.

For the space cooling application (ground heat energy delivery), the working fluid temperature of GHEs can be taken into account as the high-temperature source, and the cooling coil temperature can be taken into account as the low-temperature source in the cooling time. For the steady-state condition, the coefficient of performance of GSHPs for cooling applications is defined as:

$$COP_{\text{cooling}} = \frac{1}{1 - \frac{T_{cc}}{T_{wf}}} \tag{4.4}$$

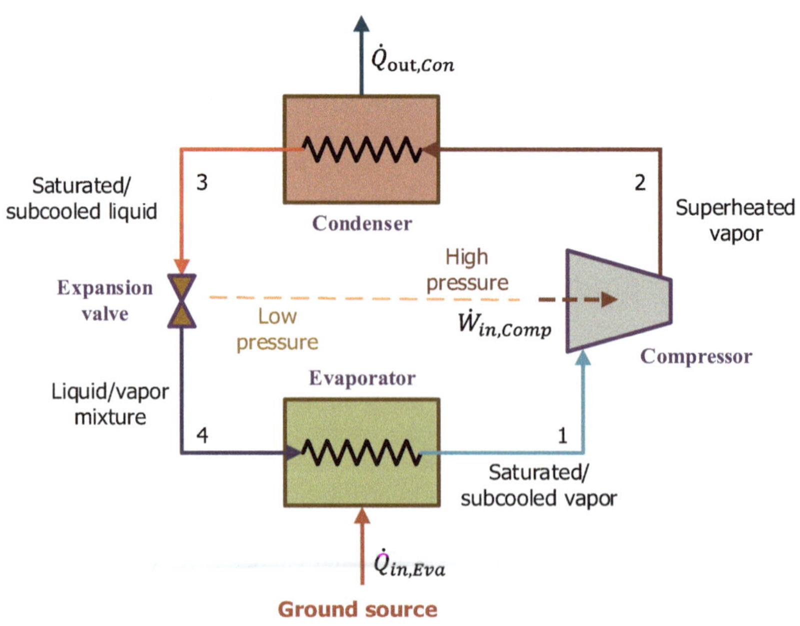

FIGURE 4.12 Schematic diagram of a typical single-stage vapor compression refrigeration plant.

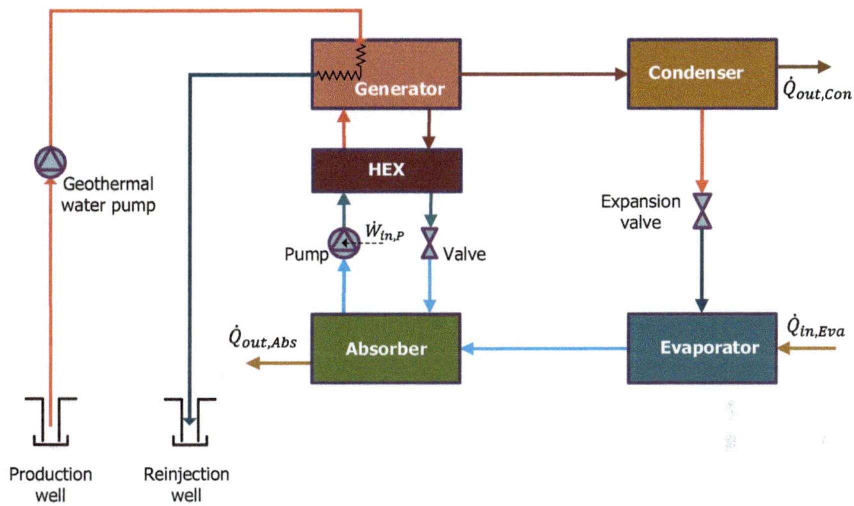

FIGURE 4.13 Schematic diagram of a single-effect absorption cooling plant.

Here, T_{cc} is the cooling coil temperature, and T_{wf} is the working fluid temperature. It can be noticed that the coefficient of performance equation just given can only be utilized for the reversible heat pump systems. Also, the energetic COP and exergetic COP terms are used to describe the cooling effects for heat pump systems for cooling applications, and these terms can be defined:

$$COP_{\text{en,cooling}} = \frac{\dot{Q}_{\text{cooling}}}{\dot{W}_{\text{Comp}}} \tag{4.5}$$

and

$$COP_{\text{ex,cooling}} = \frac{\dot{E}x^{Q}_{\text{cooling}}}{\dot{E}x^{W}_{\text{Comp}}} \tag{4.6}$$

Fig. 4.13 shows the single-effect absorption cooling plant. Ammonia and water can be used as working fluids for this cooling cycle, and in other applications, different working fluids are used in these cycles. The cooling market utilizes air conditioners, domestic-commercial fridges, coolers, precoolers, refrigerators, heat pumps, and freezers for different utilizations, ranging from food cooling to space cooling, which fundamentally utilizes cooling plants to reach the essential cooling outputs. This is performed at the designed temperature where refrigerants are used as cooling fluids. Also, these cooling fluids have great heat-absorbing qualities, making them appropriate for refrigeration. Cooling fluids are usually well-known as the refrigerants absorbing heat energy throughout vaporization. These cooling fluids, which supply a cooling impact throughout the phase change from liquid form to vapor form, are typically utilized in the cooling process, air conditioning, and heat pump systems, as well as chemical cycles. A detailed classification of cooling fluids (so-called refrigerants) are available elsewhere [14], and a brief listing is made here of the most widely used ones:

- halocarbons (e.g., R-11, R-12, R-22, R-113, R-114, R-115, etc.)
- hydrocarbons (e.g., R-50, R-170, R-290, R-600, R-600a, etc.)
- inorganic compounds (e.g., R-717, R-744, R-729, etc.)
- azeotropic mixtures (e.g., R-502, R-500, R-503, R-504, etc.)
- nonazeotropic mixtures (e.g., R-11 + R-12, R-12 + R-22, R-12 + R-114, R-13B1 + R152a, R-22 + R-114, R-114 + R-152a, etc.)

The energetic coefficients of performances (COP_{en}) and exergetic coefficients of performances (COP_{ex}) of single-effect absorption process can be written, respectively:

$$COP_{\mathrm{en}} = \frac{\dot{Q}_{\mathrm{eva}}}{\dot{Q}_{\mathrm{gen}} + \dot{W}_{\mathrm{p}}} \tag{4.7}$$

and

$$COP_{\mathrm{ex}} = \frac{\dot{E}x^{Q}_{\mathrm{eva}}}{\dot{E}x^{Q}_{\mathrm{gen}} + \dot{E}x^{W}_{\mathrm{P}}} \tag{4.8}$$

In this section, it is shown that a geothermal energy source is used to obtain a cooling product. Since the cooling product is used widely around the world, obtaining the cooling product from geothermal energy sources has great potential. For this reason, different designs are carried out in the literature for cooling by using geothermal energy sources, and performance analyses of these designs are made.

Example 4.3: A heat pump–based refrigeration system, as shown in Fig. 4.14, using R-134a as the cooling working fluid, is used for space cooling. R-134a enters the compressor subsystem at 143 kPa at a flow rate of 584 l/min as a saturated vapor. The isentropic efficiency of the compressor is 85%. The refrigerant then leaves the condenser at 47.02°C as a saturated liquid. The cold and hot reservoir temperatures are −5°C and 24°C, respectively. Find:

1. the rate of cooling provided by the evaporator, the work consumption rate of the compressor, and the rate of rejected heat by the condenser;
2. the energetic COP of the cooling cycle;
3. the exergy destruction in each component of the cycle, and the total exergy destruction in the cooling system;
4. the exergetic COP of the cooling cycle;
5. the variation of energetic and exergetic COP of the heat pump – based refrigeration system, when the reference temperature increases from 0°C to 40°C.

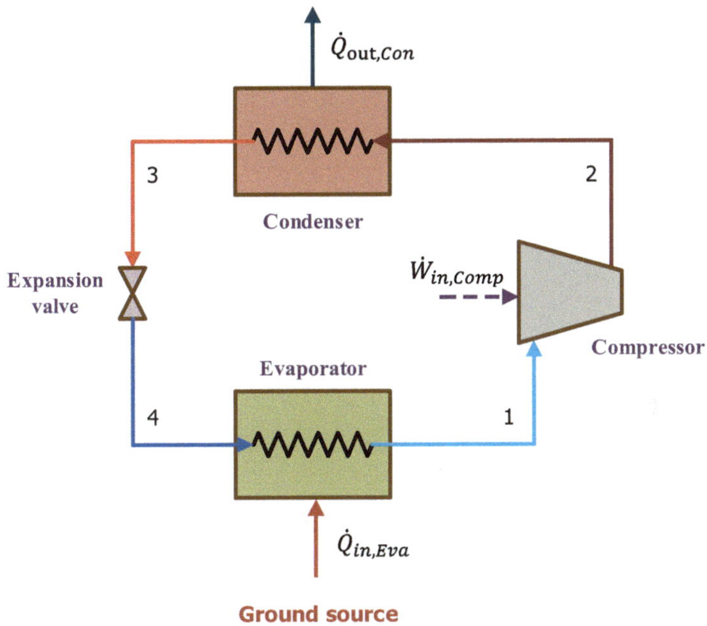

FIGURE 4.14 Schematic diagram of the heat pump – based refrigeration system.

Solution: First, some assumptions are required for cooling system analysis.

Assumptions:

- All components operate under steady-state conditions.
- The reference temperature and pressure are taken as 20°C and 101.3 kPa, respectively.
- The pressure losses in the connections between the components are negligible.
- The heat energy losses from the components are negligible.
- The changes in the kinetic and potential energies and exergies are neglected.

Analysis:

The mass, energy, entropy, and exergy balance equations for the evaporator are written:

$$\dot{m}_4 = \dot{m}_1$$

$$\dot{m}_4 h_4 + \dot{Q}_{\text{in,Eva}} = \dot{m}_1 h_1$$

$$\dot{m}_4 s_4 + \frac{\dot{Q}_{\text{in,Eva}}}{T_{\text{Eva}}} + \dot{S}_{g,\text{Eva}} = \dot{m}_1 s_1$$

$$\dot{m}_4 ex_4 + \dot{E}x^{Q}_{\text{in,Eva}} = \dot{m}_1 ex_1 + \dot{E}x_{D,\text{Eva}}$$

$\dot{E}x^{Q}_{\mathrm{in,Eva}}$ can be defined as:

$$\dot{E}x^{Q}_{\mathrm{in,Eva}} = \left(\frac{T_o}{T_{\mathrm{Eva}}} - 1\right)\dot{Q}_{\mathrm{in,Eva}}$$

The mass, energy, entropy, and exergy balance equations for the compressor are written:

$$\dot{m}_1 = \dot{m}_2$$

$$\dot{m}_1 h_1 + \dot{W}_{\mathrm{in,Cmp}} = \dot{m}_2 h_2$$

$$\dot{m}_1 s_1 + \dot{S}_{g,\mathrm{Cmp}} = \dot{m}_2 s_2$$

$$\dot{m}_1 ex_1 + \dot{E}x^{W}_{\mathrm{in,Cmp}} = \dot{m}_2 ex_2 + \dot{E}x_{D,\mathrm{Cmp}}$$

The mass, energy, entropy, and exergy balance equations for the condenser are expressed as:

$$\dot{m}_2 = \dot{m}_3$$

$$\dot{m}_2 h_2 = \dot{m}_3 h_3 + \dot{Q}_{\mathrm{out,Con}}$$

$$\dot{m}_2 s_2 + \dot{S}_{g,\mathrm{Con}} = \dot{m}_3 s_3 + \frac{\dot{Q}_{\mathrm{out,Con}}}{T_{\mathrm{Con}}}$$

$$\dot{m}_2 ex_2 = \dot{m}_3 ex_3 + \dot{E}x^{Q}_{\mathrm{out,Con}} + \dot{E}x_{D,\mathrm{Con}}$$

$\dot{E}x^{Q}_{out,Con}$ is defined as:

$$\dot{E}x^{Q}_{\mathrm{out,Con}} = \left(1 - \frac{T_o}{T_{\mathrm{Con}}}\right)\dot{Q}_{\mathrm{out,Con}}$$

The mass, energy, entropy, and exergy balance equations for the expansion valve are written:

$$\dot{m}_3 = \dot{m}_4$$

$$\dot{m}_3 h_3 = \dot{m}_4 h_4$$

$$\dot{m}_3 s_3 + \dot{S}_{g,\mathrm{EV}} = \dot{m}_4 s_4$$

$$\dot{m}_3 ex_3 = \dot{m}_4 ex_4 + \dot{E}x_{D,\mathrm{EV}}$$

1. Before calculating the heat input, heat output, and work input rates, the thermodynamic variables need to be calculated for each flow given in Fig. 4.14.

   For flow 1:

   As given in Example 4.3, $P_1 = 140$ kPa and $x_1 = 1$ (saturated vapor), and based on these data, h_1, s_1, and v_1 can be determined from the R-134a thermodynamic tables or EES as given in Table 4.5.

TABLE 4.5 Chosen and calculated thermodynamic state points of the cooling plant.

State point	$\dot{m}$(kg/s)	T (°C)	P (kPa)	h (kJ/kg)	s (kJ/kg K)	ex (kJ/kg)	$\dot{E}x$(kW)
0	—	40	101.3	289.3	1.147	—	—
1	0.07086	−18.28	143	239.5	0.9441	13.79	0.9769
2	0.07086	62.74	1223	292.3	0.9679	59.11	4.189
3	0.07086	47.02	1223	118.9	0.4278	54.88	3.889
4	0.07086	−18.28	143	118.9	0.471	41.35	2.93

For flow 2:

In case there is no pressure loss in the condenser plant, $P_2 = P_3 = P_{3,\text{sat@}T_3}$, and, as given in the question, T_3 is equal to 47.02°C. Based on these data, P_2 and P_3 can be determined from the R-134a thermodynamic tables or EES as 1223 kPa, and h_{2s} can be determined as 284.3 kJ/kg. To calculate h_2, the isentropic efficiency of compressor can be used:

$$\eta_{\text{Cmp}} = \frac{h_{2s} - h_1}{h_2 - h_1} \Rightarrow 0.85 = \frac{284.3 - 239.5}{h_2 - 239.5} \Rightarrow h_2 = 292.3 \text{ kJ/kg}$$

For flow 3:

As given in the equation, $x_3 = 0$ (saturated liquid) and P_3 is calculated as before. Based on these thermodynamically variables, h_3 and s_3 can be determined from the R-134a thermodynamic tables or EES, as given in Table 4.5.

For flow 4:

In the expansion valve, $h_3 = h_4$ and P_4 is equal to 140 kPa. Based on these data, s_4 can be determined from the R-134a thermodynamic tables or EES. Finally, the mass flow rate of R-134a can be calculated:

$$\dot{m}_{\text{wf}} = \frac{\dot{V}_1}{v_1} = \frac{(0.584/60)\text{m}^3/s}{0.1374 \text{ m}^3/\text{kg}} = 0.07086 \text{ kg}/s$$

For betting understanding, the temperature−entropy diagram of the heat pump−based refrigeration system is shown in Fig. 4.15.

The rate of cooling provided by the evaporator is:

$$\dot{Q}_{\text{in,Eva}} = \dot{m}_{\text{wf}}(h_1 - h_4) = \left(0.07086\frac{\text{kg}}{\text{s}}\right)x(239.5 - 118.9)\frac{\text{kJ}}{\text{kg}} = 8.545 \text{ kW}$$

The work consumption rate of the compressor is:

$$\dot{W}_{\text{in,Cmp}} = \dot{m}_{\text{wf}}(h_2 - h_1) = \left(0.07086\frac{\text{kg}}{\text{s}}\right)x(292.3 - 239.5)\frac{\text{kJ}}{\text{kg}} = 3.74 \text{ kW}$$

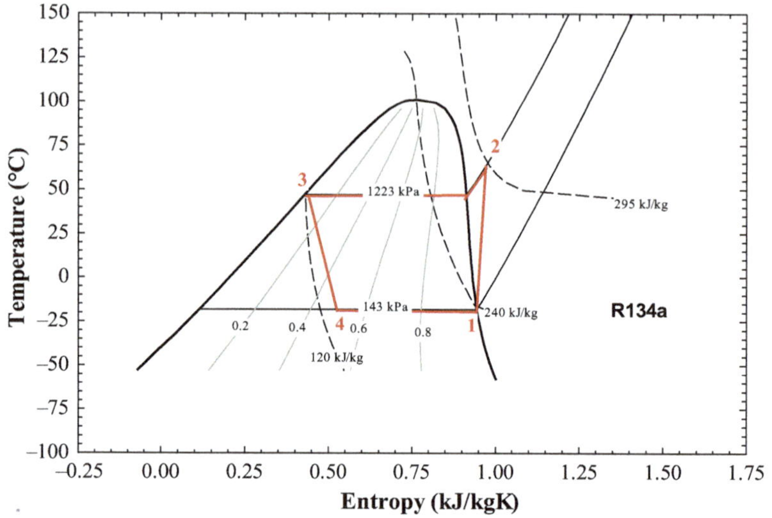

FIGURE 4.15 T-s diagram of heat pump–based refrigeration system for cooling application.

The rate of rejected heat by the condenser is:

$$\dot{Q}_{\text{out,Con}} = \dot{m}_{\text{wf}}(h_2 - h_3) = \left(0.07086\frac{\text{kg}}{\text{s}}\right) \times (292.3 - 118.9)\frac{\text{kJ}}{\text{kg}} = 12.29 \text{ kW}$$

2. The energetic COP of the cooling cycle is:

$$COP_{\text{en}} = \frac{\dot{Q}_{\text{in,Eva}}}{\dot{W}_{\text{in,Cmp}}} = \frac{8.545 \text{ kW}}{3.74 \text{ kW}} = 2.285$$

3. The exergy destruction in the evaporator is:

$$\dot{E}x_{D,\text{Eva}} = \dot{m}_{\text{wf}}(ex_4 - ex_1) + \dot{E}x^{Q}_{\text{in,Eva}} = \left(0.07086\frac{\text{kg}}{\text{s}}\right) \times (28.39 - 10.28)\frac{\text{kJ}}{\text{kg}} + \left(\frac{293}{268} - 1\right) \times 8.545 \text{ kW} = 2.0778 \text{ kW}$$

The exergy destruction in the compressor is:

$$\dot{E}x_{D,\text{Cmp}} = \dot{m}_{\text{wf}}(ex_1 - ex_2) + \dot{E}x^{W}_{\text{in,Cmp}} = \left(0.07086\frac{\text{kg}}{\text{s}}\right) \times (10.28 - 56.08)\frac{\text{kJ}}{\text{kg}} + 3.74\text{kW} = 0.4949 \text{ kW}$$

The exergy destruction in the condenser is:

$$\dot{E}x_{D,\text{Con}} = \dot{m}_{\text{wf}}(ex_2 - ex_3) - \dot{E}x^{Q}_{\text{out,Con}} = \left(0.07086\frac{\text{kg}}{\text{s}}\right) \times (56.08 - 41.05)\frac{\text{kJ}}{\text{kg}} - \left(1 - \frac{293}{297}\right) \times 12.29\ \text{kW} = 0.8996\ \text{kW}$$

The exergy destruction in the expansion valve is:

$$\dot{E}x_{D,\text{EV}} = \dot{m}_{\text{wf}}(ex_3 - ex_4) = \left(0.07086\frac{\text{kg}}{\text{s}}\right) \times (41.05 - 28.39)\frac{\text{kJ}}{\text{kg}} = 0.8977\ \text{kW}$$

The total exergy destruction in the cooling system is:

$$\dot{E}x_{D,\text{Total}} = \dot{E}x_{D,\text{Eva}} + \dot{E}x_{D,\text{Cmp}} + \dot{E}x_{D,\text{Con}} + \dot{E}x_{D,\text{EV}} = 2.0778 + 0.4949 + 0.8996 + 0.8977 = 4.37\ \text{kW}$$

4. The exergetic COP of the cooling cycle is:

$$COP_{\text{ex}} = \frac{\dot{E}x^{Q}_{\text{in,Eva}}}{\dot{E}x^{W}_{\text{in,Cmp}}} = \frac{0.7967\ \text{kW}}{3.74\ \text{kW}} = 0.213$$

5. The effect of reference temperature on the energetic and exergetic COP of the heat pump–based refrigeration system is shown in Fig. 4.16. As shown in the figure, the energetic COP of the cooling system does not change with the increasing reference temperature from 0°C to 40°C, whereas the exergetic COP of the cooling system is increased from 0.0426 to 0.3834 in the examined reference temperature change.

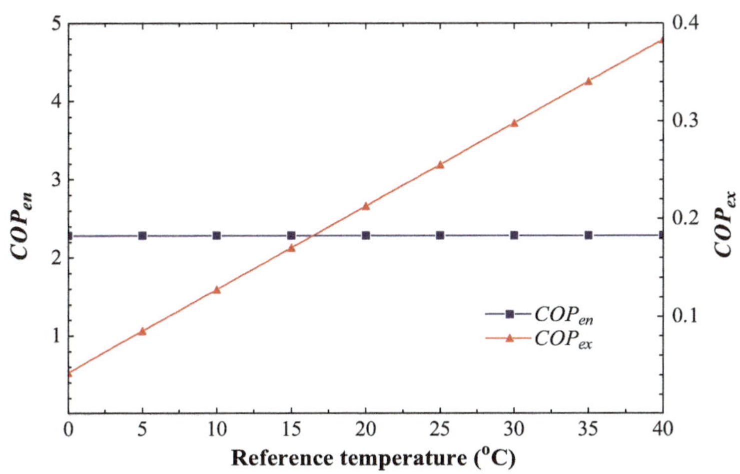

FIGURE 4.16 Effect of reference temperature on the energetic and exergetic COP of the cooling plant.

Example 4.4: A single-effect absorption refrigeration plant, as illustrated in Fig. 4.17, using ammonia–water as a working solution is used for space cooling. The strong ammonia–water solution with a mass flow rate of 2 kg/s enters the pump as a saturated liquid with a concentration of 0.6 and a pressure of 250 kPa. The strong solution leaves the pump at 400 kPa with a temperature increase of 0.4°C. The strong solution exits from the HEX and generator at points 3 and 4 at temperatures of 84°C and 99°C, respectively. Heat is supplied to the generator at a rate of 200 kW, which helps to boil the strong solution, and the weak solution and ammonia refrigerant vapor leave the generator with concentrations of 0.4 and 0.999, respectively. Consider the condenser load and evaporator temperatures to be 480.5 kW and −13.5°C, respectively. Calculate:

1. the power consumed by the pump,
2. the rate of cooling provided by the evaporator,
3. the exergy destruction rate in each component,
4. total exergy destruction rate of the cooling system,
5. the energetic and exegetic COPs,
6. the variation of exergy destruction rate, energetic and exergetic COP of the single-effect absorption cooling system, when the reference temperature increases from −10°C to 30°C.

Solution: First, some assumptions are required for geothermal energy–based single-effect absorption cooling system analysis.

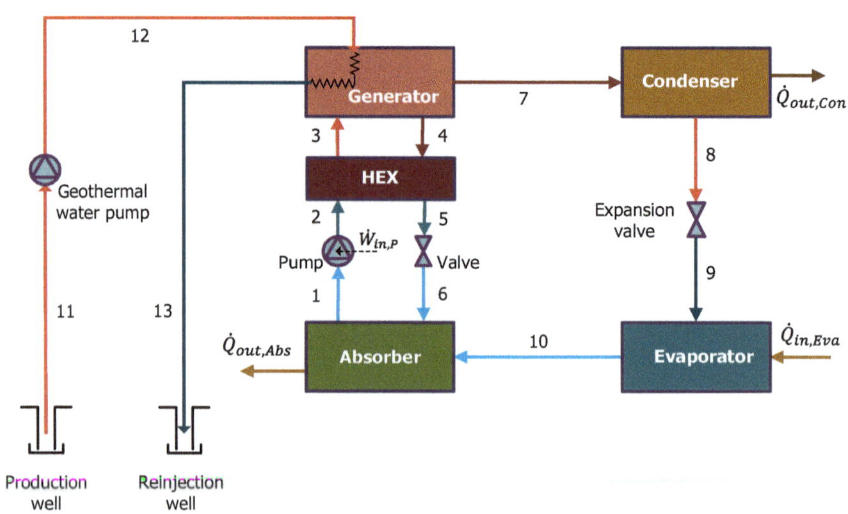

FIGURE 4.17 Schematic diagram of the geothermal energy–based single-effect absorption cooling system.

Assumptions:

- All components operate under steady-state conditions.
- The reference temperature and pressure are taken as 16°C and 101.3 kPa, respectively.
- The pressure losses in the connections between the components are negligible.
- The heat energy losses from the components are negligible.
- The changes in the kinetic and potential energies and exergies are neglected.

Analysis:

The mass, energy, entropy, and exergy balance equations are written for the generator:

$$\dot{m}_{12} = \dot{m}_{13}; \dot{m}_3 = \dot{m}_4 + \dot{m}_7$$

$$\dot{m}_3 h_3 + \dot{m}_{12} h_{12} = \dot{m}_4 h_4 + \dot{m}_7 h_7 + \dot{m}_{13} h_{13}$$

$$\dot{m}_3 s_3 + \dot{m}_{12} s_{12} + \dot{S}_{g,\mathrm{Gen}} = \dot{m}_4 s_4 + \dot{m}_7 s_7 + \dot{m}_{13} s_{13}$$

$$\dot{m}_3 ex_3 + \dot{m}_{12} ex_{12} + \dot{E}x^{Q}_{\mathrm{Gen}} = \dot{m}_4 ex_4 + \dot{m}_7 ex_7 + \dot{m}_{13} ex_{13} + \dot{E}x_{D,\mathrm{Gen}}$$

The mass, energy, entropy, and exergy balance equations can be defined for the condenser:

$$\dot{m}_7 = \dot{m}_8$$

$$\dot{m}_7 h_7 = \dot{m}_8 h_8 + \dot{Q}_{\mathrm{out,Con}}$$

$$\dot{m}_7 s_7 + \dot{S}_{g,\mathrm{Con}} = \dot{m}_8 s_8 + \dot{Q}_{\mathrm{out,Con}}/T_{\mathrm{con}}$$

$$\dot{m}_7 ex_7 = \dot{m}_8 ex_8 + \dot{E}x^{Q}_{\mathrm{out,Con}} + \dot{E}x_{D,\mathrm{Con}}$$

The mass, energy, entropy, and exergy balance equations are described for the expansion valve:

$$\dot{m}_8 = \dot{m}_9$$

$$\dot{m}_8 h_8 = \dot{m}_9 h_9$$

$$\dot{m}_8 s_8 + \dot{S}_{g,\mathrm{ev}} = \dot{m}_9 s_9$$

$$\dot{m}_8 ex_8 = \dot{m}_9 ex_9 + \dot{E}x_{D,\mathrm{ev}}$$

The mass, energy, entropy, and exergy balance equations can be written for the evaporator:

$$\dot{m}_9 = \dot{m}_{10}$$

$$\dot{m}_9 h_9 + \dot{Q}_{\mathrm{in,Eva}} = \dot{m}_{10} h_{10}$$

$$\dot{m}_9 s_9 + \dot{Q}_{\mathrm{in,Eva}}/T_{\mathrm{Eva}} + \dot{S}_{g,\mathrm{Eva}} = \dot{m}_{10} s_{10}$$

$$\dot{m}_9 ex_9 + \dot{E}x^{Q}_{\mathrm{in,Eva}} = \dot{m}_{10} ex_{10} + \dot{E}x_{D,\mathrm{Eva}}$$

The mass, energy, entropy, and exergy balance equations are defined for the absorber:

$$\dot{m}_6 + \dot{m}_{10} = \dot{m}_1$$

$$\dot{m}_6 h_6 + \dot{m}_{10} h_{10} = \dot{m}_1 h_1 + \dot{Q}_{\text{out,Abs}}$$

$$\dot{m}_6 s_6 + \dot{m}_{10} s_{10} + \dot{S}_{g,\text{Abs}} = \dot{m}_1 s_1 + \dot{Q}_{\text{out,Abs}}/T_{\text{Abs}}$$

$$\dot{m}_6 ex_6 + \dot{m}_{10} ex_{10} = \dot{m}_1 ex_1 + \dot{E}x^{Q}_{\text{out,Abs}} + \dot{E}x_{D,\text{Abs}}$$

The mass, energy, entropy, and exergy balance equations can be described for the pump as:

$$\dot{m}_1 = \dot{m}_2$$

$$\dot{m}_1 h_1 + \dot{W}_{\text{in},P} = \dot{m}_2 h_2$$

$$\dot{m}_1 s_1 + \dot{S}_{g,P} = \dot{m}_2 s_2$$

$$\dot{m}_1 ex_1 + \dot{E}x^{W}_{\text{in},P} = \dot{m}_2 ex_2 + \dot{E}x_{D,P}$$

The mass, energy, entropy, and exergy balance equations can be written for the valve:

$$\dot{m}_5 = \dot{m}_6$$

$$\dot{m}_5 h_5 = \dot{m}_6 h_6$$

$$\dot{m}_5 s_5 + \dot{S}_{g,\text{Val}} = \dot{m}_6 s_6$$

$$\dot{m}_5 ex_5 = \dot{m}_6 ex_6 + \dot{E}x_{D,\text{Val}}$$

The mass, energy, entropy, and exergy balance equations can be written for the HEX:

$$\dot{m}_2 = \dot{m}_3; \dot{m}_4 = \dot{m}_5$$

$$\dot{m}_2 h_2 + \dot{m}_4 h_4 = \dot{m}_3 h_3 + \dot{m}_5 h_5$$

$$\dot{m}_2 s_2 + \dot{m}_4 s_4 + \dot{S}_{g,\text{HEX}} = \dot{m}_3 s_3 + \dot{m}_5 s_5$$

$$\dot{m}_2 ex_2 + \dot{m}_4 ex_4 = \dot{m}_3 ex_3 + \dot{m}_5 ex_5 + \dot{E}x_{D,\text{HEX}}$$

The mass, energy, entropy, and exergy balance equations can be described for the geothermal water pump:

$$\dot{m}_{11} = \dot{m}_{12}$$

$$\dot{m}_{11} h_{11} + \dot{W}_{\text{in,GWP}} = \dot{m}_{12} h_{12}$$

$$\dot{m}_{11} s_{11} + \dot{S}_{g,\text{GWP}} = \dot{m}_{12} s_{12}$$

$$\dot{m}_{11} ex_{11} + \dot{E}x^{W}_{\text{in,GWP}} = \dot{m}_{12} ex_{12} + \dot{E}x_{D,\text{GWP}}$$

For flow 1:

As given in the equation, $P_1 = 250$ kPa, $x_1 = 0.6$, and $u_1 = 0.00$, and, based on these data, h_1, s_1, and T_1 are determined from EES and are given in Table 4.6.

For flow 2:

As given in the equation, $P_2 = 400$ kPa, $x_2 = x_1$, and $T_2 = T_1 + 0.4$, and, based on these data, h_2 and s_2 are determined from EES as written in Table 4.6.

For flow 3:

As given in the equation, $P_3 = P_2$, $x_3 = x_1$, and $T_3 = 84°C$, and, based on these data, h_3 and s_3 are determined from EES and are included in Table 4.6.

For flow 4:

As given in the equation, $P_4 = P_3$, $x_4 = 0.4$, and $T_4 = 99\ °C$, and, based on these data, h_4 and s_4 are calculated by using the EES as given in Table 4.6. Mass flow rates at points 4 and 7 can be calculated:

$$\dot{m}_3 = \dot{m}_4 + \dot{m}_7 \Rightarrow 2 = \dot{m}_4 + \dot{m}_7$$

$$\dot{m}_3 x_3 = \dot{m}_4 x_4 + \dot{m}_7 x_7 \Rightarrow 2x0.6 = \dot{m}_4 x0.4 + \dot{m}_7 x0.999$$

Solving two equations simultaneously gives us $\dot{m}_4 = 1.332\ kg/s$ and $\dot{m}_7 = 0.6678\ kg/s$, respectively.

For flow 5:

h_5 can be calculated by using the HEX energy balance equation:

$$\dot{m}_2 h_2 + \dot{m}_4 h_4 = \dot{m}_3 h_3 + \dot{m}_5 h_5$$

TABLE 4.6 Thermodynamic properties of the single-effect absorption cooling system at each state points.

1	$\dot{m}$ (kg/s)	T(°C)	P (kPa)	h (kJ/kg)	s (kJ/kg K)	x (kg/kg)	ex (kJ/kg)
0	—	45	101.3	44.96	0.6237	—	—
1	2	6.496	250	−199.7	−0.06943	0.6	−24.23
2	2	6.896	400	−197.7	−0.06301	0.6	−24.29
3	2	84	400	979.4	3.619	0.6	−18.16
4	1.332	99	400	847.7	3.046	0.4	32.51
5	1.332	−115.3	400	−919.4	−3.556	0.4	364.7
6	1.332	−115.3	250	−919.4	−3.555	0.4	364.3
7	0.6678	114.2	400	1542	5.505	0.999	−55.74
8	0.6678	−1.809	400	822.1	3.032	0.999	11.2
9	0.6678	−13.59	250	822.1	3.175	0.999	−34.19
10	0.6678	−13.5	250	1058	4.081	0.999	−86.03

Also, $P_5 = P_4$, $x_5 = x_4$, and, based on these data, T_5 and s_5 are calculated by using the EES as shown in Table 4.6.

For flow 6:

$h_6 = h_5$, $x_6 = x_5$, $P_6 = P_1$, and based on these data, T_6 and s_6 are computed by utilizing the EES as shown in Table 4.6.

For flow 7:

h_7 can be computed by using the generator energy balance equation:

$$\dot{m}_3 h_3 + \dot{Q}_{\mathrm{in,Gen}} = \dot{m}_4 h_4 + \dot{m}_7 h_7$$

Also, as given in the equation, $\dot{Q}_{\mathrm{in,Gen}} = 200$ kW, $P_7 = P_4$, $x_7 = 0.999$, and, based on these data, T_7 and s_7 are calculated by using the EES as illustrated in Table 4.6.

For flow 8:

h_8 can be calculated by utilizing the condenser energy balance equation:

$$\dot{m}_7 h_7 = \dot{m}_8 h_8 + \dot{Q}_{out,Con}$$

Also, as given in the equation $\dot{Q}_{\mathrm{out,Con}} = 480.5$ kW, $P_8 = P_7$, $x_8 = x_7$, and, based on these data, T_8 and s_8 are computed by utilizing the EES as given in Table 4.6.

For flow 9:

$P_9 = P_1$, $x_9 = x_8$, and $h_9 = h_8$, and, based on these data, T_9 and s_9 are calculated by using the EES as given in Table 4.6.

For flow 10:

$P_{10} = P_9$, $x_{10} = x_9$, and, as given in the equation, $T_{10} = -13.5\ ^\circ\mathrm{C}$, and, based on these data, h_{10} and s_{10} are computed by utilizing the EES as shown in Table 4.6.

1. The power consumed by the pump is:

$$\dot{W}_{\mathrm{in},P} = \dot{m}_2 h_2 - \dot{m}_1 h_1 = 3.967 \text{ kW}$$

2. The rate of cooling provided by the evaporator is:

$$\dot{Q}_{\mathrm{cooling}} = \dot{m}_{10} h_{10} - \dot{m}_9 h_9 = 157.8 \text{ kW}$$

3. The exergy destruction rate of the pump is:

$$\dot{E}x_{D,P} = \dot{m}_1 ex_1 - \dot{m}_2 ex_2 + \dot{E}x^{W}_{\mathrm{in},P} = 3.709 \text{ kW}$$

The exergy destruction rate of the HEX is:

$$\dot{Q}_{\mathrm{HEX}} = \dot{m}_3 h_3 - \dot{m}_2 h_2$$

$$\dot{E}x_{D,\mathrm{HEX}} = \dot{m}_2 ex_2 - \dot{m}_3 ex_3 + \dot{Q}_{\mathrm{HEX}}\left(1 - \frac{T_o}{T_{\mathrm{HEX}}}\right) = 223.4 \text{ kW}$$

The exergy destruction rate of the generator is

$$\dot{E}x_{D,\text{Gen}} = \dot{m}_3 ex_3 + \dot{Q}_{\text{in,Gen}}\left(1 - \frac{T_\text{o}}{T_\text{Gen}}\right) - \dot{m}_4 ex_4 - \dot{m}_7 ex_7 = 197.3 \text{ kW}$$

The exergy destruction rate of the condenser is:

$$\dot{E}x_{D,\text{Con}} = \dot{m}_7 ex_7 - \dot{m}_8 ex_8 - \dot{Q}_{\text{out,Con}}\left(1 - \frac{T_\text{o}}{T_\text{Con}}\right) = 15.52 \text{ kW}$$

The exergy destruction rate of the expansion valve is:

$$\dot{E}x_{D,\text{EV}} = \dot{m}_8 ex_8 - \dot{m}_9 ex_9 = 27.55 \text{ kW}$$

The exergy destruction rate of the evaporator is:

$$\dot{Q}_{\text{in,Eva}} = \dot{m}_{10} h_{10} - \dot{m}_9 h_9$$

$$\dot{E}x_{D,\text{Eva}} = \dot{m}_9 ex_9 - \dot{m}_{10} ex_{10} + \dot{Q}_{\text{in,Eva}}\left(1 - \frac{T_\text{o}}{T_\text{Eva}}\right) = 37.56 \text{ kW}$$

The exergy destruction rate of the valve is:

$$\dot{E}x_{D,\text{Val}} = \dot{m}_5 ex_5 - \dot{m}_6 ex_6 = 0.456 \text{ kW}$$

The exergy destruction rate of the absorber is:

$$\dot{Q}_{\text{out,Abs}} = \dot{m}_6 h_6 + \dot{m}_{10} h_{10} - \dot{m}_1 h_1$$

$$\dot{E}x_{D,\text{Abs}} = \dot{m}_6 ex_6 + \dot{m}_{10} ex_{10} - \dot{m}_1 ex_1 - \dot{Q}_{\text{out,Abs}}\left(1 - \frac{T_o}{T_{Abs}}\right) = 418.1 \text{ kW}$$

4. Total exergy destruction rate of the cooling system is:

$$\begin{aligned}\dot{E}x_{D,\text{Tot}} = {} & \dot{E}x_{D,P} + \dot{E}x_{D,\text{HEX}} + \dot{E}x_{D,\text{Gen}} + \dot{E}x_{D,\text{Con}} \\ & + \dot{E}x_{D,\text{EV}} + \dot{E}x_{D,\text{Eva}} + \dot{E}x_{D,\text{Val}} + \dot{E}x_{D,\text{Abs}}\end{aligned}$$

$$\begin{aligned}\dot{E}x_{D,\text{Tot}} = {} & 3.709 + 223.4 + 197.3 + 15.52 \\ & + 27.55 + 37.56 + 0.456 + 418.1 = 923.6 \text{ kW}\end{aligned}$$

5. The energetic and exegetic COPs are:

$$COP_\text{en} = \frac{\dot{Q}_\text{eva}}{\dot{Q}_\text{gen} + \dot{W}_\text{p}} = 0.7736$$

and

$$COP_\text{ex} = \frac{\dot{E}x^Q_\text{eva}}{\dot{E}x^Q_\text{gen} + \dot{E}x^W_\text{p}} = 0.3257$$

6. The effect reference temperature on the exergy destruction rate and the energetic and exergetic COP of the single-effect absorption cooling

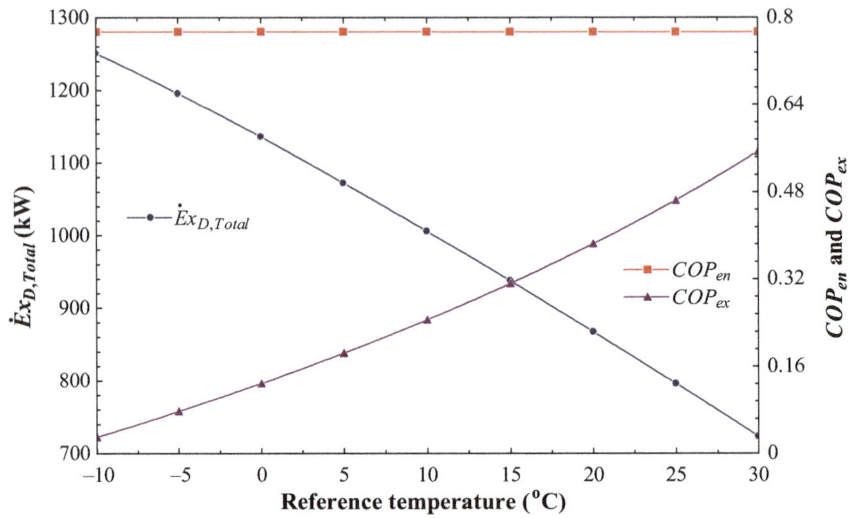

FIGURE 4.18 Effect of reference temperature on the exergy destruction rate and the energetic and exergetic COP of the single-effect absorption cooling system.

system is shown in Fig. 4.18. As shown in this figure, the exergy destruction rate of the single-effect absorption cooling system is increased from 1251 kW to 723.3 kW, and the energetic COP the cooling system does not change with the increasing reference temperature from 0°C to 40°C, whereas the exergetic COP of the cooling system is increased from 0.02986 to 0.5542 in the examined reference temperature change.

4.4 Power production

One of the most important reasons why geothermal energy sources are popular in energy systems is that electricity production can be realized with this energy source. The low- and high-temperature geothermal energy sources have a very high potential as alternative energy sources for power production.

Shown in Fig. 4.19 is the schematic diagram of power production processes based on the geothermal energy sources and some potential ways for energy storage solution, such as (1) thermal energy (storage heaters and molten salts), (2) electrostatic (capacitors and supercapacitors), (3) potential (pumped hydro and compressed air), (4) kinetic (flywheels), (5) chemicals (batteries, methanol, regenerative fuel cell, and hydrogen), and (6) electromagnetics (superconducting coils).

Power production by using geothermal energy resources is carried out with various plants. These plants are usually consisting of direct steam power production, single-flash steam power production, double-flash power production, triple-flash power production, quadruple-flash power production, binary cycle

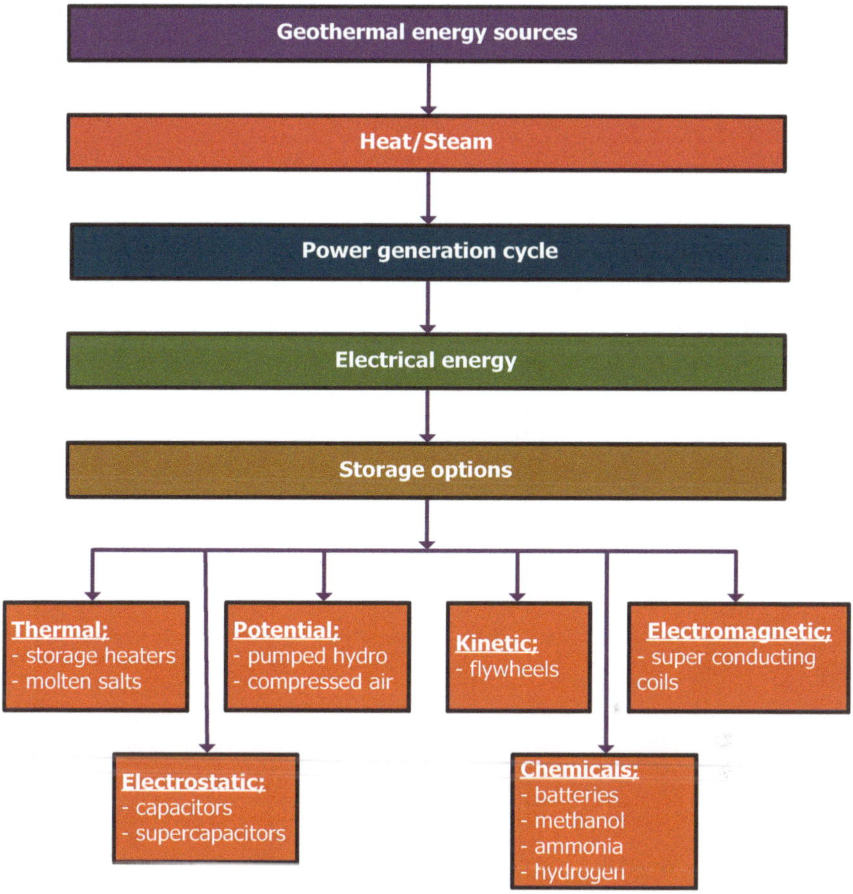

FIGURE 4.19 Geothermal energy–based power production and storage options. *Modified from [15]*

power production, combined/hybrid power production, ORC and Kalina cycles. In addition, these plants can also be classified as open cycles, as given in Fig. 4.20(a) and (b); as close cycles, as shown in Fig. 4.20(c); and as combined cycles, as illustrated in Fig. 4.20(d). The open cycle does not use a condenser. But the produced low-pressure steam needs a very–large-diameter steam turbine. This negative effect can be accomplished by using the close or the combined cycles.

4.4.1 Geothermal flashing power production

Flash steam turbines are among the most widely used components of geothermal energy systems for electricity generation. In order to use this component efficiently in electricity generation, the geothermal fluid temperature

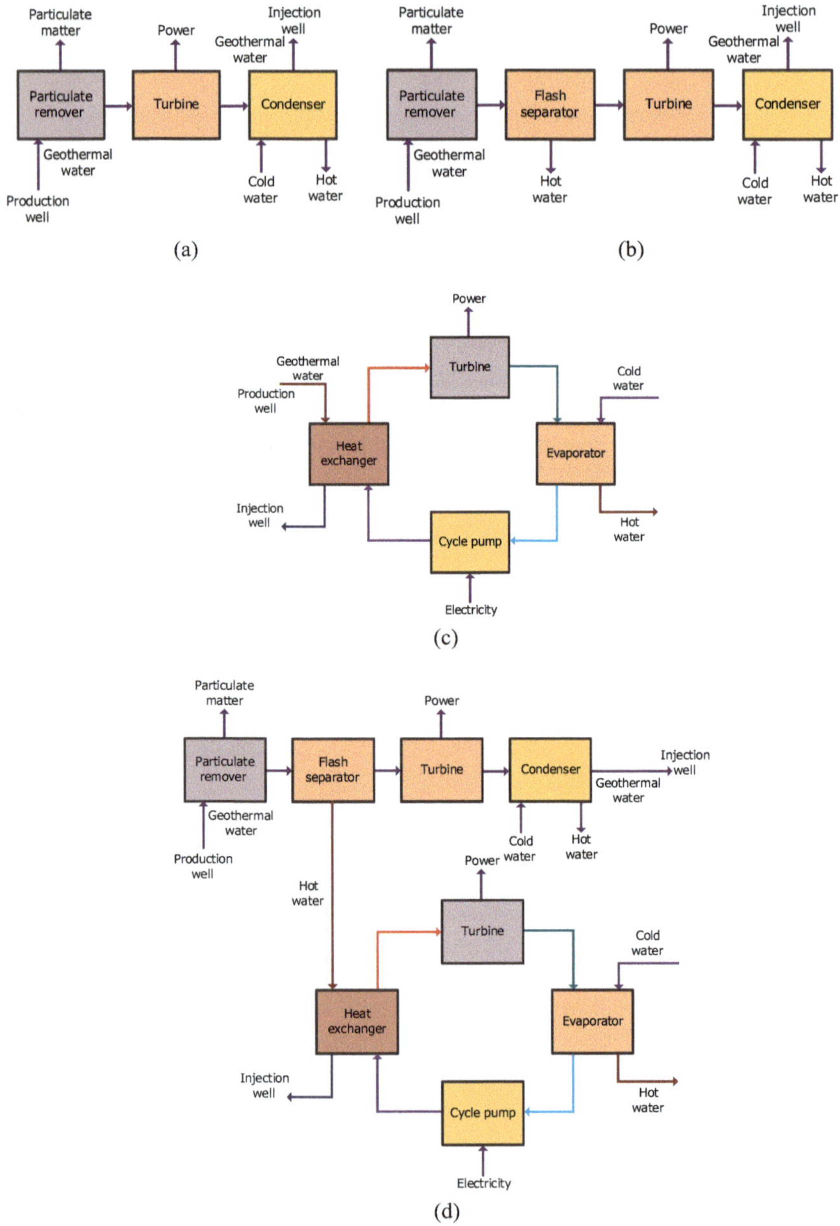

FIGURE 4.20 Geothermal power production plants include (A) direct steam cycle, (B) flash steam cycle, (C) binary cycle, and (D) combined cycle. *Modified from [15].*

must be high. This temperature value is generally accepted as above 180°C. These systems begin to work by introducing geothermal fluid as an input to the system by removing it to the Earth based on its own pressure [16,17]. As the level of geothermal fluid goes up, the pressure of the fluid decreases. This geothermal fluid then goes through a flashing process with a flashing valve. The fluid with reduced pressure is divided into two phases, liquid and steam, with the help of a separator. The vapor phase separated from the separator is sent to the turbine for electricity generation. The liquid phase separated from the separator is sent to the ground to ensure sustainability. The process described in this section is the single-flash geothermal energy generation system.

These systems have different efficiencies according to the number of flashing processes. Since the number of flashes depends on the temperature and pressure of the geothermal fluid, the temperature and pressure parameters should be taken into account when performing these flashing processes. The names of these systems also change according to the number of flashing processes. For instance, the geothermal energy generation system that includes three flashing processes is called a triple-flash geothermal generation system.

4.4.2 Binary geothermal power production

Another method of generating electricity from geothermal energy sources is the binary cycle geothermal power plant. In these systems, an organic fluid is used as the working fluid together with the geothermal fluid. Depending on the type of these organic fluids, the performance output of the dual system differs. The geothermal fluid coming from the Earth with its own pressure has heat energy.

In order for binary cycle geothermal power plants to operate, the geothermal fluid first emerges at the Earth's crust with its own pressure. Then heat energy in the geothermal fluid is transferred to the organic fluid by using a heat exchanger. After this point, the organic Rankine cycle (ORC) or Kalina cycle takes place according to the type of organic fluid. While the ammonia–water mixture is used as a working fluid in Kalina cycles, any organic fluid is used in ORC systems. After the organic fluid is evaporated by utilizing the geothermal fluid, it is sent to the turbine for electricity production. After the heat transfer between the geothermal fluid and the organic fluid takes place, the geothermal fluid is sent to the ground again for sustainability.

4.4.3 Dry steam geothermal power production

Dry steam geothermal energy systems are among the most common geothermal power plants that use the hydrothermal fluid as a working fluid. As the system's name indicates, the geothermal fluid must be in the vapor phase. This type of geothermal energy system accounts for approximately half of the installed geothermal energy systems in the world. Dry steam cycles work by transferring geothermal fluid directly to the energy production system. The geothermal fluid entering the system is directed to the turbine for electricity generation. In this way, generators provide electricity. This geothermal fluid is delivered to the turbine via pipes. Then the fluid from the turbine is sent to the condenser to obtain the heating product and then is reinjected back into the ground. These systems have some advantages over systems that use fossil fuels as an energy source. In fossil fuel systems, combustion must occur in order to obtain useful products. Dry steam geothermal energy systems do not require a combustion reaction since they provide useful products with hydrothermal fluid as fluid. Therefore, dry steam geothermal power plants do not require any fossil fuel storage and transfer of this fuel through pipes. Dry steam geothermal energy systems are advantageous in this respect compared to conventional energy generation systems. In addition, considering the environmental impact parameter, these systems are superior to traditional energy generation systems since they emit only high levels of steam and very little gas.

4.4.4 Back-pressure geothermal power production

Back-pressure-type energy production is one of the basic production methods of dry steam geothermal energy systems. An advantage of these systems is their low cost, but their low thermal efficiency is a disadvantage. These systems have the potential to be used in many areas today, with usage areas differing according to the advantages and disadvantages of the systems [15]. Back-pressure geothermal power generation, which generally provides temporary power generation, is used as a part of the systems where electricity and heating product are obtained. When we look at systems producing this type of output, it is seen that single-flash geothermal power plants are in the first place. The systems that are used most after these systems are dry steam and double-flash geothermal energy systems, respectively. There is no condenser or cooling system in such systems. Unlike the process in the dry steam geothermal energy system, the steam coming from the turbine in this system is transferred to the steam exhaust at atmospheric pressure. The pressure of the fluid coming from the back-pressure power process is above atmospheric pressure. Also, this fluid is not condensed as previously mentioned. The difference with these systems is that they use no condenser or cooling system in back-pressure geothermal power generation. The cost of back-pressure geothermal power production is lower than that of conventional condensing geothermal energy production. However, the existing power obtained from the well in these systems is not as efficient as it is in condensing systems. These systems also have the potential to be used in many areas.

4.5 Geothermal district heating and cooling

Geothermal district heating systems utilize geothermal power to supply heat energy to buildings and industries via the distribution grids. Geothermal power can also be used for space heating and cooling, domestic hot water supply, and industrial process heat requirements by district heating utilizing peaking terminals, a distribution plant, central pumping terminals, and in the building subsystems (heat exchangers, circulation pumps, etc.) [18]. 88% of energy utilization from geothermal sources are accounted for space heating (for 28 countries). China leads the geothermal district heating outputs, followed by Iceland and Turkey. Considering their populations, Iceland leads the Earth in geothermal district heating outputs per capita, and Turkey leads the individual domestic heating outputs.

4.6 Hydrogen production

Hydrogen production with a geothermal energy source has a great potential for energy production systems in terms of increasing performance and producing a new energy source. Therefore, hydrogen production with a geothermal energy source is a source of interest for academic studies.

In the near future, the use of geothermal energy plants for hydrogen generation will increase for regions with abundant geothermal power reservoirs. Geothermal energy can be used for hydrogen production in mainly four ways: (1) direct generation, (2) water electrolysis–based generation, (3) thermal energy–based thermochemical generation, and (4) hybrid thermochemical generation, as shown in Fig. 4.21.

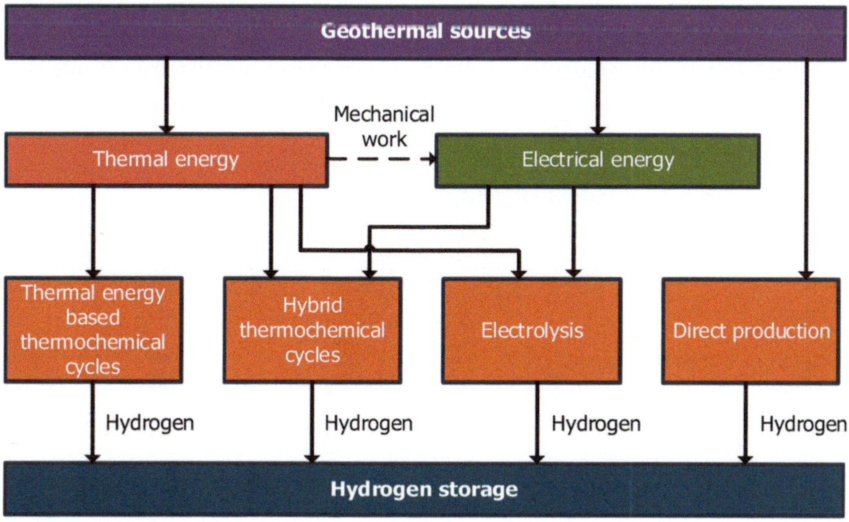

FIGURE 4.21 Illustration of the geothermal power–based hydrogen generation.

In the literature, many studies on hydrogen production with geothermal sources have been carried out. In these studies, it is seen that hydrogen production is generally classified under three headings. The first is the systems where hydrogen production is realized by making use of geothermal steam. In fact, geothermal steam is released into the atmosphere in some regions of the world with hydrogen; that is, along with geothermal gases, hydrogen gas is also released into the atmosphere. Other gases must be cleaned in order to use hydrogen. In this way, studies show that various technologies continue to benefit from hydrogen. The second method in hydrogen production utilizes thermal energy and is divided into two groups. The first group consists of systems where electricity is used as thermal energy, and the second group consists of systems that use heat as thermal energy. The third method in which hydrogen production is carried out is by means of systems utilizing electrolysis. Hydrogen can be obtained by the electrolysis of water using different technologies. The disadvantage of water electrolysis is that electricity is considered an expensive fuel. Table 4.7 compares three industrially used electrolysis methods.

There are three different types of electrolysis: (1) alkaline, (2) proton exchange membrane (PEM), and (3) high-temperature steam electrolysis (HTSE). It is seen that the electrodes used in these three electrolysis methods are different. The performances of these electrolysis methods are also different.

In the end, geothermal power combined with hydrogen generation is a hybrid thermochemical generation process. Based on the thermodynamic analysis of thermochemical hydrogen generation reactions, for some step or steps in the thermochemical processes, the required power must be provided in a mixed format of electricity and thermal energy. The temperature of heat power–based or hybrid thermochemical processes containing geothermal power–based hydrogen generation systems must be lower than 550°C. Different low-temperature thermochemical processes are listed in Table 4.8. But the temperature range of high-temperature geothermal resources are

TABLE 4.7 Kinds of industrial electrolysis.

Kinds of electrolysis	Electrolyte	Energetic performance
Alkaline	Aqueous KOH solution	78 %
Proton exchange membrane (PEM)	Hydrogen ion–conducting PEM	90 %
High-temperature steam electrolyzer (HTSE)	Oxygen-conducting ceramics	94 %

Source: Data from [19].

TABLE 4.8 Low-temperature thermochemical hydrogen generation processes.

Process	Temperature (°C)	Chemical Reactions	Reference
Cu-Cl	400–450	$2CuCl_2(s) + H_2O(g) \rightarrow CuO^*CuCl_2(s) + 2HCl(g)$	[20]
	450–500	$CuO^*CuCl_2(s) \rightarrow 2CuCl(l) + 1/2O_2(g)$	
	25–80	$4CuCl(s) + H_2O \rightarrow 2CuCl_2(aq) + 2Cu(s)$	
	>100	$CuCl_2(aq) \rightarrow CuCl_2(s)$	
	430–475	$2Cu(s) + 2HCl(g) \rightarrow 2CuCl(l) + + H_2(g)$	
Mg-Cl	450–500	$MgCl_2(s) + H_2O(g) \rightarrow 2HCl(s) + MgO(g)$	[21]
	450–500	$MgO(s) + Cl_2(g) \rightarrow MgCl_2(s) + 1/2O_2(g)$	
	80–100	$2HCl(s) \rightarrow Cl_2(s) + H_2(g)$	
Sulfuric acid	80–90	$2H_2O + SO_2 \rightarrow H_2SO_4 + H_2$	[22]
	450–500	$H_2SO_4 \rightarrow H_2O + SO_3$	
	500–550	$SO_3 \rightarrow SO_2 + 1/2O_2$	

usually between 200°C and 250°C; hence, the temperature levels of these processes must be increased to the desired 550°C–600°C using the different heat pump methods.

4.7 Ammonia production

Ammonia (NH_3) is one of the most importantly environmentally friendly fuels, power carriers, and storage medium for future generations. Non-carbon-based power technologies, in particular geothermal power–based ammonia generation, have many adequate solutions for sustainable power generation, conversion, and application. Also, ammonia is presently one of the most utilized chemicals on the earth because of numerous processes, such as fertilizers, cooling working fluids, fuel, etc. In addition, an important list of the specific properties of ammonia can be defined:

- It is formed of 1 mole of nitrogen from air separation and 3 moles of hydrogen coming from any conventional or renewable hydrogen production process.
- It is the second most generated synthetic chemical in the world, behind sulfuric acid.
- It is an important hydrogen bonding source and carrying fuel that does not carry any carbon atoms and has a high hydrogen ratio.
- Ammonia can be utilized in all kinds of combustion processes, gas turbine plants, burners as an important source with only minor revisions, and directly in fuel cells, which is an important benefit compared to other kinds of fuel sources.
- Ammonia could be utilized as a cooling working fluid for air conditioning in the vehicles and different design aims.

The Haber–Bosch system is the most effective ammonia production plant, needs very high temperatures and pressures to work, and consumes high amounts of carbon-based sources, mostly natural gas, making it a non-sustainable method in the near future. Hence, different procedures for ammonia generation are in inevitable need for improvement. Cryogenic air separation processes are currently the best performance and cost-effective techniques for producing important amounts of oxygen, nitrogen, and argon. Since ammonia is generated in large quantities, the needed nitrogen could be produced at a low cost and with high performance. The needed electrical energy can be provided either from carbon-based sources or from clean energy resources.

The chemical reaction to connect hydrogen and nitrogen thermocatalytically for the Haber–Bosch process is:

$$3/2H_2 + 1/2N_2 \rightarrow NH_3 + 45.2\ \text{kJ/mol} \tag{4.9}$$

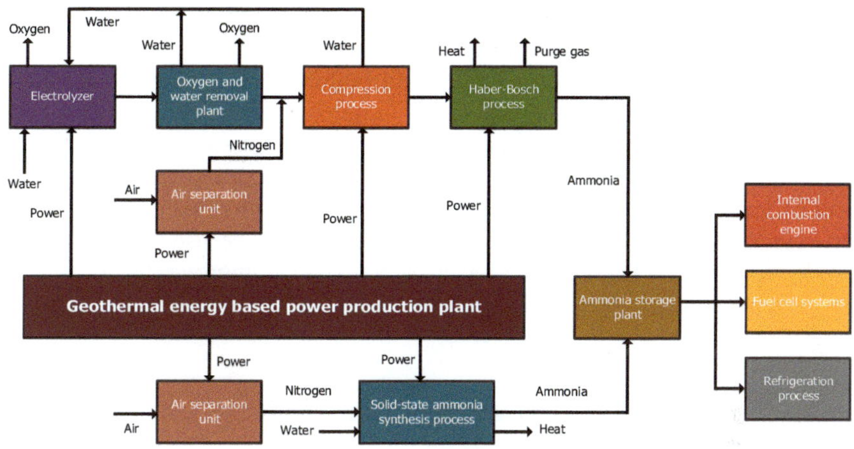

FIGURE 4.22 Block diagram of the geothermal energy–based ammonia production processes.

One of the other most important ammonia production methods in the world is the solid-state ammonia synthesis process. Similarly, for this ammonia production technique, nitrogen is produced by using the air separation process. To produce clean ammonia from the air, water and electrical energy are required for now and the near future, the block diagram of the geothermal energy–based Haber–Bosch and solid-state ammonia synthesis process is illustrated in Fig. 4.22.

4.8 Other synthetic fuels production

The process that can be realized with geothermal energy sources is synthetic fuel production. If synthetic fuels can be produced with more economical methods, these fuels have the potential to replace conventional fuels. For these reasons, different energy sources are used for the production of synthetic fuels in power generation systems. Potentially, geothermal energy sources are one of them. In order to effect synthetic fuel production efficiently, energy generation systems need to be established in the regions where geothermal energy resources are available. In the literature, various system designs have been made and continue to be made for synthetic fuel production using the geothermal energy source.

Thermal energy gained from geothermal energy resources can be utilized in several technologies. One of them is to utilize thermal energy to generate power and, later in the appropriate thermochemical cycle, to generate several synthetic fuels such as ethanol, methanol, butanol, propane, ammonia, and nonfossil methane, as given in Fig. 4.23. These synthetic fuels can be stored in the liquid phase, and then they can be transported to the final utilization.

Also, these produced fuels can be beneficial for transportation, space heating, and power production, either in fuel cells and as chemicals.

Some carbon-containing tail gases and pollutants may emerge in energy production systems. Systems producing liquid fuel are an example of such energy production systems. The production of tail gases, pollutants, and additional synthesis gases produced in these systems can occur. Liquid fuel and hydrocarbon can be produced along with these synthesis gases. Another process in which these tail gases and pollutants can be used is liquid fuel production cycles. These cycles have the potential to improve the conversion of carbon to liquid fuel or hydrocarbon.

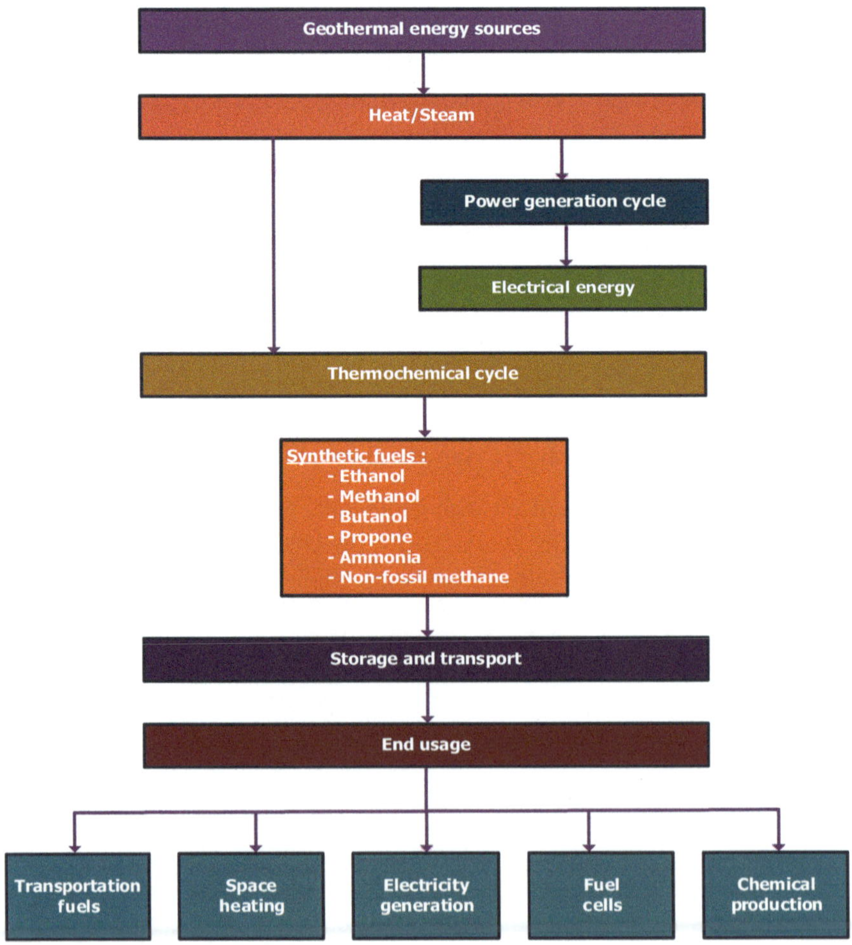

FIGURE 4.23 Geothermal energy–based synthetic fuels production and utilization options. *Modified from [23].*

4.9 Other types of applications

Geothermal energy sources are available in areas where they are used differently from the previously mentioned ways. With differences in its temperature, geothermal fluid responds to different needs in different parts of the world. In industrial systems, the fact that the supply of the fluid is costless creates great potential. Another sector in which geothermal fluid is used is the agricultural sector. In industrial applications, while the temperature of the geothermal fluid is preferred to be high, low-temperature geothermal fluid can work in agricultural activities. Some features of geothermal fluid are attractive for use in these sectors. The first of these features is its cost as an energy source. Although system installation incurs a cost for the initial investment, since the energy source itself is taken from nature, this source is attractive for these two sectors in terms of cost. The other feature is energy quality. Although the energy quality varies from region to region, it is accepted as suitable in general for these two sectors. The third feature is reliability. Reliability is one of the features that emerge after long-term operation of the system. Studies show that geothermal power generation systems are reliable and that geothermal sources are used by different sectors at different temperature ranges.

Finally, the technology of predrying and postproduction drying processes differs based on the species of dried product, such as grains, vegetables, and fruits, and on the desired results, such as moisture, shape, and further processing. Based on these results, geothermal energy resources, combined with general drying lines for grains and fruits/vegetables, should be designed, as shown in Fig. 4.24A,B.

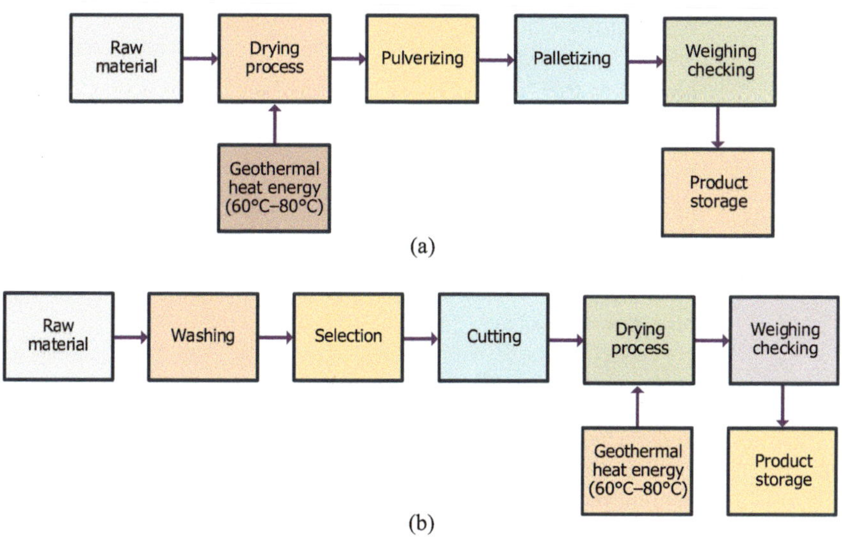

FIGURE 4.24 Process line for (A) drying grains and (B) drying fruits and vegetables by using geothermal energy. *Modified from [15].*

4.10 Closing remarks

Nowadays, the geothermal resource–based heating-cooling, power, and hydrogen production systems provide an environmentally benign alternative to the conventional source–based generation systems. This chapter has presented comprehensive case studies in order to cover energy and exergy analyses for geothermal energy–based radiator heating system, a heat pump–based heating and cooling systems, and a single-effect absorption refrigeration plant. Low- and high-temperature geothermal resources have a very high potential as an alternative energy resource for power generation. Geothermal power production plants are classified as a direct steam cycle, flash steam cycle, binary cycle, and combined cycle. Geothermal-based hydrogen generation, which basically uses geothermal energy for hydrogen generation, appears to be an environmentally conscious and sustainable avenue for countries with abundant geothermal energy resources. Geothermal energy–based hydrogen and other synthetic fuels generation options are submitted for investigating suitable geothermal hydrogen and synthetic fuel production. Also, geothermal energy based drying grains, drying fruits and vegetables are described to illustrate geothermal–based different application options. General working conditions of power production cycles, hydrogen, and other synthetic fuel production methods are defined; detailed system design options and thermodynamic analyses are given in Chapter 5.

Nomenclature

E	Energy (kJ)
e	Specific energy (kJ/kg)
$\dot{E}$	Energy rate (kW)
ex	Specific exergy (kJ/kg)
$\dot{Ex}$	Exergy rate (kW)
$\dot{Ex}_d$	Exergy destruction rate (kW)
$\dot{Ex}^Q$	Exergy transfer rate associated with heat transfer (kW)
$\dot{Ex}^W$	Exergy transfer rate associated with work (kW)
h	Specific enthalpy (kJ/kg)
m	Mass (kg)
$\dot{m}$	Mass flow rate (kg/s)
P	Pressure (kPa)
q	Specific heat transfer (kJ/kg)
Q	Heat (kJ)
$\dot{q}$	Specific heat transfer rate (kW/kg)
$\dot{Q}$	Heat rate (kW)
s	Specific entropy (kJ/kg K)
S	Entropy (kJ/K)
$\dot{S}$	Entropy rate (kW/K)
t	Time (s)
T	Temperature (°C, K)

u	Internal energy (kJ/kg)
ν	Velocity (m/s)
W	Work (kJ)
$\dot{w}$	Specific work rate (kW/kg)
$\dot{W}$	Work rate (kW)

Greek letters

Δ	Change in variable
η	Energy efficiency
ψ	Exergy efficiency

Subscript

Abs	Absorber
Cmp	Compressor
cooling	Cooling load
Con	Condenser
D	Destruction
en	Energy
Eva	Evaporator
EV	Expansion valve
ex	Exergy
Gen	Generator
heating	Heating load
p	Pump
tot	Total
Tur	Turbine
Vl	Valve
wf	Working fluid
1...74	State numbers
0	Ambient or reference condition

Superscripts

.	Rate
Ch	Chemical

Acronyms

ACS	Absorption cooling system
BHE	Borehole heat exchanger
COP	Coefficient of performance
EES	Engineering Equation Solver
GHE	Ground heat exchanger
GPS	Geothermal power system
GSHP	Ground source heat pump
HEX	Heat exchanger
HGHE	Horizontal ground heat exchanger
HTSE	High-temperature steam electrolysis

KC	Kalina cycle
KCGP	Kalina cycle geothermal plant
ORC	Organic Rankine cycle
PEM	Proton exchange membrane
RHS	Radiator heating system
SEACS	Single-effect absorption cooling system

References

[1] B. Lindal, Industrial and other applications of geothermal energy, Geotherm. Energy, Earth Sci. UNESCO, Paris. 12 (1973) 135–148. Paris, UNESCO.

[2] I. Dincer, H. Ozcan, "Geothermal energy, Comprehensive Energy Systems, Elsevier, 2018, pp. 702–732.

[3] M.H. Dickson, M. Fanelli, Geothermal energy: utilization and technology, UNESCO Publishing, Renewable Energy Series, John Wiley & Sons, 1995.

[4] A. Watson, Introduction, in Geothermal Engineering, Springer, New York, NY, 2013.

[5] M. Ozturk, Energy and exergy analysis of a combined ground source heat pump system, Appl. Therm. Eng. 73 (1) (2014) 362–370.

[6] F. Razi, I. Dincer, A new solar combined cycle integrated with heat pump system, Appl. Therm. Eng. 173 (2020) 114784.

[7] S.J. Self, B.V. Reddy, M.A. Rosen, Geothermal heat pump systems: status review and comparison with other heating options, Appl. Energy 101 (2013) 341–348.

[8] E. Atam, L. Helsen, Ground-coupled heat pumps: part 2. Literature review and research challenges in optimal design, Renew. Sustain. Energy Rev. 54 (2016) 1668–1684.

[9] M.A. Rosen, S. Koohi-Fayegh, Geothermal Energy: Sustainable Heating and Cooling Using the Ground, Wiley, United Kingdom, 2017.

[10] GHPC, Geothermal Heat Pump Consortium (2013) http://www.geoexchange.org [Accessed 25 March 2020].

[11] B. Ozcan, I. Aykurt, M. Akpak, T. Tacer, N. Yildirim, A. Hepbasli, et al., Thermodynamic analysis and assessment of a geothermal cooling system for a house, Int. J. Exergy 29 (2-4) (2019) 350–369.

[12] A. Hepbasli, Heat pumps, Comprehensive Energy Systems, Comprehensive Energy Systems, Elsevier, 2018, pp. 98–124.

[13] I. Stober, K. Bucher, History of geothermal energy use, Geothermal Energy, Springer, Berlin, Heidelberg, 2013.

[14] I. Dincer, Refrigerants, Compr. Energy Syst. 2 (2018) 435–474.

[15] M. Ozturk, I. Dincer, Geothermal energy conversion, Comprehensive Energy Systems, Elsevier, 2018, pp. 474–544.

[16] Y.E. Yuksel, M. Ozturk, I. Dincer, Analysis and performance assessment of a combined geothermal power-based hydrogen production and liquefaction system, Int. J. Hydrog. Energy 43 (22) (2018) 10268–10280.

[17] M.R. Kolahi, M. Amidpour, M. Yari, Multi-objective metaheuristic optimization of combined flash-binary geothermal and humidification dehumidification desalination systems, Desalination 490 (2020) 114456.

[18] J.W. Lund, T.L. Boyd, Direct utilization of geothermal energy 2015 worldwide review, Geothermics 60 (2016) 66–93.

[19] M.T. Balta, I. Dincer, A. Hepbasli, Thermodynamic assessment of geothermal energy use in hydrogen production, Int. J. Hydrog. Energy 34 (7) (2009) 2925–2939.

[20] M.F. Orhan, I. Dincer, M.A. Rosen, Energy and exergy assessments of the hydrogen production step of a copper–chlorine thermochemical water splitting cycle driven by nuclear-based heat, Int. J. Hydrog. Energy 33 (22) (2008) 6456–6466.
[21] M.F. Simpson, S.D. Herrmann, B.D. Boyle, A hybrid thermochemical electrolytic process for hydrogen production based on the reverse deacon reaction, Int. J. Hydrog. Energy 31 (2006) 1241–1246.
[22] IAEA, International Atomic Energy Agency "Advanced Applications of Water-Cooled Nuclear Power Plants", ISBN 978-92-0-105808-9, July 2007.
[23] I. Dincer, C. Zamfirescu, New York, NY: Elsevier Advanced Power Generation Systems, Elsevier, USA, 2014.

Study questions and problems

4.1. The most general application of geothermal energy is power production. Some other utilizations are process heating, district heating, district cooling, greenhouse heating, and heating for fish farming. From a thermodynamic point of view and considering the quality of energy, explain which of these utilizations you recommend most.

4.2. Please compare the air-source, water-source, and ground-source heat pump plants using the exergetic point of view.

4.3. Please define the difference between ground-source heat pumps and geothermal heat pumps.

4.4. In the heat pump plant, exergy destructions occur in the different parts such as the compressor, condenser, evaporator, and expansion valve. What are the causes of the exergy destruction in each of these system parts?

4.5. The COP of refrigeration and heat pump plants can be greater than 1. Can the exergy efficiency of such plants be greater than 1? Explain.

4.6. Is the exergy analysis more useful for a heating plant based on geothermal energy or for a cooling plant based on geothermal energy? Explain.

4.7. A student calculates the exergy efficiency of a geothermal energy-based heating system to be greater than 100%. Is this result reasonable? Explain.

4.8. The exergetic performance of an electric resistance heater is typically less than 10%. Explain how the same heating can be accomplished by utilizing a reversible heat pump. Consider 5 kW of heating with an outdoor air temperature of 2°C and an indoor temperature of 20°C. Draw a schematic of this heat pump unit, and calculate the COP and the exergetic performance of this heat pump.

4.9. Repeat the previous problem considering a ground source heat pump with a ground temperature of 12°C.

4.10. Compare the conventional vapor compression air conditioning plant and the single-effect absorption cooling plant by using the energetic point of view.

4.11. In the single-effect absorption cooling plant, exergy destructions occur in different parts such as the generator, heat exchanger, condenser, evaporator, pump, and expansion valve. What are the causes of exergy destruction in each of these system parts?
4.12. In Example 4.1, what are the effects of increasing the geothermal fluid temperature from 80°C to 120°C on (a) the exergy destruction rate and (b) energy and exergy efficiencies of the radiator heating system?
4.13. In Example 4.2, what are the effects of increasing the flow rate of R-134a from 600 l/min to 1500 l/min on (a) the exergy destruction rate, (b) energy and exergy efficiencies of the heat pump–based heating plant?
4.14. In Example 4.3, what are the effects of increasing the flow rate of R-134a from 450 l/min to 900 l/min on (a) the exergy destruction rate, (b) energy and exergy efficiencies of the heat pump–based refrigeration system?
4.15. In Example 4.4, what are the effects of increasing the strong solution concentration of NH_3-H_2O from 0.4 to 0.6 on (a) the exergy destruction rate, (b) energy and exergy efficiencies of the single-effect absorption cooling system?
4.16. Define the geothermal energy–based power production technologies.
4.17. Describe the geothermal energy–based hydrogen production methods.
4.18. Define the geothermal energy–based ammonia production technologies.
4.19. Obtain a published article on the exergetic analysis of heat pump plants. Using the working data provided in the article, perform a detailed exergetic analysis of the plant, and compare your results to those in the original article.
4.20. Obtain a published article on the exergetic analysis of single-effect absorption cooling systems. Using the operating data provided in the article, perform a detailed exergetic analysis of the cooling system, and compare your results to those in the original article.

Chapter 5

Basic geothermal energy systems

5.1 Introduction

Geothermal energy is an important energy source in terms of both renewable energy and the optimistic outlook in environmental impact analysis. Geothermal energy sources can appear anywhere in the world. To obtain power from this type of energy, it is necessary to provide the necessary temperature value in addition to the emergence of the geothermal energy source [1]. Geothermal energy is considered in two different categories: high enthalpy and low enthalpy. High-enthalpy geothermal energy sources are generally known to be in volcanoes and geysers. Those with low enthalpy are known to be stored in rocks on the Earth's crust. Low-enthalpy geothermal energy sources are used for heating and cooling. Generally, geothermal energy sources are used in power generation for heating and cooling. For this reason, the number of energy-producing systems using geothermal energy sources is gradually increasing. In 2016, 3567 MW of useful output was achieved with geothermal energy in the United States, and in most countries, projects capable of delivering 1272 MW output were made. By the year 2018, geothermal energy–based systems capable of generating 16.7 billion kWh of electricity are designed and 65.6 billion kWh systems are planned by 2050 [2]. While low-cost operating conditions are generally required to benefit from geothermal energy, these costs may increase during the development phase. According to a study conducted in the United States, systems using geothermal energy sources as energy sources emit 11 times less carbon dioxide than energy generation systems using coal [3]. Carbon dioxide emissions in comparison to other power systems are given in Fig. 5.1.

Many studies have appeared in the open literature for the direct use of geothermal energy source. This chapter describes the working procedures and design methods of basic, hybrid, and combined-type geothermal energy systems to produce heat and power using a single energy resource. In addition, various systems are designed using geothermal energy sources as an energy source, and thermodynamically mathematical models of these systems are created. To better investigate system performance, parametric studies and example studies are given to investigate the effects of different

Geothermal Energy Systems. DOI: https://doi.org/10.1016/B978-0-12-820775-8.00009-X

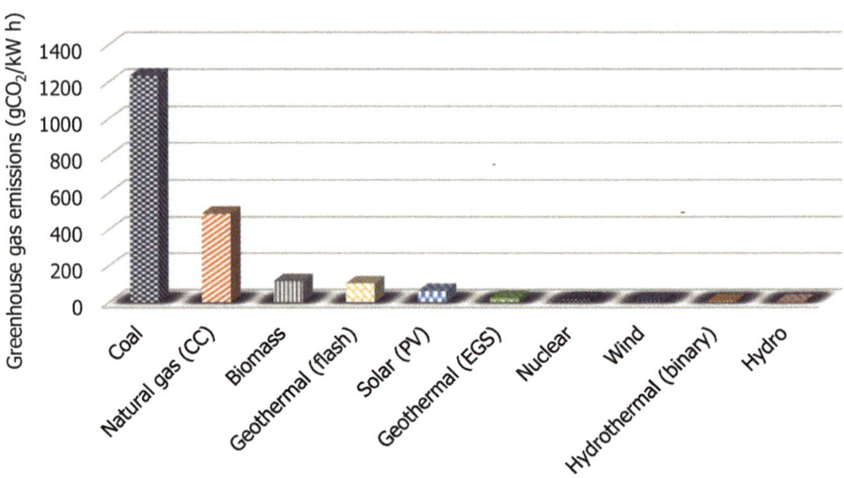

FIGURE 5.1 Greenhouse gas emissions from power generating systems. *Data from US DOE [4].*

working conditions and design parameters on energetic and exergetic performances, exergy destruction rates of system components, and power generation rates.

5.2 Basic geothermal energy systems

Although we use "basic" for geothermal energy systems here, the term goes further to cover hybrid and combined geothermal energy systems, which are alternatives to conventional power production systems. As given in Fig. 5.2, these basic systems can be classified into three categories: (1) direct steam geothermal power plants, (2) flashing geothermal power systems (GPS), and (3) binary cycle GPSs. The direct steam geothermal power plant is one of the oldest types of geothermal energy–based power production systems. The basic geothermal energy system with double turbines can be viewed as an enhanced version of the basic geothermal energy system. The design criteria and working conditions of basic flashing geothermal energy systems are described qualitatively for two different types such as the single-flash steam GPS and double-flash steam GPS. Also, geothermal energy–based binary cycle power production systems can be considered mainly in three design options: organic Rankine cycle (ORC) [single-stage (SS), SS with two turbines and double-stage (DS)], Kalina cycle (KC), and combined flash/binary cycle. In the next chapters 6 and 7, we will discuss the design methods of advanced and multigenerational geothermal energy systems, respectively, that are used to analyze integrated systems by considering the first and the second law of thermodynamics simultaneously.

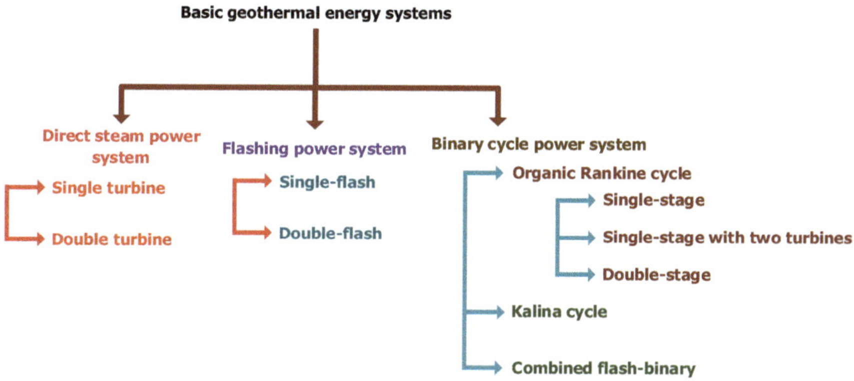

FIGURE 5.2 Schematic diagram of basic geothermal energy systems.

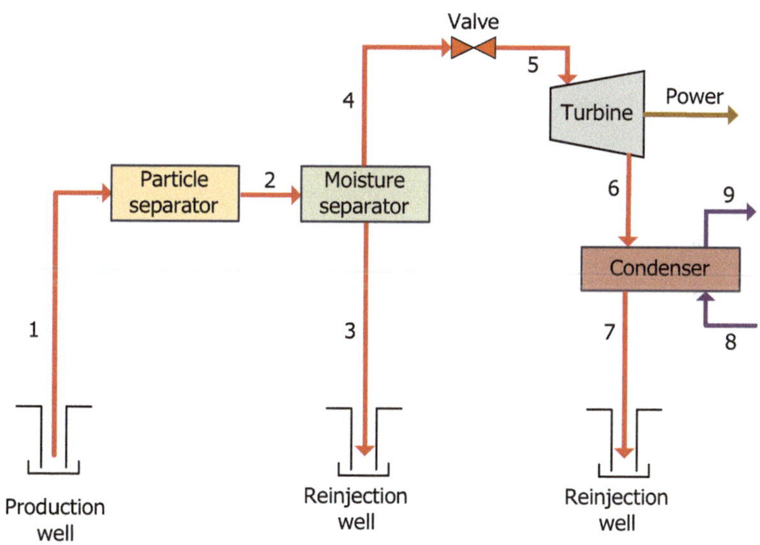

FIGURE 5.3 Simplified flow diagram of a direct steam geothermal power system.

5.3 Direct steam geothermal power plant

A direct steam GPS is the most basic system to obtain power by using geothermal energy [5]. Thermodynamic and performance analysis of this system can be carried out easily based on thermodynamic equations. In its most general form, the schematic diagram of the direct steam GPS is given in Fig. 5.3. These systems are the most basic both economically and in terms of simplicity. To benefit directly from the geothermal energy source, the temperature value of the incoming fluid must be at a certain value. These systems can be integrated into different systems since the underground fluid consists only of steam. While integrating, changes in the mathematical model of the system should also be taken into consideration.

Explaining the working principle of the system, in general, is useful for understanding how the system works. To eliminate particles in the fluid from the geothermal energy source, the fluid is sent to the particle separator by flow 1. Also, since the pressure in the geothermal energy source is higher than where the particle lifter is located, it is transmitted to the particle lifter after pressure adjustment. After the particles in the fluid are removed, the fluid coming out of the particle extractor is sent to the moisture separator with the number 2 flow to remove the moisture in the fluid. Here, the moisture released from the fluid is sent to the underground with flow number 4. For this system to perform its essential purpose, the fluid from the moisture separator is sent to the gas turbine to obtain power. With the generation of power, the fluid coming out of the turbine is sent to the condenser to be condensed with flow number 6. The fluid coming out of the condenser is sent again to the reinjection well.

Example 5.1: A direct steam GPS, as shown in Fig. 5.4, is utilized for power generation. The geothermal working fluid enters the particle separator at the temperature and pressure of 220°C and 2316 kPa, respectively, and a mass flow rate of 40 kg/s. The mass flow rate of unwanted particles ($\dot{m}_{prt}$) is equal to 10% of the input geothermal water mass flow rate. Then the clean geothermal working fluid enters the moisture separator and is

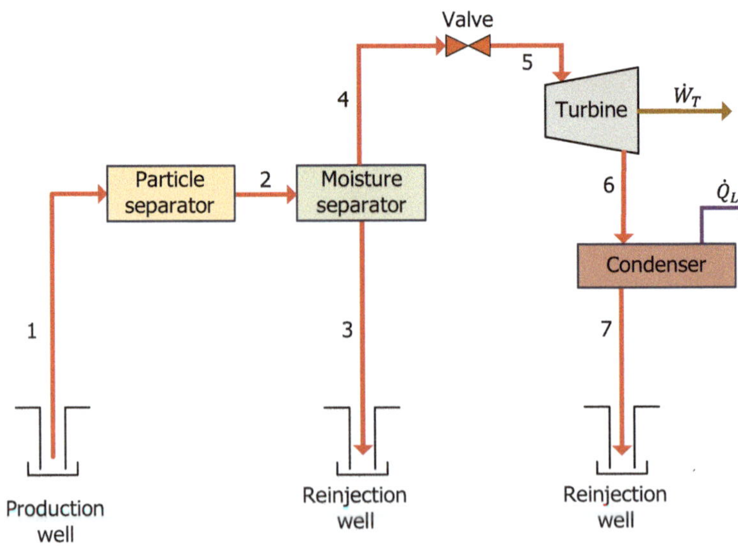

FIGURE 5.4 Schematic diagram of the direct steam geothermal power system.

separated in a saturated form in this plant; the fraction of saturated water at flow 3 is 0.4. The exit pressure from the valve is 772 kPa, and then the dry steam is utilized to produce power in a turbine. Finally, the discharged geothermal working fluid from the steam turbine is at 25 kPa with a quality of 0.84, which is then condensed and reinjected to the injection well. Calculate the:

1. power production rate by the turbine,
2. total exergy destruction rate of the direct steam GPS,
3. energy and exergy efficiencies of the whole system,
4. variation of power production rate and exergy destruction rate of the whole system, when the geothermal water mass flow rate increases from 100 to 180 kg/s, and
5. variation of energy and exergy efficiencies of the whole system when the reference temperature increases from 0°C to 40°C.

Solution: First, some assumptions are required for the thermodynamic assessment of the direct steam GPS.

Assumptions:

- All components operate under steady-state conditions.
- The reference temperature and pressure are taken as 25°C and 101.3 kPa, respectively.
- The pressure losses in the connections between the components are negligible.
- The heat energy losses from the components are negligible.
- The changes in the kinetic and potential energies and exergies are neglected.

Analysis:

The mass, energy, entropy, and exergy balance equations for the particle separator are defined:

$$\dot{m}_1 = \dot{m}_2$$
$$\dot{m}_1 h_1 = \dot{m}_2 h_2$$
$$\dot{m}_1 s_1 + \dot{S}_{g,ps} = \dot{m}_2 s_2$$
$$\dot{m}_1 ex_1 = \dot{m}_2 ex_2 + \dot{E}x_{D,ps}$$

The mass, energy, entropy, and exergy balance equations for the moisture separator can be written:

$$\dot{m}_2 = \dot{m}_3 + \dot{m}_4$$
$$\dot{m}_2 h_2 = \dot{m}_3 h_3 + \dot{m}_4 h_4$$
$$\dot{m}_2 s_2 + \dot{S}_{g,ms} = \dot{m}_3 s_3 + \dot{m}_4 s_4$$
$$\dot{m}_2 ex_2 = \dot{m}_3 ex_3 + \dot{m}_4 ex_4 + \dot{E}x_{D,ms}$$

The mass, energy, entropy, and exergy balance equations for the valve can be written:

$$\dot{m}_4 = \dot{m}_5$$
$$\dot{m}_4 h_4 = \dot{m}_5 h_5$$
$$\dot{m}_4 s_4 + \dot{S}_{g,val} = \dot{m}_5 s_5$$
$$\dot{m}_4 ex_4 = \dot{m}_5 ex_5 + \dot{E}x_{D,val}$$

The mass, energy, entropy, and exergy balance equations for the turbine can be written:

$$\dot{m}_5 = \dot{m}_6$$
$$\dot{m}_5 h_5 = \dot{m}_6 h_6 + \dot{W}_T$$
$$\dot{m}_5 s_5 + \dot{S}_{g,Tur} = \dot{m}_6 s_6$$
$$\dot{m}_5 ex_5 = \dot{m}_6 ex_6 + \dot{E}x^{W}_{Tur} + \dot{E}x_{D,Tur}$$

The mass, energy, entropy, and exergy balance equations for the condenser are written:

$$\dot{m}_6 = \dot{m}_7$$
$$\dot{m}_6 h_6 = \dot{m}_7 h_7 + \dot{Q}_L$$
$$\dot{m}_6 s_6 + \dot{S}_{g,con} = \dot{m}_7 s_7 + \dot{Q}_L / T_{con}$$
$$\dot{m}_6 ex_6 = \dot{m}_7 ex_7 + \dot{E}x^{Q}_{L} + \dot{E}x_{D,con}$$

The thermodynamic variables of all flows are defined:

For flow 1:

As given in the equation, $P_1 = 2316$ kPa and $T_1 = 220°C$, and, based on these data, h_1 and s_1 are determined from Engineering Equation Solver (EES) as 2801 kJ/kg and 6.285 kJ/kg K.

For flow 2:

$P_2 = P_1$ and $T_2 = T_1$; also, the mass flow rate at flow 2 can be determined from:

$$\dot{m}_2 = 0.9\dot{m}_1 = 36 \text{ kg/s}$$

For flow 3:

$P_3 = P_2$ and $T_3 = T_2$; also, the mass flow rate at flow 3 can be determined from:

$$\dot{m}_3 = 0.4\dot{m}_2 = 14.4 \text{ kg/s}$$

For flow 4:

$P_4 = P_3$ and $T_4 = T_3$; also, the mass flow rate at flow 4 can be calculated:

$$\dot{m}_4 = \dot{m}_2 - \dot{m}_3 = 21.6 \text{ kg/s}$$

For flow 5:

As given in the equation, $P_5 = 772$ kPa and $h_5 = h_4$, and, based on these data, T_5 and s_5 are determined by utilizing the EES program as 183.1°C and 6.75 kJ/kg K.

TABLE 5.1 State point thermodynamic data for the direct steam geothermal power system.

State point	$\dot{m}$ (kg/s)	T (°C)	P (kPa)	h (kJ/kg)	s (kJ/kg K)	x (–)	ex (kJ/kg)
0	–	40	101.3	167.6	0.5722	Liquid	–
1	40	220	2316	2801	6.285	1	844.8
2	36	220	2316	2801	6.285	1	844.8
3	14.4	220	2316	2801	6.285	1	844.8
4	21.6	220	2316	2801	6.285	1	844.7
5	21.6	183.1	772	2801	6.75	1	699.1
6	21.6	64.98	25	2242	6.72	0.84	149.4
7	21.6	64.98	25	272	0.8933	0	3.881

For flow 6:

As given in the equation, $x_6 = 0.84$ and $P_6 = 25$ kPa, and T_6, h_6, and s_6 are determined by utilizing the EES program as 64.98°C, 2242 kJ/kg, and 6.72 kJ/kg K.

For flow 7:

As given, $P_7 = P_6$ and $x_7 = 0$ (saturated liquid), and, based on these data, T_7, h_7, and s_7 are computed by utilizing the EES program as 64.98°C, 272 kJ/kg, and 0.8933 kJ/kg K. All of the thermodynamic data for every flow are given in Table 5.1.

1. The power production rate by the steam turbine is calculated:

$$\dot{W}_T = \dot{m}_5 h_5 - \dot{m}_6 h_6 = 12{,}076 \text{ kW}$$

2. The total exergy destruction rate of the direct steam GPS is found to be:

$$\dot{E}x_{D,DSGP} = \dot{m}_1 ex_1 - \dot{m}_3 ex_3 - \dot{m}_7 ex_7 - \dot{E}x^W_{out,Tur} = 11{,}562 \text{ kW}$$

3. The energy efficiency of the direct steam GPS is obtained:

$$\eta_{DSGP} = \frac{\dot{W}_T}{\dot{m}_1 h_1 - \dot{m}_3 h_3 - \dot{m}_7 h_7} = 0.1834 = 18.34\%$$

Also, the exergy efficiency of the direct steam GPS is found:

$$\psi_{SFGP} = \frac{\dot{E}x^W_{Tur}}{\dot{m}_1 ex_1 - \dot{m}_3 ex_3 - \dot{m}_7 ex_7} = 0.5109 = 51.09\%$$

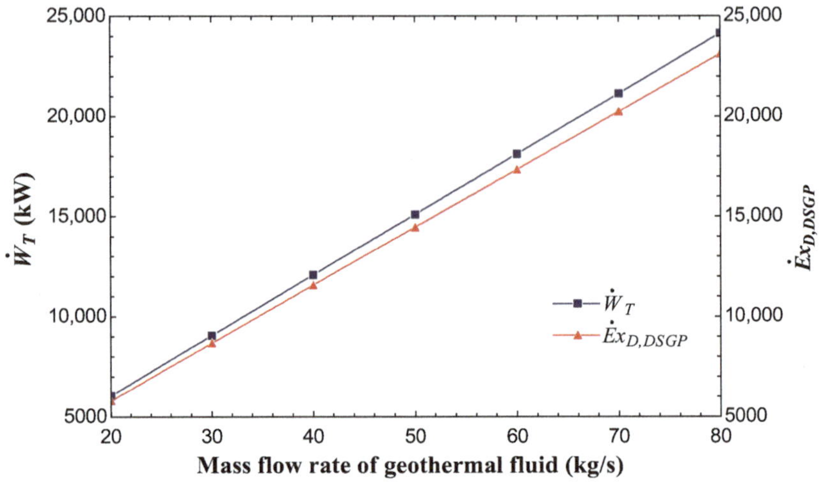

FIGURE 5.5 Effect of geothermal fluid mass flow rate on the power generation rate and exergy destruction rate for the direct steam geothermal power system.

4. The effect of geothermal water mass flow rate on the power production rate and exergy destruction rate of the direct steam GPS is illustrated in Fig. 5.5. As given in this figure, the power production rate increases from 6038 to 24,152 kW, and the exergy destruction rate increases from 5781 to 23,125 kW with the increasing geothermal mass flow rate from 20 to 80 kg/s, respectively.
5. The effect of reference temperature on both energy and exergy efficiencies of the direct steam GPS is shown in Fig. 5.6. Based on this figure, the energy efficiency of the geothermal plant does not change with the increasing reference temperature from 0°C to 40°C, whereas the exergy efficiency of the geothermal plant increases from 44.46% to 56.06% in the given reference temperature change.

5.3.1 Case study 5.1

In the first case study, the geothermal energy resource–based direct steam GPS with two turbines is analyzed by utilizing the energy and exergy analyses. The simplified flow diagram of the direct steam GPS in the first case study is illustrated in Fig. 5.7.

As reference conditions, ambient temperature and pressure are taken as 25°C and 101.3 kPa, respectively. The mass, energy, entropy, and exergy balance equations for the particle separator, moisture separator, valve, turbine 1, and condenser

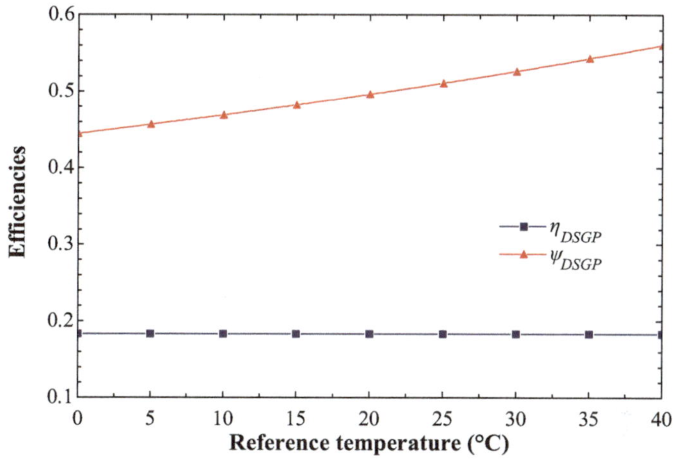

FIGURE 5.6 Effect of reference temperature on the energy and exergy efficiencies of the direct steam geothermal power system.

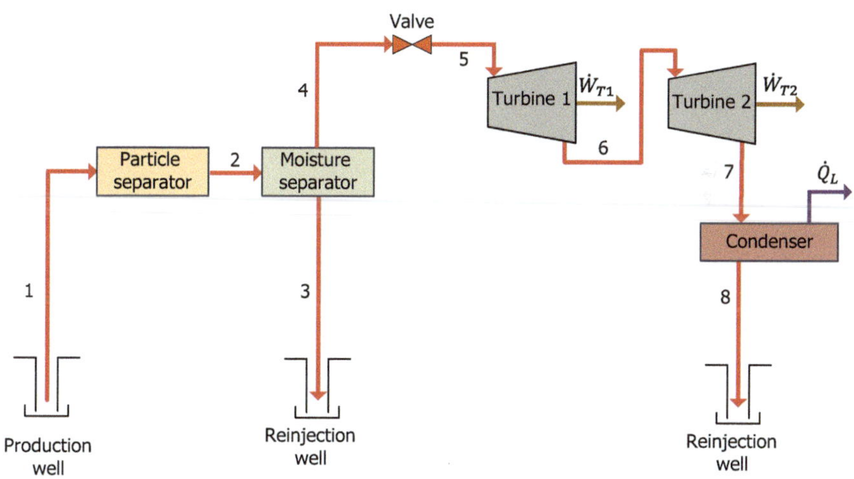

FIGURE 5.7 Schematic diagram of a direct steam geothermal power system with two turbines.

subsystems are defined in Example 5.1. In addition, the mass, energy, entropy, and exergy balance equations for the turbine 2 are defined:

$$\dot{m}_6 = \dot{m}_7$$
$$\dot{m}_6 h_6 = \dot{m}_7 h_7 + \dot{W}_{T2}$$
$$\dot{m}_6 s_6 + \dot{S}_{g,T2} = \dot{m}_7 s_7$$
$$\dot{m}_6 ex_6 = \dot{m}_7 ex_7 + \dot{E}x_{T2}^{W} + \dot{E}x_{D,T2}$$

TABLE 5.2 Operating parameters for the direct steam geothermal power system with two turbines.

Variables	Values
Geofluid source temperature, T_1	220°C
Geofluid source pressure, P_1	2316 kPa
Geofluid mass flow rate, $\dot{m}_1$	40 kg/s
Turbine 1 inlet pressure, P_5	1080 kPa
Turbine 1 inlet temperature, T_5	191°C
Turbine 2 inlet pressure, P_6	772 kPa
Turbine 2 inlet temperature, T_6	169°C
Geofluid reinjection temperature, T_8	64.98°C

TABLE 5.3 Thermodynamic assessment results for the power generation process components of the direct steam geothermal system with two turbines.

System components	Exergy destruction rate (kW)	Exergy destruction ratio (%)	Exergy efficiency (%)
Particle separator	784	8.83	84.63
Moisture separator	626	7.05	86.07
Valve	510	5.75	83.93
Turbine I	2928	32.99	36.27
Turbine II	2469	27.82	38.05
Condenser	1559	17.56	28.09
Overall system	8876	100	62.45

The assumptions used in the operating conditions of the direct steam GPS with two turbines are given in Table 5.2. The heat and work input/output rate, entropy generation rate, and exergy destruction rates, as well as energy and exergy efficiencies are calculated by using the balance equations and chosen assumptions.

The exergy destruction rate, dimensionless exergy destruction ratio, and exergy efficiency of the components of the direct steam GPS with two turbines are given in Table 5.3.

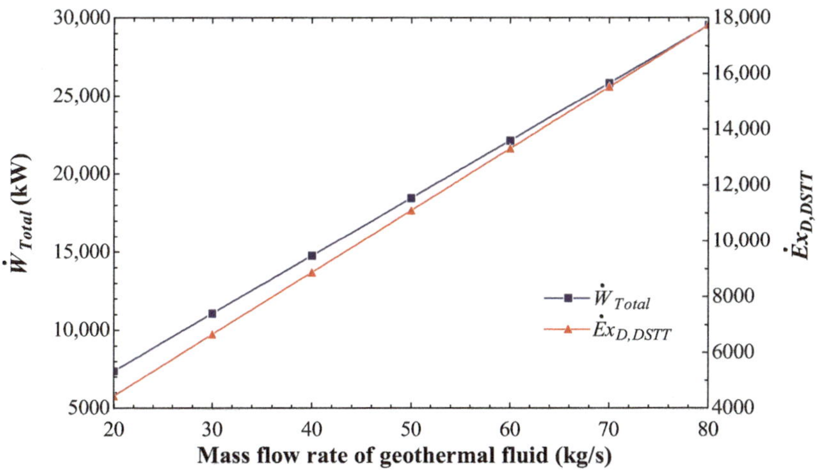

FIGURE 5.8 Effect of geothermal fluid mass flow rate on the power generation rate and exergy destruction rate for the direct steam geothermal power system with two turbines.

The effect of geothermal water mass flow rate on the power production rate and exergy destruction rate of the direct steam GPS with two turbines is shown in Fig. 5.8. As given in this figure, the power production rate increases from 7381 to 29,525 kW, and the exergy destruction rate increases from 4438 to 17,752 kW with the increasing geothermal mass flow rate from 20 to 80 kg/s.

The effects of reference temperature on the energy and exergy efficiencies of the direct steam GPS with two turbines are illustrated in Fig. 5.9. As given in this figure, the energy efficiency of the geothermal plant does not change with increasing the reference temperature from 0°C to 40°C, whereas the exergy efficiency of the geothermal plant increases from 54.35% to 68.53% in the given reference temperature change.

5.4 Basic flashing geothermal power systems

Basic flashing GPSs can be grouped into two main classes based on the number of flash chambers used to produce vapor from the geothermal working fluid. For example, one flash chamber is used in a single-flash steam GPS, and two are utilized in a double-flash steam GPS. In this section, the working and design conditions and also the energy and exergy analyses of single- and double-flash steam GPSs are provided. To investigate the effect of design and working conditions on the performance of these plants, two example studies are provided.

5.4.1 Single-flash steam geothermal power system

The single-flash steam GPSs form the basis in geothermal energy systems [6]. The fluid used in geothermal energy systems is supplied from the underground.

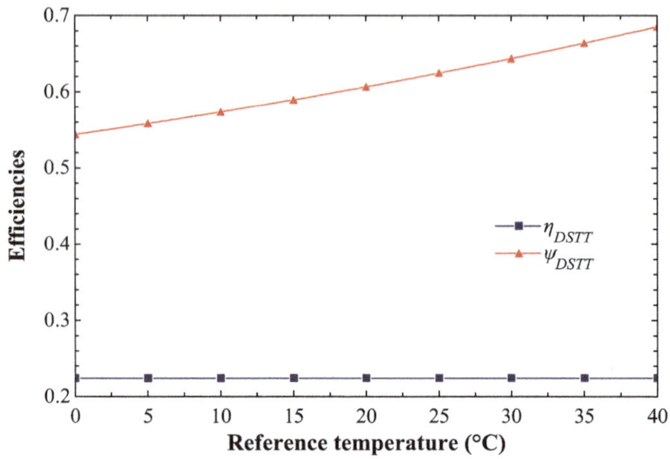

FIGURE 5.9 Effect of reference temperature on the energy and exergy efficiencies of direct steam geothermal power system with two turbines.

This fluid is available as a mixture of steam and liquid [7]. The most basic energy system that can produce electricity by making use of this mixture is the single-flash steam geothermal energy system [8]. The schematic diagram of the single-flash steam GPS is given in Fig. 5.10. In the single-flash system, the fluid in the energy system goes through a single-flash operation. This terminology can be expressed more clearly as the process of obtaining pressurized liquid and vapor from the liquid with this saturation pressure by decreasing the pressure of the underground pressurized liquid.

As can be seen in Fig. 5.10, high-pressure (HP) geothermal fluid coming from the underground is sent to the valve where flashing will take place with the number 1 flow. Here, the fluid whose pressure is reduced is transmitted to the separator with flow number 2 as a mixture of liquid and vapor. The incoming fluid is divided into two as steam and brine in the separator. The steam coming out of the separator is sent to the purifier with flow number 4 to increase the performance of the fluid. The fouling part in the fluid is removed from the system by flow number 5.

The steam fluid with increased quality is sent to the turbine with flow number 6 to obtain the electrical output. Then the steam coming out of the turbine is sent with flow number 7 to the condenser to be condensed. Studies in the literature or the application on the selection of the condenser that performs the condensation process. The condenser is used for the heating output, which is a useful output. The fluid coming out of the condenser is sent to the ground with the number 8 flow. In general, single-flash steam geothermal power generation systems operate with this logic. The activity diagram of electricity and heating generation in the single-flash steam GPS is also given in Fig. 5.11. The way the system works can be easily understood with a careful look at the activity diagram.

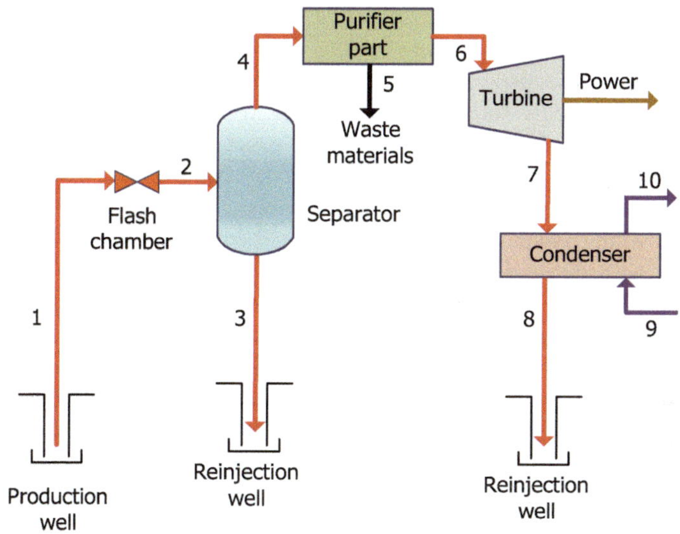

FIGURE 5.10 Simplified flow diagram of a single-flash steam geothermal power system.

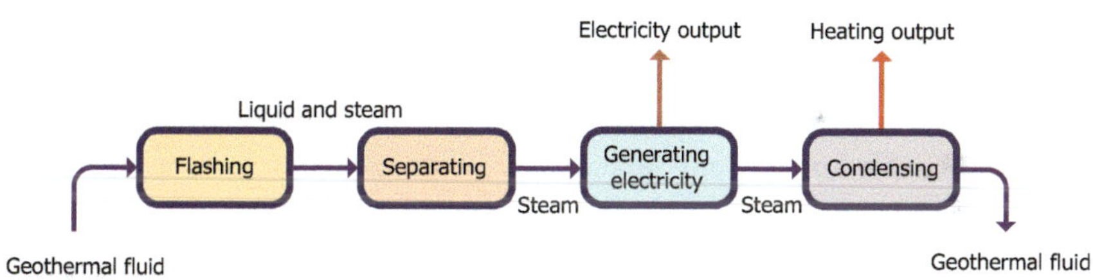

FIGURE 5.11 Activity diagram of electricity and heating generation in the single-flash steam geothermal power system.

Examining the activity process in Fig. 5.11, together with the schematic diagram of the system, increases the intelligibility of the system. The geothermal fluid coming to the system in the activity process comes from underground. Then it is carried out with the only flashing valve in the system. After flashing, liquid and vapor mixture enters the fluid separator which separates liquid as brine and vapor mixture as steam. The steam coming out of the separator is sent to the purifier for the improvement process again to ensure good performance in power production. Although this improvement process is not in the activation process, it can easily be seen in the schematic diagram of the system. The quality-enhanced steam is also sent to the turbine for power production. For the heating output, the steam coming from the turbine is condensed in the condenser, and the heating output is obtained. Then the fluid coming out of the system is sent to the ground. Since the relationship between the schematic diagram of the system and the activation process can be easily seen, it can be said that the activity process meaningfully expresses the schematic diagram of the system. Subsystems of the single-flash steam GPS are also clearly shown in Fig. 5.12.

There are generally four main subsystems in this system: the flashing process, separation process, turbine expansion process, and condensing process. How and which outputs are obtained in these subsystems have been explained.

The choice of system components is recognized as a significant task in power generating systems. For this reason, it is important to select these components in accordance with the analyses made and to plan accordingly. After the system components are selected, improvements should be continued in the context of the system and subsystems to increase system performance. For component selection, a brief explanation of the separator will be useful. The better the performance of the separator, the better the vapor–liquid mixture can be separated. Of course, in the selection of these components, besides performance, cost should also be considered. The useful outputs obtained from the overall geothermal energy system vary with the performance of the system components: the higher the components' performance, the higher the system's performance. This principle should be applied to each of the system components, including the separator. In this regard, the designer of the energy system needs to pay attention to the cost-performance analysis.

Some diagrams provided by System Modeling Language (SysML) are useful to carry out a more comfortable design and analysis of different areas of the single-flash geothermal energy system. This modeling language is particularly suitable for applications such as the design, analysis, and verification of various systems in the field of systems engineering. The diagrams provided by

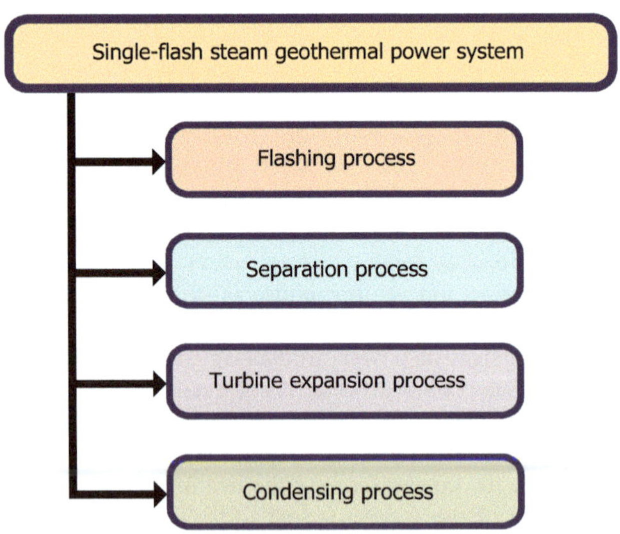

FIGURE 5.12 Subsystems of single-flash steam geothermal power system.

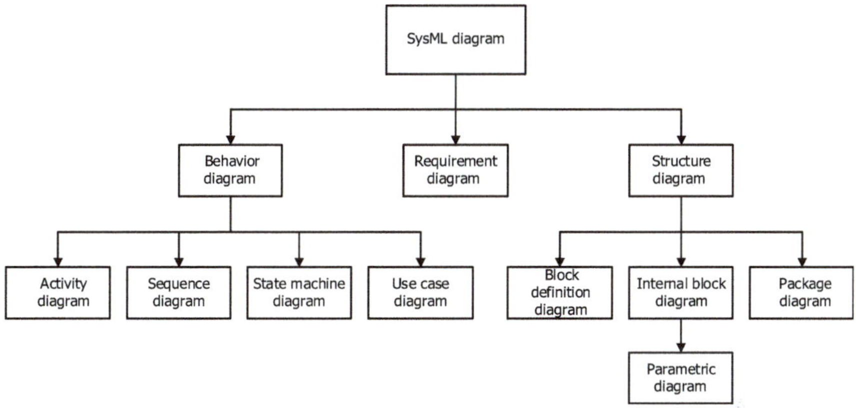

FIGURE 5.13 SysML diagram types for modeling any system. *SysML*, System Modeling Language.

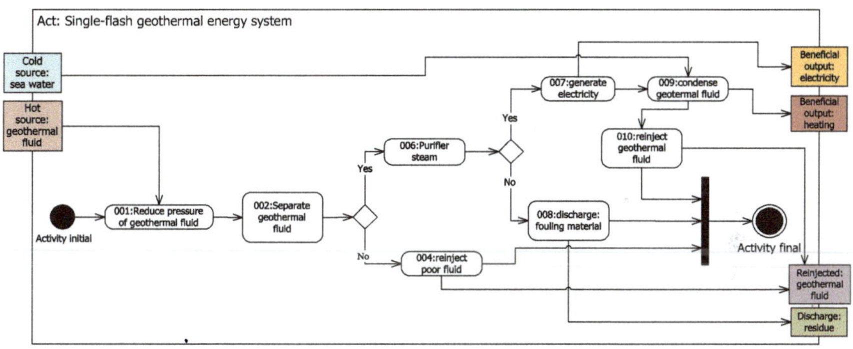

FIGURE 5.14 Activity diagram of the single-flash steam geothermal energy system.

this modeling language to model any system are given in Fig. 5.13. When all of these diagrams are created for a system, the parts of the system to be used at different times and within different systems can be easily integrated into other systems. At the same time, if the subsystems in the system are used at different times with different parameter values, the design of the system can be realized more effectively with this modeling language.

In this section, some diagrams describing the single-flash steam GPS are given by making use of using this modeling language. The first diagram given for the system is the activity diagram shown in Fig. 5.14. The activity diagram of electricity and heating generation given in Fig. 5.11 shows the steps taken while creating this diagram. However, the activity diagram in Fig. 5.14 is of the system that generally describes the system. The diagram clearly indicates the pathways occurred depending on the conditions.

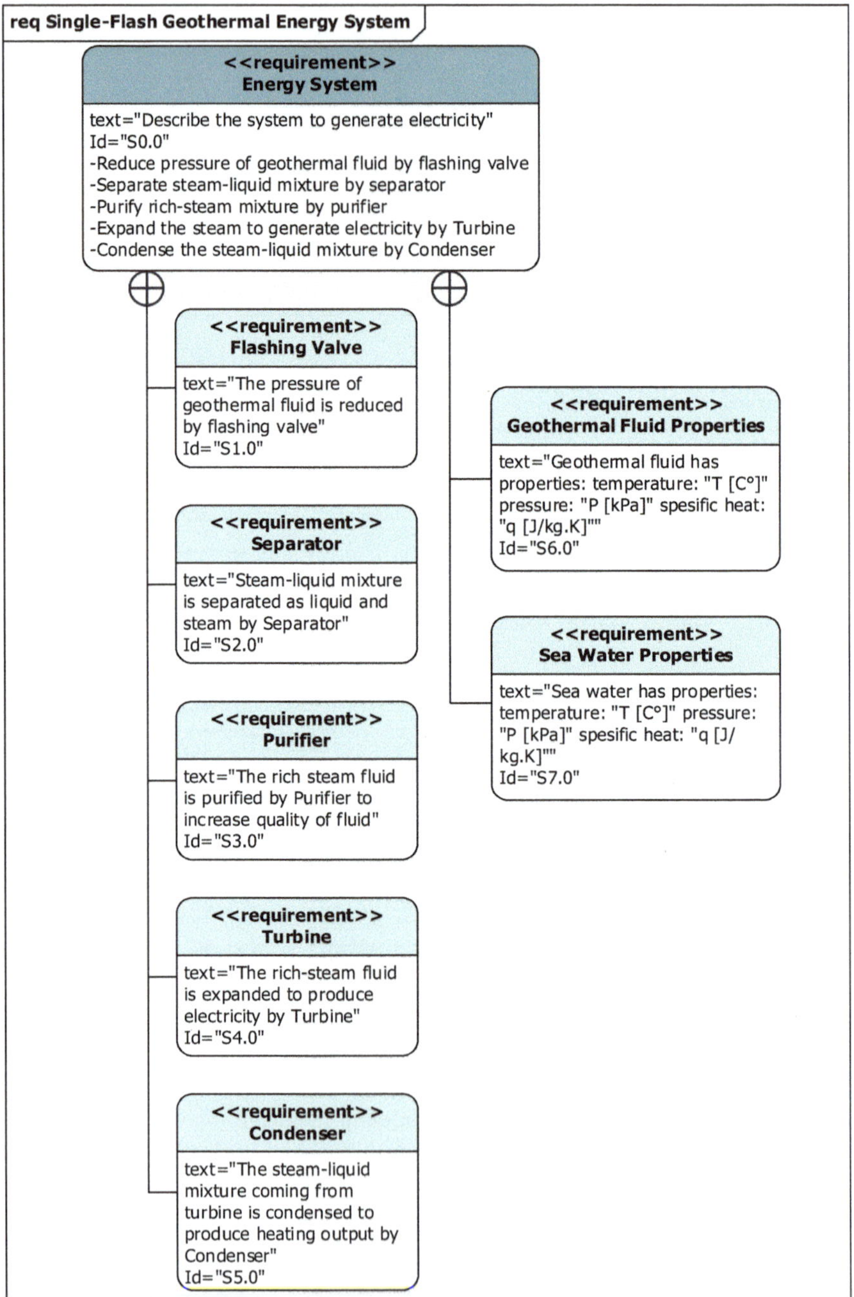

FIGURE 5.15 Process requirements diagram of the single-flash steam geothermal power plant.

Another diagram that defines the system is the requirement diagram. This diagram clearly describes the tasks performed by the system components for any system design. Unlike system components, this diagram can also be used to describe other states or phenomena within the system. In this section, the requirement diagram is used to define the tasks of the components of the single-flash system. This diagram is presented in Fig. 5.15. As can be seen from the figure, almost all of the components that function in the system are given here. This diagram can be designed in different ways given the parameters defined or desired to be defined in the system. Since the design of these systems is explained in this section, the properties of the fluids desired to be used in the system are specified in this diagram. However, since their properties can vary, the units of the parameters are written in the fluid properties section of the diagram.

The block identification diagram, one of the most basic diagrams defining the system, is used in this section to describe the electricity generation of single-flash GPS. The block definition diagram of this system is given in Fig. 5.16. As can be seen from the figure, the activities that must take place for electricity generation from the system are expressed in this diagram.

Many activities in energy systems take place at the same time or in different rows. It is useful to use the sequence diagram in the SysML modeling language to define this timing. Therefore this diagram makes it possible to see the order

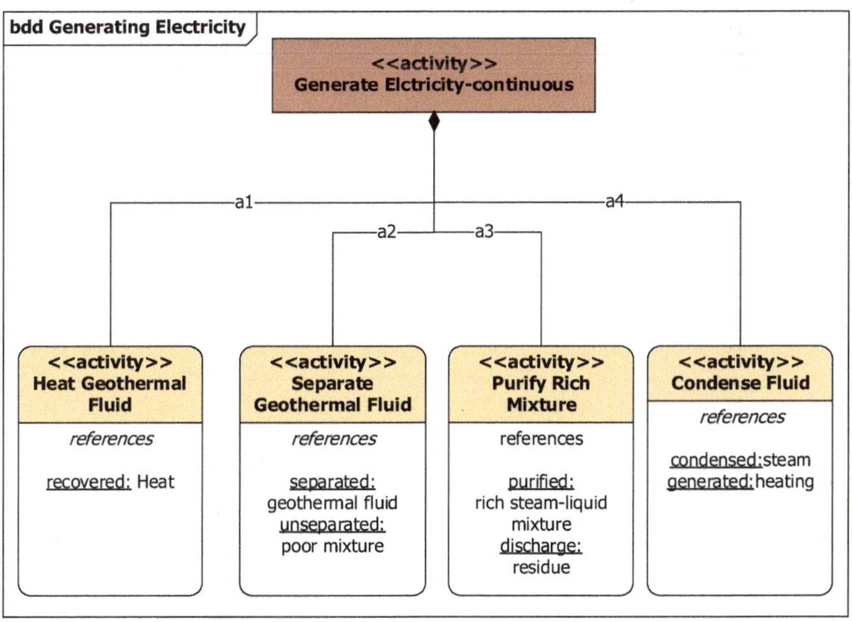

FIGURE 5.16 Block definition diagram of generating electricity from the single-flash steam geothermal plant.

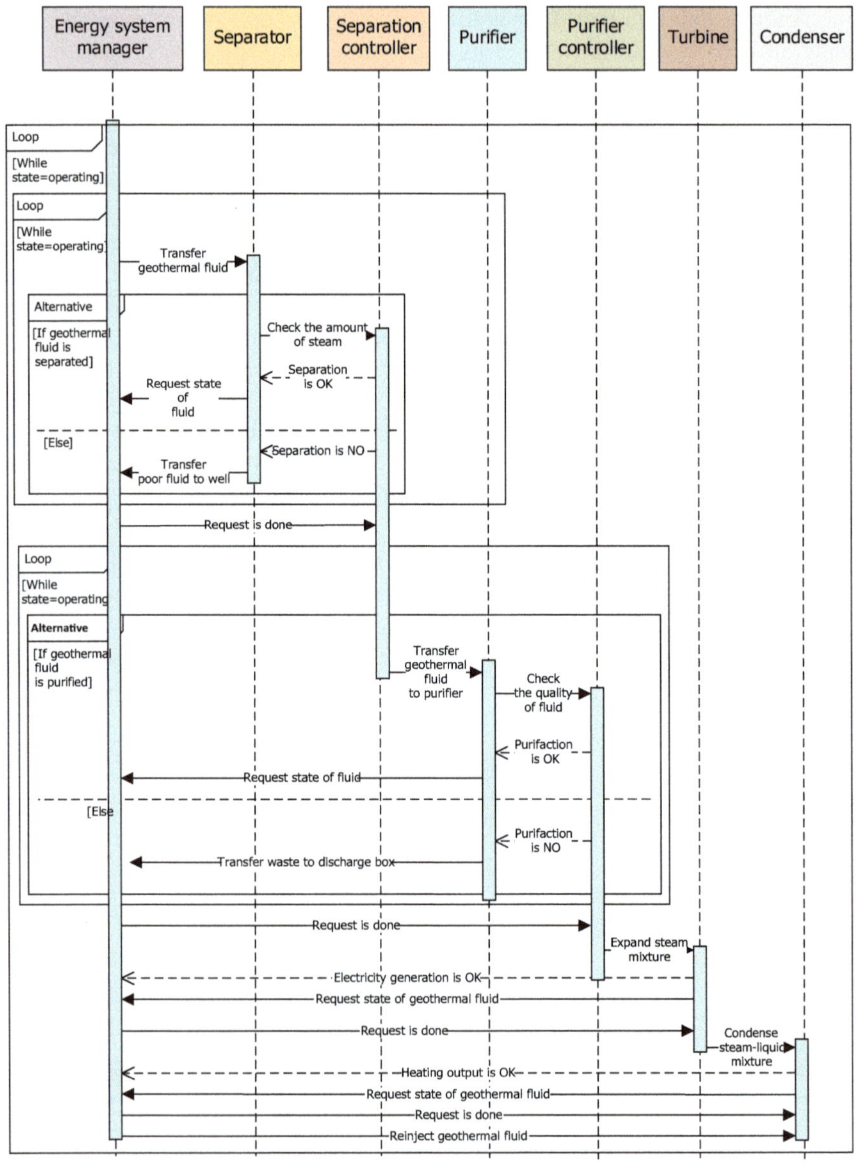

FIGURE 5.17 Sequence diagram of the single-flash steam geothermal energy plant.

of the activities taking place in a single-flash system. The sequence diagram of the single-flash system is given in Fig. 5.17. Depending on the path followed by the geothermal fluid in a single-flash system, the order in which the activities in the components are performed is given in this diagram. If other situations are realized in this diagram, these states can be expanded by add-ons.

The diagrams created using the SysML modeling language provide a clearer definition of the system. If an entire system is clearly defined using this modeling language, the design, analysis, and integration of the system and other systems into which this system is integrated become more effective. Since modeling and analysis studies in energy systems play such a large role, this modeling language can make the design of new, future systems more effective.

Example 5.2: A single-flash steam GPS, as shown in Fig. 5.18, is used for power production. In this system, 140 kg/s saturated liquid geothermal fluid at the temperature and pressure of 218°C and 2316 kPa, respectively, is isenthalpically flashed in a flash chamber at a pressure ratio of 3 and separated in the saturated form in a vapor–liquid separator. The fraction of vapor at the flash part outlet is 0.11. Then the produced vapor is used to generate power in a steam turbine. Finally, the discharged geothermal working fluid from the steam turbine is at 11 kPa with a quality of 0.84, which is then condensed and reinjected to the injection well. Calculate the:

1. work production rate by the steam turbine,
2. total exergy destruction rate of the single-flash steam GPS,
3. energy and exergy efficiencies of the single-flash steam GPS,
4. variation of power production rate and exergy destruction rate of the single-flash steam GPS when the geothermal water mass flow rate increases from 100 to 180 kg/s, and
5. variation of energy and exergy efficiencies of the single-flash steam GPS when the reference temperature increases from 0°C to 40°C.

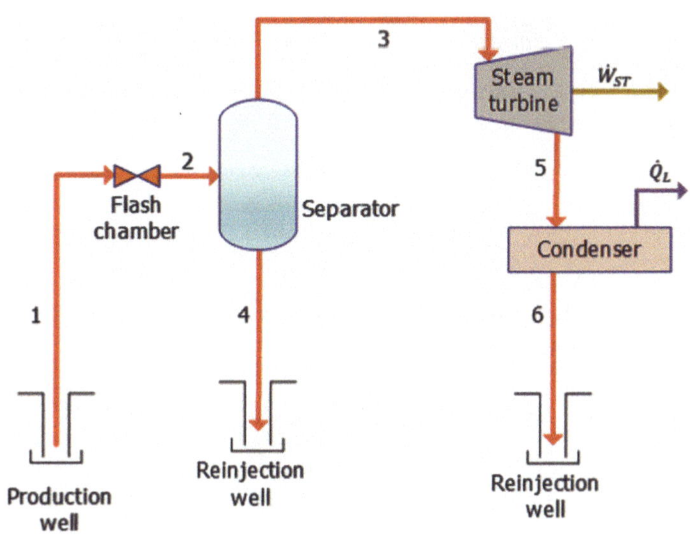

FIGURE 5.18 Schematic diagram of the single-flash steam geothermal power system.

Solution: First, some assumptions are required for the single-flash steam GPS analysis.

Assumptions:

- All components operate under steady-state conditions.
- The reference temperature and pressure are taken as 25°C and 101.3 kPa, respectively.
- The pressure losses in the connections between the components are negligible.
- The heat energy losses from the components are negligible.
- The changes in the kinetic and potential energies and exergies are neglected.

Analysis:

The mass, energy, entropy, and exergy balance equations for the flash chamber are defined:

$$\dot{m}_1 = \dot{m}_2$$
$$\dot{m}_1 h_1 = \dot{m}_2 h_2$$
$$\dot{m}_1 s_1 + \dot{S}_{g,fc} = \dot{m}_2 s_2$$
$$\dot{m}_1 ex_1 = \dot{m}_2 ex_2 + \dot{E}x_{D,fc}$$

The mass, energy, entropy, and exergy balance equations for the separator can be obtained:

$$\dot{m}_2 = \dot{m}_3 + \dot{m}_4$$
$$\dot{m}_2 h_2 = \dot{m}_3 h_3 + \dot{m}_4 h_4$$
$$\dot{m}_2 s_2 + \dot{S}_{g,sep} = \dot{m}_3 s_3 + \dot{m}_4 s_4$$
$$\dot{m}_2 ex_2 = \dot{m}_3 ex_3 + \dot{m}_4 ex_4 + \dot{E}x_{D,sep}$$

The mass, energy, entropy, and exergy balance equations for the steam turbine can be expressed:

$$\dot{m}_3 = \dot{m}_5$$
$$\dot{m}_3 h_3 = \dot{m}_5 h_5 + \dot{W}_{ST}$$
$$\dot{m}_3 s_3 + \dot{S}_{g,ST} = \dot{m}_5 s_5$$
$$\dot{m}_3 ex_3 = \dot{m}_5 ex_5 + \dot{E}x_{ST}^{W} + \dot{E}x_{D,ST}$$

The mass, energy, entropy, and exergy balance equations for the condenser can be obtained:

$$\dot{m}_5 = \dot{m}_6$$
$$\dot{m}_5 h_5 = \dot{m}_6 h_6 + \dot{Q}_L$$
$$\dot{m}_5 s_5 + \dot{S}_{g,con} = \dot{m}_6 s_6 + \dot{Q}_L / T_{con}$$
$$\dot{m}_5 ex_5 = \dot{m}_6 ex_6 + \dot{E}x_L^{Q} + \dot{E}x_{D,con}$$

Thermodynamic variables of all flows for the single-flash steam GPS are defined:

For flow 1:

As given in the equation, $P_1 = 2316$ kPa and $T_1 = 218°C$, and, based on these data, h_1 and s_1 are determined from the EES as 934.3 kJ/kg and 2.499 kJ/kg K.

For flow 2:

As given in the equation, $x_2 = 0.11$ and $P_2 = P_1/3$, and, based on these data, T_2, h_2, and s_2 are determined by using the EES as 169°C, 940.6 kJ/kg, and 2.543 kJ/kg K.

For flow 3:

$P_3 = P_2$ and $x_3 = 1$ (saturated vapor), and, based on these data, T_3, h_3, and s_3 are determined by using the EES as 169°C, 2767 kJ/kg, and 6.675 kJ/kg K. The mass flow rate at flow 3 can be determined:

$$\dot{m}_3 = 0.11 \times \dot{m}_2 = 15.4 \text{ kg/s}$$

For flow 4:

$P_4 = P_2$ and $x_4 = 0$ (saturated liquid), and also, T_4, h_4, and s_4 are determined by using the EES as 169°C, 714.8 kJ/kg, and 2.032 kJ/kg K. The mass flow rate at flow 4 can be determined:

$$\dot{m}_4 = \dot{m}_2 - \dot{m}_3 = 124.6 \text{ kg/s}$$

For flow 5:

As given in the equation, $x_5 = 0.84$ and $P_5 = 11$ kPa, and T_5, h_5, and s_5 are determined by utilizing the EES program as 47.69°C, 2205 kJ/kg, and 6.924 kJ/kg K.

For flow 6:

As given, $P_6 = P_5$ and $x_6 = 0$ (saturated liquid), and also, T_6, h_6, and s_6 are determined by utilizing the EES program as 47.69°C, 199.7 kJ/kg, and 0.6738 kJ/kg K.

All these thermodynamic data for flows in the investigated plant are evaluated by using the EES software program, and the thermodynamic results are given in Table 5.4.

1. The power production rate by the steam turbine given in Fig. 5.18 can be determined:

$$\dot{W}_{ST} = \dot{m}_3 h_3 - \dot{m}_5 h_5 = 8660 \text{ kW}$$

2. The total exergy destruction rate of the single-flash steam GPS can be calculated:

$$\dot{E}x_{D,SFGP} = \dot{m}_1 ex_1 - \dot{m}_4 ex_4 - \dot{m}_6 ex_6 - \dot{E}x_{ST}^{W} = 4283 \text{ kW}$$

TABLE 5.4 State point thermodynamic data for the single-flash steam geothermal power system.

State point	$\dot{m}$ (kg/s)	T (°C)	P (kPa)	h (kJ/kg)	s (kJ/kg K)	x (–)	ex (kJ/kg)
0	–	25	101.3	104.8	0.3669	Liquid	–
1	140	218	2316	934.3	2.499	0	193.8
2	140	169	772	940.6	2.543	0.11	187
3	15.4	169	772	2767	6.675	1	781.9
4	124.6	169	772	714.8	2.032	0	113.5
5	15.4	47.69	11	2205	6.924	0.84	145.2
6	15.4	47.69	11	199.7	0.6738	0	3.348

3. The energy efficiency of the single-flash steam GPS can be determined:

$$\eta_{SFGP} = \frac{\dot{W}_{ST}}{\dot{m}_1 h_1 - \dot{m}_4 h_4 - \dot{m}_6 h_6} = 0.224 = 22.4\%$$

Also, the exergy efficiency of the single-flash steam GPS can be computed:

$$\psi_{SFGP} = \frac{\dot{E}x_{ST}^{W}}{\dot{m}_1 ex_1 - \dot{m}_4 ex_4 - \dot{m}_6 ex_6} = 0.6691 = 66.91\%$$

4. The effect of geothermal water mass flow rate on the power production rate and exergy destruction rate of the single-flash steam GPS is shown in Fig. 5.19. As shown in this figure, the power production rate increases from 6185 to 11,134 kW, and the exergy destruction rate increases from 3059 to 5506 kW with the increasing geothermal mass flow rate from 100 to 180 kg/s.
5. The effect of reference temperature on the energy and exergy efficiencies of the single-flash steam GPS is illustrated in Fig. 5.20. It can be shown that the energy efficiency of the geothermal plant does not change with the increasing reference temperature from 0°C to 40°C, whereas the exergy efficiency of the investigated plant increases from 0.5735 to 0.7434 in the chosen reference temperature change.

5.4.2 Double-flash steam geothermal power system

The double-flash steam GPS can be defined as one that occurs when the single-flash is upgraded in terms of performance [9]. Since the double-flash steam geothermal system requires more knowledge in terms of understandability than the single-flash one, it has extra requirements in terms of both

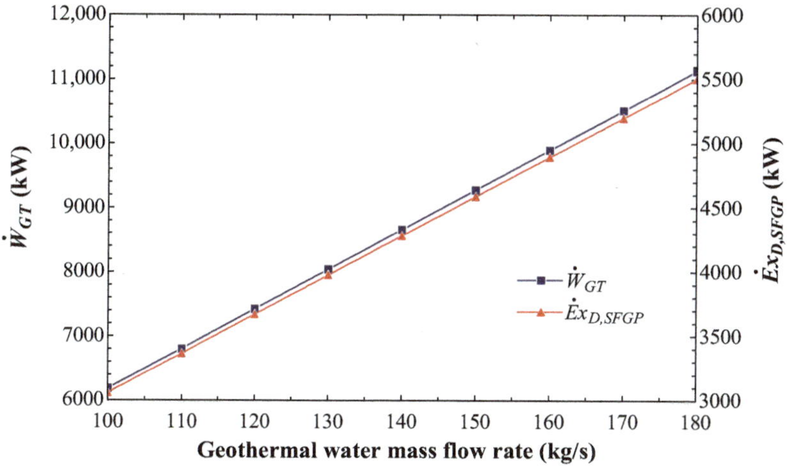

FIGURE 5.19 Effect of geothermal water mass flow rate on the power production and exergy destruction rates for the single-flash steam geothermal power system.

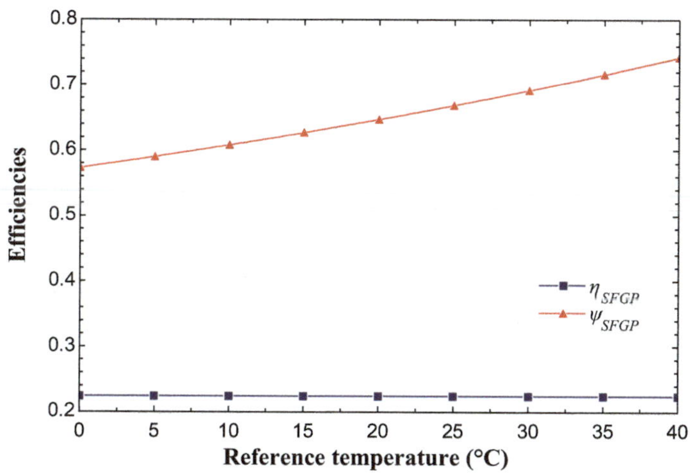

FIGURE 5.20 Effect of reference temperature on the energy and exergy efficiencies for the single-flash steam geothermal power system.

cost and maintenance [10]. As this system is the advanced version of the single-flash system, similar situations occur at some points [11]. The schematic diagram of the double-flash steam GPS in its most general form is given in Fig. 5.21. Looking at the double-flash system, the difference that can be seen directly from the single-flash steam geothermal system is that the working fluid enters the flashing process twice. In this way, considering geothermal fluid with the same amount and temperature, more power can be obtained from this system compared to a single-flash system.

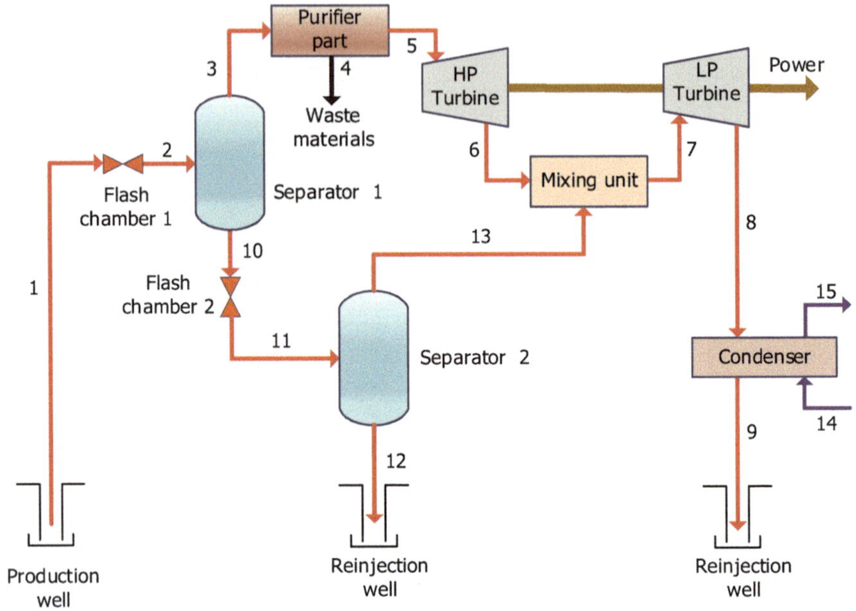

FIGURE 5.21 Simplified flow diagram of a double-flash steam geothermal power system.

As can be seen from Fig. 5.20, the geothermal fluid coming from underground is sent to flash chamber 1, where the first flashing takes place with the number 1 flow. Here, the pressure-reduced liquid–vapor mixture is delivered to separator 1 with flow number 2. After separating steam from this fluid with separator 1, this steam is sent to the purification section with flow 3 to increase the quality. The dirty part in the steam is removed from the system by flow number 4. The steam, whose quality is increased, is sent to the HP turbine with flow number 5 to produce a power output. The steam here is expanded between flow 5 and flow 6, and power can be produced. While this is happening, another process continues. Fluid leaving separator 1 with flow number 10 is sent to the flash chamber 2 for the second flashing process. Here, a second flashing takes place; that is, the liquid–vapor mixture is obtained by decreasing the pressure of the fluid again. To separate the steam in the liquid–vapor mixture, the fluid is sent to separator 2 with flow number 11. The steam coming out of separator 2 is sent to the mixing unit with flow 13 to be mixed with the steam obtained from the first flashing process. Then the steam coming from the HP turbine with flow 6 and the steam coming from separator 2 with flow 13 are mixed in the mixing unit. This steam, once mixed in the mixing unit, is sent to the low-pressure (LP) turbine with flow 7 to obtain the electrical output. Then the steam coming out of the turbine is sent to the condenser with flow 8 to be condensed. The condenser is used for the heating output, which is a beneficial output. The fluid

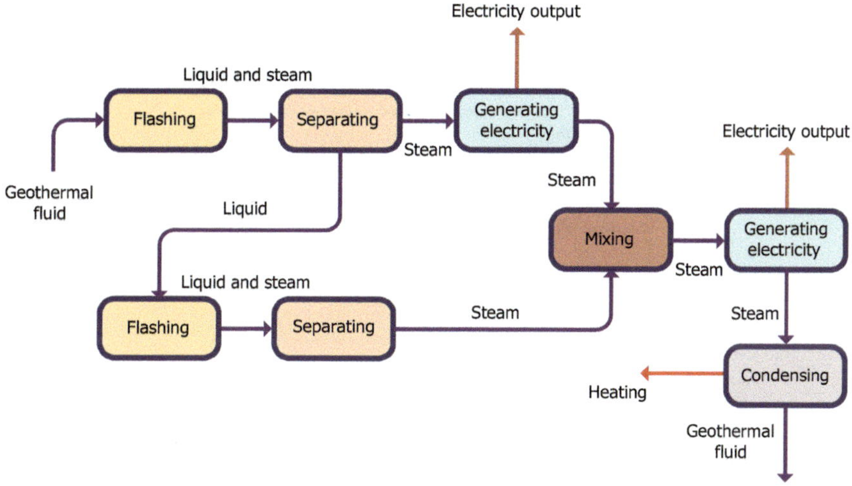

FIGURE 5.22 Activity diagram of electricity and heating generation in the double-flash steam geothermal power system.

coming out of the condenser is sent to the ground with flow 9. In general, the double-flash steam geothermal power generation systems work with this logic.

The activity diagram of the double-flash steam geothermal energy system is given in Fig. 5.22. To properly understand the activity diagram, it is useful to examine it together with the schematic diagram of the system.

It can be easily seen that this system, which was first noticed in the activity diagram in its most general form, is a two-flash system. This system can obtain more power than a single-flash system under the same conditions. As seen in the activity diagram, the fluid coming from the underground goes through the first flashing process. Then the fluid in the liquid–vapor mixture is separated from the mixture by using separator 1 to use the steam in power generation. To improve the performance of the fluid in the system, it is sent to the purifier to separate the dirty part in the fluid from separator 1. Now, with the fluid whose quality is increased, power can be produced with an HP turbine. In addition, the liquid part from separator 1 enters the flashing phase again and the liquid–vapor mixture is converted. Separator 2 separates the steam from the fluid passing through the flashing phase. Then the steam coming from here and the steam coming from the turbine are mixed, and this steam mixture is sent to the LP turbine for the second power output. For the heating product, the steam coming from the turbine is condensed in the condenser, and the heating product is obtained. Then the fluid coming out of the system is sent to the ground. In the section up to now, it can be seen that the double-flash system is more efficient and more comprehensive than the single-flash system.

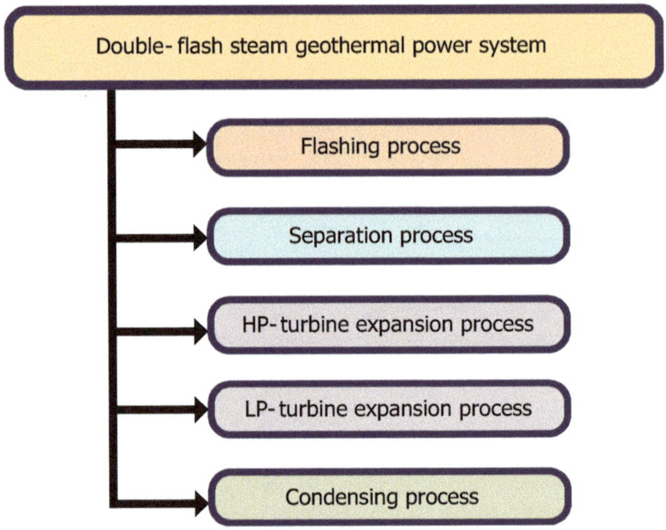

FIGURE 5.23 Subsystems of the double-flash steam geothermal power system.

Subsystems of the double-flash steam GPS are also clearly given in Fig. 5.23. There are generally five main subsystems in this system: the flashing process, separation process, HP turbine expansion process, LP-turbine expansion process, and the condensing process.

Example 5.3: A double-flash steam GPS, as shown in Fig. 5.24, is used for power production by using the HP and LP turbines. In this plant, 140 kg/s saturated liquid geothermal water at the temperature and pressure of 218°C and 2316 kPa, respectively, is isenthalpically flashed in the flash chamber 1 at a pressure ratio of 2, separated in the saturated form in the vapor–liquid separator 1; the fraction of vapor at the flash chamber outlet is 0.07623, and then the produced vapor is used to generate power in the HP turbine. The quality and pressure of working fluid exiting from the HP turbine are 0.9605 and $P_3/2$. The liquid working fluid enters flash chamber 2 and is isenthalpically flashed in this flash chamber at the same pressure level of flow 6, separated in the saturated form in the vapor–liquid separator 2, also the fraction of vapor at the flash part outlet is 0.0607, and then the produced vapor is used to generate power in the LP turbine. The discharged geothermal working fluid from the steam turbine with a quality of 0.8392 is at 10 kPa, which is then condensed and reinjected to the injection well. The saturated liquid working fluid exiting separator 2 is at the same pressure level of flow 6. Calculate the:

1. work production rate by the HP and LP turbine and total work production rate,
2. total exergy destruction rate of the double-flash steam GPS,

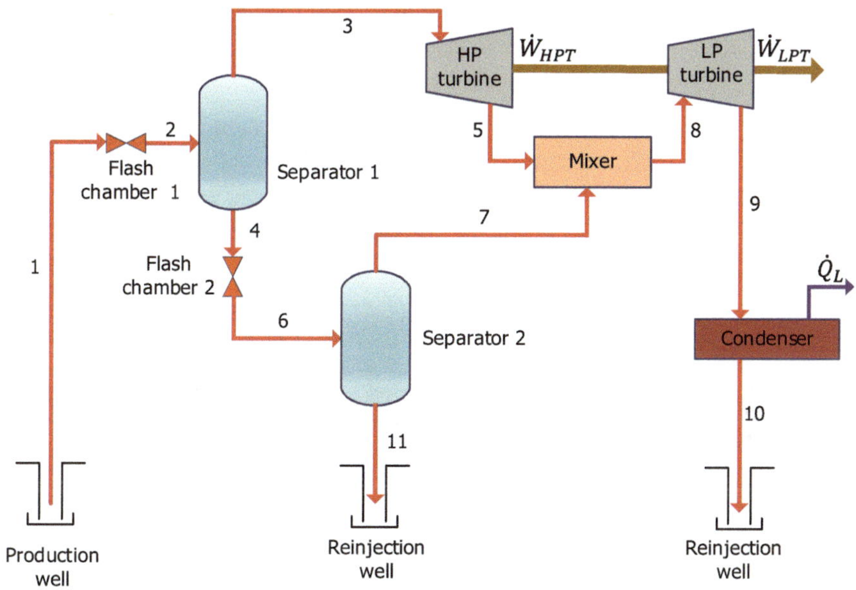

FIGURE 5.24 Schematic diagram of the double-flash steam geothermal power system.

3. energy and exergy efficiencies of the double-flash steam GPS,
4. variation of work production rate and exergy destruction rate of the double-flash GPS when the geothermal water mass flow rate increases from 100 to 180 kg/s, and
5. variation of energy and exergy efficiencies of the double-flash steam GPS when the reference temperature increases from 0°C to 40°C.

Solution: First, some assumptions are required for the double-flash steam GPS analysis.

Assumptions:

- All components operate under steady-state conditions.
- The reference temperature and pressure are taken as 25°C and 101.3 kPa, respectively.
- The pressure losses in the connections between the components are negligible.
- The heat energy losses from the components are negligible.
- The changes in the kinetic and potential energies and exergies are neglected.

Analysis:

The mass, energy, entropy, and exergy balance equations for the flash chamber 1 are defined:

$$\dot{m}_1 = \dot{m}_2$$
$$\dot{m}_1 h_1 = \dot{m}_2 h_2$$
$$\dot{m}_1 s_1 + \dot{S}_{g,fc1} = \dot{m}_2 s_2$$
$$\dot{m}_1 ex_1 = \dot{m}_2 ex_2 + \dot{E}x_{D,fc1}$$

The mass, energy, entropy, and exergy balance equations for the separator 1 can be expressed as:

$$\dot{m}_2 = \dot{m}_3 + \dot{m}_4$$
$$\dot{m}_2 h_2 = \dot{m}_3 h_3 + \dot{m}_4 h_4$$
$$\dot{m}_2 s_2 + \dot{S}_{g,sep1} = \dot{m}_3 s_3 + \dot{m}_4 s_4$$
$$\dot{m}_2 ex_2 = \dot{m}_3 ex_3 + \dot{m}_4 ex_4 + \dot{E}x_{D,sep1}$$

The mass, energy, entropy, and exergy balance equations for the HP turbine can be obtained:

$$\dot{m}_3 = \dot{m}_5$$
$$\dot{m}_3 h_3 = \dot{m}_5 h_5 + \dot{W}_{HPT}$$
$$\dot{m}_3 s_3 + \dot{S}_{g,HPT} = \dot{m}_5 s_5$$
$$\dot{m}_3 ex_3 = \dot{m}_5 ex_5 + \dot{E}x^W_{HPT} + \dot{E}x_{D,HPT}$$

The mass, energy, entropy, and exergy balance equations for the flash chamber 2 can be written:

$$\dot{m}_4 = \dot{m}_6$$
$$\dot{m}_4 h_4 = \dot{m}_6 h_6$$
$$\dot{m}_4 s_4 + \dot{S}_{g,fc2} = \dot{m}_6 s_6$$
$$\dot{m}_4 ex_4 = \dot{m}_6 ex_6 + \dot{E}x_{D,fc2}$$

The mass, energy, entropy, and exergy balance equations for the separator 2 can be defined:

$$\dot{m}_6 = \dot{m}_7 + \dot{m}_{11}$$
$$\dot{m}_6 h_6 = \dot{m}_7 h_7 + \dot{m}_{11} h_{11}$$
$$\dot{m}_6 s_6 + \dot{S}_{g,sep2} = \dot{m}_7 s_7 + \dot{m}_{11} s_{11}$$
$$\dot{m}_6 ex_6 = \dot{m}_7 ex_7 + \dot{m}_{11} ex_{11} + \dot{E}x_{D,sep2}$$

The mass, energy, entropy, and exergy balance equations for the mixer can be expressed:

$$\dot{m}_5 + \dot{m}_7 = \dot{m}_8$$
$$\dot{m}_5 h_5 + \dot{m}_7 h_7 = \dot{m}_8 h_8$$
$$\dot{m}_5 s_5 + \dot{m}_7 s_7 + \dot{S}_{g,mix} = \dot{m}_8 s_8$$
$$\dot{m}_5 ex_5 + \dot{m}_7 ex_7 = \dot{m}_8 ex_8 + \dot{E}x_{D,mix}$$

The mass, energy, entropy, and exergy balance equations for the LP turbine are defined:

$$\dot{m}_8 = \dot{m}_9$$
$$\dot{m}_8 h_8 = \dot{m}_9 h_9 + \dot{W}_{LPT}$$
$$\dot{m}_8 s_8 + \dot{S}_{g,LPT} = \dot{m}_9 s_9$$
$$\dot{m}_8 ex_8 = \dot{m}_9 ex_9 + \dot{E}x^W_{LPT} + \dot{E}x_{D,LPT}$$

The mass, energy, entropy, and exergy balance equations for the condenser are expressed:

$$\dot{m}_9 = \dot{m}_{10}$$
$$\dot{m}_9 h_9 = \dot{m}_{10} h_{10} + \dot{Q}_L$$
$$\dot{m}_9 s_9 + \dot{S}_{g,con} = \dot{m}_{10} s_{10} + \dot{Q}_L / T_{con}$$
$$\dot{m}_9 ex_9 = \dot{m}_{10} ex_{10} + \dot{E}x^Q_L + \dot{E}x_{D,con}$$

The thermodynamic variables of all flows for the double-flash GPS are defined:

For flow 1:

As given in the equation, $P_1 = 2316$ kPa and $T_1 = 218°C$, and, based on these data, h_1 and s_1 are determined from EES as 934.3 kJ/kg and 2.499 kJ/kg K.

For flow 2:

As given in the equation, $x_2 = 0.07623$ and $P_2 = P_1/2$, and, based on these data, T_2, h_2, and s_2 are determined by using the EES as 186.4°C, 943.4 kJ/kg, and 2.532 kJ/kg K.

For flow 3:

$P_3 = P_2$ and $x_3 = 1$ (saturated vapor), and, based on these data, T_3, h_3, and s_3 are determined by using the EES as 186.4°C, 2783 kJ/kg, and 6.535 kJ/kg K. The mass flow rate at flow 3 can be determined:

$$\dot{m}_3 = 0.07623 \times \dot{m}_2 = 10.67 \text{ kg/s}$$

For flow 4:

$P_4 = P_2$ and $x_4 = 0$ (saturated liquid), and T_4, h_4, and s_4 are determined by using the EES as 186.4°C, 791.5 kJ/kg, and 2.201 kJ/kg K. The mass flow rate at flow 4 can be determined:

$$\dot{m}_4 = \dot{m}_2 - \dot{m}_3 = 129.3 \text{ kg/s}$$

For flow 5:

In this flow $x_5 = 0.9605$ and $P_5 = P_3/2 = 579.5$ kPa, and T_5, h_5, and s_5 are determined by utilizing the EES program as 157.5°C, 2673 kJ/kg, and 6.58 kJ/kg K.

For flow 6:

As given in the equation, $x_6 = 0.0607$ and $P_6 = P_5$, and, based on these data, T_6, h_6, and s_6 are determined by using the EES as 157.5°C, 791.6 kJ/kg, and 2.212 kJ/kg K.

For flow 7:

$P_7 = P_6$ and $x_7 = 1$ (saturated vapor), and, based on these data, T_7, h_7, and s_7 are determined by using the EES as 157.5°C,

2755 kJ/kg, and 6.772 kJ/kg K. The mass flow rate at flow 7 can be determined:

$$\dot{m}_7 = 0.0607 \times \dot{m}_6 = 7.85\ \text{kg/s}$$

For flow 8:

$P_8 = P_7$ and $x_8 = (x_5 \times \dot{m}_5 + x_7 \times \dot{m}_7)/\dot{m}_5 + \dot{m}_7 = 0.9772$, and, based on these data, T_8, h_8, and s_8 are determined by using the EES as 157.5°C, 2708 kJ/kg, and 6.662 kJ/kg K. The mass flow rate at flow 8 can be determined:

$$\dot{m}_8 = \dot{m}_5 + \dot{m}_7 = 18.52\ \text{kg/s}$$

For flow 9:

In this flow $x_9 = 0.8392$ and $P_9 = 10$ kPa, and T_9, h_9, and s_9 are determined by utilizing the EES program as 45.82°C, 2199 kJ/kg, and 6.942 kJ/kg K.

For flow 10:

As given, $P_{10} = P_9$ and $x_{10} = 0$ (saturated liquid), and T_{10}, h_{10}, and s_{10} are determined by utilizing the EES program as 45.82°C, 191.8 kJ/kg, and 0.6493 kJ/kg K.

For flow 11:

$P_{11} = P_6$ and $x_{11} = 0$ (saturated liquid), and T_{11}, h_{11}, and s_{11} are determined by using the EES as 157.5°C, 664.7 kJ/kg, and 1.918 kJ/kg K. State point thermodynamic data for direct steam GPS are given in Table 5.5.

1. The work production rate by the HP turbine is:

$$\dot{W}_{HPT} = \dot{m}_3 h_3 - \dot{m}_5 h_5 = 1180\ \text{kW}$$

The work production rate by the LP turbine is:

$$\dot{W}_{LPT} = \dot{m}_8 h_8 - \dot{m}_9 h_9 = 9417\ \text{kW}$$

And the total work production rate is:

$$\dot{W}_{Total} = \dot{W}_{HPT} + \dot{W}_{LPT} = 10597\ \text{kW}$$

2. The total exergy destruction rate of the double-flash steam GPS is:

$$\dot{E}x_{D,DFGP} = \dot{m}_1 ex_1 - \dot{m}_{10} ex_{10} - \dot{m}_{11} ex_{11} - \dot{E}x^{W}_{Total} = 4645\ \text{kW}$$

3. The energy efficiency of the double-flash steam GPS is:

$$\eta_{DFGP} = \frac{\dot{W}_{total}}{\dot{m}_1 h_1 - \dot{m}_{10} h_{10} - \dot{m}_{11} h_{11}} = 0.2279 = 22.79\%$$

The exergy efficiency of the double-flash steam GPS is:

$$\psi_{DFGP} = \frac{\dot{E}x^{W}_{Total}}{\dot{m}_1 ex_1 - \dot{m}_{10} ex_{10} - \dot{m}_{11} ex_{11}} = 0.6953 = 69.53\%$$

4. The effect of the geothermal water mass flow rate on the total power production rate and exergy destruction rate of the double-flash steam GPS is

TABLE 5.5 State point thermodynamic data for the double-flash steam geothermal power system.

State point	$\dot{m}$ (kg/s)	T (°C)	P (kPa)	h (kJ/kg)	s (kJ/kg K)	x (−)	ex (kJ/kg)
0	–	25	101.3	104.8	0.3669	Liquid	
1	140	218	2316	934.3	2.499	0	193.8
2	140	186.4	1158	943.4	2.532	0.07623	193.1
3	10.67	186.4	1158	2783	6.535	1	839.2
4	129.3	186.4	1158	791.5	2.201	0	139.8
5	10.67	157.5	579	2673	6.58	0.9605	715.2
6	129.3	157.5	579	791.6	2.212	0.0607	136.5
7	7.85	157.5	579	2755	6.772	1	740.6
8	18.52	157.5	579	2708	6.662	0.9772	726
9	18.52	45.82	10	2199	6.942	0.8392	133.8
10	18.52	45.82	10	191.8	0.6493	0	2.814
11	121.5	157.5	579	664.7	1.918	0	97.5

illustrated in Fig. 5.25. As shown in this figure, the net power production rate increases from 7569 to 13,624 kW, and the exergy destruction rate increases from 3318 to 5972 kW with increasing the geothermal mass flow rate from 100 to 180 kg/s.

5. The effect of reference temperature on the energy and exergy efficiencies of the double-flash steam GPS is shown in Fig. 5.26. It can be shown that the energy efficiency of the geothermal system does not change with increasing the reference temperature from 0°C to 40°C, whereas the exergy efficiency of the analyzed system increases from 0.5932 to 0.7753 in the chosen reference temperature change.

5.5 Binary-type geothermal power generating system

Note that binary cycle GPSs are like systems that generate energy from energy sources such as fossil fuel but contain a working fluid other than geothermal fluid [12]. The use of renewable energy in these systems distinguishes them from systems that use energy sources such as fossil [13]. Binary cycle GPSs, which have advantages in terms of both environmental impact and cost, have an important potential in energy production [14]. This section focuses on three binary cycle GPSs. The subsystems existing and

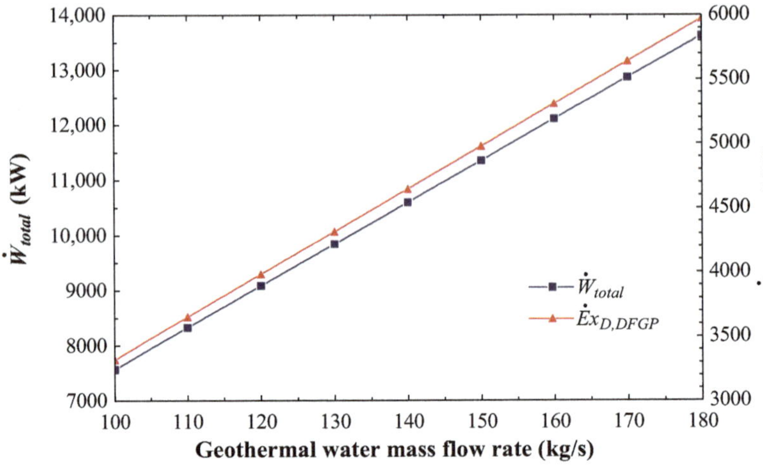

FIGURE 5.25 Effect of geothermal water mass flow rate on the power production and exergy destruction rates for the double-flash steam geothermal power system.

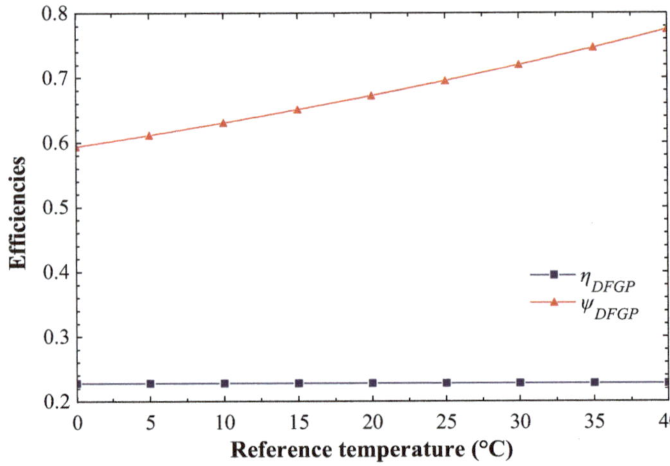

FIGURE 5.26 Effect of reference temperature on the energy and exergy efficiencies of the double-flash steam geothermal power plant.

operating in these systems and integrated with one another are generally described. The binary cycle GPS can be split into three categories:

1. ORC
   a. SSORC geothermal power generating system
   b. SSORC geothermal power generating system with two turbines
   c. DS ORC geothermal power generating system

2. KC geothermal power generating system
3. Combined flash/binary system

5.5.1 Organic Rankine cycle–based binary-type geothermal power generating system

Producing electricity from geothermal sources does not require the existence of external fuel infrastructure, as the main heat source for the electricity production purpose is located underneath the power system. Moreover, transporting, processing, and replacing fuel is not required either to extracting the necessary energy or to maintain the power plant working for power production. In addition, power production can be achieved by using medium- and high-temperature geothermal resources to operate ORC-based geothermal power plants. However, identifying the thermodynamic limits of geothermal energy conversion can determine the maximum theoretical efficiency that can be obtained from a geothermal power plant. As given previously, three traditional and commercial types of ORC plants are used for geothermal power production purposes.

5.5.1.1 Single-stage organic Rankine cycle geothermal power generating system

This section describes the operating principle of the SSORC GPS. The schematic diagram of the SSORC GPS in its most general form is given in Fig. 5.27. As can be seen from the schematic diagram of the system, the system starts to function by sending the geothermal fluid to the boiler with flow 1. By using the boiler, the thermal energy of the geothermal fluid is transferred to the working fluid used in the ORC subsystem. This transfer takes place by heat transfer between the two fluids. The energized working fluid in the ORC subsystem is sent to the turbine with flow 3. Here, the fluid expands between flow 3 and flow 4, and power is generated. Then ORC operating fluid coming out of the turbine is sent to the condenser with flow number 4 to be condensed. The condenser is used for the heating output, which is a beneficial product.

The ORC working fluid from the condenser is sent to pump 1 with flow 5 to increase its pressure. To increase the temperature of the pressure-enhanced ORC fluid, this fluid is transferred to the boiler with flow 6. The process of increasing the temperature here takes place within the geothermal fluid coming from the production well with flow 1. The operation of the SSORC GPS takes place as described here.

To better understand the working principle of the SSORC GPS, the activity diagram of electricity and heating generation in the system is given in Fig. 5.28. While examining the activity process, a schematic diagram of the system may be needed to clearly understand the system's working principle. Therefore it is useful to work on the system with two diagrams.

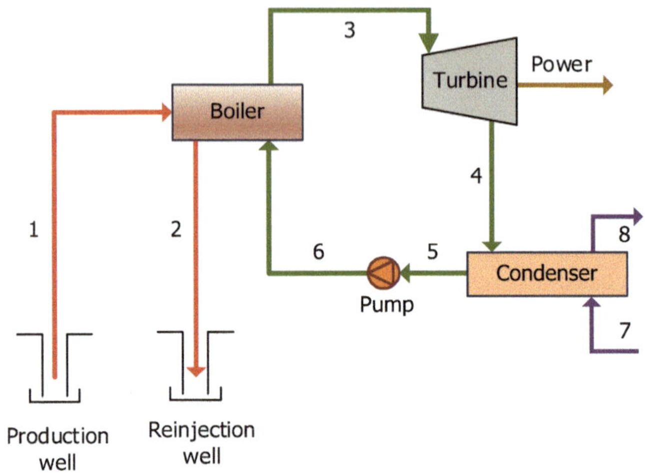

FIGURE 5.27 Simplified flow diagram of a single-stage ORC geothermal power generating system. *ORC*, Organic Rankine cycle.

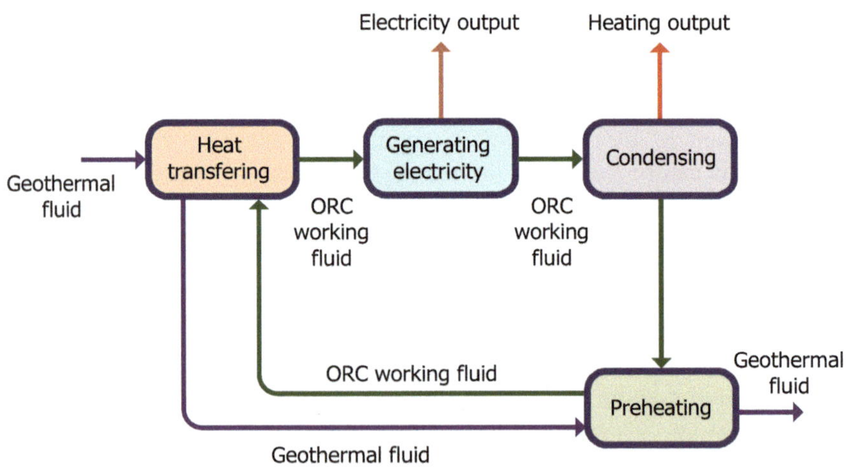

FIGURE 5.28 Activity diagram of electricity and heating generation in the single-stage ORC geothermal power generating system. *ORC*, Organic Rankine cycle.

The most common activity diagram of the SSORC GPS using SysML modeling language is given in Fig. 5.29. The activity diagram of electricity and heating generation in Fig. 5.28 shows mainly the activities taking place in the system. The activity diagram in Fig. 5.29 shows the general activity diagram of the SSORC geothermal system. It is clearly seen in the activity diagram in Fig. 5.28 that the useful outputs obtained from the system and other activities taking place in the system are interconnected. If a useful

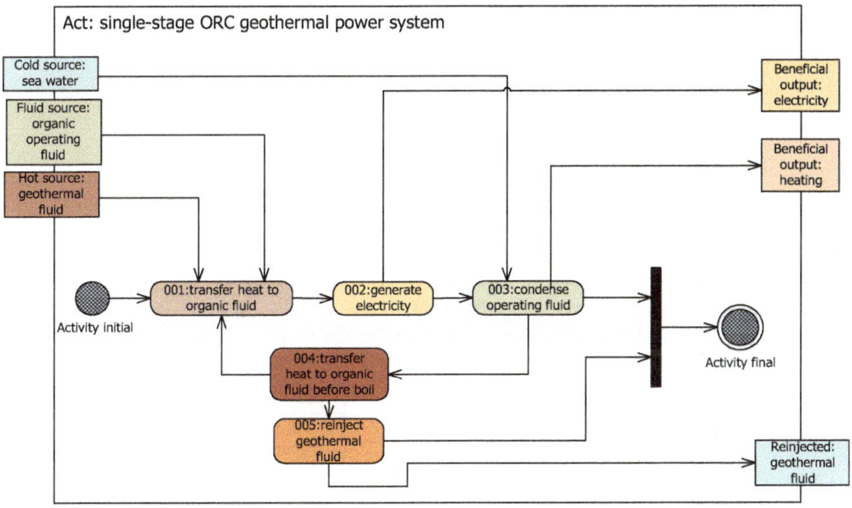

FIGURE 5.29 Activity diagram of the single-stage ORC geothermal power plant. *ORC*, Organic Rankine cycle.

product occurs with each activity in a SS system, this useful product is shown in the diagram depending on the activity.

As mentioned, requirement diagrams show the work descriptions of the components in the system. For the SSORC power plant, the requirement diagram defining system requirements is given in Fig. 5.30. As can be seen from the diagram, the functions of the components in the SS system are defined individually in this diagram. As the complexity of the designed system increases, the diagrams created with SysML modeling language become essential because, with a system containing many more components, the diagrams provided by this modeling language make defining those components more effective and understandable.

The block definition diagram of electricity and heating production is given in Fig. 5.31 in the SSORC system, which provides electricity and heating products. The activities for the production of useful electrical and heating outputs are given in the block definition diagram. Other block definition diagrams are also available for this system. However, since the aim is to show how this modeling language can be used in the design of energy systems, the most basic diagrams of the SSORC power plant are given.

The sequence of activities for the useful product in the SSORC power plant is shown in the sequence diagram in Fig. 5.32. As seen from the diagram, the part that evaluates the feedbacks in the energy system plays an important role in the working of the system. At the same time, according to the sequence diagram of the SSORC geothermal energy system, the sequence of these activities can be seen more clearly.

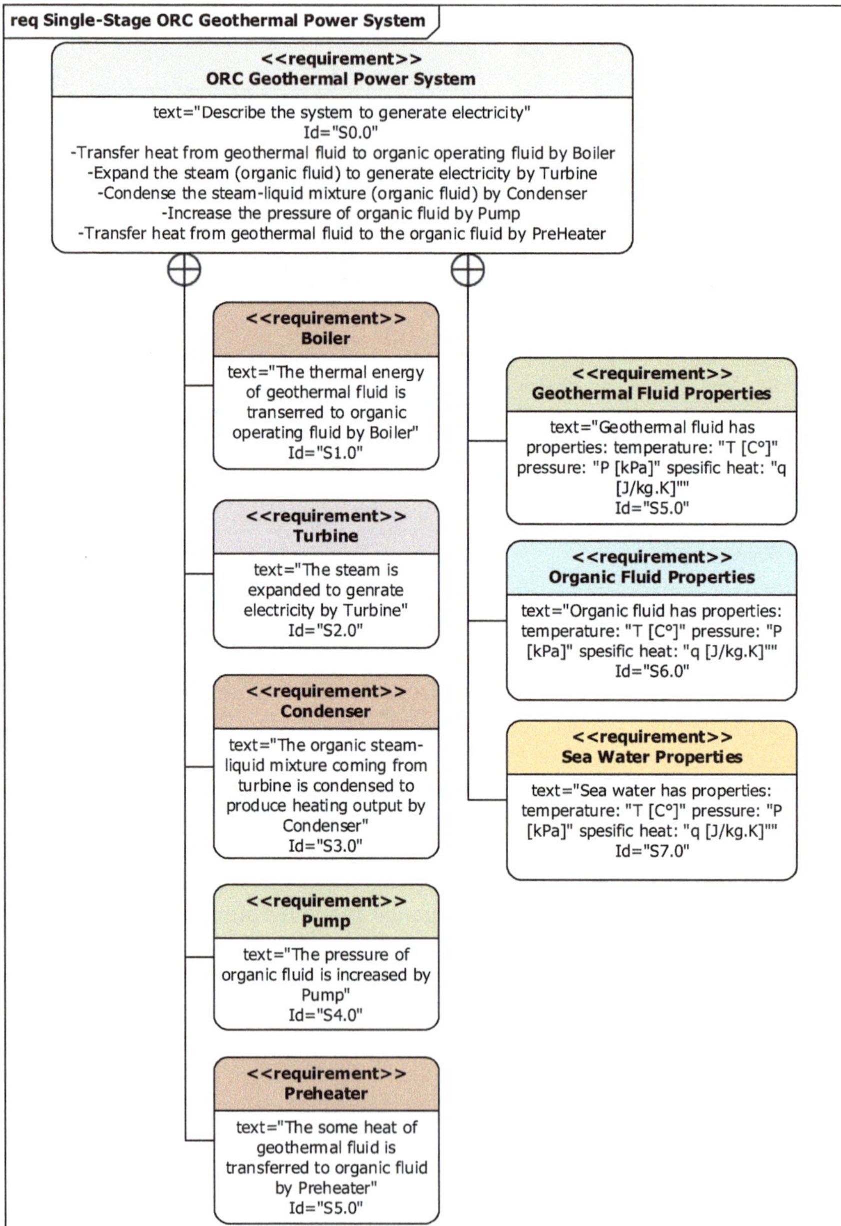

FIGURE 5.30 Requirement diagram of the single-stage ORC power plant. *ORC*, Organic Rankine cycle.

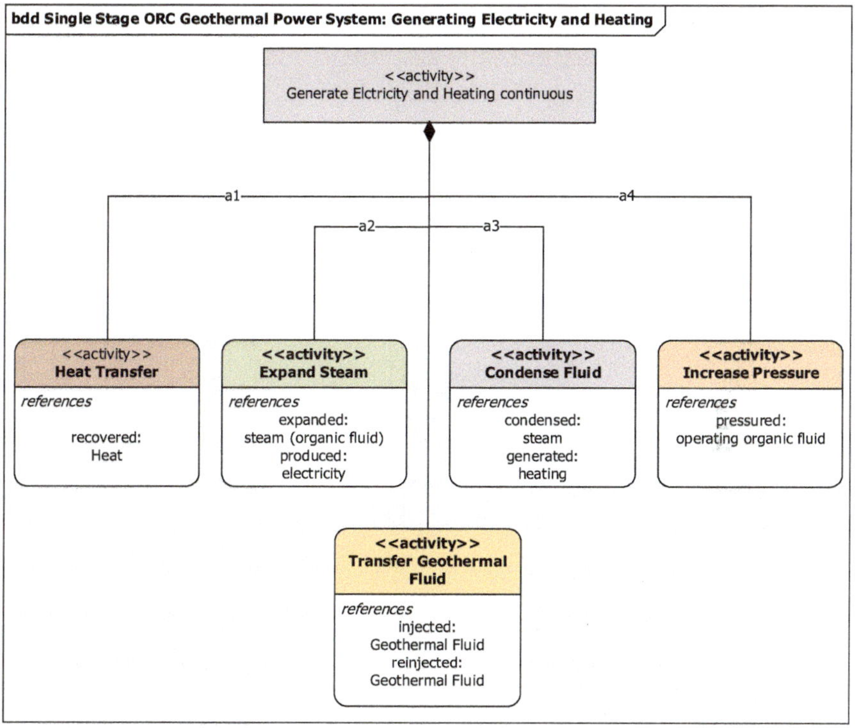

FIGURE 5.31 Block definition diagram of the single-stage ORC geothermal energy plant. *ORC*, Organic Rankine cycle.

It is seen that the system can be better defined from the diagrams created with the SysML modeling language. This system can be further defined by other diagrams offered by this modeling language. However, this section aims to give basic diagrams on how the language can be used in the design of energy systems. In addition, it is very convenient to use this modeling language for complex systems that contain different subsystems because these systems have components that are used with different parameters in different subsystems. For these systems, this modeling language can offer effective designs.

Example 5.4: A SSORC geothermal power plant, as shown in Fig. 5.33, is considered for power production. In this system, 140 kg/s saturated liquid geothermal fluid at the pressure of 772 kPa, respectively, enters a heat exchanger (HEX) to release its thermal energy to an ORC subsystem. The exit temperature of geothermal water from HEX is 47.69°C. In this subsystem, isobutane is used for the working fluid, the exiting pressure levels of the pump with a temperature increase of 0.4°C and the ORC turbine are 2700 and 450 kPa, respectively. The output temperature of the ORC turbine

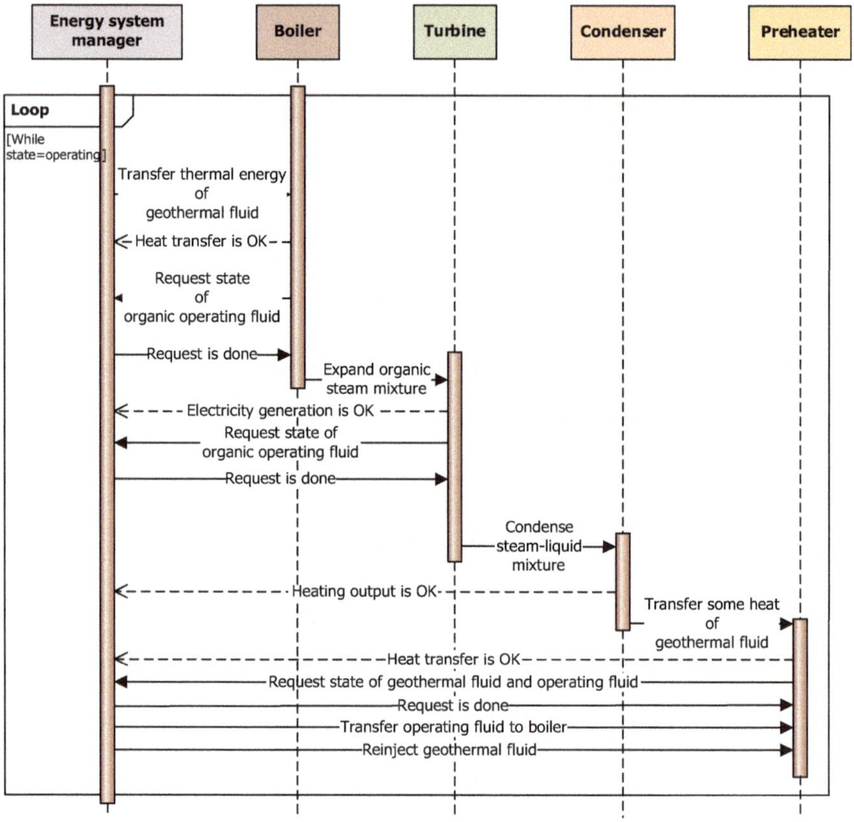

FIGURE 5.32 Sequence diagram of the single-stage ORC geothermal power plant. *ORC*, Organic Rankine cycle.

is 104°C. The pinch point temperature of HEX is assumed as 10°C. Calculate the:

1. work production rate by the ORC turbine,
2. total exergy destruction rate of the SSORC geothermal power plant,
3. energy and exergy efficiencies of the SSORC geothermal power plant,
4. variation of power production rate and exergy destruction rate of the SSORC geothermal power plant when the geothermal water mass flow rate increases from 100 to 180 kg/s, and
5. variation of energy and exergy efficiencies of the SSORC geothermal power plant when the reference temperature increases from 0°C to 40°C.

Solution: First, some assumptions are required for the thermodynamic assessment of the SSORC geothermal power plant.

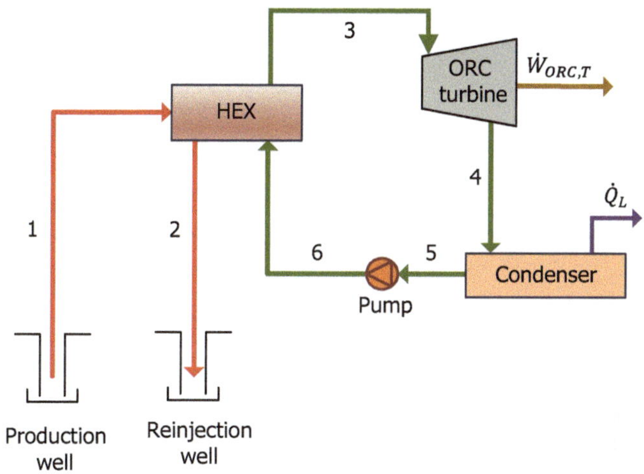

FIGURE 5.33 Schematic diagram of the single-stage ORC geothermal power plant. *ORC*, Organic Rankine cycle.

Assumptions:

- All components operate under steady-state conditions.
- The reference temperature and pressure are taken as 25°C and 101.3 kPa, respectively.
- The pressure losses in the connections between the components are negligible.
- The heat energy losses from the components are negligible.
- The changes in the kinetic and potential energies and exergies are neglected.

Analysis:

The mass, energy, entropy, and exergy balance equations for the HEX are defined:

$$\dot{m}_1 = \dot{m}_2; \dot{m}_6 = \dot{m}_3$$
$$\dot{m}_1 h_1 + \dot{m}_6 h_6 = \dot{m}_2 h_2 + \dot{m}_3 h_3$$
$$\dot{m}_1 s_1 + \dot{m}_6 s_6 + \dot{S}_{g,HEX} = \dot{m}_2 s_2 + \dot{m}_3 s_3$$
$$\dot{m}_1 ex_1 + \dot{m}_6 ex_6 = \dot{m}_2 ex_2 + \dot{m}_3 ex_3 + \dot{E}x_{D,HEX}$$

The mass, energy, entropy, and exergy balance equations for the ORC turbine can be expressed:

$$\dot{m}_3 = \dot{m}_4$$
$$\dot{m}_3 h_3 = \dot{m}_4 h_4 + \dot{W}_{ORC,T}$$
$$\dot{m}_3 s_3 + \dot{S}_{g,ORCT} = \dot{m}_4 s_4$$
$$\dot{m}_3 ex_3 = \dot{m}_4 ex_4 + \dot{E}x^{W}_{ORC,T} + \dot{E}x_{D,ORCT}$$

The mass, energy, entropy, and exergy balance equations for the condenser can be obtained:

$$\dot{m}_4 = \dot{m}_5$$
$$\dot{m}_4 h_4 = \dot{m}_5 h_5 + \dot{Q}_L$$
$$\dot{m}_4 s_4 + \dot{S}_{g,con} = \dot{m}_5 s_5 + \dot{Q}_L / T_{con}$$
$$\dot{m}_4 ex_4 = \dot{m}_5 ex_5 + \dot{E}x_L^Q + \dot{E}x_{D,con}$$

The mass, energy, entropy, and exergy balance equations for the pump are defined:

$$\dot{m}_5 = \dot{m}_6$$
$$\dot{m}_5 h_5 + \dot{W}_{in,P} = \dot{m}_6 h_6$$
$$\dot{m}_5 s_5 + \dot{S}_{g,P} = \dot{m}_6 s_6$$
$$\dot{m}_5 ex_5 + \dot{E}x_{in,P}^W = \dot{m}_6 ex_6 + \dot{E}x_{in,P}^W + \dot{E}x_{D,P}$$

Thermodynamic variables of all flows for the SSORC geothermal power plant are defined:

For flow 1:

As given in the equation, $P_1 = 772$ kPa and $x_1 = 0$, and, based on these data, T_1, h_1, and s_1 are determined from EES as 169°C, 714.8 kJ/kg, and 2.032 kJ/kg K.

For flow 2:

As given in the equation, $T_2 = 47.69$ and $P_2 = P_1$, and, based on these data, h_2 and s_2 are determined by using the EES as 200.3 kJ/kg and 0.6734 kJ/kg K.

For flow 3:

As given, $T_3 = T_1 - 10°C$ and $P_3 = P_6$, and h_3 and s_3 are determined by utilizing the EES program as 814 kJ/kg and 2.687 kJ/kg K.

For flow 4:

As given, $P_4 = 450$ kPa and $T_4 = 104°C$, and h_4 and s_4 are determined by using the EES software program as 738.9 kJ/kg and 2.723 kJ/kg K.

For flow 5:

$P_5 = P_4$ and $x_5 = 0$, and T_5, h_5, and s_5 are determined by utilizing the EES program as 33.84°C, 281.1 kJ/kg, and 1.279 kJ/kg K.

For flow 6:

As given in the equation, $T_6 = T_5 + 0.4°C$ and $P_6 = 2700$ kPa, and h_6 and s_6 are determined by utilizing the EES program as 283.4 kJ/kg and 1.272 kJ/kg K. The state point thermodynamic data for the SSORC geothermal power plant are given in Table 5.6.

1. The work production rate by the ORC turbine is found to be:

$$\dot{W}_{ORC,T} = \dot{m}_4 h_4 - \dot{m}_5 h_5 = 10199 \text{ kW}$$

The work consumption rate by the pump becomes:

$$\dot{W}_{in,P} = \dot{m}_6 h_6 - \dot{m}_5 h_5 = 275.4 \text{ kW}$$

And the total work production rate is found to be:

$$\dot{W}_{Total} = \dot{W}_{ORC,T} - \dot{W}_{in,P} = 9923.6\ \text{kW}$$

2. The total exergy destruction rate of the SSORC geothermal power plant is:

$$\dot{E}x_{D,BGP} = \dot{m}_1 ex_1 - \dot{m}_2 ex_2 - \dot{E}x^{W}_{out,Total} = 5393\ \text{kW}$$

3. The energy efficiency of the SSORC geothermal power plant is:

$$\eta_{SSORC} = \frac{\dot{W}_{Total}}{\dot{m}_1 h_1 - \dot{m}_2 h_2} = 0.1378 = 13.78\%$$

The exergy efficiency of the SSORC geothermal power plant is:

$$\psi_{SSORC} = \frac{\dot{E}x^{W}_{Total}}{\dot{m}_1 ex_1 - \dot{m}_2 ex_2} = 0.6479 = 64.79\%$$

4. The effect of the geothermal water mass flow rate on the power production rate and exergy destruction rate of the SSORC geothermal power plant is shown in Fig. 5.34. As shown in this figure, the power production rate increases from 7088 to 12,759 kW, and the exergy destruction rate increases from 3852 to 6933 kW with increasing the geothermal mass flow rate from 100 to 180 kg/s.
5. The effect of reference temperature on the energy and exergy efficiencies of the SSORC geothermal power plant is illustrated in Fig. 5.35. It can be shown that the energy efficiency of the geothermal plant does not change with increasing the reference temperature from 0°C to 40°C, whereas the exergy efficiency of the investigated plant increases from 0.4944 to 0.7962 in the chosen reference temperature change.

5.5.1.2 *Single-stage organic Rankine cycle geothermal power generating system with two turbines*

This section describes the SSORC GPS with two turbines. Configuration of this system is given in the diagram in Fig. 5.36. The ORC GPS starts to work by transferring the geothermal fluid with energy potential to the system. The pressure level of the geothermal fluid also plays a role in this transfer. The geothermal fluid coming to the system is first transferred to the HEX with flow 1. In the HEX, the energy transfer takes place between the geothermal fluid and the working fluid in the ORC subsystem with two turbines. The energy potential of the geothermal fluid plays an active role in the energy transfer in the HEX. After the energy transfer, geothermal fluid is reinjected to the ground with flow 2 to ensure sustainability in energy resources. The working fluid at a certain pressure in the ORC subsystem with two turbines is transferred to the HEX with flow 7 for energy transfer. In the HEX, the working fluid, which obtains a certain energy potential by utilizing the energy potential in the geothermal fluid,

TABLE 5.6 State point thermodynamic data for the single-stage organic Rankine cycle geothermal power plant.

State point	Fluid	$\dot{m}$ (kg/s)	T (°C)	P (kPa)	h (kJ/kg)	s (kJ/kg K)	x (–)	ex (kJ/kg)
0	Water	–	–	–	598.9	2.513	–	–
00	Isobutane	–	25	101.3	104.8	0.3669	–	–
1	Water	140	169	772	714.8	2.032	0	113.5
2	Water	140	47.69	772	200.3	0.6734	0	4.109
3	Isobutane	135.8	159	2700	814	2.687	Superheated	163.2
4	Isobutane	135.8	104	450	738.9	2.723	Superheated	77.32
5	Isobutane	135.8	33.84	450	281.4	1.279	0	50.55
6	Isobutane	135.8	34.24	2700	283.4	1.272	Compressed liquid	54.65

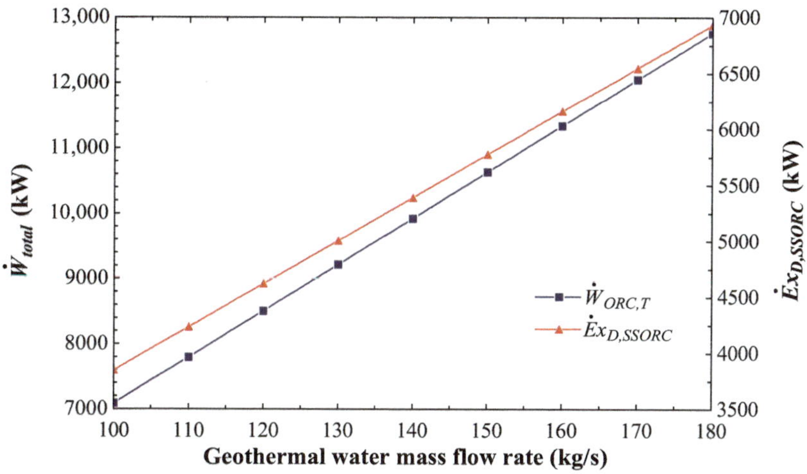

FIGURE 5.34 Effect of geothermal water mass flow rate on the power production and exergy destruction rates for the single-stage ORC geothermal power plant. *ORC*, Organic Rankine cycle.

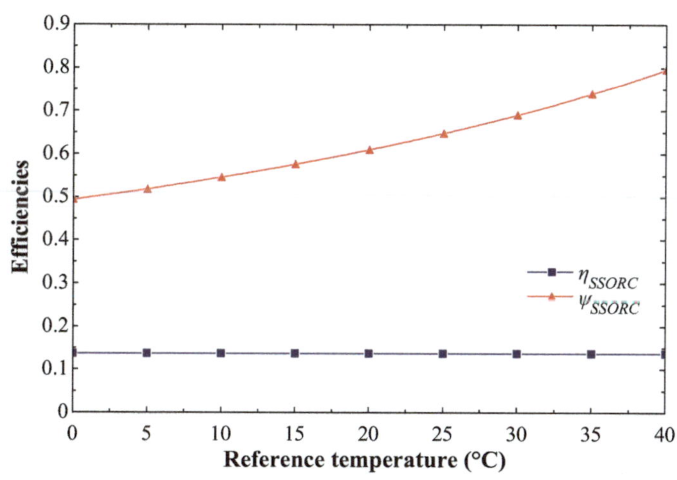

FIGURE 5.35 Effect of reference temperature on the energy and exergy efficiencies for the single-stage ORC geothermal power plant. *ORC*, Organic Rankine cycle.

is transferred to the ORC turbine 1 with flow 3. In ORC turbine 1, power generation is realized by expanding the working fluid between flows 3 and 4. After the power generation in ORC turbine 1 is realized, the working fluid is transferred to ORC turbine 2 with flow 4 to benefit from the energy potential of the working fluid again. In ORC turbine 2, the working fluid is expanded between flows 4 and 5, and power generation is realized again. After the power generation in the ORC turbine 2 is also realized, the working fluid with reduced pressure but still having a certain enthalpy is transferred to the condenser with flow

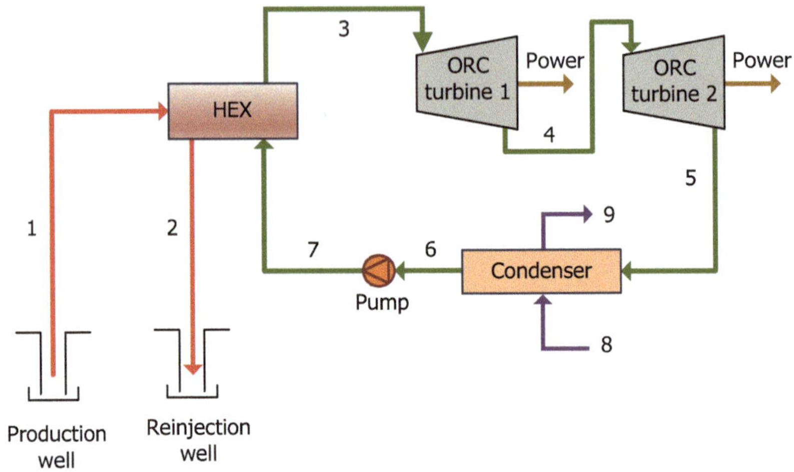

FIGURE 5.36 Simplified flow diagram of a single-stage ORC geothermal power system with two turbines. *ORC*, Organic Rankine cycle.

5 to produce heating output. After the power generation in the condenser is effected, the working fluid with reduced enthalpy is transferred to the pump with flow 6 to increase its pressure. The working fluid, reaching a certain pressure by using the pump, is transferred back to the HEX with flow 7 to meet the energy requirement for power generation. A cycle of the SSORC GPS with two turbines is completed in this way.

Example 5.5: A SSORC geothermal power plant with two turbines, as shown in Fig. 5.37, is considered for power production. In this system, 58 kg/s geothermal fluid at the pressure of 240 kPa, respectively, enters a HEX to release its thermal energy to an ORC subsystem with two turbines. The exit temperature of geothermal water from the HEX is 88.69°C. In this subsystem, R245fa is used for the working fluid, the exiting pressure levels of the pump with a temperature increase of 0.4°C, and the ORC turbine 1 and ORC turbine 2 are 1930, 748, and 724 kPa, respectively. Also, the output temperature of the ORC turbine 1 and ORC turbine 2 are 89.3°C and 78.4°C. Calculate the:

1. work production rate by ORC turbine 1 and ORC turbine 2,
2. total exergy destruction rate of SS an ORC geothermal power plant with two turbines,
3. energy and exergy efficiencies of the SSORC geothermal power plant with two turbines,
4. variation of power production rate and exergy destruction rate of the SSORC geothermal power plant with two turbines when the geothermal water mass flow rate increases from 30 to 80 kg/s, and

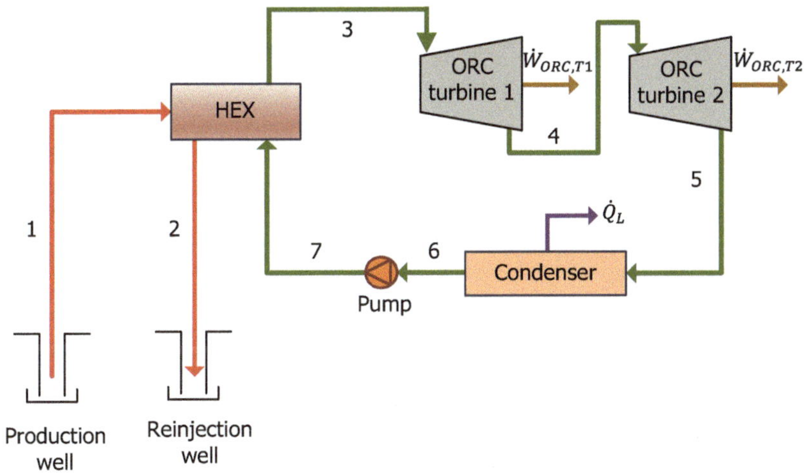

FIGURE 5.37 Schematic diagram of the single-stage ORC geothermal power plant with two turbines. *ORC*, Organic Rankine cycle.

5. variation of energy and exergy efficiencies of the SSORC geothermal power plant with two turbines when the reference temperature increases from 0°C to 40°C.

Solution: First, some assumptions are required for the thermodynamic assessment of the SSORC geothermal power plant with two turbines.

Assumptions:

- All components operate under steady-state conditions.
- The reference temperature and pressure are taken as 25°C and 101.3 kPa, respectively.
- The pressure losses in the connections between the components are negligible.
- The heat energy losses from the components are negligible.
- The changes in the kinetic and potential energies and exergies are neglected.

Analysis:

The mass, energy, entropy, and exergy balance equations for the HEX are defined:

$$\dot{m}_1 = \dot{m}_2; \dot{m}_7 = \dot{m}_3$$
$$\dot{m}_1 h_1 + \dot{m}_7 h_7 = \dot{m}_2 h_2 + \dot{m}_3 h_3$$
$$\dot{m}_1 s_1 + \dot{m}_7 s_7 + \dot{S}_{g,HEX} = \dot{m}_2 s_2 + \dot{m}_3 s_3$$
$$\dot{m}_1 ex_1 + \dot{m}_7 ex_7 = \dot{m}_2 ex_2 + \dot{m}_3 ex_3 + \dot{E}x_{D,HEX}$$

The mass, energy, entropy, and exergy balance equations for the ORC turbine 1 can be expressed:

$$\dot{m}_3 = \dot{m}_4$$
$$\dot{m}_3 h_3 = \dot{m}_4 h_4 + \dot{W}_{ORC,T1}$$
$$\dot{m}_3 s_3 + \dot{S}_{g,ORCT1} = \dot{m}_4 s_4$$
$$\dot{m}_3 ex_3 = \dot{m}_4 ex_4 + \dot{E}x^W_{ORC,T1} + \dot{E}x_{D,ORCT1}$$

The mass, energy, entropy, and exergy balance equations for the ORC turbine 2 are defined as:

$$\dot{m}_4 = \dot{m}_5$$
$$\dot{m}_4 h_4 = \dot{m}_5 h_5 + \dot{W}_{ORC,T2}$$
$$\dot{m}_4 s_4 + \dot{S}_{g,ORCT2} = \dot{m}_5 s_5$$
$$\dot{m}_4 ex_4 = \dot{m}_5 ex_5 + \dot{E}x^W_{ORC,T2} + \dot{E}x_{D,ORCT2}$$

The mass, energy, entropy, and exergy balance equations for the condenser can be obtained:

$$\dot{m}_5 = \dot{m}_6$$
$$\dot{m}_5 h_5 = \dot{m}_6 h_6 + \dot{Q}_L$$
$$\dot{m}_5 s_5 + \dot{S}_{g,con} = \dot{m}_6 s_6 + \dot{Q}_L / T_{con}$$
$$\dot{m}_5 ex_5 = \dot{m}_6 ex_6 + \dot{E}x^Q_L + \dot{E}x_{D,con}$$

The mass, energy, entropy, and exergy balance equations for the pump are defined:

$$\dot{m}_6 = \dot{m}_7$$
$$\dot{m}_6 h_6 + \dot{W}_{in,P} = \dot{m}_7 h_7$$
$$\dot{m}_6 s_6 + \dot{S}_{g,P} = \dot{m}_7 s_7$$
$$\dot{m}_6 ex_6 + \dot{E}x^W_{in,P} = \dot{m}_7 ex_7 + \dot{E}x^W_{in,P} + \dot{E}x_{D,P}$$

Thermodynamic variables of all flows for the SSORC geothermal power plant are defined:

For flow 1:

As given in the equation, $P_1 = 240$ kPa and $x_1 = 0$, and, based on these data, T_1, h_1, and s_1 are determined from EES as 126.1°C, 529.8 kJ/kg, and 1.593 kJ/kg K.

For flow 2:

As given in the equation, $T_2 = 88.69$ and $P_2 = P_1$, and, based on these data, h_2 and s_2 are determined by using the EES as 371.6 kJ/kg and 1.177 kJ/kg K.

For flow 3:

As given, $T_3 = 120$°C and $P_3 = P_7$, and h_3 and s_3 are determined by utilizing the EES program as 372.6 kJ/kg and 1.515 kJ/kg K.

For flow 4:

As given, $P_4 = 748$ kPa and superheated, and T_4, h_4, and s_4 are determined by using the EES software program as 89.3°C, 367.7 kJ/kg, and 1.514 kJ/kg K.

For flow 5:

As given, $P_5 = 724$ kPa and superheated, and T_5, h_5, and s_5 are determined by using the EES software program as 78.4°C, 360.6 kJ/kg, and 1.476 kJ/kg K.

For flow 6:

$P_6 = P_5$ and $x_6 = 0$, and T_6, h_6, and s_6 are determined by utilizing the EES program as 76.53°C, 304.1 kJ/kg, and 1.333 kJ/kg K.

For flow 7:

As given in the equation, $T_7 = T_6 + 0.4°C$ and $P_7 = 1930$ kPa, and h_7 and s_7 are determined by utilizing the EES program as 304.7 kJ/kg and 1.332 kJ/kg K. The state point thermodynamic data for the SSORC geothermal power plant are given in Table 5.7.

1. The work production rate by the ORC turbine 1 is:

$$\dot{W}_{ORC,T1} = \dot{m}_3 h_3 - \dot{m}_4 h_4 = 664.4\ \text{kW}$$

The work production rate by the ORC turbine 2 is:

$$\dot{W}_{ORC,T2} = \dot{m}_4 h_4 - \dot{m}_5 h_5 = 826.4\ \text{kW}$$

The work consumption rate by the pump is:

$$\dot{W}_{in,P} = \dot{m}_7 h_7 - \dot{m}_6 h_6 = 73.81\ \text{kW}$$

And the total work production rate is:

$$\dot{W}_{Total} = \dot{W}_{ORC,T1} + \dot{W}_{ORC,T2} - \dot{W}_{in,P} = 1417\ \text{kW}$$

2. The total exergy destruction rate of the SSORC geothermal power plant with two turbines is:

$$\dot{E}x_{D,ORCTT} = \dot{m}_1 ex_1 - \dot{m}_2 ex_2 - \dot{E}x^W_{Total} = 564.6\ \text{kW}$$

3. The energy efficiency of the SSORC geothermal power plant with two turbines is:

$$\eta_{ORC,TT} = \frac{\dot{W}_{Total}}{\dot{m}_1 h_1 - \dot{m}_2 h_2} = 0.1544 = 15.44\%$$

The exergy efficiency of the SSORC geothermal power plant with two turbines is:

$$\psi_{ORC,TT} = \frac{\dot{E}x^W_{Total}}{\dot{m}_1 ex_1 - \dot{m}_2 ex_2} = 0.7151 = 71.51\%$$

4. The effect of the geothermal water mass flow rate on the power production rate and exergy destruction rate of the SSORC geothermal power plant with two turbines is shown in Fig. 5.38. As shown in the figure, the power production rate increases from 733 to 1955 kW, and the exergy destruction rate increases from 292.1 to 778.8 kW with the increasing geothermal mass flow rate from 30 to 80 kg/s.

TABLE 5.7 State point thermodynamic data for the single-stage organic Rankine cycle geothermal power plant with two turbines.

State point	Fluid	$\dot{m}$ (kg/s)	T (°C)	P (kPa)	h (kJ/kg)	s (kJ/kg K)	x (–)	ex (kJ/kg)
0	Water		25	101.3	104.8	0.3669	–	–
00	R245fa				424.6	1.78	–	–
1	Water	58	126.1	240	529.8	1.593	0	113.5
2	Water	58	88.69	240	371.6	1.177	0	4.109
3	R245fa	135	120	1930	372.6	1.515	Superheated	163.2
4	R245fa	135	89.3	748	367.7	1.514	Superheated	77.32
5	R245fa	135	88.7	724	361.6	1.498	Superheated	50.55
6	R245fa	135	76.53	724	304.1	1.333	0	54.65
7	R245fa	135	76.93	1930	304.7	1.332	Compressed liquid	

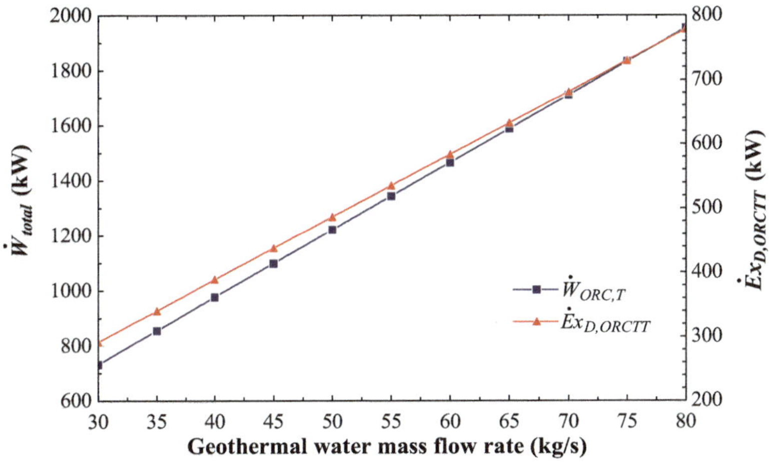

FIGURE 5.38 Effect of geothermal water mass flow rate on the power production and exergy destruction rates for the single-stage ORC geothermal power plant with two turbines. *ORC*, Organic Rankine cycle.

5. The effect of the reference temperature on the energy and exergy efficiencies of the SSORC geothermal power plant with two turbines is illustrated in Fig. 5.39. It can be shown that the energy efficiency of the geothermal plant does not change with the increasing reference temperature from 0°C to 40°C, whereas the exergy efficiency of the investigated plant increases from 0.5482 to 0.8749 in the chosen reference temperature change.

5.5.1.3 Double-stage organic Rankine cycle geothermal power generating system

The DS ORC geothermal power generating systems are considered a stage-enhanced version of their SS counterparts [5]. The schematic diagram of the DS ORC GPS is given in Fig. 5.40 in its most general form. As can be seen from Fig. 5.40, the system starts to function with the inclusion of geothermal fluid into the system. Geothermal fluid coming from the ground is sent to boiler 1 with flow 1. Here, with the help of the boiler, the energy in the geothermal fluid is transferred to the working fluid used in the ORC process. The fluid in the ORC process, loaded with energy from the geothermal fluid, is sent to the turbine with the number 9 flow. Here, the fluid expands between flow 9 and flow 10, and electricity is produced. Then ORC operating fluid coming out of the turbine is sent to the condenser with flow 10 to be condensed. The condenser is used for the heating output, which is a useful product. The ORC working fluid from the condenser is sent to pump 1 with flow 11 to increase its

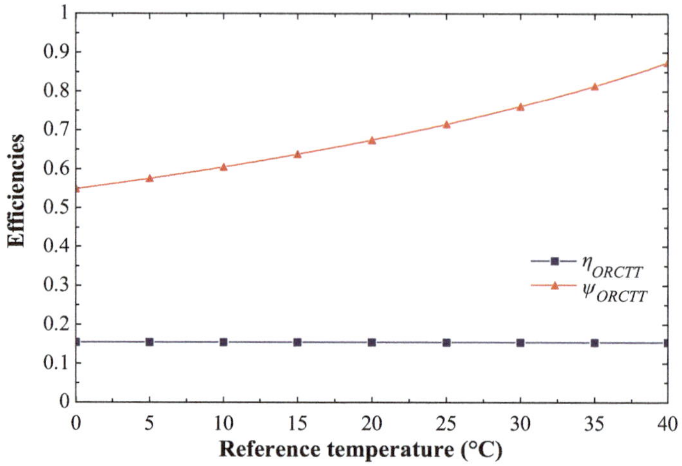

FIGURE 5.39 Effect of reference temperature on the energy and exergy efficiencies for the single-stage ORC geothermal power plant with two turbines. *ORC*, Organic Rankine cycle.

pressure. To increase its temperature, the pressure-enhanced ORC fluid is sent to the preheater with flow 12. The process of increasing the temperature here takes place within the geothermal fluid coming from the three-way valve 1 with flow 4. As seen in Fig. 5.40, geothermal fluid is supplied from the three-way valve to the preheater 1 with flow 4. Thus heat transfer can take place between the geothermal fluid and the ORC working fluid. Similarly, the same situation applies to the other part of the system. When the geothermal fluid reaches boiler 2 with flow 2, there is an energy transfer between the geothermal fluid and the working fluid in the other ORC process. The energized ORC working fluid is sent to the turbine with flow 16 for electricity generation. Here, the fluid expands between flow 16 and flow 17, and electricity is generated. Then the ORC working fluid coming out of the ORC turbine is sent to condenser 2 with flow 17 to be condensed. The condenser is used for the heating output, which is a beneficial product. The ORC working fluid from the condenser is sent to pump 2 with flow 18 to increase its pressure. To increase the temperature of the pressure-enhanced ORC operating fluid, this fluid is sent to preheater 2 with flow number 19. The process of increasing the temperature here takes place within the geothermal fluid coming from three-way valve 1 with flow 6. Flows 5 and 7 and geothermal fluid coming from preheater 1 and preheater 2 are collected with three-way valve 2 and sent to the ground with flow 8. This is a summary of the logic of the DS ORC GPS in its most general form.

To better understand the DS ORC geothermal system, the activity diagram of the system is helpful. In general, the system activity diagram is given in

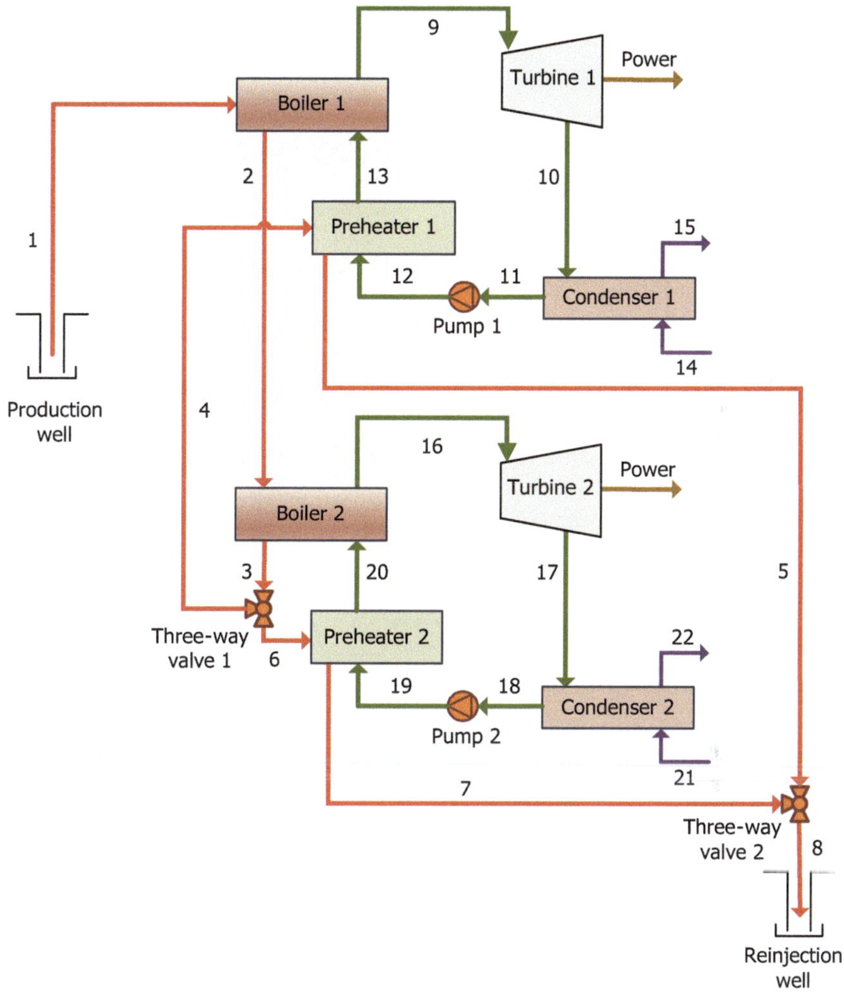

FIGURE 5.40 Simplified flow diagram of a double-stage ORC geothermal power system. *ORC*, Organic Rankine cycle.

Fig. 5.41. The diagram clearly shows the activities that take place in the system, and if this activity diagram is studied together with the schematic diagram of the system, the operating logic of the system can be better understood.

5.5.2 Case study 5.2

In this case study, the geothermal energy resource–based DS ORC GPS is investigated by utilizing energy and exergy analyses. The simplified flow diagram of the DS ORC GPS is illustrated in Fig. 5.40. As reference

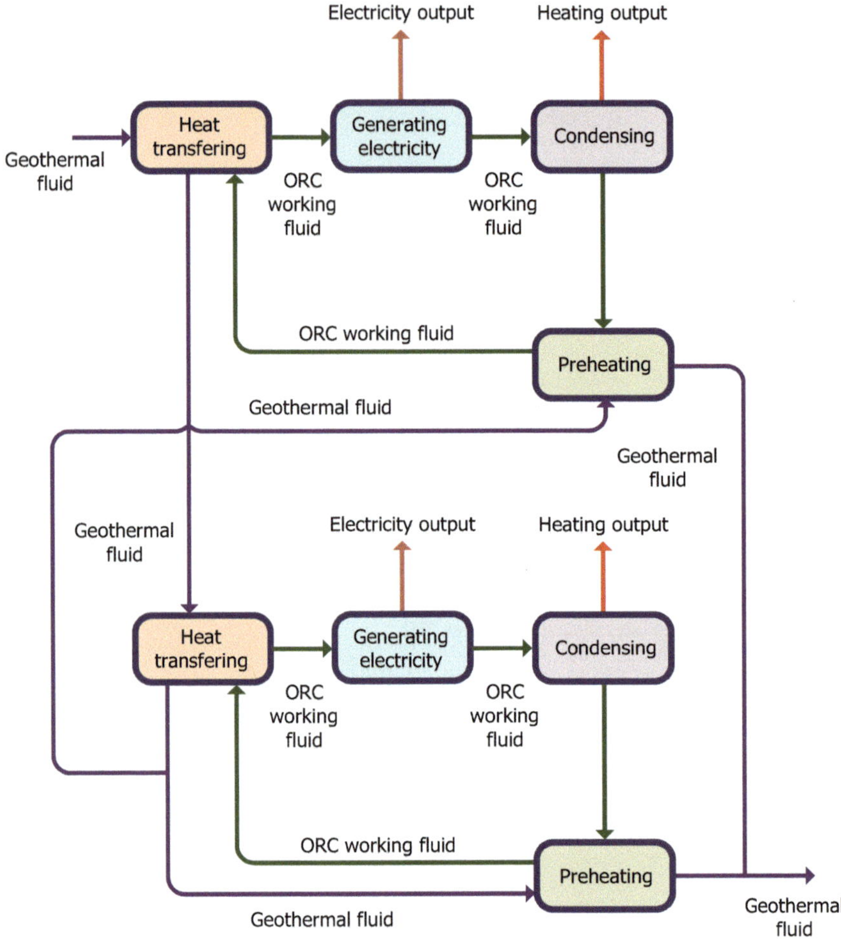

FIGURE 5.41 Activity diagram of electricity and heating generation in the double-stage ORC geothermal power system. *ORC*, Organic Rankine cycle.

conditions, ambient temperature and pressure are taken as 25°C and 101.3 kPa, respectively. The assumptions used in the operating conditions of the DS ORC GPS are given in Table 5.8.

The mass, energy, entropy, and exergy balance equations for the boiler 1 are defined:

$$\dot{m}_1 = \dot{m}_2; \dot{m}_{13} = \dot{m}_9$$
$$\dot{m}_1 h_1 + \dot{m}_{13} h_{13} = \dot{m}_2 h_2 + \dot{m}_9 h_9$$
$$\dot{m}_1 s_1 + \dot{m}_{13} s_{13} + \dot{S}_{g,bl1} = \dot{m}_2 s_2 + \dot{m}_9 s_9$$
$$\dot{m}_1 ex_1 + \dot{m}_{13} ex_{13} = \dot{m}_2 ex_2 + \dot{m}_9 ex_9 + \dot{E}x_{D,bl1}$$

TABLE 5.8 Assumptions for the double-stage organic Rankine cycle (ORC) geothermal power system.

Variables	Values
Geofluid source temperature, T_1	210°C
Geofluid source pressure, P_1	1500 kPa
Geofluid mass flow rate, $\dot{m}_1$	140 kg/s
Working fluid of ORC subsystem 1	R113
Working fluid of ORC subsystem 2	R123
ORC turbine 1 inlet pressure, P_9	2000 kPa
ORC turbine 2 inlet pressure, P_{16}	733 kPa
ORC turbine 1 inlet temperature, T_9	179°C
ORC turbine 2 inlet temperature, T_{16}	97°C
ORC turbine 1 output pressure, P_{10}	21.43 kPa
ORC turbine 2 output pressure, P_{17}	98.46 kPa
ORC turbine 1 output temperature, T_{10}	85°C
ORC turbine 2 output temperature, T_{17}	42°C
Geofluid reinjection temperature, T_8	49.73°C

The mass, energy, entropy, and exergy balance equations for the ORC turbine 1 can be written:

$$\dot{m}_9 = \dot{m}_{10}$$
$$\dot{m}_9 h_9 = \dot{m}_{10} h_{10} + \dot{W}_{ORC,T1}$$
$$\dot{m}_9 s_9 + \dot{S}_{g,ORCT1} = \dot{m}_{10} s_{10}$$
$$\dot{m}_9 ex_9 = \dot{m}_{10} ex_{10} + \dot{E}x^W_{ORC,T1} + \dot{E}x_{D,ORCT1}$$

The mass, energy, entropy, and exergy balance equations for the condenser 1 can be defined:

$$\dot{m}_{10} = \dot{m}_{11}; \dot{m}_{14} = \dot{m}_{15}$$
$$\dot{m}_{10} h_{10} + \dot{m}_{14} h_{14} = \dot{m}_{11} h_{11} + \dot{m}_{15} h_{15}$$
$$\dot{m}_{10} s_{10} + \dot{m}_{14} s_{14} + \dot{S}_{g,con1} = \dot{m}_{11} s_{11} + \dot{m}_{15} s_{15}$$
$$\dot{m}_{10} ex_{10} + \dot{m}_{14} ex_{14} = \dot{m}_{11} ex_{11} + \dot{m}_{15} ex_{15} + \dot{E}x_{D,con1}$$

The mass, energy, entropy, and exergy balance equations for the pump 1 are expressed as:

$$\dot{m}_{11} = \dot{m}_{12}$$
$$\dot{m}_{11} h_{11} + \dot{W}_{P1} = \dot{m}_{12} h_{12}$$
$$\dot{m}_{11} s_{11} + \dot{S}_{g,P1} = \dot{m}_{12} s_{12}$$
$$\dot{m}_{11} ex_{11} + \dot{E}x^W_{P1} = \dot{m}_{12} ex_{12} + \dot{E}x_{D,P1}$$

The mass, energy, entropy, and exergy balance equations for the preheater 1 are obtained:

$$\dot{m}_4 = \dot{m}_5; \dot{m}_{12} = \dot{m}_{13}$$
$$\dot{m}_4 h_4 + \dot{m}_{12} h_{12} = \dot{m}_5 h_5 + \dot{m}_{13} h_{13}$$
$$\dot{m}_4 s_4 + \dot{m}_{12} s_{12} + \dot{S}_{g,PH1} = \dot{m}_5 s_5 + \dot{m}_{13} s_{13}$$
$$\dot{m}_4 ex_4 + \dot{m}_{12} ex_{12} = \dot{m}_5 ex_5 + \dot{m}_{13} ex_{13} + \dot{E}x_{D,PH1}$$

The mass, energy, entropy, and exergy balance equations for the boiler 2 are written:

$$\dot{m}_2 = \dot{m}_3; \dot{m}_{20} = \dot{m}_{16}$$
$$\dot{m}_2 h_2 + \dot{m}_{20} h_{20} = \dot{m}_3 h_3 + \dot{m}_{16} h_{16}$$
$$\dot{m}_2 s_2 + \dot{m}_{20} s_{20} + \dot{S}_{g,bl2} = \dot{m}_3 s_3 + \dot{m}_{16} s_{16}$$
$$\dot{m}_2 ex_2 + \dot{m}_{20} ex_{20} = \dot{m}_3 ex_3 + \dot{m}_{16} ex_{16} + \dot{E}x_{D,bl2}$$

The mass, energy, entropy, and exergy balance equations for the ORC turbine 2 are defined:

$$\dot{m}_{16} = \dot{m}_{17}$$
$$\dot{m}_{16} h_{16} = \dot{m}_{17} h_{17} + \dot{W}_{ORC,T2}$$
$$\dot{m}_{16} s_{16} + \dot{S}_{g,ORCT2} = \dot{m}_{17} s_{17}$$
$$\dot{m}_{16} ex_{16} = \dot{m}_{17} ex_{17} + \dot{E}x^{W}_{ORC,T2} + \dot{E}x_{D,ORCT2}$$

The mass, energy, entropy, and exergy balance equations for the condenser 2 can be expressed as:

$$\dot{m}_{17} = \dot{m}_{18}; \dot{m}_{21} = \dot{m}_{22}$$
$$\dot{m}_{17} h_{17} + \dot{m}_{21} h_{21} = \dot{m}_{18} h_{18} + \dot{m}_{22} h_{22}$$
$$\dot{m}_{17} s_{17} + \dot{m}_{21} s_{21} + \dot{S}_{g,con2} = \dot{m}_{18} s_{18} + \dot{m}_{22} s_{22}$$
$$\dot{m}_{17} ex_{17} + \dot{m}_{21} ex_{21} = \dot{m}_{18} ex_{18} + \dot{m}_{22} ex_{22} + \dot{E}x_{D,con2}$$

The mass, energy, entropy, and exergy balance equations for the pump 2 are defined:

$$\dot{m}_{18} = \dot{m}_{19}$$
$$\dot{m}_{18} h_{18} + \dot{W}_{P2} = \dot{m}_{19} h_{19}$$
$$\dot{m}_{18} s_{18} + \dot{S}_{g,P2} = \dot{m}_{19} s_{19}$$
$$\dot{m}_{18} ex_{18} + \dot{E}x^{W}_{P2} = \dot{m}_{19} ex_{19} + \dot{E}x_{D,P2}$$

The mass, energy, entropy, and exergy balance equations for the preheater 2 can be obtained:

$$\dot{m}_6 = \dot{m}_7; \dot{m}_{19} = \dot{m}_{20}$$
$$\dot{m}_6 h_6 + \dot{m}_{19} h_{19} = \dot{m}_7 h_7 + \dot{m}_{20} h_{20}$$
$$\dot{m}_6 s_6 + \dot{m}_{19} s_{19} + \dot{S}_{g,PH2} = \dot{m}_7 s_7 + \dot{m}_{20} s_{20}$$
$$\dot{m}_6 ex_6 + \dot{m}_{19} ex_{19} = \dot{m}_7 ex_7 + \dot{m}_{20} ex_{20} + \dot{E}x_{D,PH2}$$

The mass, energy, entropy, and exergy balance equations for the three-way valve 1 can be written:

$$\dot{m}_3 = \dot{m}_4 + \dot{m}_6$$
$$\dot{m}_3 h_3 = \dot{m}_4 h_4 + \dot{m}_6 h_6$$
$$\dot{m}_3 s_3 + \dot{S}_{g,twv1} = \dot{m}_4 s_4 + \dot{m}_6 s_6$$
$$\dot{m}_3 ex_3 = \dot{m}_4 ex_4 + \dot{m}_6 ex_6 + \dot{E}x_{D,twv1}$$

The mass, energy, entropy, and exergy balance equations for the three-way valve 2 are defined:

$$\dot{m}_5 + \dot{m}_7 = \dot{m}_8$$
$$\dot{m}_5 h_5 + \dot{m}_7 h_7 = \dot{m}_8 h_8$$
$$\dot{m}_5 s_5 + \dot{m}_7 s_7 + \dot{S}_{g,twv2} = \dot{m}_8 s_8$$
$$\dot{m}_5 ex_5 + \dot{m}_7 ex_7 = \dot{m}_8 ex_8 + \dot{E}x_{D,twv2}$$

To investigate the efficiency of DS ORC GPS more comprehensively, the parametric studies are given here to examine the impacts of different indicator variables on the total power production rate, exergy destruction rate, and exergy efficiency. As the first parametric study, the effect of the geothermal water mass flow rate on the total power production rate, exergy destruction rate, and exergy efficiency of the DS ORC GPS are shown in Fig. 5.42. It can be shown that the total power production rate increases from 8263 to 9235 kW, the exergy destruction rate increases from 6726 to 7458 kW, and the exergy efficiency increases from 0.4665 to 0.5132, with increasing the geothermal water mass flow rate from 100 to 180 kg/s.

The effects of the reference temperature on the total power production rate, exergy destruction rate, and exergy efficiency of the DS ORC GPS are analyzed as another case study, and study outputs are illustrated in Fig. 5.43. As shown in this figure, the total power production rate increases from 8312 to 9000 kW, the exergy destruction rate increases from 6772 to 7275 kW, and the exergy efficiency increases from 0.4701 to 0.5011, with increasing the reference temperature from 0°C to 40°C.

The effects of the reference temperature on the energy and exergy efficiencies and the total power production rate of the DS ORC GPS are analyzed as another case study, and study outputs are illustrated in Fig. 5.44. As shown in this figure, the energy efficiency of the DS ORC GPS does not change with increasing the reference temperature from 0°C to 40°C, whereas the exergy efficiency of the DS ORC GPS increases from 0.645 to 0.8371, and the total power production rate increases from 12,358 to 19,163 kW in the chosen reference temperature change.

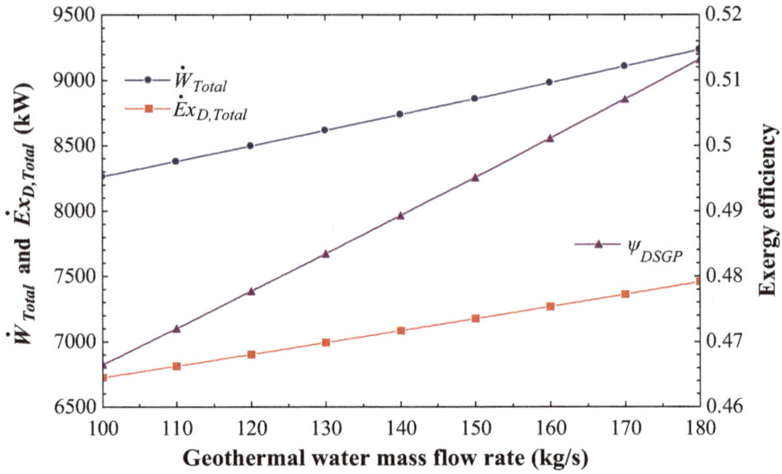

FIGURE 5.42 Effect of the mass flow rate of geothermal working fluid on net power generation and exergy efficiency for the double-stage ORC geothermal power system. *ORC*, Organic Rankine cycle.

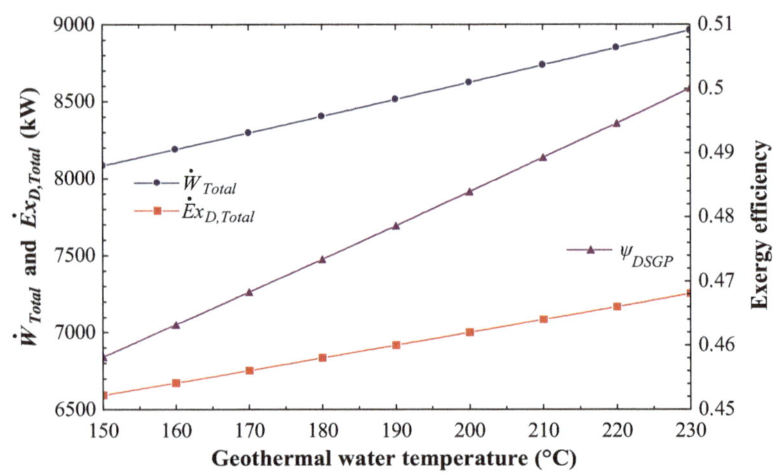

FIGURE 5.43 Effect of geothermal working fluid temperature on net power generation and exergy efficiency for the double-stage ORC geothermal power system. *ORC*, Organic Rankine cycle.

5.5.3 Kalina cycle geothermal power generating system

In this section, power generation with a binary cycle geothermal energy system including the KC is explained. Generating electricity with double-cycle GPSs containing the KC is similar to the systems containing the ORC subsystem [15]. The working fluid of the ORC subsystem is an organic fluid,

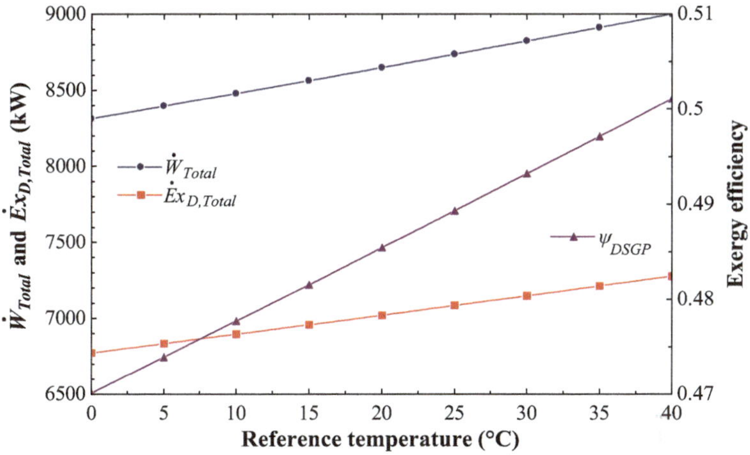

FIGURE 5.44 Effect of ambient temperature on net power generation and exergy efficiency for the double-stage ORC geothermal power system. *ORC*, Organic Rankine cycle.

while the working fluid of the KC is a mixture of water and ammonia of certain percentages. The most general schematic diagram of the KC GPS is given in Fig. 5.45. As can be seen from the schematic diagram of the system, the system starts to function by sending the geothermal fluid into the boiler with the number 1 flow. Here, by utilizing the boiler, heat energy contained in the geothermal fluid is transferred to the working fluid in the KC. The energized working fluid is sent to the separator with flow 3. Here, by increasing the percentage of ammonia in the working fluid, a rich mixture is obtained. The rich mixture is sent to the turbine with flow 4 to produce electricity. In the turbine, the fluid is expanded from stream 4 to stream 5, and electricity is produced. Weak fluid in the separator is sent to the valve with flow 6, and from there, it is sent to the three-way valve with flow 7. The fluid from flow 7 and the fluid from flow 5 are combined with a three-way valve. The operating fluid coming out of the three-way valve is sent to the condenser with flow 8 to be condensed. The condenser is used for the heating output, which is a beneficial product. Then the KC working fluid is sent from the condenser to the pump with flow 9 to increase its pressure. The pressure-enhanced KC working fluid is then sent to the boiler with flow 10 to increase its temperature. The geothermal fluid, which performs its function in one cycle in the boiler, is then sent back to the ground with flow number 2. In general, the KC GPS works this way.

The activity diagram of the energy generation systems is one of the optimum options to describe and understand the working principle of the system. To make the working principle of the system understandable, the activity diagram of electricity and heating generation in the KC GPS is shown in Fig. 5.46 in its most general form.

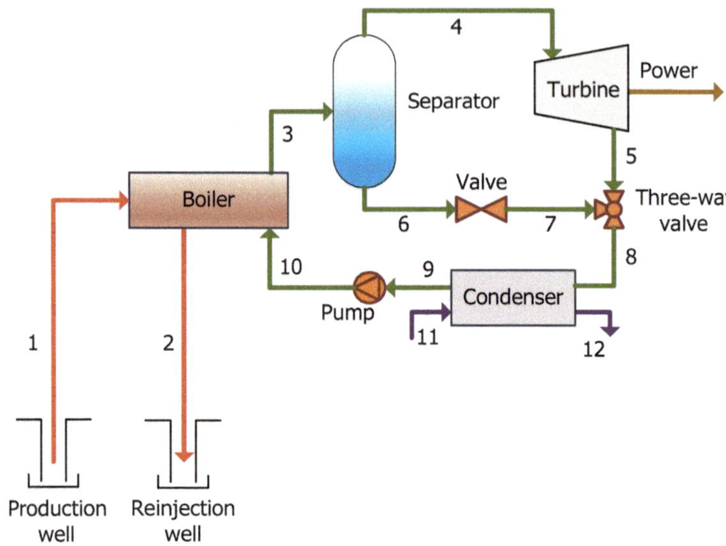

FIGURE 5.45 Simplified flow diagram of a Kalina cycle geothermal power system.

The general activity diagram of the KC geothermal power plant is given in Fig. 5.47. As can be seen from the figure, many activities are linked to each other during the production of useful outputs obtained from the system, and the activity diagram clearly shows these connections. At the same time, conditional activities can be seen more clearly in this activity diagram. The activities in the system begin with the heat transfer between the geothermal fluid included in the system and the working fluid. Then conditional fluid transfer is seen in the separator. In the part where the separation process takes place for the working fluid, the rich fluid mixture is sent to the turbine for electricity generation, the other mixture is directed to the condenser for heating output. With this diagram, it is possible to see the activities in the system directly.

The requirement diagram of the KC geothermal energy system is given in Fig. 5.48. From this diagram, the functions of the components in the KC energy system can be clearly seen. The diagram also includes definitions for fluids in the energy system. However, since the parameter values of these fluids may differ for different designs, in this diagram, the values of these fluid parameters are shown in parameter units. Predefining the system requirements for any system makes the design of the system more effective. The requirement diagram of the KC geothermal energy system can potentially contribute to the integration of this system into different systems or to the design of this system for different applications.

The block definition diagram of the activities carried out to obtain the useful outputs of electricity and heating from the KC geothermal system is

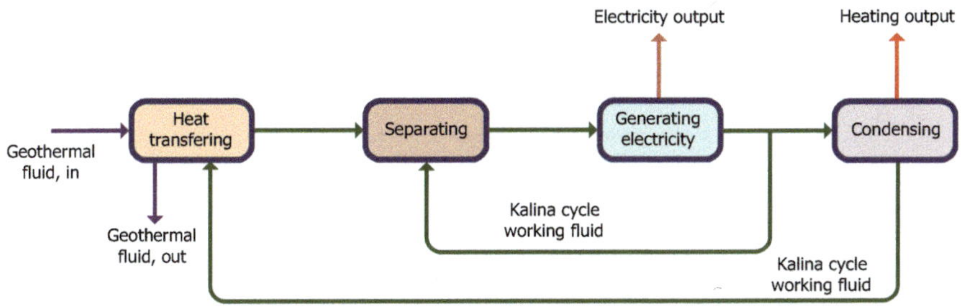

FIGURE 5.46 Activity diagram of electricity and heating generation in the Kalina cycle geothermal power generating system.

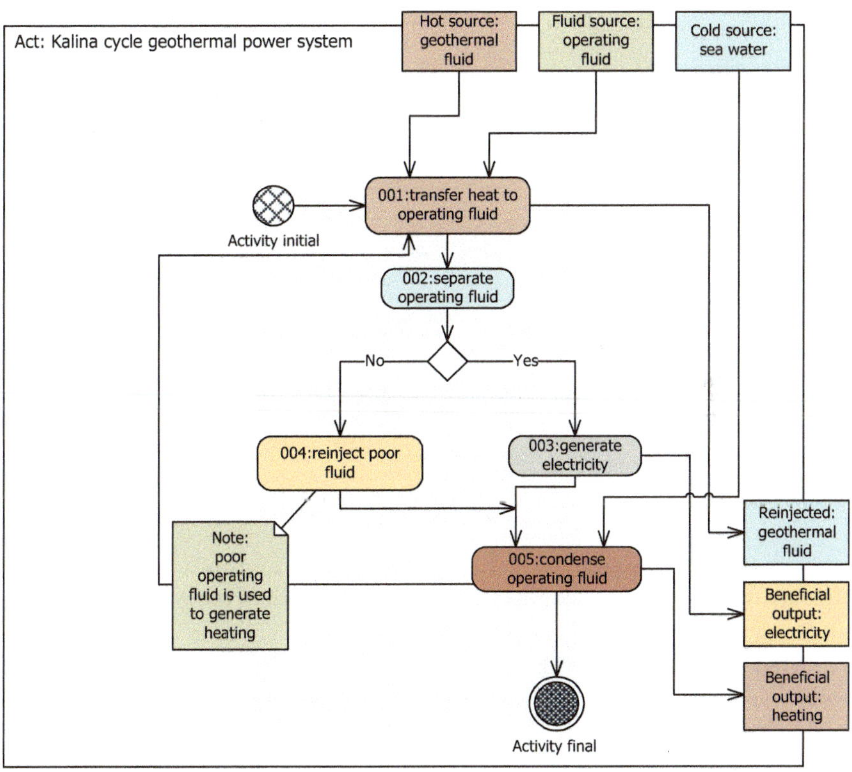

FIGURE 5.47 Activity diagram of the Kalina cycle geothermal power plant.

given in Fig. 5.49. The states of the geothermal fluid and the operating fluid are given in each activity. At the same time, the beneficial outcomes obtained depending on the activities in the system are shown in the blocks with these activities. As can be seen in the block definition diagram, in general, six activities mainly occur for the electricity and heating outputs from this system. If there are subsystems, another block definition diagram can be

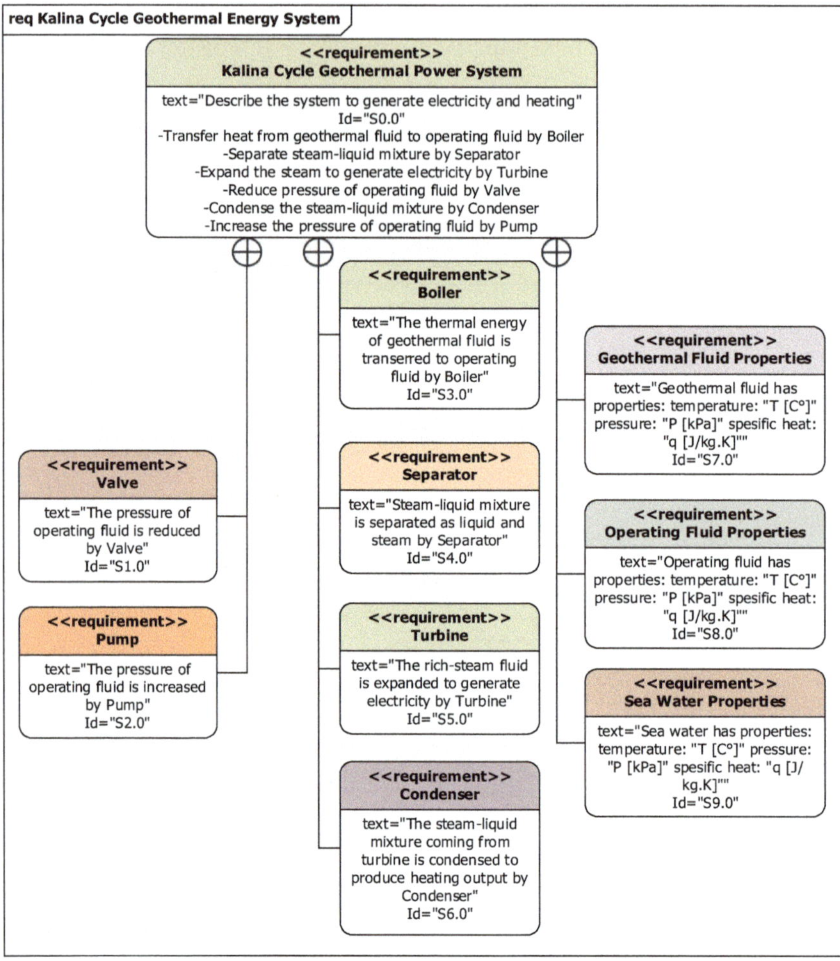

FIGURE 5.48 Requirement diagram of the Kalina cycle geothermal energy system.

prepared to show them. The system designer can detail these diagrams to better describe the system. Since the KC, a geothermal energy system, is a basic energy system, basic diagrams are given in this section. Generally, the block definition diagram is used to describe the system structure.

The sequence diagram of the system is shown in Fig. 5.50 to show the order of the activities taking place in the KC geothermal energy system. Since conditional activities take place in the system, these activities are defined in the sequence diagram to meet this situation. When looking at the separation process, the conditional activity can be seen.

In the separation process that takes place in the separator, a separation controller that controls this process is added to the sequence diagram. This section sends feedback to determine the component to which fluids from the

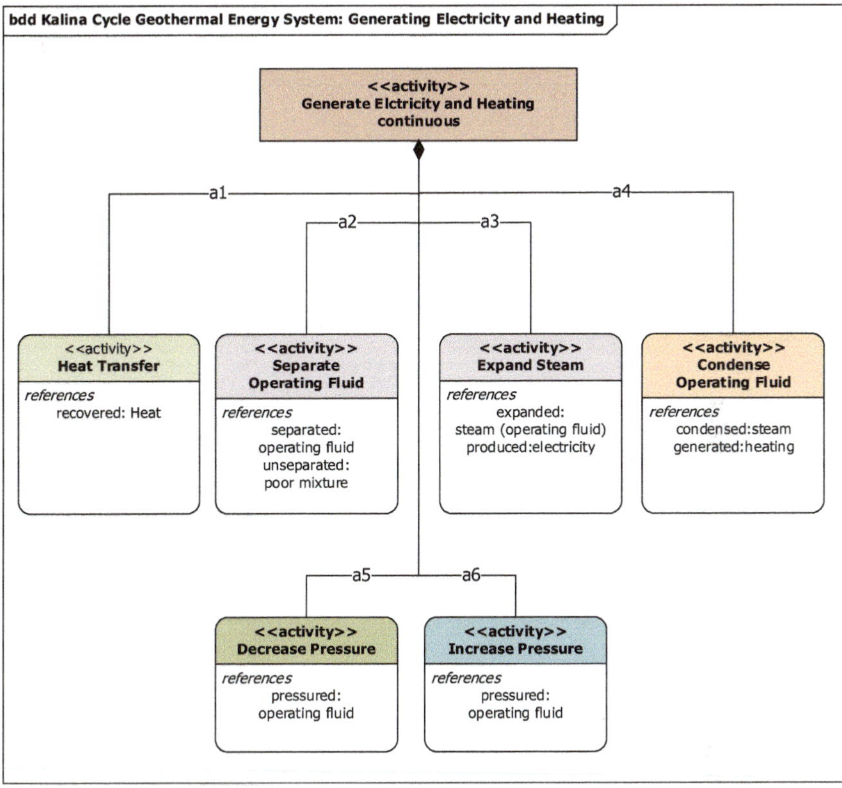

FIGURE 5.49 Block definition diagram of the Kalina cycle geothermal energy system.

separation process are to be transferred. Depending on the response to the feedback, the working fluid is sent either to the turbine or to the condenser. This situation is clearly shown in the sequence diagram. At the same time, feedback is sent after each useful output generation in the sequence diagram. After this feedback, the component to which the geothermal fluid in the system will be transferred is determined. In this way, a general sequence diagram of an SSORC geothermal power plant is created.

It can be easily seen from these diagrams that the diagrams created using the SysML modeling language have the potential for system definition. The diagrams created with this modeling language, as just shown, clearly describe the activities taking place in the system, the system structure, and the order of the activities. Therefore, designs made with this modeling language have the potential for future designs.

Example 5.6: A KC geothermal energy system, as shown in Fig. 5.51, is considered for power production. In this system, 140 kg/s saturated liquid geothermal fluid at the temperature and pressure of 170°C and 1800 kPa,

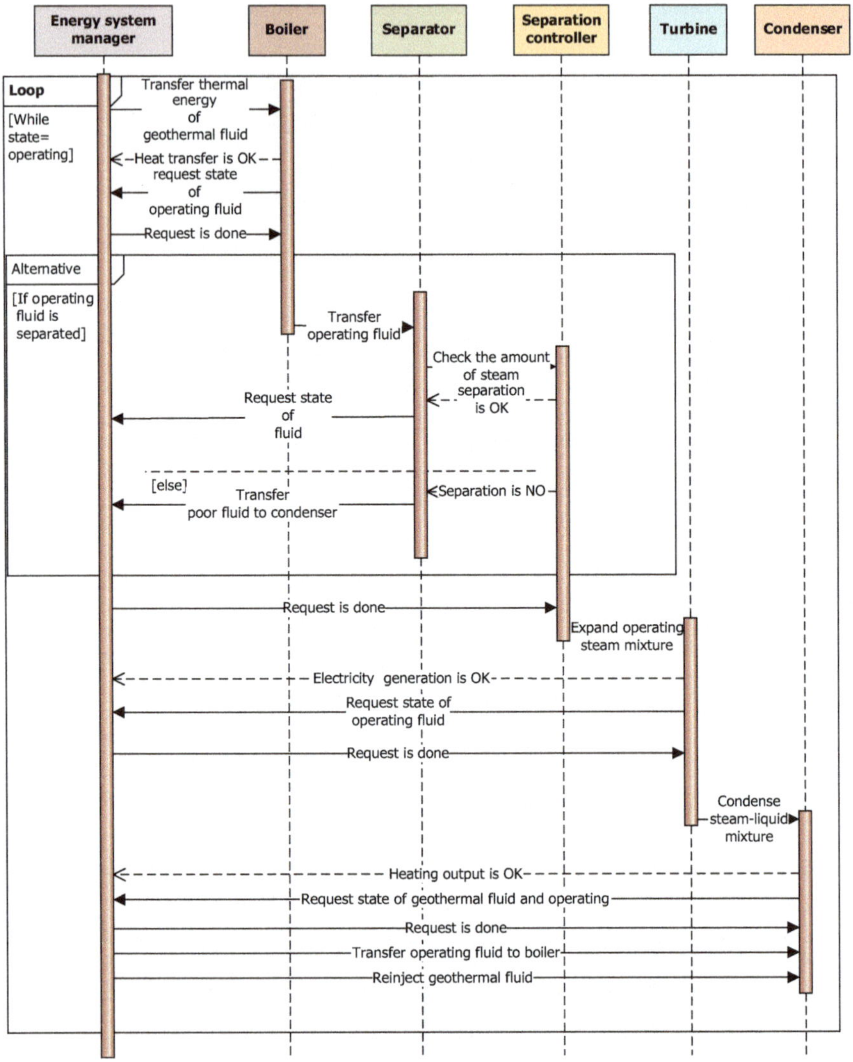

FIGURE 5.50 Sequence diagram of the Kalina cycle geothermal energy system.

respectively, enters a vapor generator to release its thermal energy to a Kalina subsystem. In this subsystem, ammonia–water is used for the working fluid, the exiting pressure levels of the pump with a temperature increase of 0.36°C and the Kalina turbine are 2323 and 516.2 kPa, respectively. The output temperature of the regenerator and Kalina turbine are 57°C and 104°C, the output pressure of the valve is 516.2 kPa. The pinch point temperature of the HEX is assumed to be 9°C. The ammonia concentration of

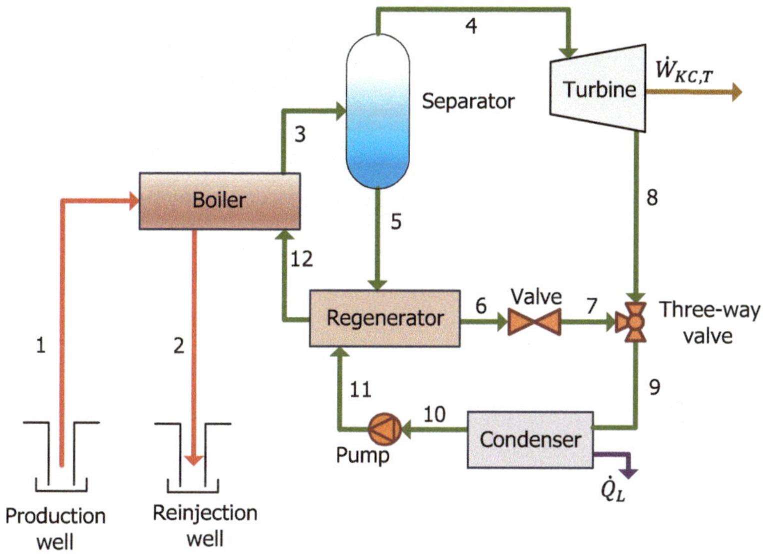

FIGURE 5.51 Schematic diagram of the Kalina cycle–based geothermal energy system.

the KC is given as $x_3 = x_9 = x_{10} = x_{11} = x_{12} = 0.6$, $x_4 = x_8 = 0.7649$, and $x_5 = x_6 = x_7 = 0.2319$. Calculate the:

1. total work production rate,
2. total exergy destruction rate of the Kalina system,
3. energy and exergy efficiencies of the Kalina power system,
4. variations of the power production rate and the exergy destruction rate of the Kalina system when the geothermal water mass flow rate increases from 100 to 180 kg/s,
5. variations of energy and exergy efficiencies of the Kalina power system when the reference temperature increases from 0°C to 40°C.

Solution: First, some assumptions are required for the KC geothermal energy system analysis.

Assumptions:

- All components operate under steady-state conditions.
- The reference temperature and pressure are taken as 25°C and 101.3 kPa, respectively.
- The pressure losses in the connections between the components are negligible.
- The heat energy losses from the components are negligible.
- The changes in the kinetic and potential energies and exergies are neglected.

Analysis:

The mass, energy, entropy, and exergy balance equations for the boiler are defined:

$$\dot{m}_1 = \dot{m}_2; \dot{m}_{12} = \dot{m}_3$$
$$\dot{m}_1 h_1 + \dot{m}_{12} h_{12} = \dot{m}_2 h_2 + \dot{m}_3 h_3$$
$$\dot{m}_1 s_1 + \dot{m}_{12} s_{12} + \dot{S}_{g,Bo} = \dot{m}_2 s_2 + \dot{m}_3 s_3$$
$$\dot{m}_1 ex_1 + \dot{m}_{12} ex_{12} = \dot{m}_2 ex_2 + \dot{m}_3 ex_3 + \dot{E}x_{D,Bo}$$

The mass, energy, entropy, and exergy balance equations for the separator can be defined:

$$\dot{m}_3 = \dot{m}_4 + \dot{m}_5$$
$$\dot{m}_3 h_3 = \dot{m}_4 h_4 + \dot{m}_5 h_5$$
$$\dot{m}_3 s_3 + \dot{S}_{g,sep} = \dot{m}_4 s_4 + \dot{m}_5 s_5$$
$$\dot{m}_3 ex_3 = \dot{m}_4 ex_4 + \dot{m}_5 ex_5 + \dot{E}x_{D,sep}$$

The mass, energy, entropy, and exergy balance equations for the turbine are expressed as:

$$\dot{m}_4 = \dot{m}_8$$
$$\dot{m}_4 h_4 = \dot{m}_8 h_8 + \dot{W}_{KC,T}$$
$$\dot{m}_4 s_4 + \dot{S}_{g,KCT} = \dot{m}_8 s_8$$
$$\dot{m}_4 ex_4 = \dot{m}_8 ex_8 + \dot{E}x^W_{KC,T} + \dot{E}x_{D,KCT}$$

The mass, energy, entropy, and exergy balance equations for the regenerator are obtained:

$$\dot{m}_5 = \dot{m}_6; \dot{m}_{11} = \dot{m}_{12}$$
$$\dot{m}_5 h_5 + \dot{m}_{11} h_{11} = \dot{m}_6 h_6 + \dot{m}_{12} h_{12}$$
$$\dot{m}_5 s_5 + \dot{m}_{11} s_{11} + \dot{S}_{g,Reg} = \dot{m}_6 s_6 + \dot{m}_{12} s_{12}$$
$$\dot{m}_5 ex_5 + \dot{m}_{11} ex_{11} = \dot{m}_6 ex_6 + \dot{m}_{12} ex_{12} + \dot{E}x_{D,Reg}$$

The mass, energy, entropy, and exergy balance equations for the valve are defined:

$$\dot{m}_6 = \dot{m}_7$$
$$\dot{m}_6 h_6 = \dot{m}_7 h_7$$
$$\dot{m}_6 s_6 + \dot{S}_{g,val} = \dot{m}_7 s_7$$
$$\dot{m}_6 ex_6 = \dot{m}_7 ex_7 + \dot{E}x_{D,val}$$

The mass, energy, entropy, and exergy balance equations for the three-way valve are written:

$$\dot{m}_7 + \dot{m}_8 = \dot{m}_9$$
$$\dot{m}_7 h_7 + \dot{m}_8 h_8 = \dot{m}_9 h_9$$
$$\dot{m}_7 s_7 + \dot{m}_8 s_8 + \dot{S}_{g,twv} = + \dot{m}_9 s_9$$
$$\dot{m}_7 ex_7 + \dot{m}_8 ex_8 = \dot{m}_9 ex_9 + \dot{E}x_{D,twv}$$

The mass, energy, entropy, and exergy balance equations for the condenser are defined:

$$\dot{m}_9 = \dot{m}_{10}$$
$$\dot{m}_9 h_9 = \dot{m}_{10} h_{10} + \dot{Q}_L$$
$$\dot{m}_9 s_9 + \dot{S}_{g,con} = \dot{m}_{10} s_{10} + \dot{Q}_L / T_{con}$$
$$\dot{m}_9 ex_9 = \dot{m}_{10} ex_{10} + \dot{E}x_L^Q + \dot{E}x_{D,con}$$

The mass, energy, entropy, and exergy balance equations for the pump can be expressed as:

$$\dot{m}_{10} = \dot{m}_{11}$$
$$\dot{m}_{10} h_{10} + \dot{W}_{in,P} = \dot{m}_{11} h_{11}$$
$$\dot{m}_{10} s_{10} + \dot{S}_{g,P} = \dot{m}_{11} s_{11}$$
$$\dot{m}_{10} ex_{10} + \dot{E}x_{in,P}^W = \dot{m}_{11} ex_{11} + \dot{E}x_{D,P}$$

For flow 1:

As given in the equation, $P_1 = 1800$ kPa and $T_1 = 170$°C, and, based on these data, h_1 and s_1 are determined from EES as 719.6 kJ/kg and 2.04 kJ/kg K.

For flow 2:

As given in the equation, $P_2 = P_1$ and $T_2 = 96$°C, and based on these data, h_2 and s_2 are determined by using the EES as 403.6 kJ/kg and 1.261 kJ/kg K.

For flow 3:

$P_3 = P_{11}$, $x_3 = 0.6$ and $T_3 = 161$°C, and, based on these data, h_3 and s_3 are determined from EES as 1457 kJ/kg and 4.328 kJ/kg K. The mass flow rate at flow 3 can be determined:

$$\dot{m}_1 (h_1 - h_2) = \dot{m}_3 (h_3 - h_{12})$$

For flow 4:

$P_4 = P_3$, $x_4 = 0.7649$ and $T_4 = T_3$, and, based on these data, h_4 and s_4 are determined from EES as 1852 kJ/kg and 5.355 kJ/kg K.

For flow 5:

$P_5 = P_3$, $x_5 = 0.2319$ and $T_5 = T_3$, and based on these data, h_5 and s_5 are determined from EES as 560 kJ/kg and 2.003 kJ/kg K.

For flow 6:

$P_6 = P_3$, $x_6 = x_5$ and $T_6 = 57$°C, and, based on these data, h_6 and s_6 are determined from EES as 90.8 kJ/kg and 0.7719 kJ/kg K.

For flow 7:

$P_7 = 516.2$, $x_7 = x_6$, and $h_7 = h_6$, and, based on these data, T_7 and s_7 are determined from EES as 57.34°C and 0.778 kJ/kg K.

For flow 8:

$P_8 = P_7$, $x_8 = x_4$ and $T_8 = 104$°C, and based on these data, h_8 and s_8 are determined from EES as 1634 kJ/kg and 5.475 kJ/kg K.

For flow 9:

$P_9 = P_8$, $x_9 = x_3$ and $T_9 = 99°C$, and based on these data, h_9 and s_9 are determined from EES as 1144 kJ/kg and 4 kJ/kg K.

For flow 10:

$P_{10} = P_9$, $x_{10} = x_3$ and $T_{10} = 27°C$, and based on these data, h_{10} and s_{10} are determined from EES as -105.6 kJ/kg and 0.2536 kJ/kg K.

For flow 11:

$P_{11} = 2323$ kPa, $x_{11} = x_3$ and $T_{11} = T_{10} + 0.36°C$, and also based on these data, h_{11} and s_{11} are determined by utilizing the EES program as -102.6 kJ/kg and 0.2565 kJ/kg K.

For flow 12:

$P_{12} = P_{11}$, $x_{12} = x_3$, and $T_{12} = 59°C$, and, based on these data, h_{12} and s_{12} are determined from the EES as 42.88 kJ/kg and 0.7166 kJ/kg K. The state points thermodynamic data for the KC geothermal energy system are given in Table 5.9.

1. The work production rate is:

$$\dot{W}_{KC,T} = \dot{m}_4 h_4 - \dot{m}_8 h_8 = 4699 \text{ kW}$$

The work consumption rate by the pump is:

$$\dot{W}_{in,P} = \dot{m}_{11} h_{11} - \dot{m}_{10} h_{10} = 99 \text{ kW}$$

And the total work production rate is:

$$\dot{W}_{Total} = \dot{W}_{KC,T} - \dot{W}_{in,P} = 4600 \text{ kW}$$

2. The total exergy destruction rate of the KC geothermal energy system is:

$$\dot{E}x_{D,KCGP} = \dot{m}_1 ex_1 - \dot{m}_2 ex_2 - \dot{E}x^W_{Total} = 7092 \text{ kW}$$

3. The energy efficiency of the KC geothermal energy system is:

$$\eta_{KCGP} = \frac{\dot{W}_{Total}}{\dot{m}_1 h_1 - \dot{m}_2 h_2} = 0.104 = 10.4\%$$

The exergy efficiency of the KC geothermal energy system is:

$$\psi_{KCGP} = \frac{\dot{E}x^W_{Total}}{\dot{m}_1 ex_1 - \dot{m}_2 ex_2} = 0.3934 = 39.34\%$$

4. The effects of geothermal water mass flow rate on the power production rate and exergy destruction rate of the KC geothermal energy system are shown in Fig. 5.52. As shown in this figure, the power production rate increases from 3286 to 5915 kW, and the exergy destruction rate increases from 5066 to 9119 kW with increasing the geothermal mass flow rate from 100 to 180 kg/s.

TABLE 5.9 State points thermodynamic data for the Kalina cycle geothermal energy system.

State point	Working fluid	$\dot{m}$ (kg/s)	T (°C)	P (kPa)	h (kJ/kg)	s (kJ/kg K)	x (–)	ex (kJ/kg)
0	H_2O	–	25	101.3	−92.38	0.2829	–	–
00	NH_3–H_2O	–	25	101.3	104.9	0.3672	–	–
1	H_2O	140	170	1800	719.6	2.04	1	115.9
2	H_2O	140	96	1800	403.6	1.261	1	32.34
3	NH_3–H_2O	31.3	161	2323	1457	4.328	0.6	342.7
4	NH_3–H_2O	21.61	161	2323	1852	5.355	0.7649	431.6
5	NH_3–H_2O	9.687	161	2323	560	2.003	0.2319	139.6
6	NH_3–H_2O	9.687	57	2323	90.8	0.7719	0.2319	37.38
7	NH_3–H_2O	9.687	57.34	516.2	90.8	0.778	0.2319	35.57
8	NH_3–H_2O	21.61	104	516.2	1634	5.475	0.7649	178.4
9	NH_3–H_2O	31.3	99	516.2	1144	4	0.6	128.5
10	NH_3–H_2O	31.3	27	516.2	−105.7	0.2536	0.6	−4.62
11	NH_3–H_2O	31.3	27.36	2323	−102.6	0.2565	0.6	−2.316
12	NH_3–H_2O	31.3	59	2323	42.88	0.7166	0.6	5.959

5. The effects of reference temperature on the energy and exergy efficiencies of the KC geothermal energy system are illustrated in Fig. 5.53. It can be shown that the energy efficiency of the geothermal plant does not change with increasing the reference temperature from 0°C to 40°C, whereas the exergy efficiency of the investigated plant increases from 0.319 to 0.4575 in the chosen reference temperature change.

5.5.4 Combined flash/binary geothermal power generating system

The combined flash/binary geothermal energy systems produce electricity and other useful outputs using energy from a renewable energy source [16]. This section describes the working principle of a combined flash/binary GPS. The most common schematic diagram of such a system is shown in Fig. 5.54.

As can be seen from the schematic diagram of the system in Fig. 5.54, the geothermal fluid is transferred to the flash chamber with flow 1, and the system starts to function. The pressure of the geothermal fluid coming to the flash chamber is reduced, and the geothermal fluid is converted into a mixture of liquid and steam. Then this liquid–steam mixture is sent to the separator with flow 2. The incoming fluid is divided into two as steam and brine in the separator. The steam coming out of the separator is sent to the purifier with flow 3 to increase the performance of the fluid. The fouling part in the fluid is removed from the system by flow 4. The steam fluid with increased quality is sent to the turbine with flow 5 to obtain the electrical output. The incoming steam is expanded between flow 5 and flow 6 to generate

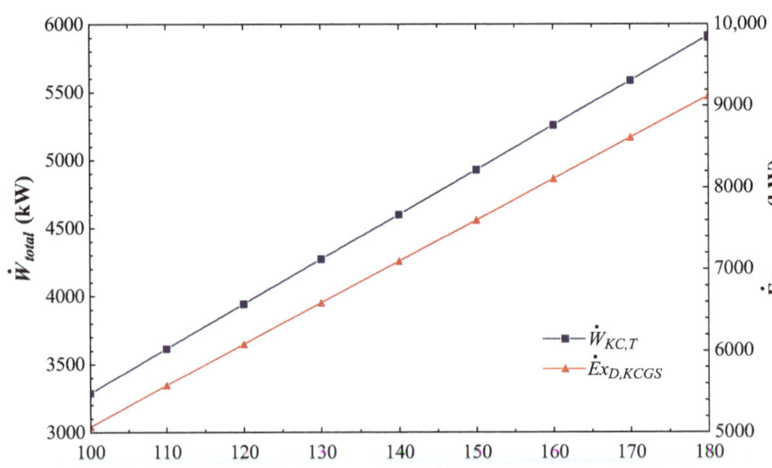

FIGURE 5.52 Effect of geothermal water mass flow rate on the power production and exergy destruction rates for the Kalina cycle geothermal energy system.

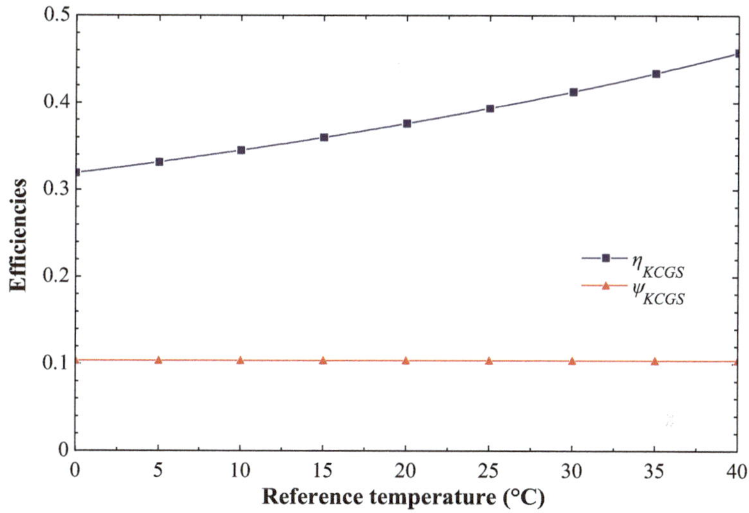

FIGURE 5.53 Effect of reference temperature on the energy and exergy efficiencies for the Kalina cycle geothermal energy system.

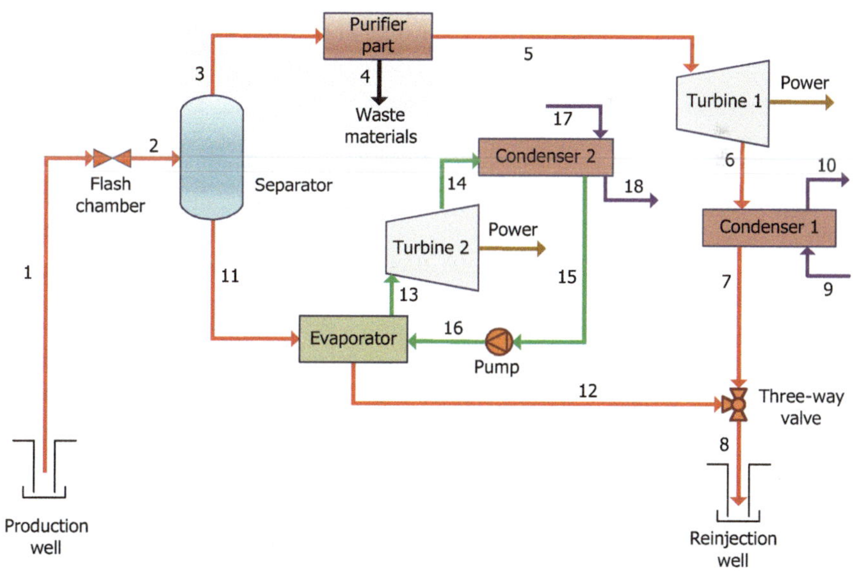

FIGURE 5.54 Simplified flow diagram of a combined flash/binary geothermal power system.

electricity. Then the steam coming out of the turbine is sent to the condenser with flow 6 to be condensed. The condenser is used for the heating output, which is a beneficial product. The fluid coming out of the condenser is sent to the three-way valve with the number 7 flow. While the geothermal fluid is circulating in the system, a binary cycle is also performed in the system. While the process up to this point is taking place, at the same time, the other fluid leaving the separator is sent to the evaporator with flow 11. By using the evaporator, heat energy in the geothermal fluid is transferred to the organic working fluid in the ORC cycle. This energy transfer is carried out by heat transfer between the geothermal fluid and the ORC working fluid. The energy-laden organic working fluid is sent to the turbine with flow 13 for electricity generation. The incoming steam is expanded between flow 13 and flow 14 to produce electricity. Then ORC operating fluid coming out of the turbine is sent to condenser 2 with flow 14 to be condensed. The condenser is used for the heating output, which is a useful product. The ORC working fluid from condenser 2 is sent to the pump with flow 15 to increase its pressure. To increase the temperature of the pressure-enhanced ORC fluid, this fluid is sent to the evaporator with flow 16. The process of increasing the temperature here takes place within the geothermal fluid coming from a separator with flow 11. Then the geothermal fluid coming out of the evaporator is sent to the three-way valve with flow 12. Fluids combined with the three-way valve are sent to the ground with flow 8. The combined flash/binary GPS generally works this way.

The operating logic of combined flash/binary GPSs can be better understood with the activity diagram of the system, which is given in Fig. 5.55. As seen from the activity diagram, an organic fluid is used together with the geothermal fluid in the system. With the geothermal fluid, both beneficial outputs are obtained, and energy transfer to the working fluid of another cycle is performed. By using the geothermal fluid and power obtained, the useful output of heating is obtained. With the transfer of heat energy in the geothermal fluid to the organic fluid in the ORC subsystem, the heating output is also obtained along with electrical output from this subsystem.

The electrical outputs in the system are produced by the expansion of the fluids coming into the turbine. Heating outputs are obtained by condensing the fluid coming to the condensers. Since the selection of the components used in the system is a subject in itself, these components should be selected on the basis of both cost and performance analyses. While making such cost and performance analyses, the thermodynamic equations also help to form the mathematical model of the system. Therefore, if the mathematical models of these and similar systems are prepared clearly, the selection of the components in the system is also facilitated. The combined energy systems yield useful outputs in terms of both efficiency and performance. This section focuses on the working principle of a basic combined flash/binary GPS. To make the operating logic of the system understandable, the schematic diagram of the system and the activity diagram of the system have been included.

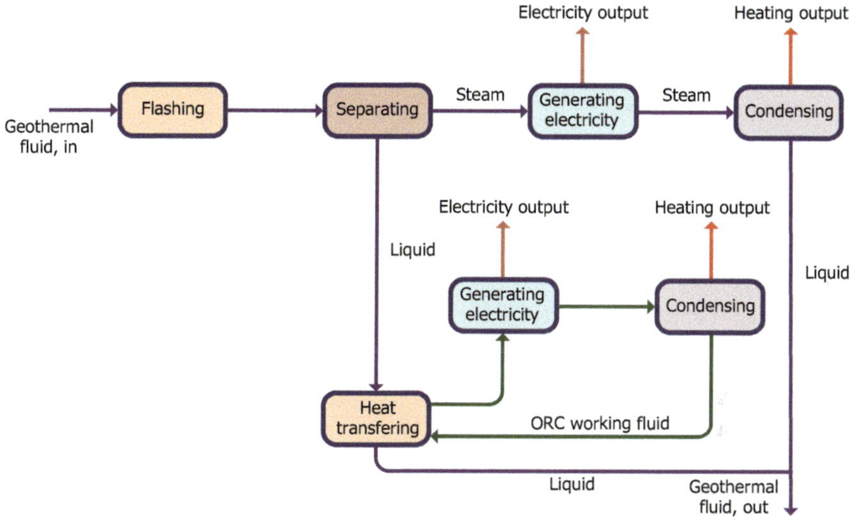

FIGURE 5.55 Activity diagram of electricity and heating generation in the combined flash/binary geothermal power system.

Example 5.7: A combined flash/binary GPS, as shown in Fig. 5.54, is considered for power production. In this system, 140 kg/s saturated liquid geothermal fluid at the temperature and pressure of 218°C and 2316 kPa, respectively, is isenthalpically flashed in a flashing part at a pressure ratio of 3, separated in the saturated form in a vapor–liquid separator; the fraction of vapor at the flash chamber outlet is 0.11136. The produced vapor is then used to generate power in a steam turbine. The discharged geothermal working fluid from the steam turbine is at 11 kPa with a quality of 0.8445 and is then condensed and reinjected to the injection well. To increase geothermal plant efficiency, the saturated water exiting from the separator enters a HEX to release its thermal energy to an ORC subsystem. In this subsystem, R600a is used for the working fluid, the exiting pressure levels of the pump undergo a temperature increase of 0.4°C, and the ORC turbine are 2700 and 450 kPa, respectively. The output temperature of ORC is 104°C. The pinch point temperature of the HEX is assumed to be 10°C. Calculate the:

1. total work production rate,
2. total exergy destruction rate of the combined flash/binary GPS,
3. energy and exergy efficiencies of the combined flash/binary GPS,
4. variations of power production rate and the exergy destruction rate of the combined flash/binary GPS when the geothermal water mass flow rate increases from 100 to 180 kg/s,
5. variations of energy and exergy efficiencies of the combined flash/binary GPS when the reference temperature increases from 0°C to 40°C.

Solution: First, some assumptions are required for the combined flash/binary GPS analysis.

Assumptions:

- All components operate under steady-state conditions.
- The reference temperature and pressure are taken as 25°C and 101.3 kPa, respectively.
- The pressure losses in the connections between the components are negligible.
- The heat energy losses from the components are negligible.
- The changes in the kinetic and potential energies and exergies are neglected.

Analysis:

The mass, energy, entropy, and exergy balance equations for the combined flash/binary geothermal power plant components are given in Examples 5.2 and 5.4.

For flow 1:

As given in the equation, $P_1 = 2316$ kPa and $T_1 = 218°C$, and, based on these data, h_1 and s_1 are determined from EES as 934.3 kJ/kg and 2.499 kJ/kg K.

For flow 2:

As given in the equation, $x_2 = 0.1114$ and $P_2 = P_1/3$, and, based on these data, T_2, h_2, and s_2 are determined by using the EES as 169°C, 943.4 kJ/kg, and 2.549 kJ/kg K.

For flow 3:

$P_3 = P_2$ and $x_3 = 1$ (saturated vapor), and, based on these data, T_3, h_3, and s_3 are determined by using the EES as 169°C, 2767 kJ/kg, and 6.675 kJ/kg K. The mass flow rate at flow 3 can be determined:

$$\dot{m}_3 = 0.1114 \times \dot{m}_2 = 15.59 \text{ kg/s}$$

For flow 4:

$P_4 = P_2$ and $x_4 = 0$ (saturated liquid), and T_4, h_4, and s_4 are determined by using the EES as 169°C, 714.8 kJ/kg, and 2.032 kJ/kg K. The mass flow rate at flow 4 can be determined

$$\dot{m}_4 = \dot{m}_2 - \dot{m}_3 = 124.41 \text{ kg/s}$$

For flow 5:

As given in the equation, $x_5 = 0.8445$ and $P_1 = 11$ kPa, and T_5, h_5, and s_5 are determined by utilizing the EES program as 47.69°C, 2216 kJ/kg, and 6.958 kJ/kg K.

For flow 6:

As given, $P_6 = P_5$ and $x_6 = 0$ (saturated liquid), and T_6, h_6, and s_6 are determined by utilizing the EES program as 47.69°C, 199.7 kJ/kg, and 0.6738 kJ/kg K.

For flow 7:

As given, $T_7 = T_6$ and $P_7 = P_4$, and h_6 and s_6 are determined by utilizing the EES program as 47.69°C, 200.3 kJ/kg, and 0.6735 kJ/kg K.

For flow 8:

As given, $T_8 = T_4 - 10°C$ and $P_8 = P_{11}$, and h_8 and s_8 are determined by utilizing the EES program as 937.3 kJ/kg and 4.141 kJ/kg K.

For flow 9:

As given, $P_9 = 450$ kPa and $T_9 = 104°C$, and h_9 and s_9 are determined by using the EES software program as 738.9 kJ/kg and 2.723 kJ/kg K.

For flow 10:

$P_{10} = P_9$ and $x_{10} = 0$, and T_{10}, h_{10}, and s_{10} are determined by utilizing the EES program as 33.92°C, 411.2 kJ/kg, and 2.751 kJ/kg K.

For flow 11:

As given in the equation, $T_{11} = T_{10} + 0.4°C$ and $P_{11} = 2700$ kPa, and h_{11} and s_{11} are determined by utilizing the EES program as 416.4 kJ/kg and 2.754 kJ/kg K. The state points thermodynamic data for the combined GPS are given in Table 5.10.

1. The work production rate by the turbine is found:

$$\dot{W}_T = \dot{m}_3 h_3 - \dot{m}_5 h_5 = 8599 \text{ kW}$$

The work production rate by the ORC turbine is written as:

$$\dot{W}_{ORC,T} = \dot{m}_9 h_9 - \dot{m}_{10} h_{10} = 9114 \text{ kW}$$

The work consumption rate by the pump is:

$$\dot{W}_{in,P} = \dot{m}_{11} h_{11} - \dot{m}_{10} h_{10} = 608 \text{ kW}$$

The net work production rate is:

$$\dot{W}_{Total} = \dot{W}_T + \dot{W}_{ORC,T} - \dot{W}_{in,P} = 17{,}106 \text{ kW}$$

2. The total exergy destruction rate of the combined flash/binary GPS is:

$$\dot{E}x_{D,BGP} = \dot{m}_1 ex_1 - \dot{m}_6 ex_6 - \dot{m}_7 ex_7 - \dot{E}x^W_{Total} = 9468 \text{ kW}$$

3. The energy efficiency of the combined flash/binary GPS is:

$$\eta_{BGP} = \frac{\dot{W}_{Total}}{\dot{m}_1 h_1 - \dot{m}_6 h_6 - \dot{m}_7 h_7} = 0.1665 = 16.65\%$$

The exergy efficiency of the combined flash/binary GPS is:

$$\psi_{BGP} = \frac{\dot{E}x^W_{Total}}{\dot{m}_1 ex_1 - \dot{m}_6 ex_6 - \dot{m}_7 ex_7} = 0.6437 = 64.37\%$$

4. The effects of geothermal water mass flow rate on the power production rate and exergy destruction rate of the combined flash/binary GPS are

TABLE 5.10 State points thermodynamic data for the combined flash/binary geothermal power system.

State point	Working fluid	$\dot{m}$ (kg/s)	T (°C)	P (kPa)	h (kJ/kg)	s (kJ/kg K)	x (–)	ex (kJ/kg)
0	Water	–	25	101.3	598.9	2.513	–	–
00	Isobutane	–	25	101.3	104.8	0.3669	–	–
1	Water	140	218	2316	934.3	2.499	0	193.8
2	Water	140	169	772	943.4	2.549	0.1114	187.9
3	Water	15.59	169	772	2767	6.675	1	781.9
4	Water	124.4	169	772	714.8	2.032	0	113.5
5	Water	15.59	47.69	11	2216	6.958	0.8445	146
6	Water	15.59	47.69	11	199.7	0.6738	0	3.348
7	Water	124.4	47.69	772	200.3	0.6735	0	4.11
8	Isobutane	121.3	159	2700	814	2.687	Superheated	163.2
9	Isobutane	121.3	104	450	738.9	2.723	Superheated	77.32
10	Isobutane	121.3	33.84	450	281.4	1.279	0	50.55
11	Isobutane	121.3	35.44	2700	286.4	1.282	Compressed liquid	54.74

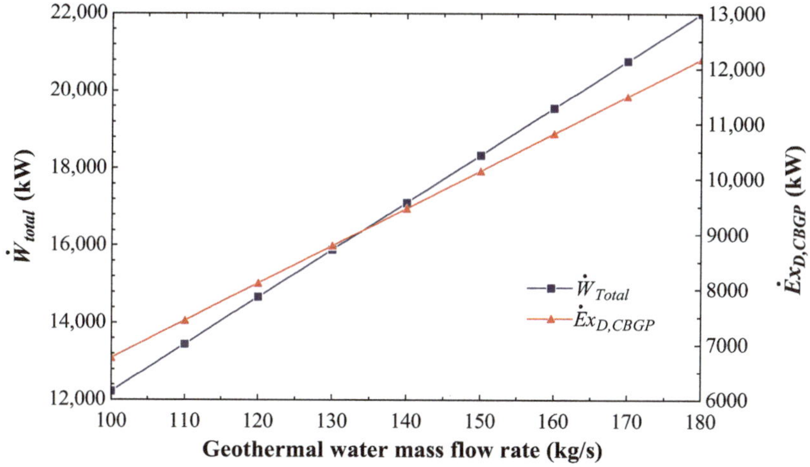

FIGURE 5.56 Effect of geothermal water mass flow rate on the power production and exergy destruction rates for the combined flash/binary geothermal power system.

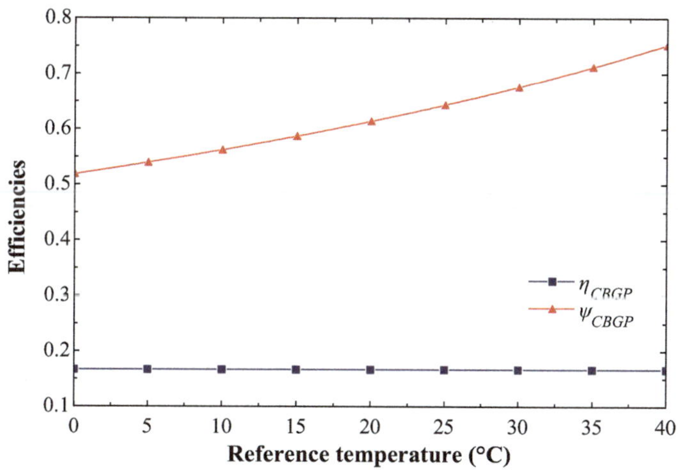

FIGURE 5.57 Effect of reference temperature on the energy and exergy efficiencies for the combined flash/binary geothermal power system.

shown in Fig. 5.56. As shown in the figure, the power production rate increases from 12,218 to 21,993 kW, and the exergy destruction rate increases from 6763 to 12,173 kW with increasing the geothermal mass flow rate from 100 to 180 kg/s.

5. The effects of reference temperature on the energy and exergy efficiencies of the combined flash/binary GPS is illustrated in Fig. 5.57. It can be shown that the energy efficiency of the geothermal plant does not change with increasing the reference temperature from 0°C to 40°C,

whereas the exergy efficiency of the investigated plant increases from 0.5189 to 0.7522 in the chosen reference temperature change.

5.6 Closing remarks

This chapter has comprehensively discussed the basic, hybrid, and combined geothermal energy systems, such as the direct steam geothermal power generating system, single-flash steam geothermal power generating system, double-flash steam geothermal power generating system, SSORC geothermal power generating system, SSORC geothermal power generating system with two turbines, DS ORC geothermal power generating system, KC geothermal power generating system, and combined flash/binary power generating systems.

It has also presented the designs of the systems for thermodynamic analysis and assessment using energy and exergy approaches. Both energy and exergy efficiencies have been considered in the studies. The exergy destruction rates in the overall geothermal energy–based basic plants and subcomponents are quantified. The great majority of the overall exergy losses for the geothermal energy plants are associated with the high-temperature difference between the components of the geothermal energy plants and their surroundings. The examples and case studies have illustrated that results obtained through exergy analysis are clearer and more meaningful than those obtained by energy analysis and help indicate potential modifications to improve efficiency. In other words, the use of exergy is important because it clearly takes into account the loss of availability and the temperature of the heat in geothermal energy–based basic plants, and hence it more correctly reflects the thermodynamic value of the power and heating producing plants.

Nomenclature

A	Area (m^2)
E	Energy (kJ)
e	Specific energy (kJ/kg)
$\dot{E}$	Energy rate (kW)
ex	Specific exergy (kJ/kg)
$\dot{E}x$	Exergy rate (kW)
$\dot{E}x_D$	Exergy destruction rate (kW)
$\dot{E}x^Q$	Exergy transfer rate associated with heat transfer (kW)
$\dot{E}x^W$	Exergy transfer rate associated with work (kW)
h	Specific enthalpy (kJ/kg)
L	Length (m)
m	Mass (kg)
$\dot{m}$	Mass flow rate (kg/s)
P	Pressure (kPa)
q	Specific heat transfer (kJ/kg)

Q	Heat (kJ)
$\dot{q}$	Specific heat transfer rate (kW/kg)
$\dot{Q}$	Heat rate (kW)
s	Specific entropy (kJ/kg K)
S	Entropy (kJ/K)
$\dot{S}$	Entropy rate (kW/K)
t	Time (s)
T	Temperature (°C, K)
u	Internal energy (kJ/kg)
V	Volume (m^3)
W	Work (kJ)
$\dot{w}$	Specific work rate (kW/kg)
$\dot{W}$	Work rate (kW)

Greek letters

Δ	Change in variable
η	Energy efficiency
ψ	Exergy efficiency

Subscript

a	Air
con	Condenser
D	Destruction
e	Exit condition
en	Energy
eva	Evaporator
ev	Expansion valve
ex	Exergy
f	Fuel
fc	Flash chamber
g	Generation
gen	Generator
heating	Heating load
HP	High pressure
l	Liquid
L	Loss
LP	Low pressure
i	Inlet condition
MP	Medium pressure
ms	Moisture separator
mx	Mixer
p	Pump
ps	Particle separator
sep	Separator
ST	Steam turbine
tot	Total
tur	Turbine

val	Valve
wf	Working fluid
1...74	State numbers
0	Ambient or reference condition

Superscripts

.	Rate
Ch	Chemical

Acronyms

CC	Combined cycle
CHP	Combined heat and power
DFGP	Double-flash geothermal plant
DS	Double-stage
DSGP	Direct steam geothermal plant
EES	Engineering Equation Solver
EGS	Enhanced geothermal system
GPS	Geothermal power system
HEX	Heat exchanger
KC	Kalina cycle
KCGP	Kalina cycle geothermal plant
LCA	Life cycle assessment
LCI	Life cycle inventory
ORC	Organic Rankine cycle
SFGP	Single-flash geothermal plant
SS	Single-stage
SysML	System Modeling Language

References

[1] I. Dincer, C. Zamfirescu, Renewable Energies. Sustainable Energy Systems and Applications, Springer US, Boston, MA, 2011.

[2] GEA, Annual US & Global Geothermal Power Production Report, Geothermal Energy Association, 2016.

[3] GEA, Geothermal Basics—Environmental Benefits, Geothermal Energy Association, 2014.

[4] US DOE, Life Cycle Analysis Results of Geothermal Systems in Comparison to Other Power Systems, Argonne National Laboratory, 2010.

[5] M. Ozturk, I. Dincer, Geothermal energy conversion, Comprehensive Energy Systems, Elsevier, 2018, pp. 474–544.

[6] Y.E. Yuksel, M. Ozturk, Thermodynamic and thermoeconomic analyses of a geothermal energy based integrated system for hydrogen production, Int. J. Hydrogen Energy 42 (4) (2017) 2530–2546.

[7] M. Yari, Exergetic analysis of various types of geothermal power plants, Renew. Energy 35 (2010) 112–121.

[8] Y.E. Yuksel, M. Ozturk, I. Dincer, Development of a geothermal based integrated plant for generating clean hydrogen and other useful commodities, J. Energy Resour. Technol. 142 (9) (2020) 1–13.

[9] I. Dincer, H. Ozcan, Geothermal energy, Comprehensive Energy Systems, Elsevier, 2018, pp. 702–732.
[10] T.A.H. Ratlamwala, I. Dincer, Comparative efficiency assessment of novel multi-flash integrated geothermal systems for power and hydrogen production, Appl. Therm. Eng. 48 (2012) 359–366.
[11] T.A.H. Ratlamwala, I. Dincer, Energetic and exergetic investigation of novel multi-flash geothermal systems integrated with electrolyzers, J. Power Sources 254 (2014) 306–315.
[12] R.S. El-Emam, I. Dincer, Exergy and exergoeconomic analyses and optimization of geothermal organic Rankine cycle, Appl. Therm. Eng. 59 (2013) 435–444.
[13] T.A.H. Ratlamwala, I. Dincer, M.A. Gadalla, Performance analysis of a novel integrated geothermal-based system for multi-generation applications, Appl. Therm. Eng. 40 (2012) 71–79.
[14] I. Dincer, M.A. Rosen, Exergy: Energy, Environment and Sustainable Development, second ed., Elsevier, Oxford, 2013.
[15] I. Dincer, C. Zamfirescu, Advanced Power Generation Systems, Elsevier, New York, 2014.
[16] Y.E. Yuksel, M. Ozturk, I. Dincer, Energetic and exergetic performance evaluations of a geothermal power plant based integrated system for hydrogen production, Int. J. Hydrogen Energy 43 (1) (2018) 78–90.

Study questions and problems

5.1. Describe geothermal energy–based electricity generation technologies.
5.2. Describe the direct steam geothermal power plant technologies.
5.3. Describe the single-, double-, triple-, and quadruple-flash steam power plant technologies.
5.4. Describe the binary cycle power plant technologies.
5.5. Describe the combined flash/binary power plant technologies.
5.6. Define the energy and exergy efficiencies for various geothermal power plants. How can you express the energy and exergy of a geothermal reservoir?
5.7. Explain the importance of using exergy analysis for assessing and designing geothermal energy–based power plants.
5.8. Explain how exergy analysis can help improve the performance of geothermal energy–based power plants.
5.9. Identify the main causes of exergy destruction in geothermal power plants, and propose methods for reducing or minimizing them.
5.10. Compare the energy and exergy efficiencies of a geothermal power plant. Which one is greater?
5.11. How do you define that the exergy efficiency of the geothermal power plant is generally higher than its energy efficiency?
5.12. What is the effect of ambient air temperature on the exergetic performance of any geothermal power plant?

5.13. A student calculates the exergy efficiency of geothermal energy–based power production plant to be greater than 100%. Is this result reasonable? Explain.

5.14. Do you support utilizing geothermal power sources below 160°C for electricity generation? If such an application is to occur, what is the most appropriate cycle? Explain.

5.15. In a geothermal power system utilizing a 150°C source, geothermal fluid is reinjected into the ground at nearly 80°C. What is the ratio of the exergy of the brine reinjected to the exergy of brine in the reservoir? How can you utilize this brine further before its reinjection?

5.16. A single-flash geothermal power plant, as illustrated in Fig. 5.18, is utilizing hot geothermal water at 240°C. The mass flow rate of the geothermal water is 28 kg/s. Calculate the mass flow rate of the steam entering the steam turbine, the power generated from the steam turbine, the isentropic efficiency of the steam turbine, the energy and exergy efficiencies of the single-flash geothermal power plant. Also show the effect of increasing the geothermal mass flow rate from 18 to 38 kg/s on the power production rate and exergy efficiency of the plant.

5.17. A double-flash geothermal power plant, as illustrated in Fig. 5.24, is utilizing hot geothermal water at 240°C. The mass flow rate of the geothermal water is 28 kg/s. Calculate the temperature of the steam that will enter the second steam turbine, the power generation rate from the second turbine, and the overall energy and exergy efficiencies of the double-flash geothermal power plant. Also show the effect of increasing geothermal mass flow rate from 18 to 38 kg/s on the power production rate and exergy efficiency of the plant.

5.18. A single-stage ORC geothermal power plant, as shown in Fig. 5.33, is utilizing 85 kg/s saturated liquid geothermal fluid at a pressure of 786 kPa. This working fluid enters the HEX to release its thermal energy to an ORC subsystem. The exit temperature of geothermal water from the HEX is 51.24°C. Isobutane is used for the working fluid; with a temperature increase of 0.42°C, the exiting pressure levels of the pump and the ORC turbine are 2850 and 462 kPa, respectively. The output temperature of the ORC turbine is 98°C. The pinch point temperature of the HEX is assumed to be 12°C. Calculate the total power production rate, and determine the overall energy and exergy efficiencies of the single-stage ORC geothermal power plant. Also show the effect of increasing ORC turbine pressure from 2100 to 3100 kPa on the power production rate and exergy efficiency of the geothermal plant.

5.19. A Kalina cycle geothermal energy system, as shown in Fig. 5.51, is utilizing 85 kg/s saturated liquid geothermal fluid at the temperature and pressure of 184°C and 1950 kPa, respectively, enters a vapor generator to release its thermal energy to a Kalina subsystem. Ammonia–water is used for the working fluid; with a temperature increase of 0.4°C, the exiting pressure levels of the pump and of the Kalina turbine are 2325 and 518 kPa, respectively. Also the output temperature of the regenerator and Kalina turbine are 58°C and 98°C; the output pressure of the valve is 518 kPa. The pinch point temperature of the HEX is assumed to be 12°C. The ammonia concentration of the Kalina cycle is given as $x_3 = x_9 = x_{10} = x_{11} = x_{12} = 0.62$, $x_4 = x_8 = 0.766$, and $x_5 = x_6 = x_7 = 0.232$. Calculate the total power production rate and the overall energy and exergy efficiencies of the Kalina system geothermal energy system. Also show the effect of increasing the Kalina turbine pressure from 2000 to 2500 kPa on the power production rate and exergy efficiency of the geothermal plant.

5.20. A combined flash/binary geothermal power system, as shown in Fig. 5.54, is utilizing 85 kg/s saturated liquid geothermal fluid at the temperature and pressure of 209°C and 2324 kPa, respectively. The fluid is isenthalpically flashed in a flashing part at a pressure ratio of 3 and separated in a saturated form in a vapor–liquid separator. The fraction of vapor at the flash chamber outlet is 0.11136, and the produced vapor is then used to generate power in a steam turbine. The discharged geothermal working fluid from the steam turbine is at 11.2 kPa with a quality of 0.8476, which is then condensed and reinjected to the injection well. To increase geothermal plant efficiency, the saturated water exiting from the separator enters a HEX to release its thermal energy to an ORC subsystem. R600a is used for the working fluid, and the exiting pressure levels of the pump with a temperature increase of 0.4°C and the ORC turbine are 2750 and 475 kPa, respectively. Also, the output temperature of ORC is 98°C. The pinch point temperature of the HEX is assumed to be 12°C. Calculate the total power production rate and the overall energy and exergy efficiencies of the combined flash/binary geothermal power system. Also show the effect on the power production rate and exergy efficiency of the geothermal plant of increasing ORC turbine pressure from 2250 to 3250 kPa.

Chapter 6

Advanced geothermal energy systems

6.1 Introduction

Power producing plants are usually treated as big heat engines to convert thermal energy (in the form of heat) input into mechanical work, which is then used to generate electrical energy at a sustained rate. Thermal energy input is supplied by the burning of fossil fuels (coal, oil, and natural gas) and biomass by processing nuclear energy, or by harvesting thermal power from renewable energy resources. Generally, conventional power plants contain multiple generating components that are designed to work at their nominal load when they operate optimally [1].

Cogeneration, which is the simultaneous generation of thermal energy and electric energy, has important potential for transforming existing energy supply plants so that they become more sustainable. In any type of thermal power system, a portion, generally 20%–45%, of the input thermal power should be converted to electrical energy, and the remainder is released to the environment as waste thermal energy. As geothermal energy–based power production plants can generate high-pressure and -temperature working fluids, these energy resources should be used to produce heating, cooling, or freshwater with a topping cycle.

Geothermal energy is an important renewable energy source that has shown rapid growth in recent years [2]. A lot of thermodynamic cycles have been developed for geothermal energy–based power production aim [3–5]. The selection of the cycle depends on the kind of geothermal working fluid, its flow rate, and the level of temperature. The development of geothermal energy–based power production plants that are more sustainable and effective concerning economics and environment requires new design methods. Geothermal energy technology has evolved extremely fast in the last decades, with the discovery of new methods of power conversion. New concepts and developments concerning plant combinations for improved effectiveness, decreased cost and environmental impact, and increased energy and exergy efficiencies are required [6]. It should be noted that the geothermal energy–based power production system design must be such that it extracts as much exergy as possible from the geothermal working fluid. Further advantages should be achieved from the plant combination, which entails

Geothermal Energy Systems. DOI: https://doi.org/10.1016/B978-0-12-820775-8.00006-4

integrating different subplants into a larger plant in which the subplants operate together for better performance. Therefore, the main objective of combination is to achieve better energy usage indicators and better effectiveness with respect to separate plants. In this chapter, we try to address the geothermal energy−based advanced systems briefly. There are five types: (1) multistaged direct systems, (2) multiflashing systems, (3) multistaged binary systems, (4) multiflashing binary systems, and (5) combined/integrated systems. Besides power, these advanced systems coproduce heating, freshwater, cooling, hydrogen, process heat, synthetic fuels, or other commodities with market value. The next chapter will take another perspective on geothermally driven cogeneration plants in addition to discussing multigeneration geothermal energy systems.

6.2 Classification of advanced geothermal energy systems

Many studies are carried out on different thermodynamic cycles where useful products are obtained by using geothermal energy resources. When utilizing a geothermal energy source, certain parameters depend on which thermodynamic cycle is performed. Some of these parameters are the type of geothermal fluid, the flow rate of the geothermal fluid, and the temperature of the geothermal fluid. The geothermal fluid can be found in dry steam, low-pressure brine, and high-pressure brine forms. Designs for geothermal energy−based systems should be developed in such a way that they can benefit more from the potential of the geothermal energy source. In Chapter 5, we overviewed the commonly used basic geothermal energy systems design strategies by using the viewpoint of thermodynamic analysis. In the current chapter, we will discuss the design fundamentals of advanced geothermal energy systems with multigeneration.

A comprehensive attempt to classify combination options for advanced geothermal energy−based power generation plants is illustrated in Fig. 6.1. As given in this figure, five main options of plant integration are contained: multistaging, multiflashing, multistaging with binary, multiflashing with binary, and combining/integrating.

Geothermal fluid in the form of dry steam is used in multistaged direct systems. Since the geothermal fluid is in the form of dry steam, it can be used for power generation in turbines and then for the heating output in the condenser by setting a certain pressure after passing through the particle and moisture separator. In multiflashing systems, geothermal fluid in the form of high-pressure brine is used. In these systems, first of all, a flashing process is performed to convert the geothermal fluid in liquid form into saturated steam. The quality of the saturated vapor phase geothermal fluid is increased with a separator, and after a purification process, this geothermal fluid is transferred to the turbine for power generation. Depending on the moisture content of the steam expanding in the turbine, double, triple, and quadruple-flashing processes can be performed. Multistaged binary systems are

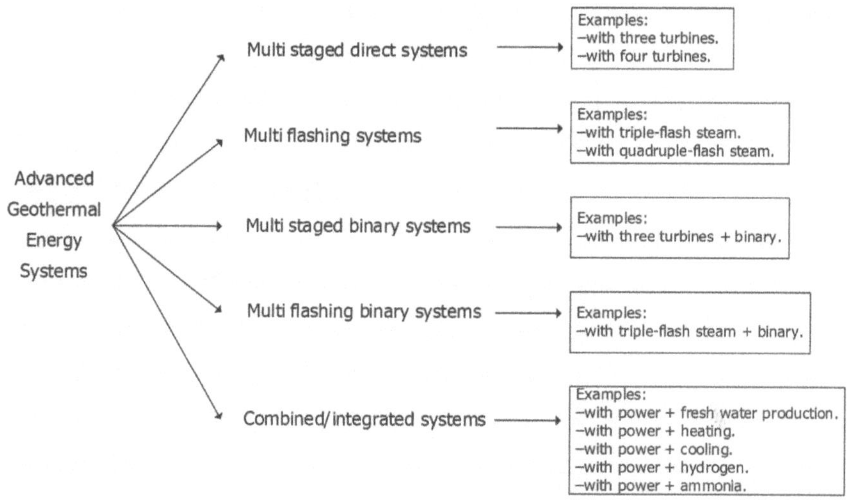

FIGURE 6.1 Main options of advanced geothermal energy systems.

designed by integrating a power cycle into multistaged direct systems. In these systems, an energy transfer is realized between the expanding steam coming from the turbine and the working fluid used in the power cycle. With this energy transfer, the energy required for the power cycle integrated into multistaged direct systems is met. Multiflashing binary systems are modeled by integrating a power cycle into multiflashing systems. In these systems, an energy transfer is performed between the expanding steam coming from the turbine and the working fluid used in the power cycle. With this energy transfer, the energy required for the power cycle integrated into multiflashing systems is provided. In geothermal-based combined/integrated systems, besides power cycles, there are subsystems where useful outputs such as heating, cooling, freshwater, hydrogen, and ammonia can be produced. Since the geothermal energy source can be used in the production of different useful outputs in these systems, these systems have an important potential in terms of performance and sustainability.

6.3 Multistaged direct geothermal energy systems

The schematic diagram of a multistaged direct system with three turbines is illustrated in Fig. 6.2. Explaining the working principle of the system, in general, will be useful for understanding how the system works.

To eliminate particles in the fluid from the geothermal energy source, the fluid is sent to the particle separator by flow 1. Since the pressure in the geothermal energy source is higher than where the particle lifter is located, it is transmitted to the particle lifter after pressure adjustment. After the particles in the fluid are removed, the fluid coming out of the particle extractor is sent to the moisture

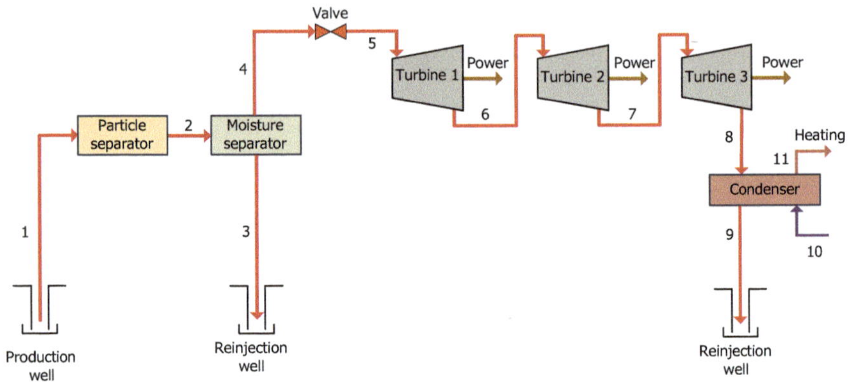

FIGURE 6.2 Simplified flow diagram of a multistaged direct system with three turbines.

separator with the number 2 flow to remove the moisture in the fluid. Here, the moisture released from the fluid is sent to the underground with flow 3. Quality fluid coming from the moisture separator is transferred to the valve with flow 4 to adjust its pressure. For this system to serve its essential purpose, the fluid from valve is sent to turbine 1 to obtain power by flow 5. After the power generation in turbine 1, the expanding steam in turbine 1 is transferred to turbine 2 for power generation with flow 6 in order to benefit from dry steam again. After the power generation in turbine 2 is realized, the expanding steam in turbine 2 is transferred to turbine 3 with flow 7 in order to realize power generation again. The fluid coming out of turbine 3 is sent to the condenser to be condensed with flow 8 to obtain heating output. The fluid coming out of the condenser is sent again to the reinjection well by flow 9. A cycle of the multistage direct geothermal system is completed in this way.

6.3.1 Case study 6.1

In the first case study, the geothermal energy source–based direct steam geothermal power system with three turbines is investigated by using thermodynamic analysis. The simplified flow diagram of the direct steam geothermal power system with three turbines in the first case study is shown in Fig. 6.3.

The mass, energy, entropy, and exergy balance equations for the particle separator can be written:

$$\dot{m}_1 = \dot{m}_2$$

$$\dot{m}_1 h_1 = \dot{m}_2 h_2$$

$$\dot{m}_1 s_1 + \dot{S}_{g,ps} = \dot{m}_2 s_2$$

$$\dot{m}_1 ex_1 = \dot{m}_2 ex_2 + \dot{E}x_{D,ps}$$

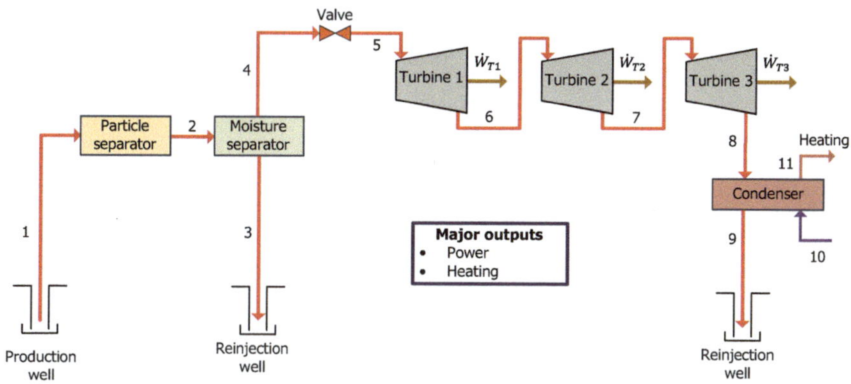

FIGURE 6.3 Schematic diagram of the multistaged direct system with three turbines.

The mass, energy, entropy, and exergy balance equations for the moisture separator can be written:

$$\dot{m}_2 = \dot{m}_3 + \dot{m}_4$$

$$\dot{m}_2 h_2 = \dot{m}_3 h_3 + \dot{m}_4 h_4$$

$$\dot{m}_2 s_2 + \dot{S}_{g,ms} = \dot{m}_3 s_3 + \dot{m}_4 s_4$$

$$\dot{m}_2 ex_2 = \dot{m}_3 ex_3 + \dot{m}_4 ex_4 + \dot{E}x_{D,ms}$$

The mass, energy, entropy, and exergy balance equations for the valve can be written:

$$\dot{m}_4 = \dot{m}_5$$

$$\dot{m}_4 h_4 = \dot{m}_5 h_5$$

$$\dot{m}_4 s_4 + \dot{S}_{g,val} = \dot{m}_5 s_5$$

$$\dot{m}_4 ex_4 = \dot{m}_5 ex_5 + \dot{E}x_{D,val}$$

The mass, energy, entropy, and exergy balance equations for the turbine 1 can be written:

$$\dot{m}_5 = \dot{m}_6$$

$$\dot{m}_5 h_5 = \dot{m}_6 h_6 + \dot{W}_{T1}$$

$$\dot{m}_5 s_5 + \dot{S}_{g,T1} = \dot{m}_6 s_6$$

$$\dot{m}_5 ex_5 = \dot{m}_6 ex_6 + \dot{E}x_{T1}^{W} + \dot{E}x_{D,T1}$$

The mass, energy, entropy, and exergy balance equations for the turbine 2 can be written:

$$\dot{m}_6 = \dot{m}_7$$

$$\dot{m}_6 h_6 = \dot{m}_7 h_7 + \dot{W}_{T2}$$

$$\dot{m}_6 s_6 + \dot{S}_{g,T2} = \dot{m}_7 s_7$$

$$\dot{m}_6 ex_6 = \dot{m}_7 ex_7 + \dot{E}x^W_{T2} + \dot{E}x_{D,T2}$$

The mass, energy, entropy, and exergy balance equations for the turbine 3 can be written:

$$\dot{m}_7 = \dot{m}_8$$

$$\dot{m}_7 h_7 = \dot{m}_8 h_8 + \dot{W}_{T3}$$

$$\dot{m}_7 s_7 + \dot{S}_{g,T3} = \dot{m}_8 s_8$$

$$\dot{m}_7 ex_7 = \dot{m}_8 ex_8 + \dot{E}x^W_{T3} + \dot{E}x_{D,T3}$$

The mass, energy, entropy, and exergy balance equations for the condenser are defined:

$$\dot{m}_8 = \dot{m}_9, \; \dot{m}_{10} = \dot{m}_{11}$$

$$\dot{m}_8 h_8 + \dot{m}_{10} h_{10} = \dot{m}_9 h_9 + \dot{m}_{11} h_{11}$$

$$\dot{m}_8 s_8 + \dot{m}_{10} s_{10} + \dot{S}_{g,con} = \dot{m}_9 s_9 + \dot{m}_{11} s_{11}$$

$$\dot{m}_8 ex_8 + \dot{m}_{10} ex_{10} = \dot{m}_9 ex_9 + \dot{m}_{11} ex_{11} + \dot{E}x_{D,con}$$

In order to model the geothermal energy—based combined system, some parameters are selected as input data for the simulation. The operating indicators utilized in the multistaged direct geothermal energy system with three turbines are listed in Table 6.1. The heat and work input/output rate, entropy generation rate, and exergy destruction rates, as well as energy and exergy effectiveness, are computed by utilizing the balance equalities and chosen assumptions.

The exergy destruction rate, dimensionless exergy destruction ratio, and exergy efficiency of the multistaged direct geothermal energy system with three turbines are written in Table 6.2. The turbines and condenser subsystems have the maximum exergy destruction rate for the investigated geothermal energy—based power and heating production system.

Power production rates from turbine 1, turbine 2, and turbine 3 and the heating producing rate from the condenser subsystem are given in Table 6.3. These data show that the net power output from turbine 2 is much higher than those of turbine 1 and turbine 3 and that the heating production rate is equal to 15,425 kW.

TABLE 6.1 Operating parameters for the multistaged direct geothermal energy system with three turbines.

Variables	Values
Reference temperature, T_0	25°C
Reference pressure, P_0	101.3 kPa
Geofluid source temperature, T_1	260°C
Geofluid source pressure, P_1	2816 kPa
Geofluid mass flow rate, $\dot{m}_1$	40 kg/s
Turbine 1 inlet pressure, P_5	1580 kPa
Turbine 1 inlet temperature, T_5	204.4°C
Turbine 2 inlet pressure, P_6	972 kPa
Turbine 2 inlet temperature, T_6	178.7°C
Turbine 3 inlet pressure, P_7	525 kPa
Turbine 3 inlet temperature, T_7	153.7°C
Heating fluid exit pressure, P_{11}	110 kPa
Heating fluid exit temperature, T_{11}	54°C
Geofluid reinjection temperature, T_9	52.73°C

TABLE 6.2 Thermodynamic analysis results for the multistaged direct geothermal energy system with three turbines.

System components	Exergy destruction rate (kW)	Exergy destruction ratio (%)	Exergy efficiency (%)
Particle separator	902	5.14	82.57
Moisture separator	758	4.32	85.64
Valve	693	3.95	88.53
Turbine 1	5396	30.77	37.86
Turbine 2	4456	25.41	38.93
Turbine 3	3591	20.47	39.74
Condenser	1743	9.94	34.38
Overall system	17,539	100	63.34

TABLE 6.3 Useful outputs from the multistaged direct geothermal energy system with three turbines.

Useful outputs	Value (kW)
Power production from turbine 1, $\dot{W}_{T1}$	7114
Power production from turbine 2, $\dot{W}_{T2}$	9910
Power production from turbine 3, $\dot{W}_{T3}$	7788
Useful heating production, $\dot{Q}_{Heating}$	15,425

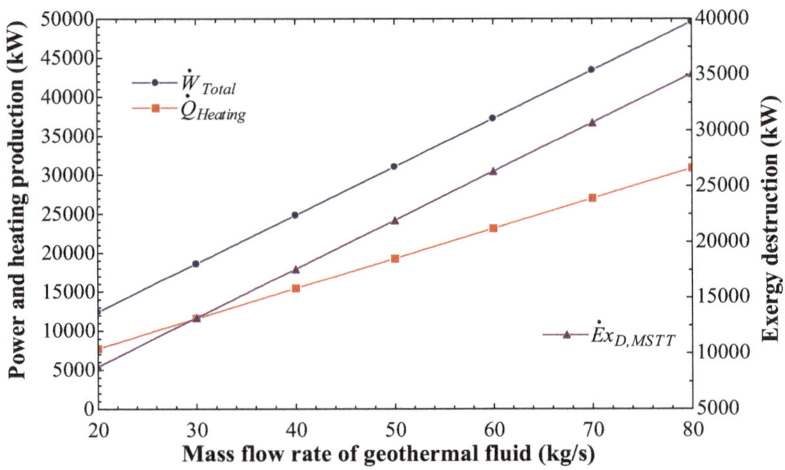

FIGURE 6.4 Effect of geothermal fluid mass flow rate on the power, heating generation, and exergy destruction rate for the multistaged direct system with three turbines.

The effects of geothermal water mass flow rate on the total power and heating production, as well as the exergy destruction rate of the multistaged direct system with three turbines, are shown in Fig. 6.4. As illustrated in this figure, the power production rate increases from 12,406 to 49,625 kW, the heating production rate increases from 7712 to 30,850 kW, and the exergy destruction rate increases from 8769 to 35,078 kW with increasing the geothermal mass flow rate from 20 to 80 kg/s, respectively.

The effect of the reference temperature on the energy and exergy efficiencies of the multistaged direct system with three turbines is shown in Fig. 6.5. As illustrated in this figure, the energy efficiency of the geothermal energy–based power plant does not change with increasing the reference temperature from 0°C to 40°C, and it is equal to 35.87%, whereas the exergy efficiency of this geothermal plant increases from 57.76% to 72.68% in the given reference temperature change.

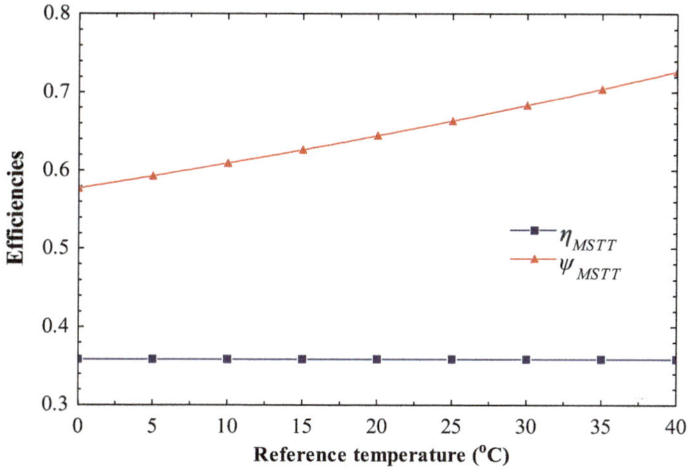

FIGURE 6.5 Effect of reference temperature on the energy and exergy efficiencies of multi-staged direct system with three turbines.

6.3.2 Case study 6.2

In the second case study, to investigate the effect of increasing the number of turbines on system performance, the geothermal energy resource–based multistaged direct system with four turbines is examined by utilizing energy and exergy analyses. The simplified flow diagram of the multistaged direct system with four turbines in the second case study is illustrated in Fig. 6.6.

In this case study, the ambient temperature and ambient pressure are chosen as 25°C and 101.3 kPa, respectively. The mass, energy, entropy, and exergy balance equations for the particle separator, moisture separator, valve, turbine 1, turbine 2, turbine 3, and condenser subsystems are described in Case Study 6.1. Also, the mass, energy, entropy, and exergy balance equations for the turbine 4 can be given as:

$$\dot{m}_8 = \dot{m}_9$$

$$\dot{m}_8 h_8 = \dot{m}_9 h_9 + \dot{W}_{T4}$$

$$\dot{m}_8 s_8 + \dot{S}_{g,T4} = \dot{m}_9 s_9$$

$$\dot{m}_8 ex_8 = \dot{m}_9 ex_9 + \dot{E}x_{T4}^{W} + \dot{E}x_{D,T4}$$

The mass, energy, entropy, and exergy balance equations for the condenser are defined:

$$\dot{m}_9 = \dot{m}_{10},\ \dot{m}_{11} = \dot{m}_{12}$$

$$\dot{m}_9 h_9 + \dot{m}_{11} h_{11} = \dot{m}_{10} h_{10} + \dot{m}_{12} h_{12}$$

$$\dot{m}_9 s_9 + \dot{m}_{11} s_{11} + \dot{S}_{g,con} = \dot{m}_{10} s_{10} + \dot{m}_{12} s_{12}$$

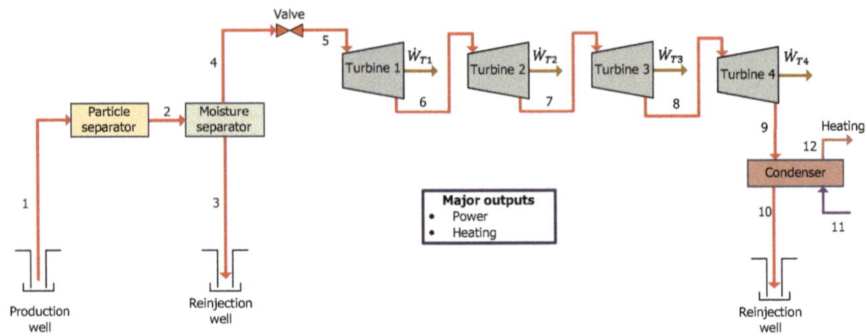

FIGURE 6.6 Schematic diagram of the multistaged direct system with four turbines.

The assumptions utilized in the working conditions of the multistaged direct system with four turbines are shown in Table 6.4 and Fig. 6.6. The heat and work input/output rate, entropy generation rate, and exergy destruction rates, and also energy and exergy efficiencies can be computed by using these balance equalities and chosen conditions.

The exergy destruction rate, dimensionless exergy destruction ratio, and exergy efficiency of the multistaged direct geothermal energy system with four turbines are given in Table 6.5. As given in this table, the valve, moisture separator, and particle separator subsystems have the maximum exergy efficiency for the analyzed geothermal energy source–based power and heating generating plant. Turbines in this geothermal energy–based combined plant have the maximum exergy destruction rates. Power production rates from turbines and heating producing rate are illustrated in Table 6.6. It should be noted that the power production rate from turbine 3 is higher than those of the other turbines.

The effects of geothermal water mass flow rate on the total power and heating production and the exergy destruction rate of the multistaged direct system with four turbines are illustrated in Fig. 6.7. As presented in this figure, the power production rate increases from 14,787 to 59,148, the heating production rate increases from 8785 to 35,139 kW, and the exergy destruction rate increases from 7579 to 30,136 kW with increasing the geothermal mass flow rate from 20 to 80 kg/s.

The effects of reference temperature on the energy and exergy efficiencies of the multistaged direct system with four turbines are illustrated in Fig. 6.8. As given in this figure, the energy efficiency of the geothermal energy–based power plant does not change with increasing the reference temperature from 0°C to 40°C, and it is equal to 64.73%, whereas the exergy efficiency of this geothermal energy source–based power production plant increases from 71.98% to 86.53% in the given reference temperature change.

TABLE 6.4 Operating parameters for the multistaged direct geothermal energy system with four turbines.

Variables	Values
Reference temperature, T_0	25°C
Reference pressure, P_0	101.3 kPa
Geofluid source temperature, T_1	280°C
Geofluid source pressure, P_1	3146 kPa
Geofluid mass flow rate, $\dot{m}_1$	40 kg/s
Turbine 1 inlet pressure, P_5	2056 kPa
Turbine 1 inlet temperature, T_5	215.1°C
Turbine 2 inlet pressure, P_6	1452 kPa
Turbine 2 inlet temperature, T_6	196.8°C
Turbine 3 inlet pressure, P_7	825 kPa
Turbine 3 inlet temperature, T_7	171.7°C
Turbine 4 inlet pressure, P_8	540 kPa
Turbine 4 inlet temperature, T_8	154.8°C
Heating fluid exit pressure, P_{12}	110 kPa
Heating fluid exit temperature, T_{12}	54°C
Geofluid reinjection temperature, T_{10}	48.67°C

6.4 Multiflashing systems

In multiflash geothermal-based systems, geothermal fluid in the form of high-pressure brine is used as the energy source. Since the moisture content of the geothermal fluid used in these systems is high, double-, triple-, or quadruple-flashing processes are performed.

6.4.1 Triple-flash steam geothermal power system

Triple-flash steam geothermal power system is defined as the development of three-flash steam geothermal power systems [7]. In this system, three flashing processes are carried out in order to increase the benefit of geothermal energy. Adding another flashing process to the system with double flashes adds cost and component requirements to the system [8]. In general, this system has common points with a double-flash steam geothermal power system. The schematic diagram with the most common form of triple-flash

TABLE 6.5 Thermodynamic analysis results for the multistaged direct geothermal energy system with four turbines.

System components	Exergy destruction rate (kW)	Exergy destruction ratio (%)	Exergy efficiency (%)
Particle separator	857	4.52	80.93
Moisture separator	726	3.83	84.56
Valve	653	3.45	87.99
Turbine 1	4827	25.99	38.57
Turbine 2	4013	21.18	39.94
Turbine 3	3324	17.56	40.15
Turbine 4	2967	15.66	40.98
Condenser	1509	7.82	36.21
Overall system	13,963	100	80.54

TABLE 6.6 Useful outputs from the multistaged direct geothermal energy system with four turbines.

Useful outputs	Value (kW)
Power production from turbine 1, $\dot{W}_{T1}$	6600
Power production from turbine 2, $\dot{W}_{T2}$	8239
Power production from turbine 3, $\dot{W}_{T3}$	9561
Power production from turbine 4, $\dot{W}_{T4}$	6706
Useful heating production, $\dot{Q}_{Heating}$	17,570

steam geothermal power system is given in Fig. 6.9. When looking at the schematic diagram of the system, the triple-flashing process can be clearly seen. The flashing process, which is completed with separator 2 in the double-flash steam geothermal power system, is completed in separator 3 by continuing one more step in this system. With this system, more power output is obtained from the geothermal energy source of the same amount and temperature compared to one with a double-flashing process.

As seen from the schematic diagram of the system, the geothermal fluid coming from underground is sent to flashing chamber 1 where the first flashing will take place with the number 1 flow. Here, the pressure-reduced

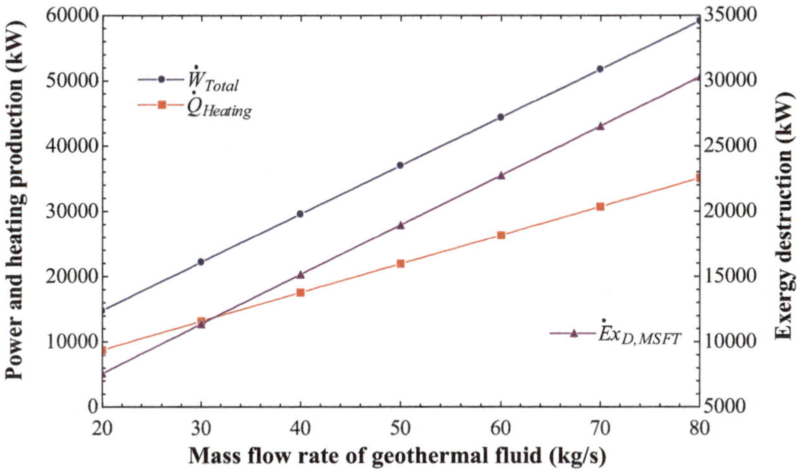

FIGURE 6.7 Effect of geothermal fluid mass flow rate on the power, heating generation, and exergy destruction rate for the multistaged direct system with four turbines.

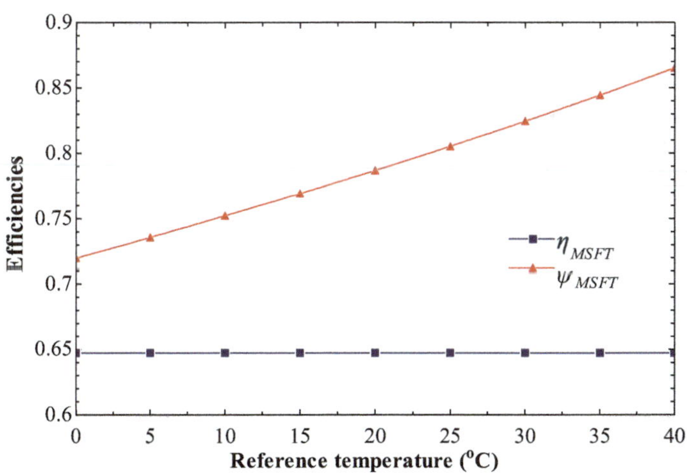

FIGURE 6.8 Effect of reference temperature on the energy and exergy efficiencies of the multistaged direct system with four turbines.

liquid–vapor mixture is delivered to separator 1 with flow 2. After separating steam from this fluid with separator 1, this steam is sent to the purification section with flow 3 to increase the quality. The fouling part in the steam is removed from the system by flow 4. The steam, whose quality is increased, is sent to the high-pressure turbine with flow 5 in order to produce a power output. The steam here is expanded between flow 5 and flow 6, and

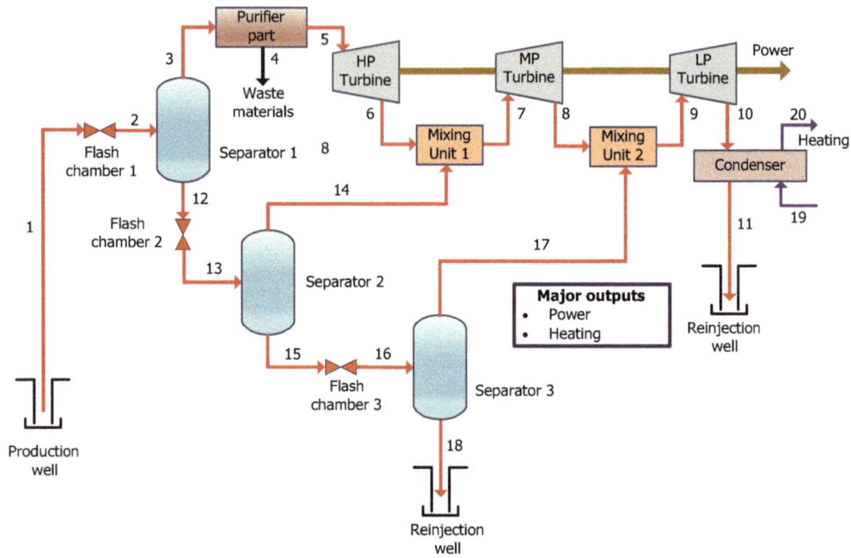

FIGURE 6.9 Simplified flow diagram of the triple-flash steam geothermal power system.

power can be produced. While the process so far is carried out, the fluid coming out of the separator 1 is also transmitted to flash chamber 2 with flow 12. Here, the fluid, which is reduced to pressure and turned into a vapor–liquid mixture, is sent to separator 2 with flow 13.

The steam coming out of separator 2 is sent to the mixing unit with flow 14 to be mixed with the steam obtained from the first flashing process. Then the steam coming from the high-pressure turbine with flow 6 and the steam coming from separator 2 with flow 14 are mixed in the mixing unit. This steam, which is mixed in the mixing unit, is sent to the medium pressure turbine with flow 7 in order to obtain the power output. The steam here is expanded between flow 7 and flow 8, and power is generated. Again, while the process up to this point is taking place, the liquid fluid coming out of separator 2 is sent to flash chamber 3 with flow 15 for the third flashing process. Here, the fluid, which is reduced to pressure and turned into a vapor–liquid mixture, is sent to separator 3 with flow 16. The steam coming out of separator 3 is sent to the mixing unit with flow 17 to be mixed with the steam obtained from the third flashing process. In separator 3, the fluid separated from the mixture as a liquid is sent to the ground with flow 18. Then the steam coming from the medium pressure turbine with flow 8 and the steam coming from separator 3 with flow 17 are mixed in the mixing unit. This steam, which is mixed in the mixing unit, is sent to the low-pressure turbine with flow 9 in order to obtain a power product. The steam here is expanded between flow 9 and flow 10, and power is produced. Then the steam coming out of the turbine is sent to the condenser with flow 10 to be condensed. The condenser is used for the heating output, which is a useful product. The fluid coming out of the condenser is sent to the ground with the number 11 flow. In general, the triple-flash steam geothermal power system works with this flow.

Subplants of the triple-flash steam geothermal power system are also clearly shown in Fig. 6.10. There are generally six main subplants in this system: the flashing process, separation process, HP-turbine expansion process, MP-turbine expansion process, LP-turbine expansion process, and condensing process.

In order to understand the operating logic of the system, the activity diagram of the triple-flash steam geothermal power system is shown in Fig. 6.11. The three flashing processes in the system can be easily seen from the activity diagram of the system. If the activity diagram of the system is also examined with the schematic diagram, the operating logic of the system can be better understood.

6.4.2 Case study 6.3

In this case study, the geothermal energy resource integrated triple-flash steam geothermal power system is analyzed by using thermodynamic assessment. The simplified flow diagram of the triple-flash steam geothermal

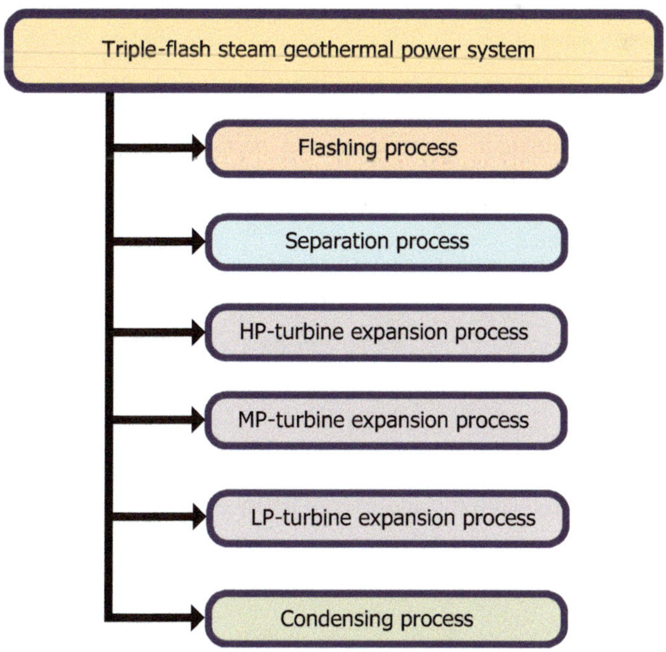

FIGURE 6.10 The subsystems of the triple-flash steam geothermal power system.

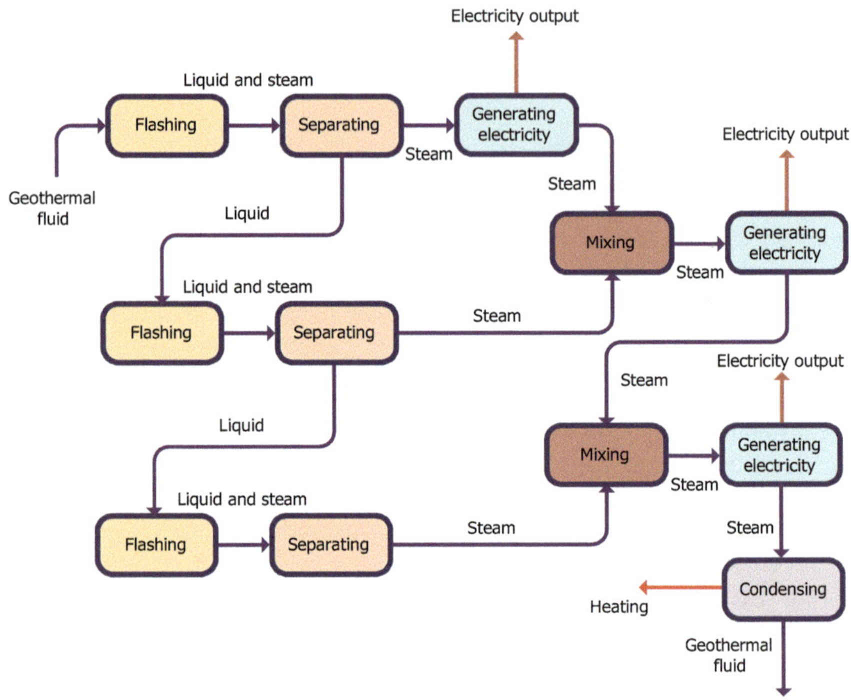

FIGURE 6.11 Activity diagram of electricity and heating generation in the triple-flash steam geothermal power system.

power generation system in this case study is illustrated in Fig. 6.9. As reference conditions, ambient temperature and pressure are taken as 25°C and 101.3 kPa, respectively.

The mass, energy, entropy, and exergy balance equations for the flash chamber 1 are defined:

$$\dot{m}_1 = \dot{m}_2$$

$$\dot{m}_1 h_1 = \dot{m}_2 h_2$$

$$\dot{m}_1 s_1 + \dot{S}_{g,fc1} = \dot{m}_2 s_2$$

$$\dot{m}_1 ex_1 = \dot{m}_2 ex_2 + \dot{E}x_{D,fc1}$$

The mass, energy, entropy, and exergy balance equations for the separator 1 can be defined:

$$\dot{m}_2 = \dot{m}_3 + \dot{m}_{12}$$

$$\dot{m}_2 h_2 = \dot{m}_3 h_3 + \dot{m}_{12} h_{12}$$

$$\dot{m}_2 s_2 + \dot{S}_{g,sep1} = \dot{m}_3 s_3 + \dot{m}_{12} s_{12}$$

$$\dot{m}_2 ex_2 = \dot{m}_3 ex_3 + \dot{m}_{12} ex_{12} + \dot{E}x_{D,sep1}$$

The mass, energy, entropy, and exergy balance equations for the flash chamber 2 can be written:

$$\dot{m}_{12} = \dot{m}_{13}$$

$$\dot{m}_{12} h_{12} = \dot{m}_{13} h_{13}$$

$$\dot{m}_{12} s_{12} + \dot{S}_{gen,fc2} = \dot{m}_{13} s_{13}$$

$$\dot{m}_{12} ex_{12} = \dot{m}_{13} ex_{13} + \dot{E}x_{D,fc2}$$

The mass, energy, entropy, and exergy balance equations for the separator 2 can be obtained:

$$\dot{m}_{13} = \dot{m}_{14} + \dot{m}_{15}$$

$$\dot{m}_{13} h_{13} = \dot{m}_{14} h_{14} + \dot{m}_{15} h_{15}$$

$$\dot{m}_{13} s_{13} + \dot{S}_{g,sep2} = \dot{m}_{14} s_{14} + \dot{m}_{15} s_{15}$$

$$\dot{m}_{13} ex_{13} = \dot{m}_{14} ex_{14} + \dot{m}_{15} ex_{15} + \dot{E}x_{D,sep2}$$

The mass, energy, entropy, and exergy balance equations for the flash chamber 3 are defined:

$$\dot{m}_{15} = \dot{m}_{16}$$

$$\dot{m}_{15} h_{15} = \dot{m}_{16} h_{16}$$

$$\dot{m}_{15} s_{15} + \dot{S}_{g,fc3} = \dot{m}_{16} s_{16}$$

$$\dot{m}_{15} ex_{15} = \dot{m}_{16} ex_{16} + \dot{E}x_{D,fc3}$$

The mass, energy, entropy, and exergy balance equations for the separator 3 can be expressed:

$$\dot{m}_{16} = \dot{m}_{17} + \dot{m}_{18}$$

$$\dot{m}_{16} h_{16} = \dot{m}_{17} h_{17} + \dot{m}_{18} h_{18}$$

$$\dot{m}_{16} s_{16} + \dot{S}_{g,sep3} = \dot{m}_{17} s_{17} + \dot{m}_{18} s_{18}$$

$$\dot{m}_{16} ex_{16} = \dot{m}_{17} ex_{17} + \dot{m}_{18} ex_{18} + \dot{E}x_{D,sep3}$$

The mass, energy, entropy, and exergy balance equations for the purifier part are defined:

$$\dot{m}_3 = \dot{m}_4 + \dot{m}_5$$

$$\dot{m}_3 h_3 = \dot{m}_4 h_4 + \dot{m}_5 h_5$$

$$\dot{m}_3 s_3 + \dot{S}_{g,pp} = \dot{m}_4 s_4 + \dot{m}_5 s_5$$

$$\dot{m}_3 ex_3 = \dot{m}_4 ex_4 + \dot{m}_5 ex_5 + \dot{E}x_{D,pp}$$

The mass, energy, entropy, and exergy balance equations for the high-pressure turbine can be defined:

$$\dot{m}_5 = \dot{m}_6$$

$$\dot{m}_5 h_5 = \dot{m}_6 h_6 + \dot{W}_{HPT}$$

$$\dot{m}_5 s_5 + \dot{S}_{g,HPT} = \dot{m}_6 s_6$$

$$\dot{m}_5 ex_5 = \dot{m}_6 ex_6 + \dot{E}x^W_{HPT} + \dot{E}x_{D,HPT}$$

The mass, energy, entropy, and exergy balance equations for the mixing unit 1 are expressed:

$$\dot{m}_6 + \dot{m}_{14} = \dot{m}_7$$

$$\dot{m}_6 h_6 + \dot{m}_{14} h_{14} = \dot{m}_7 h_7$$

$$\dot{m}_6 s_6 + \dot{m}_{14} s_{14} + \dot{S}_{g,mu1} = \dot{m}_7 s_7$$

$$\dot{m}_6 ex_6 + \dot{m}_{14} ex_{14} = \dot{m}_7 ex_7 + \dot{E}x_{D,mu1}$$

The mass, energy, entropy, and exergy balance equations for the middle-pressure turbine can be obtained:

$$\dot{m}_7 = \dot{m}_8$$

$$\dot{m}_7 h_7 = \dot{m}_8 h_8 + \dot{W}_{MPT}$$

$$\dot{m}_7 s_7 + \dot{S}_{g,MPT} = \dot{m}_8 s_8$$

$$\dot{m}_7 ex_7 = \dot{m}_8 ex_8 + \dot{E}x^W_{MPT} + \dot{E}x_{D,MPT}$$

The mass, energy, entropy, and exergy balance equations for the mixing unit 2 are defined:

$$\dot{m}_8 + \dot{m}_{17} = \dot{m}_9$$

$$\dot{m}_8 h_8 + \dot{m}_{17} h_{17} = \dot{m}_9 h_9$$

$$\dot{m}_8 s_8 + \dot{m}_{17} s_{17} + \dot{S}_{g,mu2} = \dot{m}_9 s_9$$

$$\dot{m}_8 ex_8 + \dot{m}_{17} ex_{17} = \dot{m}_9 ex_9 + \dot{E}x_{D,mu2}$$

The mass, energy, entropy, and exergy balance equations for the low-pressure turbine are defined:

$$\dot{m}_9 = \dot{m}_{10}$$

$$\dot{m}_9 h_9 = \dot{m}_{10} h_{10} + \dot{W}_{LPT}$$

$$\dot{m}_9 s_9 + \dot{S}_{g,LPT} = \dot{m}_{10} s_{10}$$

$$\dot{m}_9 ex_9 = \dot{m}_{10} ex_{10} + \dot{E}x^W_{LPT} + \dot{E}x_{D,LPT}$$

The mass, energy, entropy, and exergy balance equations for the condenser are defined:

$$\dot{m}_{10} = \dot{m}_{11};\ \dot{m}_{19} = \dot{m}_{20}$$

$$\dot{m}_{10} h_{10} + \dot{m}_{19} h_{19} = \dot{m}_{11} h_{11} + \dot{m}_{20} h_{20}$$

$$\dot{m}_{10} s_{10} + \dot{m}_{19} s_{19} + \dot{S}_{g,con} = \dot{m}_{11} s_{11} + \dot{m}_{20} s_{20}$$

$$\dot{m}_{10} ex_{10} + \dot{m}_{19} ex_{19} = \dot{m}_{11} ex_{11} + \dot{m}_{20} ex_{20} + \dot{E}x_{D,con}$$

The assumptions used in the operating conditions of the triple-flash steam geothermal power system are given in Table 6.7. Software code in Engineering Equation Solver (ESS) is also written and used for this geothermal energy–based combined plant in order to analyze its corresponding baseline model using the same methods previously described. Also, the heat and work input/output rate, entropy generation rate, and exergy destruction rates, and also the energy and exergy efficiencies are calculated by using the balance equations and given assumptions.

The exergy destruction rate, dimensionless exergy destruction ratio and exergy efficiency of the triple-flash steam geothermal power generation components are given in Table 6.8. It can be seen that the purifier, mixing rooms and separators have the maximum exergy efficiency. The high-pressure turbine, medium-pressure turbine, low-pressure turbine, flash chamber 1, and flash chamber 2 have the maximum exergy destruction rates. The components that have high-grade heat interaction result in higher exergy destruction rates.

The exergy analysis results showed that the HP, MP, and LP turbines, as well as the flashing subcomponents, are the main sources of irreversibility. The purifier subcomponent has the maximum exergetic performance among the other system components. The exergy efficiencies of separators and mixing rooms in the geothermal process vary between 81.95% and 84.76% and

TABLE 6.7 Operating parameters for the triple-flash steam geothermal power system.

Variables	Values
Geofluid source temperature, T_1	210°C
Geofluid source pressure, P_1	1500 kPa
Geofluid mass flow rate, $\dot{m}_1$	140 kg/s
Separator I inlet pressure, P_2	530 kPa
Separator II inlet pressure, P_{13}	95 kPa
Separator III inlet pressure, P_{16}	50 kPa
HP turbine output pressure, P_6	95 kPa
MP turbine output pressure, P_8	50 kPa
LP turbine output pressure, P_{10}	10 kPa
Geofluid reinjection temperature, T_{11}	45.82°C

between 87.25% and 88.24%, respectively. These exergy efficiencies can be observed to be higher than those of other system components.

To investigate the effectiveness of a triple-flash steam geothermal power system more comprehensively, the parametric studies are given next in order to examine the effects of some significant parameter variables on the power production rate, exergy destruction rate, and exergetic performance. According to the findings, the predominant parameter affecting total power production is the mass flow rate. As seen from Fig. 6.12, as the mass flow rate triples from 100 to 180 kg/s, the power production rate, exergy destruction rate, and exergy efficiency of the investigated system increase from 9591 kW to about 17,263 kW, from 4159 to 7487 kW, and from 52% to 64%, respectively. The increment in geothermal mass flow rate makes the turbine produce more work.

Because higher-temperature fluid transfers more energy to the turbines of the investigated system, the increment in geothermal water temperature has positive effects on both power generation rate and exergetic performance. As seen from Fig. 6.13, the power production rate increases from 8552 to 15,133 kW and exergy efficiency increases from 58.84% to 78.48%. Also, the exergy destruction rate increases from 4480 to 7924 kW.

According to the findings of this case study, the reference temperature is the most significant factor affecting exergy efficiency. The definition of exergy clarifies this increment. As seen from Fig. 6.14, as the reference temperature changes from 0°C to 40°C, the exergy efficiency increases from about 61.91% to 80.83%. Proportionally, the produced power increases about

TABLE 6.8 Thermodynamic assessment results for the triple-flash steam geothermal power generation process components.

System components	Exergy destruction rate (kW)	Exergy destruction ratio (%)	Exergy efficiency (%)
Flash chamber 1	1018.6	12.7	74.52
Separator 1	42.22	0.53	84.76
Flash chamber 2	1110.7	13.8	73.28
Separator 2	75.9	0.95	83.26
Flash chamber 3	407.7	5.08	72.52
Separator 3	87.7	1.09	81.95
Purifier	65.39	0.82	97.25
HP turbine	1564.1	19.5	44.62
Mixing room 1	71.02	0.89	88.24
MP turbine	1413	17.6	41.86
Mixing room 2	69.52	0.87	87.25
LP turbine	1277.4	15.9	40.28
Condenser	818.5	10.2	31.28

6500 kW with the same temperature change. The energy efficiency of the investigated plant is calculated as 22.91%, but energetic performance does not change with increasing the reference temperature.

6.4.3 Quadruple-flash steam geothermal power system

The quadruple-flash steam geothermal power system is created by adding a flashing process to the triple-flash steam geothermal system [9]. In this system, there are four flashing processes in total. Since adding another flashing process to the system requires cost and component requirements in addition to the system, the quadruple-flash steam geothermal power system is more comprehensive and has greater needs than others. The schematic diagram of the quadruple-flash steam geothermal system is shown in Fig. 6.15 in its most general form. Looking at the schematic diagram, the four flashing processes presented in the system can be clearly seen. In the triple-flash steam geothermal power system, the flashing process is completed in separator 3, while in this system it is completed in separator 4. With this system, more

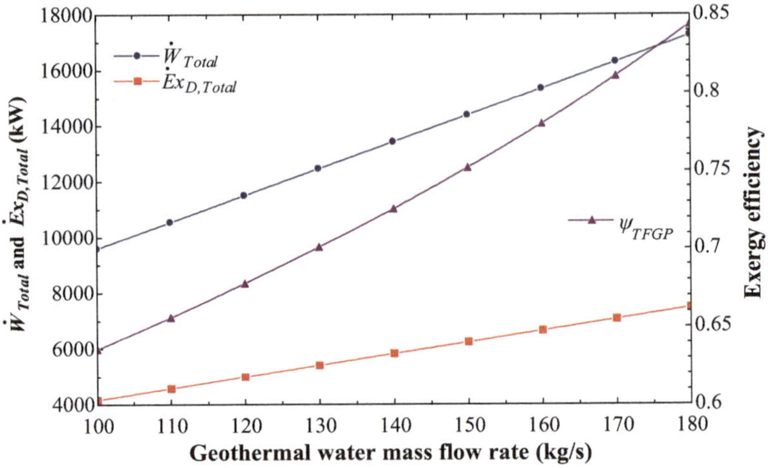

FIGURE 6.12 Effect of the mass flow rate of geothermal working fluid on net power generation and exergy efficiency.

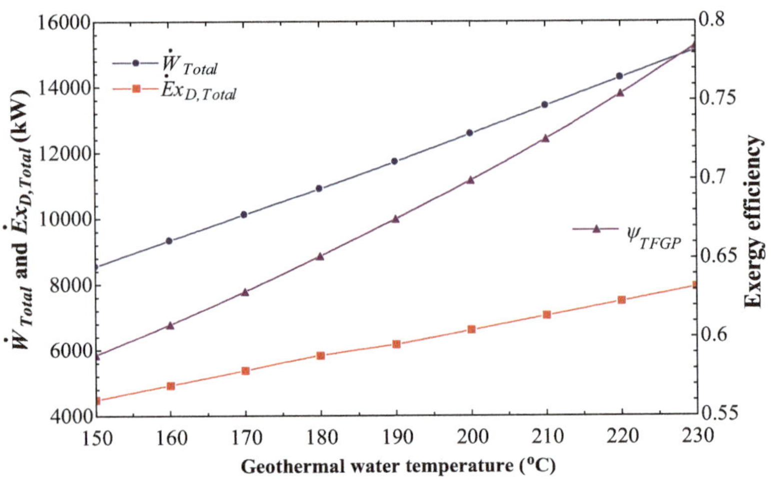

FIGURE 6.13 Effect of geothermal working fluid temperature on net power generation and exergy efficiency.

power is obtained from systems containing other flashing processes from the same amount and temperature of geothermal fluid.

As seen from the schematic diagram of the quadruple-flash steam geothermal power system, the geothermal fluid is sent to flash chamber 1, where the first flashing process will take place with flow 1. Here, the fluid that is depressurized is sent to separator 1 as a liquid–vapor mixture. By using

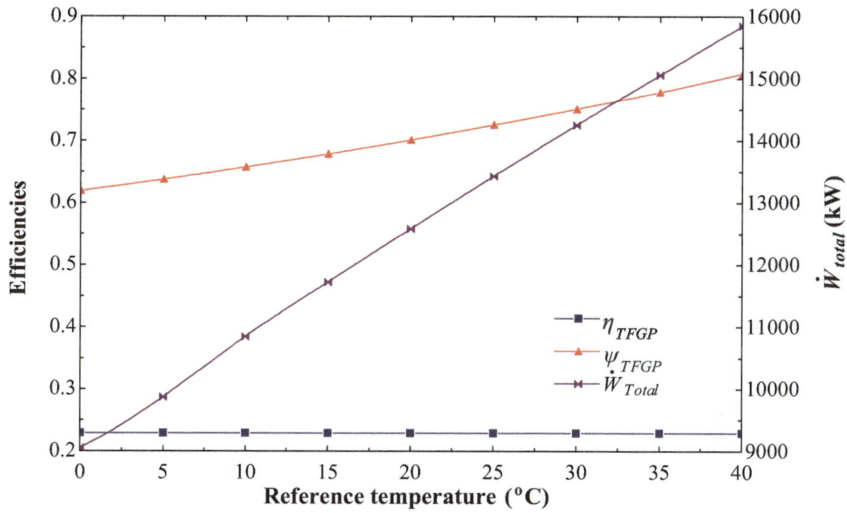

FIGURE 6.14 Effect of ambient temperature on net power generation and exergy efficiency.

separator 1, it is sent to the purifier with flow 3 to increase the steam quality separated from this mixture. Here, the dirty part in the fluid is removed from the system by flow 4. The steam, whose quality is increased, is sent to a very high-pressure turbine in order to produce power with flow 5. The steam here is expanded between flow 5 and flow 6, and power is generated. While the process so far is carried out, the fluid coming out of separator 1 is also transmitted to flash chamber 2 with flow 14. The fluid, which is converted to the liquid–vapor mixture by decreasing its pressure, is sent to separator 2 with flow 15.

The steam coming out of separator 2 is sent to the mixing unit with flow 16 to be mixed with the steam obtained from the first flashing process. Then the steam coming from the very high-pressure turbine with flow 6 and the steam coming from separator 2 with flow 16 are mixed in the mixing unit. This steam, which is mixed in the mixing unit, is sent to the high-pressure turbine with flow 7 in order to obtain a power product. The steam here is expanded between flow 7 and flow 8, and power output is produced. Again, while the process up to this point is taking place, the liquid fluid coming out of separator 2 is sent to flash chamber 3 with flow 17 for the third flashing process. Here, the liquid–vapor mixture is obtained by decreasing the pressure of the fluid. Then this mixture is sent to separator 3 with flow 18 to separate the steam part from this mixture.

The steam coming out of separator 3 is sent to the mixing unit with flow 19 to be mixed with the steam obtained from the third flashing process. Then the steam coming from the high-pressure turbine with flow 8 and the

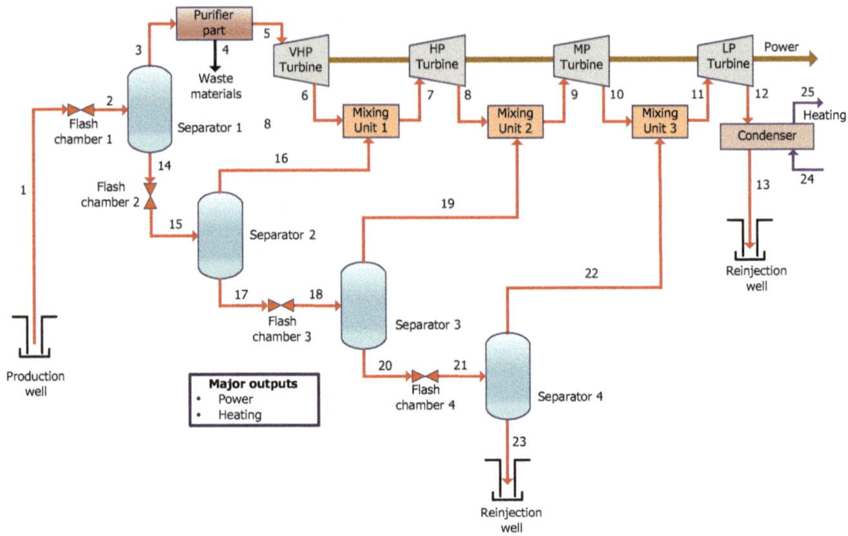

FIGURE 6.15 Simplified flow diagram of the quadruple-flash steam geothermal power system.

steam coming from separator 3 with flow 19 are mixed in the mixing unit. This steam, which is mixed in the mixing unit, is sent to the medium pressure turbine with flow 9 in order to obtain power. The steam here is expanded between flow 9 and 10, and power is generated. Again, while the process up to this point is taking place, the liquid fluid coming out of separator 3 is sent to flash chamber 4 with flow 20 for the fourth flashing process. Here, the liquid–vapor mixture is obtained by decreasing the pressure of the fluid. Then this mixture is sent to separator 4 with flow 21 to separate the steam part from this mixture. The steam coming out of separator 4 is sent to the mixing unit with flow 22 to be mixed with the steam obtained from the fourth flashing process. Then the steam coming from the medium pressure turbine with flow 10 and the steam coming from separator 4 with flow 22 are mixed in the mixing unit. This steam, which is mixed in the mixing unit, is sent to the low-pressure turbine with flow 11 in order to produce power. The steam here is expanded between flow 11 and flow 12, and power is produced. The steam coming out of the turbine is sent to the condenser with flow 12 to be condensed.

The condenser is used for the heating output, which is a useful product. The fluid coming out of the condenser is sent to the ground with the number 13 flow. The logic of the quadruple-flash steam geothermal power system in its most general form is summarized in this way.

Since the quadruple-flash steam geothermal power system is more comprehensive than geothermal power systems with fewer flashing processes, the system may require additional work in its design. Therefore, reviewing studies in the literature and applications while designing the system can provide useful technical information for system design.

The activity diagram of the quadruple-flash steam geothermal power system is useful in order to better understand its working logic. The general activity diagram of this system is given in Fig. 6.16.

The subsystems of the quadruple-flash steam geothermal power system are clearly given in Fig. 6.17. There are generally seven main subplants in this system: the flashing process, separation process, VHP-turbine expansion process, HP-turbine expansion process, MP-turbine expansion process, LP-turbine expansion process, and condensing process.

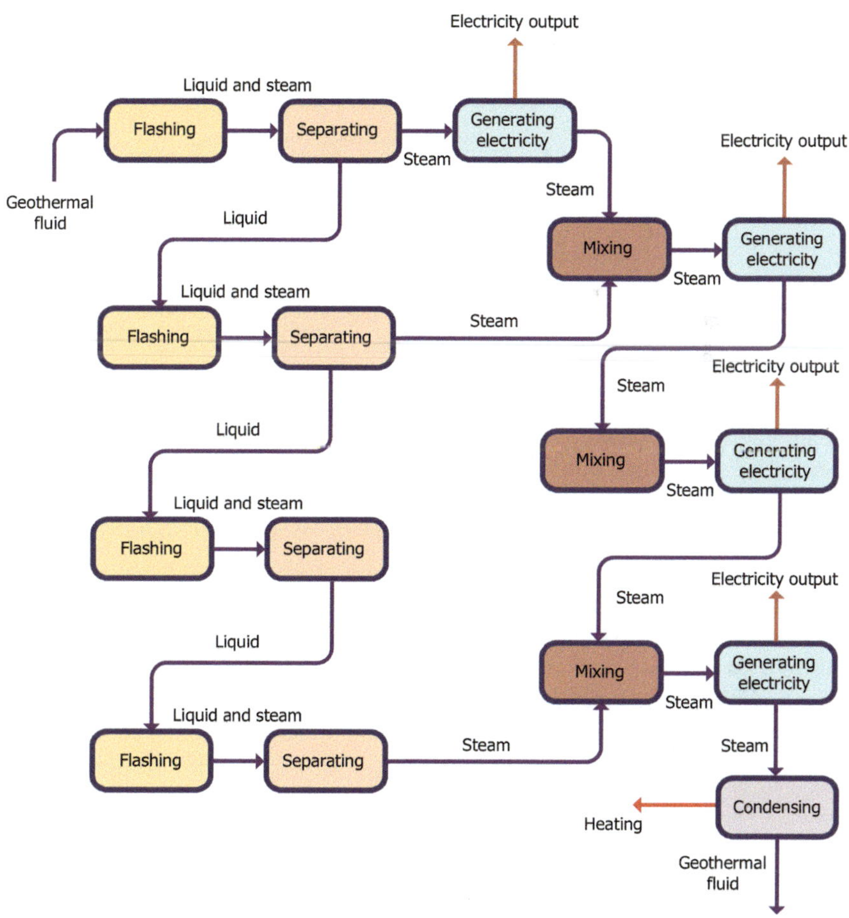

FIGURE 6.16 Activity diagram of electricity and heating generation in the quadruple-flash steam geothermal power system.

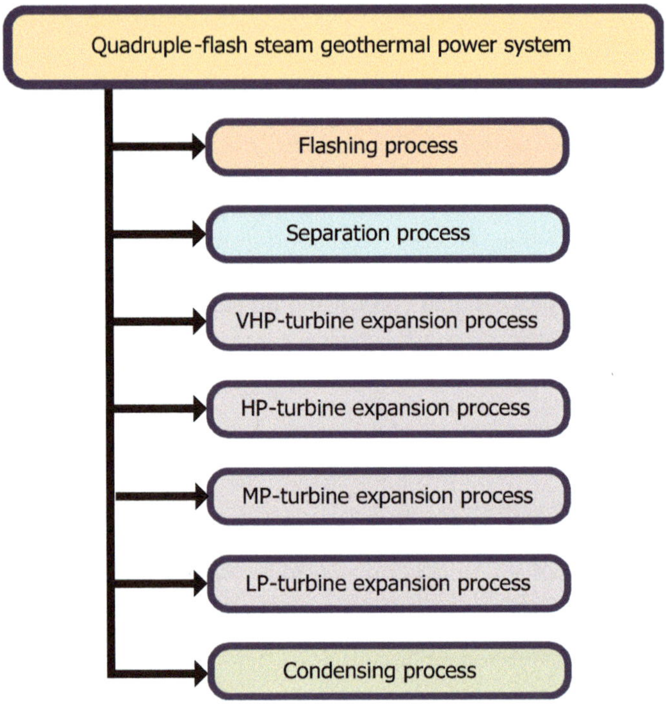

FIGURE 6.17 Subsystems of the quadruple-flash steam geothermal power system.

6.4.4 Case study 6.4

In this case study, the geothermal energy resource integrated quadruple-flash steam geothermal power system is analyzed by using thermodynamic assessment. The simplified flow diagram of the quadruple-flash steam geothermal power generation system is illustrated in Fig. 6.15. As reference conditions, ambient temperature and pressure are taken to be 25°C and 101.3 kPa, respectively. The assumptions used in the operating conditions of the quadruple-flash steam geothermal power system are given in Table 6.9.

The heat and work input/output rate, entropy generation rate, and exergy destruction rates, as well as the energy and exergy efficiencies, are calculated by using the balance equations and given assumptions. The mass, energy, entropy, and exergy balance equations for the flash chamber 1 are defined:

$$\dot{m}_1 = \dot{m}_2$$

$$\dot{m}_1 h_1 = \dot{m}_2 h_2$$

$$\dot{m}_1 s_1 + \dot{S}_{g,fc1} = \dot{m}_2 s_2$$

$$\dot{m}_1 ex_1 = \dot{m}_2 ex_2 + \dot{E}x_{D,fc1}$$

TABLE 6.9 Operating parameters for the quadruple-flash steam geothermal power system.

Variables	Values
Geofluid source temperature, T_1	210°C
Geofluid source pressure, P_1	1500 kPa
Geofluid mass flow rate, $\dot{m}_1$	140 kg/s
Separator 1 inlet pressure, P_2	667 kPa
Separator 2 inlet pressure, P_{15}	119 kPa
Separator 3 inlet pressure, P_{18}	62.92 kPa
Separator 4 inlet pressure, P_{21}	45.86 kPa
VHP turbine output pressure, P_6	400 kPa
HP turbine output pressure, P_8	300 kPa
MP turbine output pressure, P_{10}	200 kPa
LP turbine output pressure, P_{12}	100 kPa
Geofluid reinjection temperature, T_{13}	45.82°C

The mass, energy, entropy, and exergy balance equations for the separator 1 can be expressed:

$$\dot{m}_2 = \dot{m}_3 + \dot{m}_{14}$$

$$\dot{m}_2 h_2 = \dot{m}_3 h_3 + \dot{m}_{14} h_{14}$$

$$\dot{m}_2 s_2 + \dot{S}_{g,sep1} = \dot{m}_3 s_3 + \dot{m}_{14} s_{14}$$

$$\dot{m}_2 ex_2 = \dot{m}_3 ex_3 + \dot{m}_{14} ex_{14} + \dot{E}x_{D,sep1}$$

The mass, energy, entropy, and exergy balance equations for the flash chamber 2 can be obtained:

$$\dot{m}_{14} = \dot{m}_{15}$$

$$\dot{m}_{14} h_{14} = \dot{m}_{15} h_{15}$$

$$\dot{m}_{14} s_{14} + \dot{S}_{gen,fc2} = \dot{m}_{15} s_{15}$$

$$\dot{m}_{14} ex_{14} = \dot{m}_{15} ex_{15} + \dot{E}x_{D,fc2}$$

The mass, energy, entropy, and exergy balance equations for the separator 2 can be written:

$$\dot{m}_{15} = \dot{m}_{16} + \dot{m}_{17}$$

$$\dot{m}_{15} h_{15} = \dot{m}_{16} h_{16} + \dot{m}_{17} h_{17}$$

$$\dot{m}_{15}s_{15} + \dot{S}_{g,sep2} = \dot{m}_{16}s_{16} + \dot{m}_{17}s_{17}$$

$$\dot{m}_{15}ex_{15} = \dot{m}_{16}ex_{16} + \dot{m}_{17}ex_{17} + \dot{E}x_{D,sep2}$$

The mass, energy, entropy, and exergy balance equations for the flash chamber 3 can be defined:

$$\dot{m}_{17} = \dot{m}_{18}$$

$$\dot{m}_{17}h_{17} = \dot{m}_{18}h_{18}$$

$$\dot{m}_{17}s_{17} + \dot{S}_{g,fc3} = \dot{m}_{18}s_{18}$$

$$\dot{m}_{17}ex_{17} = \dot{m}_{18}ex_{18} + \dot{E}x_{D,fc3}$$

The mass, energy, entropy, and exergy balance equations for the separator 3 are expressed:

$$\dot{m}_{18} = \dot{m}_{19} + \dot{m}_{20}$$

$$\dot{m}_{18}h_{18} = \dot{m}_{19}h_{19} + \dot{m}_{20}h_{20}$$

$$\dot{m}_{18}s_{18} + \dot{S}_{g,sep3} = \dot{m}_{19}s_{19} + \dot{m}_{20}s_{20}$$

$$\dot{m}_{18}ex_{18} = \dot{m}_{19}ex_{19} + \dot{m}_{20}ex_{20} + \dot{E}x_{D,sep3}$$

The mass, energy, entropy, and exergy balance equations for the flash chamber 4 are obtained:

$$\dot{m}_{20} = \dot{m}_{21}$$

$$\dot{m}_{20}h_{20} = \dot{m}_{21}h_{21}$$

$$\dot{m}_{20}s_{20} + \dot{S}_{g,fc4} = \dot{m}_{21}s_{21}$$

$$\dot{m}_{20}ex_{20} = \dot{m}_{21}ex_{21} + \dot{E}x_{D,fc4}$$

The mass, energy, entropy, and exergy balance equations for the separator 4 are written:

$$\dot{m}_{21} = \dot{m}_{22} + \dot{m}_{23}$$

$$\dot{m}_{21}h_{21} = \dot{m}_{22}h_{22} + \dot{m}_{23}h_{23}$$

$$\dot{m}_{21}s_{21} + \dot{S}_{g,sep4} = \dot{m}_{22}s_{22} + \dot{m}_{23}s_{23}$$

$$\dot{m}_{21}ex_{21} = \dot{m}_{22}ex_{22} + \dot{m}_{23}ex_{23} + \dot{E}x_{D,sep4}$$

The mass, energy, entropy, and exergy balance equations for the purifier part are defined:

$$\dot{m}_3 = \dot{m}_4 + \dot{m}_5$$

$$\dot{m}_3 h_3 = \dot{m}_4 h_4 + \dot{m}_5 h_5$$

$$\dot{m}_3 s_3 + \dot{S}_{g,pp} = \dot{m}_4 s_4 + \dot{m}_5 s_5$$

$$\dot{m}_3 ex_3 = \dot{m}_4 ex_4 + \dot{m}_5 ex_5 + \dot{E}x_{D,pp}$$

The mass, energy, entropy, and exergy balance equations for the very high-pressure turbine can be expressed:

$$\dot{m}_5 = \dot{m}_6$$

$$\dot{m}_5 h_5 = \dot{m}_6 h_6 + \dot{W}_{VHPT}$$

$$\dot{m}_5 s_5 + \dot{S}_{g,VHPT} = \dot{m}_6 s_6$$

$$\dot{m}_5 ex_5 = \dot{m}_6 ex_6 + \dot{E}x^W_{VHPT} + \dot{E}x_{D,VHPT}$$

The mass, energy, entropy, and exergy balance equations for the mixing unit 1 can be obtained:

$$\dot{m}_6 + \dot{m}_{16} = \dot{m}_7$$

$$\dot{m}_6 h_6 + \dot{m}_{16} h_{16} = \dot{m}_7 h_7$$

$$\dot{m}_6 s_6 + \dot{m}_{16} s_{16} + \dot{S}_{g,mu1} = \dot{m}_7 s_7$$

$$\dot{m}_6 ex_6 + \dot{m}_{16} ex_{16} = \dot{m}_7 ex_7 + \dot{E}x_{D,mu1}$$

The mass, energy, entropy, and exergy balance equations for the high-pressure turbine can be written:

$$\dot{m}_7 = \dot{m}_8$$

$$\dot{m}_7 h_7 = \dot{m}_8 h_8 + \dot{W}_{HPT}$$

$$\dot{m}_7 s_7 + \dot{S}_{g,HPT} = \dot{m}_8 s_8$$

$$\dot{m}_7 ex_7 = \dot{m}_8 ex_8 + \dot{E}x^W_{HPT} + \dot{E}x_{D,HPT}$$

The mass, energy, entropy, and exergy balance equations for the mixing unit 2 can be defined:

$$\dot{m}_8 + \dot{m}_{19} = \dot{m}_9$$

$$\dot{m}_8 h_8 + \dot{m}_{19} h_{19} = \dot{m}_9 h_9$$

$$\dot{m}_8 s_8 + \dot{m}_{19} s_{19} + \dot{S}_{g,mu2} = \dot{m}_9 s_9$$

$$\dot{m}_8 ex_8 + \dot{m}_{19} ex_{19} = \dot{m}_9 ex_9 + \dot{E}x_{D,mu2}$$

The mass, energy, entropy, and exergy balance equations for the middle-pressure turbine are expressed:

$$\dot{m}_9 = \dot{m}_{10}$$

$$\dot{m}_9 h_9 = \dot{m}_{10} h_{10} + \dot{W}_{MPT}$$

$$\dot{m}_9 s_9 + \dot{S}_{g,MPT} = \dot{m}_{10} s_{10}$$

$$\dot{m}_9 ex_9 = \dot{m}_{10} ex_{10} + \dot{E}x^W_{MPT} + \dot{E}x_{D,MPT}$$

The mass, energy, entropy, and exergy balance equations for the mixing unit 3 are obtained:

$$\dot{m}_{10} + \dot{m}_{22} = \dot{m}_{11}$$

$$\dot{m}_{10} h_{10} + \dot{m}_{22} h_{22} = \dot{m}_{11} h_{11}$$

$$\dot{m}_{10} s_{10} + \dot{m}_{22} s_{22} + \dot{S}_{g,mu3} = \dot{m}_{11} s_{11}$$

$$\dot{m}_{10} ex_{10} + \dot{m}_{22} ex_{22} = \dot{m}_{11} ex_{11} + \dot{E}x_{D,mu3}$$

The mass, energy, entropy, and exergy balance equations for the low-pressure turbine are written:

$$\dot{m}_{11} = \dot{m}_{12}$$

$$\dot{m}_{11} h_1 = \dot{m}_{12} h_{12} + \dot{W}_{LPT}$$

$$\dot{m}_{11} s_{11} + \dot{S}_{g,LPT} = \dot{m}_{12} s_{12}$$

$$\dot{m}_{11} ex_{11} = \dot{m}_{12} ex_{12} + \dot{E}x^W_{LPT} + \dot{E}x_{D,LPT}$$

The mass, energy, entropy, and exergy balance equations for the condenser are defined:

$$\dot{m}_{12} = \dot{m}_{13};\ \dot{m}_{24} = \dot{m}_{25}$$

$$\dot{m}_{12} h_{12} + \dot{m}_{24} h_{24} = \dot{m}_{13} h_{13} + \dot{m}_{25} h_{25}$$

$$\dot{m}_{12} s_{12} + \dot{m}_{24} s_{24} + \dot{S}_{g,con} = \dot{m}_{13} s_{13} + \dot{m}_{25} s_{25}$$

$$\dot{m}_{12} ex_{12} + \dot{m}_{24} ex_{24} = \dot{m}_{13} ex_{13} + \dot{m}_{25} ex_{25} + \dot{E}x_{D,con}$$

In order to study the performance of the quadruple-flash steam geothermal power system more comprehensively through the energy and exergy approaches, parametric studies are undertaken here to examine the impacts of different indicator variables on the exergy destruction rate and exergy efficiency. The effect of the geothermal water mass flow rate on the total power production rate, exergy destruction rate, and exergy efficiency of the quadruple-flash steam geothermal power system is shown in Fig. 6.18. It can be shown that the total power production rate increases from 12,076 to 21,736 kW, the exergy destruction rate increases from 5167 to 9138 kW, and the exergy efficiency increases from 0.6583 to 0.8739, with increasing the geothermal water mass flow rate from 100 to 180 kg/s.

The effects of geothermal water temperature on the total power production rate, exergy destruction rate, and exergy efficiency of the quadruple-flash steam geothermal power system are analyzed in this case study, and the results are shown in Fig. 6.19. As given in this figure, the total power production rate increases from 11,161 to 18,662 kW, the exergy destruction rate increases from 4781 to 7847 kW, and the exergy efficiency increases from 0.6497 to 0.7973, with increasing the geothermal water temperature from 150°C to 230°C, respectively.

The effects of reference temperature on the energy and exergy efficiencies, as well as the total power production rate, of the quadruple-flash steam geothermal power system are analyzed in this case study, and study outputs are illustrated in Fig. 6.20. As shown in this figure, the energy efficiency of the quadruple-flash steam geothermal power system does not change with increasing the reference temperature from 0°C to 40°C, whereas the exergy efficiency of the system increases from 0.645 to 0.8371, and the total power production rate increases from 12,358 to 19,163 kW in the chosen reference temperature change.

6.5 Geothermal energy–based multistaged with binary systems

The schematic diagram of a geothermal energy based multistaged with binary system is shown in Fig. 6.21. Explaining the working principle of the system, in general, will be useful for understanding how the system works. In order to eliminate particles in the fluid from the geothermal energy source, the fluid is sent to the particle separator by flow 1. Also, since the pressure in the geothermal energy source is higher than where the particle lifter is located, it is transmitted to the particle lifter after a pressure adjustment. After the particles in the fluid are removed, the fluid coming out of the particle extractor is sent to the moisture separator with the number 2 flow to remove the moisture in the fluid. Here, the moisture released from the fluid is sent to the underground with flow 3. Quality fluid coming from the moisture separator is transferred to the valve with flow 4 to adjust its pressure. For this system to serve its essential purpose, the fluid from the

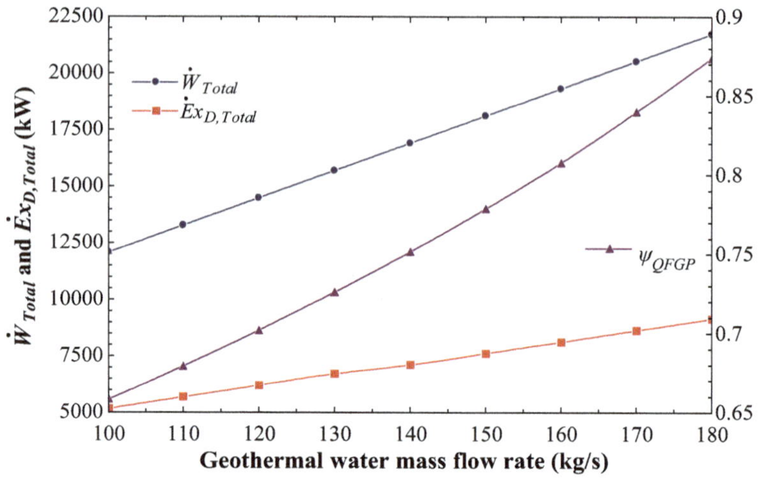

FIGURE 6.18 Effect of the mass flow rate of geothermal working fluid on net power generation and exergy efficiency for the quadruple-flash steam geothermal power system.

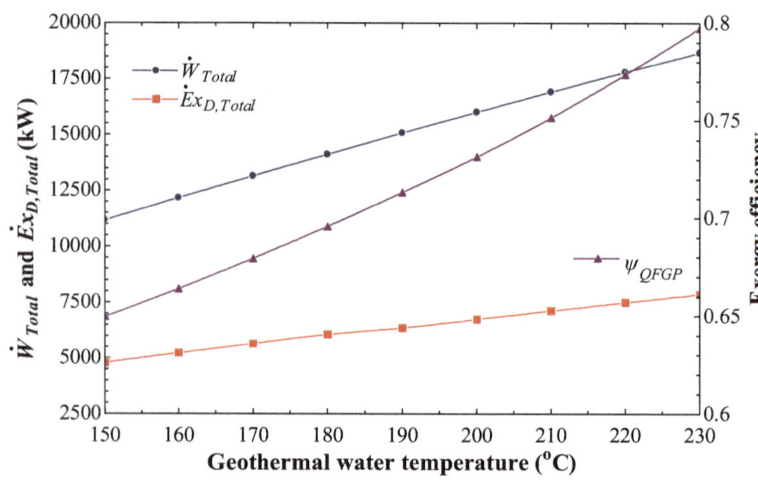

FIGURE 6.19 Effect of geothermal working fluid temperature on net power generation and exergy efficiency for the quadruple-flash steam geothermal power system.

valve is sent to turbine 1 to obtain power by flow 5. After power generation in turbine 1, the expanding steam in turbine 1 is transferred to turbine 2 for electricity generation with flow 6 in order to benefit from dry steam again. After the power generation in turbine 2 is realized, the expanding steam in turbine 2 is transferred to turbine 3 with flow 7 in order to realize electricity generation again. The fluid coming out of turbine 3 is sent to the evaporator for energy transfer with flow 8. The fluid coming out of the evaporator is

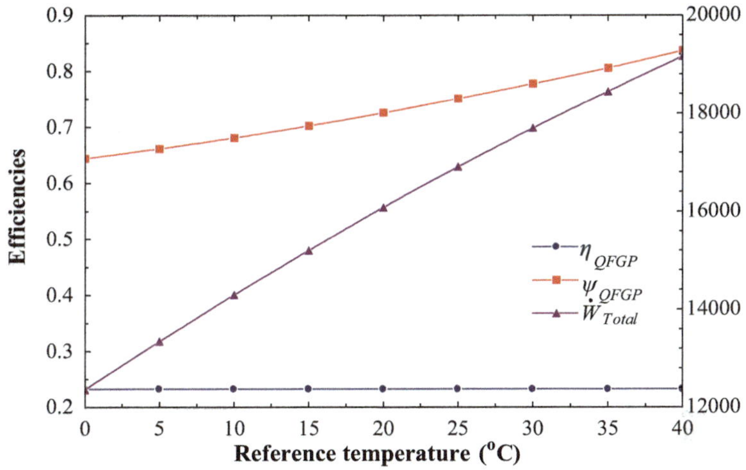

FIGURE 6.20 Effect of ambient temperature on net power generation and exergy efficiency for the quadruple-flash steam geothermal power system.

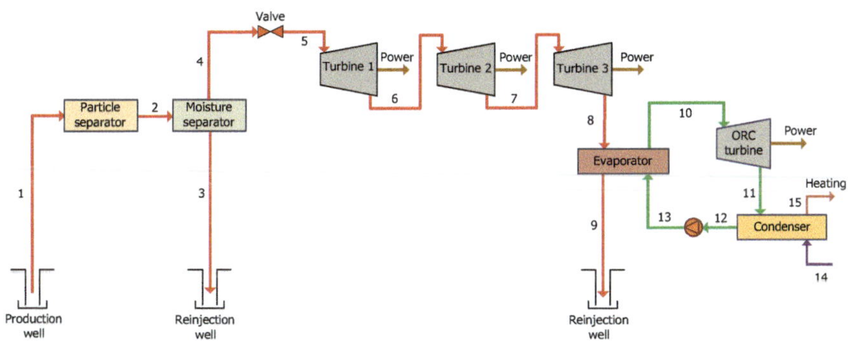

FIGURE 6.21 Simplified flow diagram of a geothermal energy based multistaged with binary system.

sent again to the reinjection well by flow 9. A cycle of the geothermal fluid in a multistaged binary system is completed in this way. The power cycle in which electricity generation is realized in the multistaged binary system is the organic Rankine cycle (ORC) subsystem. The geothermal fluid coming out of turbine 3 is sent to the evaporator with flow 8. Here, the geothermal fluid transfers some of its energy to the ORC working fluid. For this, heat transfer takes place between the geothermal fluid and ORC working fluid. The energized ORC working fluid is sent to the ORC turbine with flow 10 for electricity generation. In the ORC turbine, working fluid is expanded between flows 10 and 11, producing electrical output, a useful product. The working fluid from the ORC turbine is transferred to the condenser with flow 11 for heating output. The working fluid condensed in the condenser is sent to the pump with flow 12 to increase its pressure. The working fluid, whose pressure is increased, is transferred to the evaporator with flow 13 for energy reload. In this way, the ORC subsystem, the power cycle of the multistaged binary system, is also completed. Generally, a cycle of the geothermal based multistaged binary system is completed in this way.

6.5.1 Case study 6.5

In this case study, the geothermal energy resource–based multistaged with binary system is analyzed by using thermodynamic analysis. The simplified flow diagram of the geothermal energy resource–based multistaged with binary system is illustrated in Fig. 6.22. To perform thermodynamic analysis, the ambient temperature and pressure are taken as 25°C and 101.3 kPa, respectively.

The mass, energy, entropy, and exergy balance equations for the parts of the multistaged direct system with three turbines are previously given in Case Study 6.1. In addition to them, the mass, energy, entropy, and exergy balance equations for the ORC subsystem are given in this case study.

The mass, energy, entropy, and exergy balance equations for the evaporator are defined:

$$\dot{m}_8 = \dot{m}_9;\ \dot{m}_{13} = \dot{m}_{10}$$

$$\dot{m}_8 h_8 + \dot{m}_{13} h_{13} = \dot{m}_9 h_9 + \dot{m}_{10} h_{10}$$

$$\dot{m}_8 s_8 + \dot{m}_{13} s_{13} + \dot{S}_{g,eva} = \dot{m}_9 s_9 + \dot{m}_{10} s_{10}$$

$$\dot{m}_8 ex_8 + \dot{m}_{13} ex_{13} = \dot{m}_9 ex_9 + \dot{m}_{10} ex_{10} + \dot{E}x_{D,eva}$$

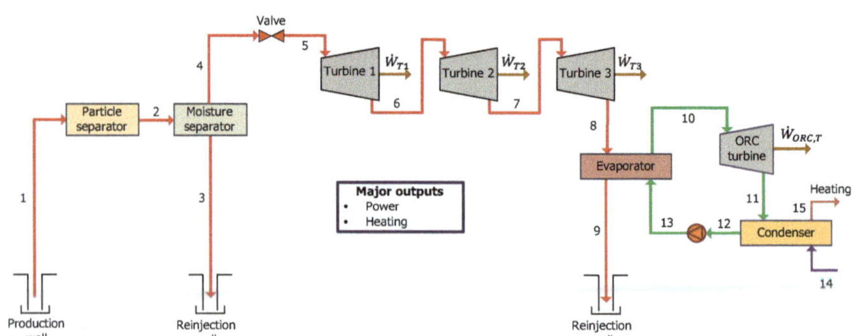

FIGURE 6.22 Schematic diagram of a geothermal based multistaged with binary system.

The mass, energy, entropy, and exergy balance equations for the ORC turbine are described:

$$\dot{m}_{10} = \dot{m}_{11}$$

$$\dot{m}_{10}h_{10} = \dot{m}_{11}h_{11} + \dot{W}_{ORC,T}$$

$$\dot{m}_{10}s_{10} + \dot{S}_{g,ORCT} = \dot{m}_{11}s_{11}$$

$$\dot{m}_{10}ex_{10} = \dot{m}_{11}ex_{11} + \dot{E}x^{W}_{ORC,T} + \dot{E}x_{D,ORCT}$$

The mass, energy, entropy, and exergy balance equations for the condenser are defined:

$$\dot{m}_{11} = \dot{m}_{12};\ \dot{m}_{14} = \dot{m}_{15}$$

$$\dot{m}_{11}h_{11} + \dot{m}_{14}h_{14} = \dot{m}_{12}h_{12} + \dot{m}_{15}h_{15}$$

$$\dot{m}_{11}s_{11} + \dot{m}_{14}s_{14} + \dot{S}_{g,con} = \dot{m}_{12}s_{12} + \dot{m}_{15}s_{15}$$

$$\dot{m}_{11}ex_{11} + \dot{m}_{14}ex_{14} = \dot{m}_{12}ex_{12} + \dot{m}_{15}ex_{15} + \dot{E}x_{D,con}$$

The mass, energy, entropy, and exergy balance equations for the pump can be defined:

$$\dot{m}_{12} = \dot{m}_{13}$$

$$\dot{m}_{12}h_{12} + \dot{W}_{P} = \dot{m}_{13}h_{13}$$

$$\dot{m}_{12}s_{12} + \dot{S}_{g,P} = \dot{m}_{13}s_{13}$$

$$\dot{m}_{12}ex_{12} + \dot{E}x^{W}_{P} = \dot{m}_{13}ex_{13} + \dot{E}x_{D,P}$$

The assumptions used in the working parameters of the geothermal energy—based multistaged with binary system are written in Table 6.10. The working fluid of the ORC subplant is chosen as R32. Also, the heat and work input/output rate, entropy generation rate, exergy destruction rates, and energetic and exergetic efficiencies can be calculated by using the mass, energy, entropy and exergy balance equations and given assumptions.

The energy and exergy efficiency equations for the multistaged direct system with three turbines subsystem are given:

$$\eta_{MSTT} = \frac{\dot{W}_{T1} + \dot{W}_{T2} + \dot{W}_{T3} + \dot{Q}_{Eva}}{\dot{m}_1h_1 - \dot{m}_3h_3 - \dot{m}_9h_9}$$

and

$$\psi_{MSTT} = \frac{\dot{E}x^{W}_{T1} + \dot{E}x^{W}_{T2} + \dot{E}x^{W}_{T3} + \dot{E}x^{Q}_{Eva}}{\dot{m}_1ex_1 - \dot{m}_3ex_3 - \dot{m}_9ex_9}$$

TABLE 6.10 Operating parameters for the geothermal energy–based multistaged with binary system.

Variables	Values
Reference temperature, T_0	25°C
Reference pressure, P_0	101.3 kPa
Geofluid source temperature, T_1	280°C
Geofluid source pressure, P_1	3024 kPa
Geofluid mass flow rate, $\dot{m}_1$	40 kg/s
Turbine 1 inlet pressure, P_5	1680 kPa
Turbine 1 inlet temperature, T_5	215.6°C
Turbine 2 inlet pressure, P_6	1021 kPa
Turbine 2 inlet temperature, T_6	185.4°C
Turbine 3 inlet pressure, P_7	634 kPa
Turbine 3 inlet temperature, T_7	155.8°C
ORC turbine inlet pressure, P_{10}	5780 kPa
ORC turbine inlet temperature, T_{10}	78.2°C
Heating fluid exit pressure, P_{11}	122 kPa
Heating fluid exit temperature, T_{11}	48°C
Geofluid reinjection temperature, T_9	46.27°C

The energy and exergy efficiency equations for the ORC subsystem can be defined:

$$\eta_{ORC} = \frac{\dot{W}_{net,ORC} + \dot{Q}_{Heating}}{\dot{m}_8 h_8 - \dot{m}_9 h_9}$$

and

$$\psi_{ORC} = \frac{\dot{E}x^{W}_{net,ORC} + \dot{E}x^{Q}_{Heating}}{\dot{m}_8 ex_8 - \dot{m}_9 ex_9}$$

where $\dot{W}_{net,ORC}$ and $\dot{Q}_{Heating}$ can be defined respectively:

$$\dot{W}_{net,ORC} = \dot{W}_{ORC} - \dot{W}_P$$

and

$$\dot{Q}_{Heating} = \dot{m}_{15} h_{15} - \dot{m}_{14} h_{14}$$

The energy and exergy efficiency equations for the geothermal energy based multistaged with binary system are given:

$$\eta_{MSBS} = \frac{\dot{W}_{net,MSBS} + \dot{Q}_{Heating}}{\dot{m}_1 h_1 - \dot{m}_3 h_3 - \dot{m}_9 h_9}$$

and

$$\psi_{MSBS} = \frac{\dot{E}x^{W}_{net,MSBS} + \dot{E}x^{Q}_{Heating}}{\dot{m}_1 ex_1 - \dot{m}_3 ex_3 - \dot{m}_9 ex_9}$$

where $\dot{W}_{net,MSBS}$ can be defined as:

$$\dot{W}_{net,MSBS} = \dot{W}_{T1} + \dot{W}_{T2} + \dot{W}_{T3} + \dot{W}_{ORC} - \dot{W}_{P}$$

Based on the balance equations, including energy and exergy efficiency equations, the results of the energy and exergy analyses for the geothermal energy–based multistaged with binary system and its subplants are written in Table 6.11. As given in the table, the energetic and exergetic efficiencies of the overall geothermal energy system are computed as 0.3819 and 0.6524, respectively.

The power and heating production rates by using the turbines of multistaged system with three turbines subsystem and condenser subsystem are written in Table 6.12. Also, the electrical power consumption rate of the ORC pump is calculated as 208 kW. Based on these data, the net power production rate from the geothermal energy–based multistaged system with binary cycle is equal to 26,991 kW.

The effects of geothermal water mass flow rate on the power and heating production rates of the geothermal energy–based multistaged system with binary cycle are shown in Fig. 6.23. As illustrated in the figure, the power production rates for turbine 1, turbine 2, turbine 3, and the ORC turbine increase from 7291 to 9716 kW, from 9410 to 12,184 kW, from 7282 to 9268 kW, and from 954 to 1186 kW, respectively, and the heating

TABLE 6.11 Energy and exergy efficiencies of the geothermal energy–based multistaged system with binary cycle.

Subsystems/overall system	Energy efficiency (%)	Exergy efficiency (%)
Multistaged direct system with three turbines	35.24	61.68
ORC subsystem	26.34	22.95
Overall system	37.68	63.07

TABLE 6.12 Useful output rates from the geothermal energy–based multistaged system with binary cycle.

Useful outputs	Value (kW)
Power production from turbine 1, $\dot{W}_{T1}$	8024
Power production from turbine 2, $\dot{W}_{T2}$	10,257
Power production from turbine 3, $\dot{W}_{T3}$	7892
Power production from ORC turbine, $\dot{W}_{ORC,T}$	1026
Heating production, $\dot{Q}_{Heating}$	9157

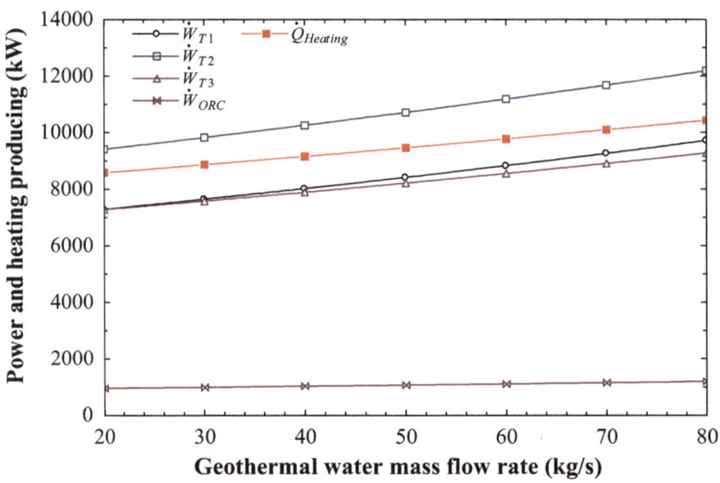

FIGURE 6.23 Impact of the mass flow rate of geothermal working fluid on the power and heating generation rates.

production rate increases from 8581 to 10,426 kW, with increasing the geothermal mass flow rate from 20 to 80 kg/s.

The effect of geothermal water mass flow rate on the energy and exergy efficiencies of the geothermal energy–based multistaged system with binary cycle is illustrated in Fig. 6.24. As given in this figure, the energy efficiencies of the multistaged direct system with three turbines subsystem, the ORC subsystem, and overall geothermal energy plant increase from 34.95% to 35.81%, from 26.18% to 26.65%, and from 37.31% to 38.43%, respectively, as well as the exergy efficiencies of these subsystems and overall plant increase from 60.58% to 63.93%, from 22.58% to 23.69%, and 61.71% to 65.89%, respectively, with increasing the geothermal mass flow rate from 20 to 80 kg/s.

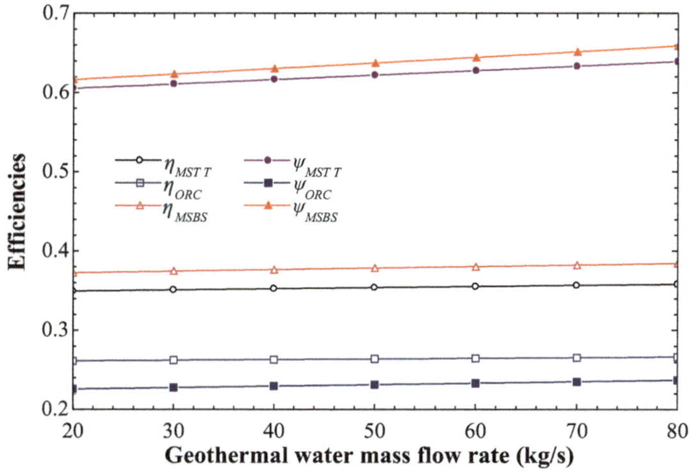

FIGURE 6.24 Impact of the mass flow rate of geothermal working fluid on the energy and exergy efficiencies of the whole system and its subsystems.

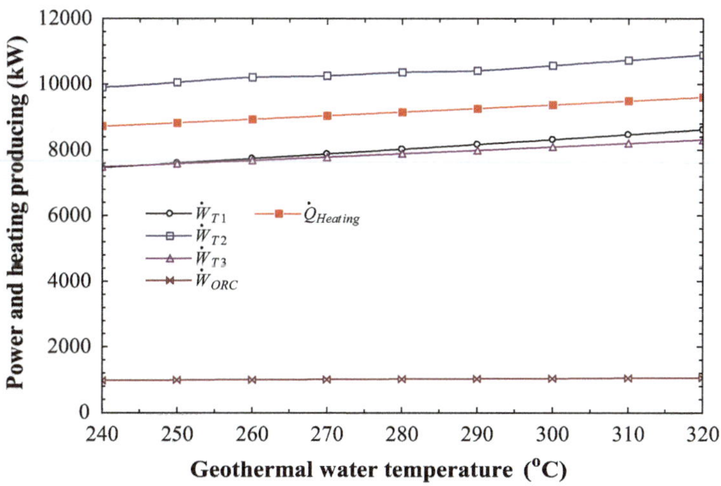

FIGURE 6.25 Impact of the geothermal working fluid temperature on the power and heating generation rates.

The effects of geothermal water temperature on the power production rate and heating production rate of the geothermal energy–based multistaged system with binary cycle are shown in Fig. 6.25. It can be shown that the power production rate for turbine 1, turbine 2, turbine 3, and the ORC turbine increase from 7471 to 8617 kW, from 9909 to 10,886 kW, from 7494 to 8310 kW, and from 989 to 1063 kW, respectively, and the heating

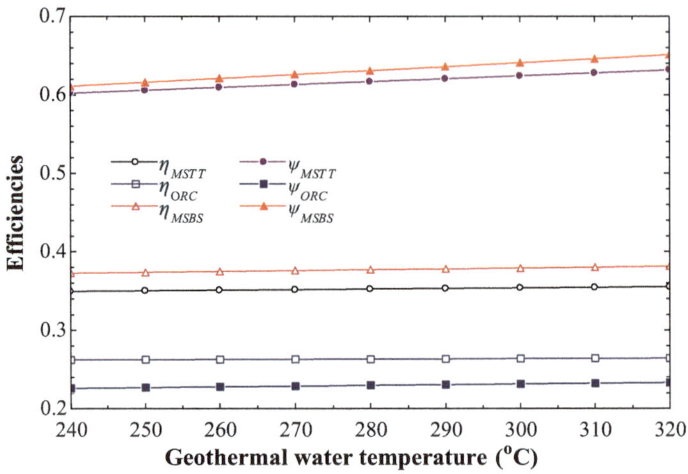

FIGURE 6.26 Impact of the geothermal working water temperature on the energy and exergy efficiencies of the whole system and its subsystems.

production rate increases from 8730 to 9604 kW, with increasing the geothermal water temperature rate from 240°C to 320°C.

The influences of the geothermal working water temperature on the energetic and exergetic efficiencies of the geothermal energy–based multistaged system with binary cycle are better illustrated in Fig. 6.26. The energetic efficiencies of the multistaged direct system with three turbines and ORC subplants, as well as the overall geothermal plant increase from 34.96% to 35.52%, from 26.23% to 26.44%, and from 37.23% to 38.14%, respectively, and the exergy efficiencies of these subplants and overall plant increase from 60.22% to 63.17%, from 22.59% to 23.31%, and 61.09% to 65.11%, respectively, with increasing the geothermal water temperature rate from 240°C to 320°C.

6.6 Geothermal energy–based multiflashing with binary systems

The schematic diagram of a geothermal energy–based triple-flashing with binary system is illustrated in Fig. 6.27. As seen from the schematic diagram of the system, the geothermal fluid coming from underground is sent to flashing chamber 1 where the first flashing takes place with the number 1 flow. Here, the pressure-reduced liquid–vapor mixture is delivered to separator 1 with flow 2. After separating steam from this fluid with separator 1, this steam is sent to the purification section with flow 3 to increase the quality. The fouling part in the steam is removed from the system by flow 4. The steam, whose quality is increased, is sent to the high-pressure turbine with flow 5 in order to produce a power output. The steam here is expanded between flow 5 and flow

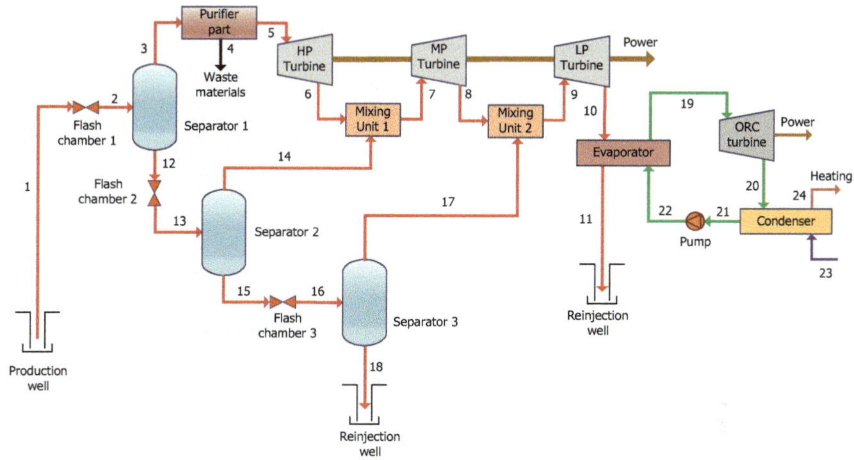

FIGURE 6.27 Simplified flow diagram of a geothermal energy–based triple-flashing with binary system.

6, and power can be produced. While the process so far is carried out, the fluid coming out of separator 1 is also transmitted to flash chamber 2 with flow 12. Here, the fluid, which is reduced in pressure and turned into a vapor–liquid mixture, is sent to separator 2 with flow 13. The steam coming out of separator 2 is sent to the mixing unit with flow 14 to be mixed with the steam obtained from the first flashing process. Then the steam coming from the high-pressure turbine with flow 6 and the steam coming from separator 2 with flow 14 are mixed in the mixing unit. This steam, which is mixed in the mixing unit, is sent to the medium-pressure turbine with flow 7 in order to obtain the power output. The steam here is expanded between flow 7 and flow 8, and power is generated. Again, while the process up to this point is taking place, the liquid fluid coming out of separator 2 is sent to flash chamber 3 with flow 15 for the third flashing process. Here, the fluid, which is reduced in pressure and turned into a vapor–liquid mixture, is sent to separator 3 with flow 16. The steam coming out of separator 3 is sent to the mixing unit with flow 17 to be mixed with the steam obtained from the third flashing process. In separator 3, the fluid separated from the mixture as a liquid is sent to the ground with flow 18. Then the steam coming from the medium-pressure turbine with flow 8 and the steam coming from separator 3 with flow 17 are mixed in the mixing unit. This steam, which is mixed in the mixing unit, is sent to the low-pressure turbine with flow 9 in order to obtain a power product. The steam here is expanded between flow 9 and flow 10, and power is produced. Then, the steam coming out of the turbine is sent to the evaporator for energy transfer with flow 10. In the evaporator, there is an energy transfer between the geothermal fluid and the ORC operating fluid. The fluid coming out of the evaporator is sent to the ground with the 11 flow. In general, a cycle of the geothermal fluid in the geothermal energy–based triple-flashing with binary system is completed in this way.

The power cycle in which electricity generation is realized in the geothermal energy–based triple-flashing with binary system is the ORC subsystem. The geothermal fluid coming out of the LP turbine is sent to the evaporator with flow 10. Here, the geothermal fluid transfers some of its energy to the ORC working fluid. For this purpose, heat transfer takes place between the geothermal fluid and ORC working fluid. The energized ORC working fluid is sent to the ORC turbine with flow 19 for electricity generation. In the ORC turbine, the working fluid is expanded between flows 19 and 20, producing electrical output, a useful product. The working fluid from the ORC turbine is transferred to the condenser with flow 20 for heating output. The working fluid condensed in the condenser is sent to the pump with flow 21 to increase its pressure. The working fluid, whose pressure is increased, is transferred to the evaporator with flow 22 for energy reload. In this way, the ORC subsystem, the power cycle of the geothermal energy–based triple-flashing with binary system, is also completed. Generally, a cycle of the geothermal energy–based triple-flashing with binary system is completed in this way.

6.6.1 Case study 6.6

In this case study, the geothermal energy resource based multiflashing with binary system is modeled by utilizing the energetic and exergetic viewpoints. The simplified flow diagram of this geothermal energy integrated is given in Fig. 6.28. As reference conditions, the ambient temperature and ambient pressure are chosen as 25°C and 101.3 kPa, respectively.

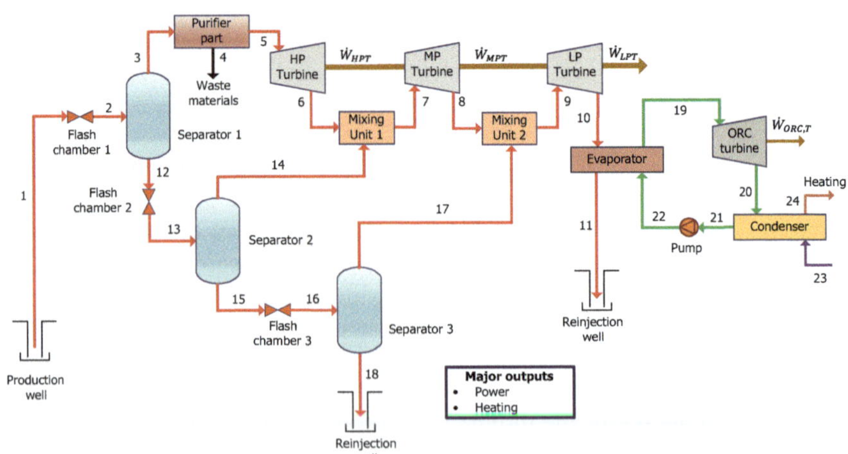

FIGURE 6.28 Schematic diagram of the geothermal energy–based triple-flashing with binary system.

The mass, energy, entropy, and exergy balance equations for the triple-flash steam geothermal power system component are previously given in Case Study 6.3. In addition to them, the mass, energy, entropy, and exergy balance equations for the ORC turbine are defined:

$$\dot{m}_{19} = \dot{m}_{20}$$

$$\dot{m}_{19}h_{19} = \dot{m}_{20}h_{20} + \dot{W}_{ORC,T}$$

$$\dot{m}_{19}s_{19} + \dot{S}_{g,ORC,T} = \dot{m}_{20}s_{20}$$

$$\dot{m}_{19}ex_{19} = \dot{m}_{20}ex_{20} + \dot{E}x^{W}_{ORC,T} + \dot{E}x_{D,ORCT}$$

The mass, energy, entropy, and exergy balance equations for the condenser are defined:

$$\dot{m}_{20} = \dot{m}_{21};\ \dot{m}_{23} = \dot{m}_{24}$$

$$\dot{m}_{20}h_{20} + \dot{m}_{23}h_{23} = \dot{m}_{21}h_{21} + \dot{m}_{24}h_{24}$$

$$\dot{m}_{20}s_{20} + \dot{m}_{23}s_{23} + \dot{S}_{g,con} = \dot{m}_{21}s_{21} + \dot{m}_{24}s_{24}$$

$$\dot{m}_{20}ex_{20} + \dot{m}_{23}ex_{23} = \dot{m}_{21}ex_{21} + \dot{m}_{24}ex_{24} + \dot{E}x_{D,con}$$

The mass, energy, entropy, and exergy balance equations for the pump are described:

$$\dot{m}_{21} = \dot{m}_{22}$$

$$\dot{m}_{21}h_{21} + \dot{W}_{P} = \dot{m}_{22}h_{22}$$

$$\dot{m}_{21}s_{21} + \dot{S}_{g,P} = \dot{m}_{22}s_{22}$$

$$\dot{m}_{21}ex_{21} + \dot{E}x^{W}_{P} = \dot{m}_{22}ex_{22} + \dot{E}x_{D,P}$$

The assumptions used in the operating conditions of the geothermal energy–based triple-flashing with binary system are shown in Table 6.13. The working fluid of the ORC subplant is R245fa. Also, the heat and work input/output rate, entropy generation rate, exergy destruction rates, and energetic and exergetic efficiencies are computed by utilizing the balance equations and chosen assumptions.

The energy and exergy efficiency equations for the triple-flashing steam geothermal energy plant are given:

$$\eta_{TFGP} = \frac{\dot{W}_{HPT} + \dot{W}_{MPT} + \dot{W}_{LPT} + \dot{Q}_{Eva}}{\dot{m}_{1}h_{1} - \dot{m}_{11}h_{11} - \dot{m}_{18}h_{18}}$$

TABLE 6.13 Operating parameters for the geothermal energy–based triple-flashing with binary system.

Variables	Values
Geofluid source temperature, T_1	230°C
Geofluid source pressure, P_1	1800 kPa
Geofluid mass flow rate, $\dot{m}_1$	140 kg/s
Separator I inlet pressure, P_2	630 kPa
Separator II inlet pressure, P_{13}	125 kPa
Separator III inlet pressure, P_{16}	65 kPa
HP turbine output pressure, P_6	105 kPa
MP turbine output pressure, P_8	55 kPa
LP turbine output pressure, P_{10}	15 kPa
ORC turbine inlet temperature, P_{13}	1000 kPa
ORC turbine inlet pressure, T_{13}	84.65°C
Geofluid reinjection temperature, T_{11}	41.96°C

and

$$\psi_{TFGP} = \frac{\dot{E}x^W_{HPT} + \dot{E}x^W_{MPT} + \dot{E}x^W_{LPT} + \dot{E}x^Q_{Eva}}{\dot{m}_1 ex_1 - \dot{m}_{11} ex_{11} - \dot{m}_{18} ex_{18}}$$

The energy and exergy efficiency equations for the ORC subsystem can be defined:

$$\eta_{ORC} = \frac{\dot{W}_{net,ORC} + \dot{Q}_{Heating}}{\dot{m}_{10} h_{10} - \dot{m}_{11} h_{11}}$$

and

$$\psi_{ORC} = \frac{\dot{E}x^W_{net,ORC} + \dot{E}x^Q_{Heating}}{\dot{m}_{10} ex_{10} - \dot{m}_{11} ex_{11}}$$

where $\dot{W}_{net,ORC}$ and $\dot{Q}_{Heating}$ can be defined respectively:

$$\dot{W}_{net,ORC} = \dot{W}_{ORC} - \dot{W}_P$$

and

$$\dot{Q}_{Heating} = \dot{m}_{24} h_{24} - \dot{m}_{23} h_{23}$$

The energy and exergy efficiency equations for the geothermal energy–based triple-flashing steam with binary system are given:

$$\eta_{TFBS} = \frac{\dot{W}_{net,TFBS} + \dot{Q}_{Heating}}{\dot{m}_1 h_1 - \dot{m}_{11} h_{11} - \dot{m}_{18} h_{18}}$$

and

$$\psi_{TFBS} = \frac{\dot{E}x^W_{net,TFBS} + \dot{E}x^Q_{Heating}}{\dot{m}_1 ex_1 - \dot{m}_{11} ex_{11} - \dot{m}_{18} ex_{18}}$$

where $\dot{W}_{net,TFBS}$ can be defined as:

$$\dot{W}_{net,TFBS} = \dot{W}_{HPT} + \dot{W}_{MPT} + \dot{W}_{LPT} + \dot{W}_{ORC} - \dot{W}_P$$

Based on this procedure, the results of the energy and exergy analyses for the geothermal energy–based triple-flashing with binary system and its subsystems are given in Table 6.14. As shown in this table, the energy efficiency and exergy efficiency of the overall system are calculated as 0.3819 and 0.6524, respectively.

The power and heating production rates by utilizing the turbines and condenser subplants are illustrated in Table 6.15. The electrical energy consumption of the pump is computed as 243 kW. Based on these data, the net power production rate from the geothermal energy–based triple-flashing with binary system is equal to 25,382 kW.

To investigate the effectiveness of the geothermal energy–based triple-flashing with binary system more comprehensively, the parametric studies are given here to examine the effects of some important indicators on the power and heating production rates, energy and exergy efficiencies of the overall system and its subsystems. As seen from Fig. 6.29, when the mass flow rate of the geothermal energy–based triple-flashing with binary system increases from 100 to 180 kg/s, the power production rate of the high-pressure turbine, the medium-pressure turbine, the low-pressure turbine, and

TABLE 6.14 Energy and exergy efficiencies of the geothermal energy–based triple-flashing with binary system

Subsystems/overall system	Energy efficiency (%)	Exergy efficiency (%)
Triple-flashing steam geothermal plant	31.57	59.05
ORC subsystem	28.16	24.39
Overall system	38.19	65.24

TABLE 6.15 Useful output rates from the geothermal energy–based triple-flashing with binary system.

Useful outputs	Value (kW)
Power production from high-pressure turbine, $\dot{W}_{HPT}$	10,928
Power production from medium-pressure turbine, $\dot{W}_{MPT}$	8146
Power production from low-pressure turbine, $\dot{W}_{LPT}$	4725
Power production from ORC turbine, $\dot{W}_{ORC,T}$	1826
Heating production, $\dot{Q}_{Heating}$	12,584

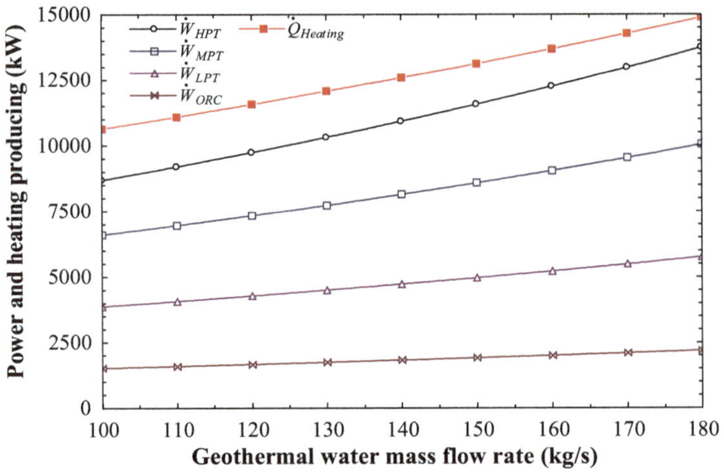

FIGURE 6.29 Effect of the mass flow rate of geothermal working fluid on the power and heating generation rates.

the ORC turbine of the investigated plant increase from 8688 kW to about 13,744 kW, from 6600 to 10,053 kW, from 3872 to 5765 kW, and from 1519 to 2194 kW, respectively, and the heating producing rate is increased from 10,633 to 14,892 kW. Based on these results, it can be said that the increment in geothermal mass flow rate makes the turbine produce more work.

The effects of geothermal mass flow rate on the energy and exergy efficiencies of the overall plant and its subsystems are illustrated accordingly in Fig. 6.30. As seen in this figure, when the mass the flow rate of the geothermal energy–based triple-flashing with binary system increases from 100 to 180 kg/s, the energy efficiencies of the triple-flash subsystem, the ORC subsystem, and the overall system increase from 31.06% to 32.07%, from 27.82% to 28.49%, and from 37.43% to 38.95%, respectively. In addition,

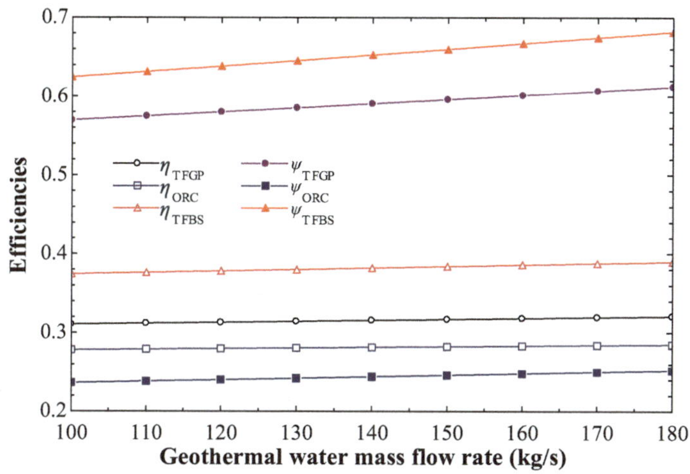

FIGURE 6.30 Effect of the mass flow rate of geothermal working fluid on the energy and exergy efficiencies of whole system and its subsystems.

the exergy efficiencies of the triple-flash subsystem, the ORC subsystem, and the overall system increase from 56.97% to 61.2%, from 23.62% to 25.17%, and from 62.44% to 68.15%, respectively.

Because higher-temperature geothermal water transfers more energy to the turbines of the system, the increment in geothermal water temperature has positive effects on the power and heating generation rates. As seen from Fig. 6.31, the power production rates of the high-pressure turbine, medium-pressure turbine, low-pressure turbine, and ORC turbine of the investigated plant increase from 9706 kW to about 11,733 kW, from 7342 to 8670 kW, from 4321 to 4984 kW, and from 1702 to 1904 kW, respectively, and the heating production rate is increased from 11,623 to 13,197 kW.

The effects of geothermal temperature on the energy and exergy efficiencies of the overall plant and its subsystems are illustrated in Fig. 6.32. As observed here, when the geothermal water temperature of the geothermal energy–based triple-flashing with binary system increases from 180°C to 260°C kg/s, the energy efficiencies of the triple-flash subsystem, the ORC subsystem, and the overall system increase from 31.25% to 31.75%, from 28.01% to 28.24% and from 37.62% to 38.53%, respectively. In addition, the exergy efficiencies of the triple-flash subsystem, the ORC subsystem, and the overall system increase from 57.31% to 60.11%, from 23.9% to 24.68%, and from 62.69% to 66.81%, respectively.

6.7 Geothermal energy–based combined/integrated system

In geothermal-based combined/integrated systems, besides power cycles, there are subsystems where useful outputs such as heating, cooling,

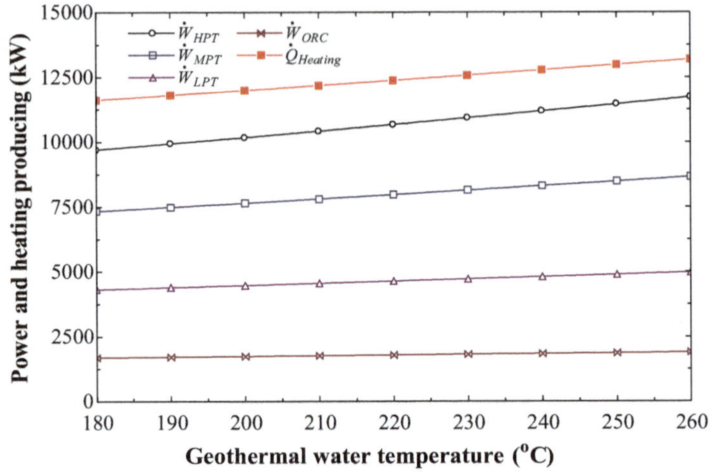

FIGURE 6.31 Effect of the geothermal water temperature on the power and heating generation rates.

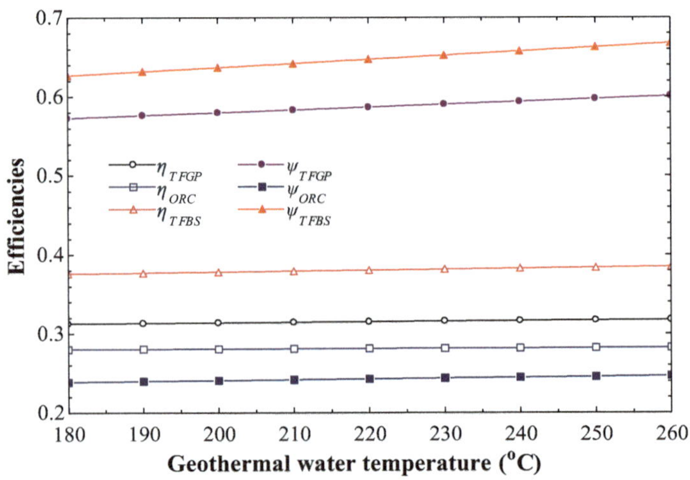

FIGURE 6.32 Effect of the geothermal water temperature on the energy and exergy efficiencies of the whole system and its subsystems.

freshwater, hydrogen, and ammonia can be produced. Since the geothermal energy source can be used in the production of different useful outputs in these systems, these systems have an important potential in terms of performance and sustainability.

6.7.1 Combined/integrated system for power and freshwater production

Geothermal-based combined/integrated systems can provide different useful outputs such as heating, cooling, freshwater, hydrogen, and ammonia. The desalination of seawater by utilizing geothermal energy sources could provide the potential to meet increasing freshwater demand. Studies on systems developed using geothermal energy to desalinate seawater are ongoing. This section describes desalination processes and various possible combinations.

Desalination processes can be grouped under two headings: processes in which either water or salt is collected first (Fig. 6.33). The processes in which water is collected can also be grouped under two headings: single-phase and phase-change processes. Phase-change processes are the multistage-flash (MSF), multieffect distillation (MED), mechanical vapor compressor, thermal vapor compressor, solar still, membrane distillation, passive vacuum desalination, humidification–dehumidification, freezing–melting, absorption heat pump desalination, and adsorption heat pump desalination. Single-phase processes can be classified as reverse osmosis (RO) and forward osmosis. The processes by which salt is collected are electrodialysis, capacitive deionization, and ion exchange. Global distribution of the installed seawater desalination capacities by utilizing technology is shown in Fig. 6.34. Among these desalination processes, MED, MSF, and RO account for about 94% of global desalination capacity, as can be seen from Fig. 6.34.

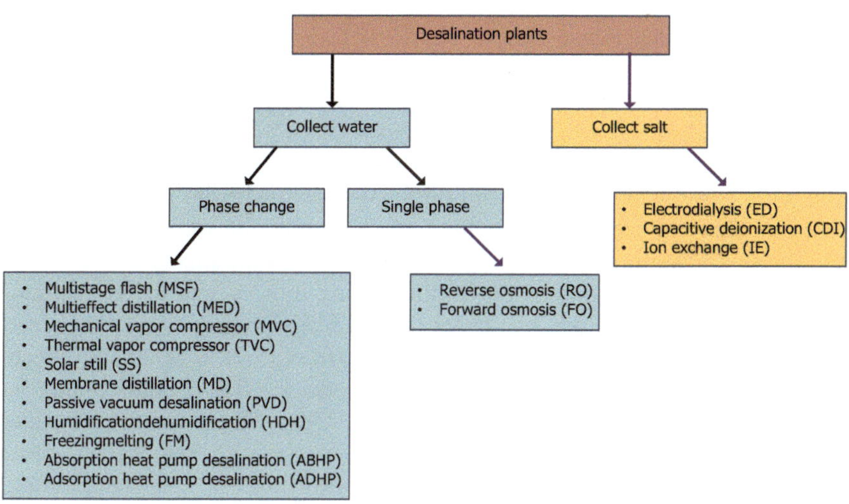

FIGURE 6.33 Grouped potential desalination plants.

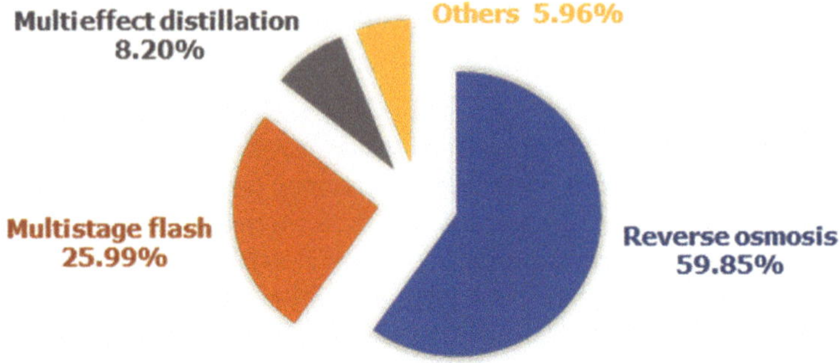

FIGURE 6.34 Global distribution of installed seawater desalination capacity by technology. *Data from World Bank [10].*

6.7.1.1 Reverse osmosis desalination unit combined with a geothermal energy system

As shown in Fig. 6.35, the RO desalination unit occurs with an evaporator, low- and high-pressure pumps, a filter, chemical processing, two three-way valves, RO, a valve, a Pelton turbine, and a freshwater storage tank. As seen from the schematic diagram of the system, the geothermal fluid coming from underground is sent to flashing chamber 1 with the number 1 flow, where the first flashing takes place. Here, the pressure-reduced liquid–vapor mixture is delivered to separator 1 with flow 2. After separating steam from this fluid with separator 1, the steam is sent to the purification section with flow 3 to increase the quality. The fouling part in the steam is removed from the system by flow 4. The steam, whose quality is increased, is sent to the high-pressure turbine with flow 5 in order to produce the power output. The steam here is expanded between flow 5 and flow 6, and power can be produced. While the process so far is carried out, the fluid coming out of separator 1 is also transmitted to flash chamber 2 with flow 12. Here, the fluid, which is reduced in pressure and turned into a vapor–liquid mixture, is sent to separator 2 with flow 13. The steam coming out of separator 2 is sent to the mixing unit with flow 14 to be mixed with the steam obtained from the first flashing process. Then the steam coming from the high-pressure turbine with flow 6 and the steam coming from separator 2 with flow 14 are mixed in the mixing unit. This steam, which is mixed in the mixing unit, is sent to the medium-pressure turbine with flow 7 in order to obtain the power output. The steam here is expanded between flow 7 and flow 8, and power is generated. Again, while the process up to this point is taking place, the liquid fluid coming out of separator 2 is sent to flash chamber 3 with flow 15 for the third flashing process. Here, the fluid, which is reduced in pressure and turned into a vapor–liquid mixture, is sent to separator 3 with flow 16. The steam coming out of separator 3 is sent to the mixing unit with flow 17 to be mixed with

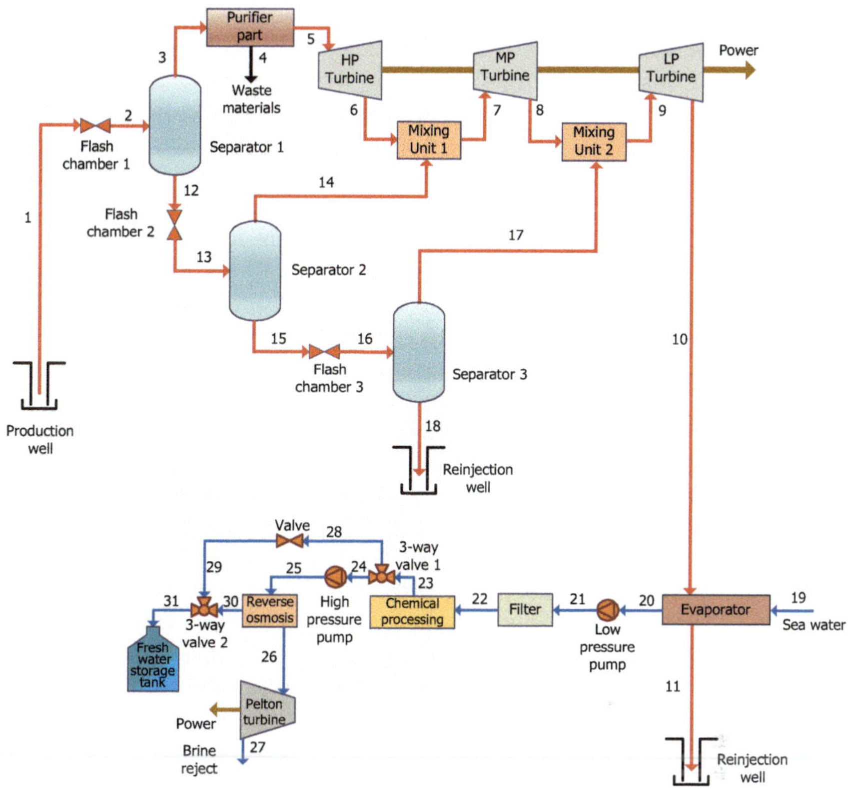

FIGURE 6.35 Simplified flow diagram of a geothermal energy–based power production including freshwater through a reverse osmosis desalination subsystem.

the steam obtained from the third flashing process. In separator 3, the fluid separated from the mixture as a liquid is sent to the ground with flow 18. Then the steam coming from the medium-pressure turbine with flow 8 and the steam coming from separator 3 with flow 17 are mixed in the mixing unit. This steam, once mixed in the mixing unit, is sent to the low-pressure turbine with flow 9 in order to obtain a power product. The steam here is expanded between flow 9 and flow 10, and power is produced. Then the steam coming out of the turbine is sent to the evaporator for energy transfer with flow 10. In the evaporator, an energy transfer occurs between geothermal fluid and seawater. The fluid coming out of the evaporator is sent to the ground with the number 11 flow. In general, a cycle of geothermal fluid in geothermal energy–based power production, including freshwater through the RO desalination subsystem, is completed in this way.

The cycle in which freshwater production is realized in the system is the RO desalination subsystem. In this subsystem, seawater is first transferred to the evaporator with flow 19. By using the low-pressure pump, the seawater coming to this subsystem is sent to the filter component where the seawater will be filtered with flow 20. Various particles that can bring a load to the components in this subsystem are separated from the seawater by filtration. Seawater, which passes through the filtering process, is sent to the chemical process with flow 21. Freshwater from this chemical treatment component is sent to three-way valve 2 with flow 22. A certain amount of freshwater from chemical processing is transferred to the high-pressure pump with flow 24. Here, the freshwater, with increased pressure, is sent to the RO component with flow 25. The high-pressure freshwater coming to the RO component is passed through semipermeable membranes. Almost all impurities in the water are purified. To take advantage of the residue remaining from the treatment, this impurities-containing residue fluid is sent to the Pelton turbine with flow 26 for electrical output. Electricity is produced by expanding this fluid between flows 26 and 27 in the Pelton turbine. In this way, the fluid containing impurities coming from the purification process is driven from the turbine to the outside of the system. While the chemical processing up to this point is taking place, the remaining part of the freshwater from chemical processing is sent to the valve with flow 28. By using this valve, the pressure of the water is reduced. Freshwater, with reduced pressure, is sent to three-way valve 2 with flow 29. This freshwater is mixed with the freshwater coming from the RO component with flow 30 and sent to the freshwater storage tank with flow 31. Freshwater in the storage tank can be used for various needs. Generally, a cycle of the geothermal energy–based power production including freshwater through a RO desalination subsystem is completed in this way.

6.7.2 Case study 6.7

In this case study, the geothermal energy resource–based power production plant including freshwater through RO desalination subsystem is analyzed according to energy and exergy analyses. The simplified flow diagram of this geothermal energy–based power and freshwater production plant is shown in Fig. 6.36. For this case study, the ambient temperature and pressure are taken as 25°C and 101.3 kPa, respectively.

The mass, energy, entropy, and exergy balance equations for the evaporator are defined:

$$\dot{m}_{10} = \dot{m}_{11};\ \dot{m}_{19} = \dot{m}_{20}$$

$$\dot{m}_{10}h_{10} + \dot{m}_{19}h_{19} = \dot{m}_{11}h_{11} + \dot{m}_{20}h_{20}$$

$$\dot{m}_{10}s_{10} + \dot{m}_{19}s_{19} + \dot{S}_{g,eva} = \dot{m}_{11}s_{11} + \dot{m}_{20}s_{20}$$

$$\dot{m}_{10}ex_{10} + \dot{m}_{19}ex_{19} = \dot{m}_{11}ex_{11} + \dot{m}_{20}ex_{20} + \dot{E}x_{D,eva}$$

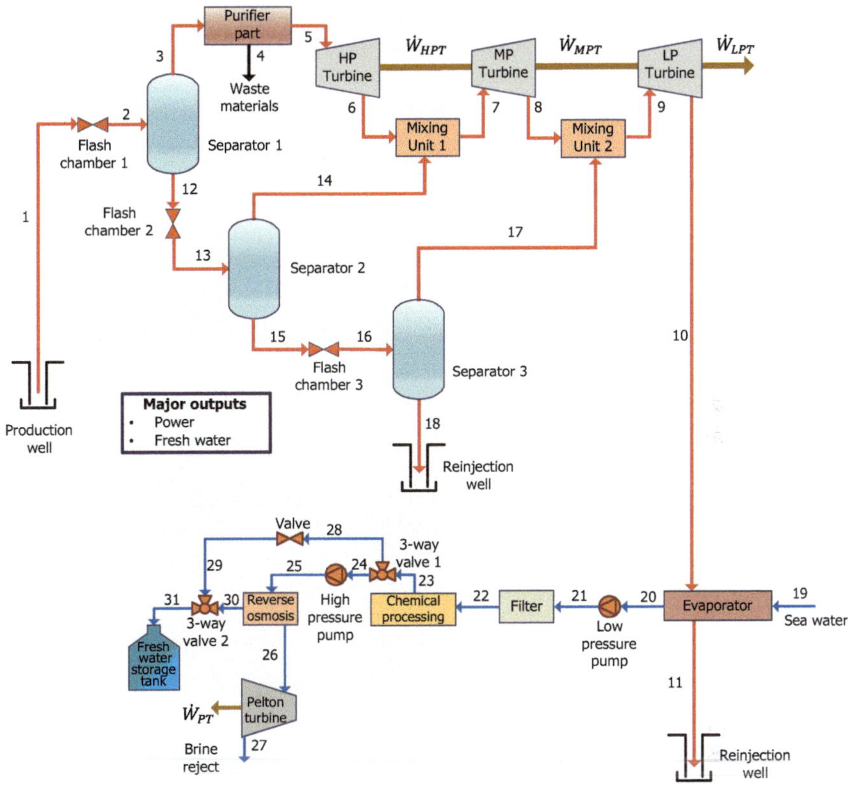

FIGURE 6.36 Schematic diagram of the geothermal energy–based power production including freshwater through a reverse osmosis desalination subsystem.

The mass, energy, entropy, and exergy balance equations for the low-pressure pump are defined:

$$\dot{m}_{20} = \dot{m}_{21}$$

$$\dot{m}_{20}h_{20} + \dot{W}_{LPP} = \dot{m}_{21}h_{21}$$

$$\dot{m}_{20}s_{20} + \dot{S}_{g,LPP} = \dot{m}_{21}s_{21}$$

$$\dot{m}_{20}ex_{20} + \dot{E}x_{LPP}^{W} = \dot{m}_{21}ex_{21} + \dot{E}x_{D,LPP}$$

The mass, energy, entropy, and exergy balance equations for the filter are defined:

$$\dot{m}_{21} = \dot{m}_{22}$$

$$\dot{m}_{21}h_{21} = \dot{m}_{22}h_{22}$$

$$\dot{m}_{21}s_{21} + \dot{S}_{g,fl} = \dot{m}_{22}s_{22}$$

$$\dot{m}_{21}ex_{21} = \dot{m}_{22}ex_{22} + \dot{E}x_{D,fl}$$

The mass, energy, entropy, and exergy balance equations for the chemical processing can be given as:

$$\dot{m}_{22} = \dot{m}_{23}$$

$$\dot{m}_{22}h_{22} = \dot{m}_{23}h_{23}$$

$$\dot{m}_{22}s_{22} + \dot{S}_{g,cf} = \dot{m}_{23}s_{23}$$

$$\dot{m}_{22}ex_{22} = \dot{m}_{23}ex_{23} + \dot{E}x_{D,cf}$$

The mass, energy, entropy, and exergy balance equations for the three-way valve 1 can be defined:

$$\dot{m}_{23} = \dot{m}_{24} + \dot{m}_{28}$$

$$\dot{m}_{23}h_{23} = \dot{m}_{24}h_{24} + \dot{m}_{28}h_{28}$$

$$\dot{m}_{23}s_{23} + \dot{S}_{g,3wv1} = \dot{m}_{24}s_{24} + \dot{m}_{28}s_{28}$$

$$\dot{m}_{23}ex_{23} = \dot{m}_{24}ex_{24} + \dot{m}_{28}ex_{28} + \dot{E}x_{D,3wv1}$$

The mass, energy, entropy, and exergy balance equations for the valve can be written:

$$\dot{m}_{28} = \dot{m}_{29}$$

$$\dot{m}_{28}h_{28} = \dot{m}_{29}h_{29}$$

$$\dot{m}_{28}s_{28} + \dot{S}_{g,val} = \dot{m}_{29}s_{29}$$

$$\dot{m}_{28}ex_{28} = \dot{m}_{29}ex_{29} + \dot{E}x_{D,val}$$

The mass, energy, entropy, and exergy balance equations for the high-pressure pump are defined:

$$\dot{m}_{24} = \dot{m}_{25}$$

$$\dot{m}_{24}h_{24} + \dot{W}_{HPP} = \dot{m}_{25}h_{25}$$

$$\dot{m}_{24}s_{24} + \dot{S}_{g,HPP} = \dot{m}_{25}s_{25}$$

$$\dot{m}_{24}ex_{24} + \dot{E}x^{W}_{HPP} = \dot{m}_{25}ex_{25} + \dot{E}x_{D,HPP}$$

The mass, energy, entropy, and exergy balance equations for the RO are written:

$$\dot{m}_{25} = \dot{m}_{26} + \dot{m}_{30}$$

$$\dot{m}_{25}h_{25} = \dot{m}_{26}h_{26} + \dot{m}_{30}h_{30}$$

$$\dot{m}_{25}s_{25} + \dot{S}_{g,ro} = \dot{m}_{26}s_{26} + \dot{m}_{30}s_{30}$$

$$\dot{m}_{25}ex_{25} = \dot{m}_{26}ex_{26} + \dot{m}_{30}ex_{30} + \dot{E}x_{D,ro}$$

The mass, energy, entropy, and exergy balance equations for the Pelton turbine can be defined:

$$\dot{m}_{26} = \dot{m}_{27}$$

$$\dot{m}_{26}h_{26} = \dot{m}_{27}h_{27} + \dot{W}_{PT}$$

$$\dot{m}_{26}s_{26} + \dot{S}_{g,PT} = \dot{m}_{27}s_{27}$$

$$\dot{m}_{26}ex_{26} = \dot{m}_{27}ex_{27} + \dot{E}x_{PT}^{W} + \dot{E}x_{D,PT}$$

The mass, energy, entropy, and exergy balance equations for the three-way valve 2 are written:

$$\dot{m}_{29} + \dot{m}_{30} = \dot{m}_{31}$$

$$\dot{m}_{29}h_{29} + \dot{m}_{30}h_{30} = \dot{m}_{31}h_{31}$$

$$\dot{m}_{29}s_{29} + \dot{m}_{30}s_{30} + \dot{S}_{g,3wv2} = \dot{m}_{31}s_{31}$$

$$\dot{m}_{29}ex_{29} + \dot{m}_{30}ex_{30} = \dot{m}_{31}ex_{31} + \dot{E}x_{D,3wv2}$$

The several assumptions utilized in the working conditions of the geothermal energy–based power production including freshwater through a RO desalination subsystem are written in Table 6.16.

The energy and exergy efficiency equations for the triple-flashing steam geothermal energy plant can be given as:

$$\eta_{TFGP} = \frac{\dot{W}_{HPT} + \dot{W}_{MPT} + \dot{W}_{LPT} + \dot{Q}_{Eva}}{\dot{m}_1h_1 - \dot{m}_{11}h_{11} - \dot{m}_{18}h_{18}}$$

and

$$\psi_{TFGP} = \frac{\dot{E}x_{HPT}^{W} + \dot{E}x_{MPT}^{W} + \dot{E}x_{LPT}^{W} + \dot{E}x_{Eva}^{Q}}{\dot{m}_1ex_1 - \dot{m}_{11}ex_{11} - \dot{m}_{18}ex_{18}}$$

TABLE 6.16 Operating parameters for the geothermal energy–based power production including freshwater through a reverse osmosis desalination subsystem.

Variables	Values
Geofluid source temperature, T_1	210°C
Geofluid source pressure, P_1	1750 kPa
Geofluid mass flow rate, $\dot{m}_1$	90 kg/s
Separator I inlet pressure, P_2	620 kPa
Separator II inlet pressure, P_{13}	115 kPa
Separator III inlet pressure, P_{16}	60 kPa
HP turbine output pressure, P_6	100 kPa
MP turbine output pressure, P_8	54 kPa
LP turbine output pressure, P_{10}	14 kPa
Seawater entering temperature, T_{19}	18°C
Seawater exiting temperature from evaporator, T_{19}	42°C
Geofluid reinjection temperature, T_{11}	52.73°C

The energy and exergy efficiency equations for the RO desalination subsystem are defined:

$$\eta_{RODS} = \frac{\dot{m}_{fw}h_{31} + \dot{W}_{net,RODS}}{\dot{m}_{19}h_{19} - \dot{m}_{27}h_{27}}$$

and

$$\psi_{RODS} = \frac{\dot{m}_{fw}ex_{31} + \dot{E}x^{W}_{net,RODS}}{\dot{m}_{19}ex_{19} - \dot{m}_{27}ex_{27}}$$

Here, net power production from the distillation subsystem can be calculated as:

$$\dot{W}_{net,RODS} = \dot{W}_{PT} - \dot{W}_{LPT} - \dot{W}_{HPT}$$

The energy and exergy efficiency equations for the geothermal energy–based power production including freshwater through the RO desalination subsystem are given:

$$\eta_{TFDP} = \frac{\dot{W}_{net,TFDP} + \dot{m}_{fw}h_{31}}{\dot{m}_1h_1 - \dot{m}_{11}h_{11} - \dot{m}_{18}h_{18}}$$

and

$$\psi_{TFDP} = \frac{\dot{E}x^{W}_{net,TFDP} + \dot{m}_{fw}ex_{31}}{\dot{m}_1 ex_1 - \dot{m}_{11} ex_{11} - \dot{m}_{18} ex_{18}}$$

where $\dot{W}_{net,TFDP}$ can be defined as:

$$\dot{W}_{net,TFDP} = \dot{W}_{HPT} + \dot{W}_{MPT} + \dot{W}_{LPT} + \dot{W}_{PT} - \dot{W}_{HPP} + \dot{W}_{LPP} - \dot{W}_{P}$$

Based on the thermodynamic analysis balance equations, the results of the energy and exergy analyses for the geothermal energy–based power production plant including freshwater through a RO desalination subsystem are listed in Table 6.17. As illustrated in this table, the energy and exergy efficiencies of the whole system are computed as 0.3819 and 0.6524, respectively.

The power production rates and mass flow rate of producing freshwater by utilizing high-pressure, medium-pressure, low-pressure, and Penton turbines and RO desalination subsystem are listed in Table 6.18. Also, the electrical energy consumption of low- and high-pressure pumps are computed as 28.7 and 204.4 kW, respectively. Based on these data, the net power production rate from this system is equal to 21,110 kW.

Some energy and exergy values depend on the intensive properties of the dead-state condition, such as reference temperature. Therefore, the energy and exergy analysis results generally are sensitive to variations in these properties. In this case study, it is seen that the reference temperature has a significant effect on both efficiencies, especially on the exergy efficiency of the geothermal energy–based system. As given in Fig. 6.37, the energy and exergy efficiencies of the geothermal energy resource–based power production plant including freshwater through a RO desalination subsystem are increased from 36.88% to 38.99% and from 43.25% to 46.47%, respectively.

TABLE 6.17 Energy and exergy efficiencies of the geothermal energy–based power production plant including freshwater through a reverse osmosis desalination subsystem.

Subsystems/overall system	Energy efficiency (%)	Exergy efficiency (%)
Triple-flashing steam geothermal plant	30.24	43.86
Reverse osmosis desalination subsystem	28.16	24.39
Overall system	38.19	45.24

TABLE 6.18 Useful output rates from the geothermal energy–based power production including freshwater through a reverse osmosis desalination subsystem.

Useful outputs	Value
Power production by using high-pressure turbine, $\dot{W}_{HPT}$	8976 kW
Power production by using medium-pressure turbine, $\dot{W}_{MPT}$	6904 kW
Power production by using low-pressure turbine, $\dot{W}_{LPT}$	4438 kW
Power production by using Penton turbine, $\dot{W}_{PT}$	1025 kW
Mass flow rate of producing freshwater, $\dot{m}_{FW}$	18.43 kg/s

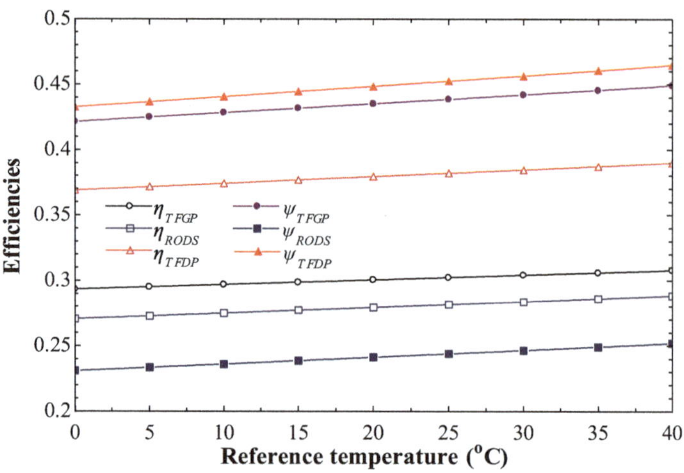

FIGURE 6.37 Effect of the reference temperature on the energy and exergy efficiencies of the whole system and its subsystems.

The effect of reference temperature on the power production and freshwater production rates from the geothermal energy resource–based power production plant including freshwater through a RO desalination subsystem are shown in Fig. 6.38. It is seen from the figure that, by increasing dead-state temperature from 0°C to 40°C, as expected, the power and freshwater production rate increases.

The geothermal water mass flow rate is an important parameter that affects the energy and exergy efficiencies of the geothermal energy resource–based power production plant including freshwater through a RO desalination subsystem. As seen from Fig. 6.39, the energy and exergy efficiencies

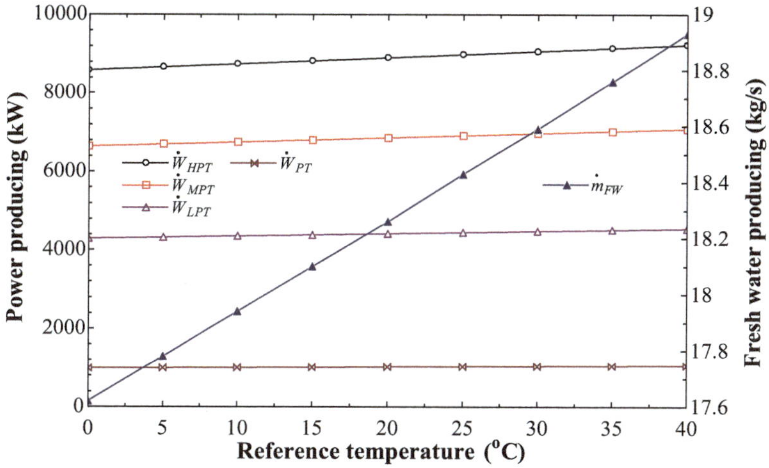

FIGURE 6.38 Effect of the reference temperature on the power and freshwater production rates.

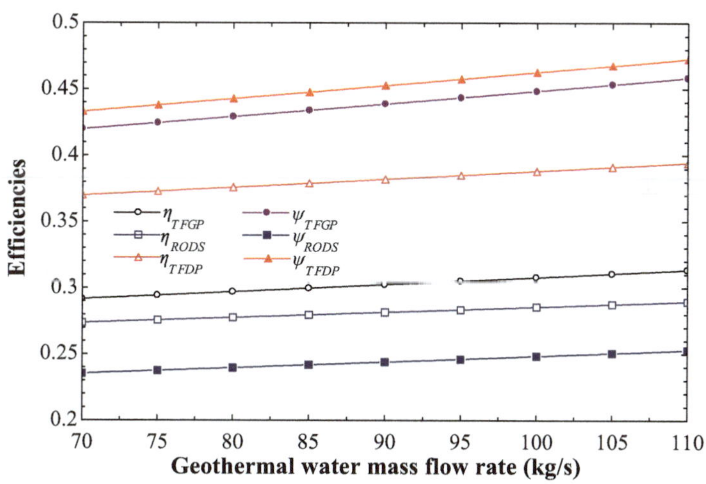

FIGURE 6.39 Effect of the geothermal water mass flow rate on the energy and exergy efficiencies of the whole system and its subsystems.

of the geothermal energy plant are increased from 36.99% to 39.42% and from 43.29% to 47.36%, respectively.

From Fig. 6.40, it can be interpreted that increasing the geothermal water mass flow rate increases the power production and freshwater production rates. The total power production rate increases from 20,307 to 22,428 kW, and freshwater production increases from 17.85 to 19.02 kg/s in the range of 70–110 kg/s.

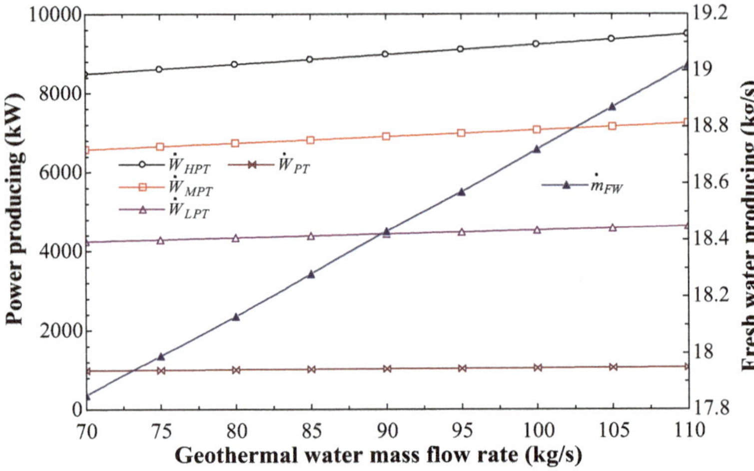

FIGURE 6.40 Effect of the geothermal water mass flow rate on the power and freshwater production rates.

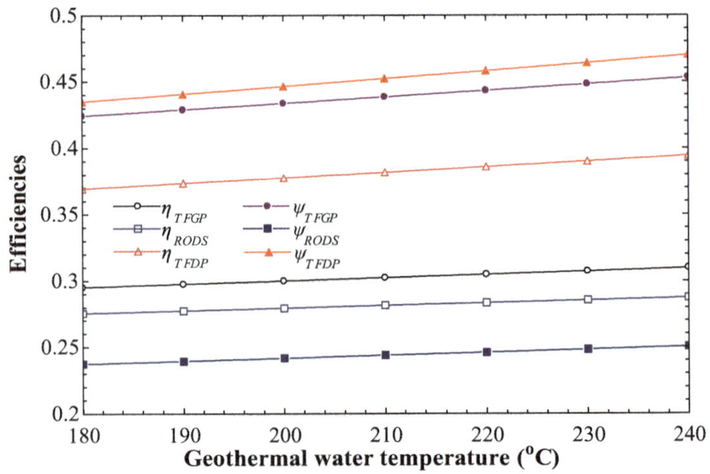

FIGURE 6.41 Effect of the geothermal water temperature on the energy and exergy efficiencies of the whole system and its subsystems.

In Fig. 6.41, the variation of energy and exergy efficiencies of geothermal energy–based combined plant and its subplants are given with respect to geothermal water temperature. As given in this figure, the energy and exergy efficiencies of the geothermal energy resource–based power production plant including freshwater through a RO desalination subsystem increase from 36.95% to 39.46% and from 43.52% to 47.03%, respectively.

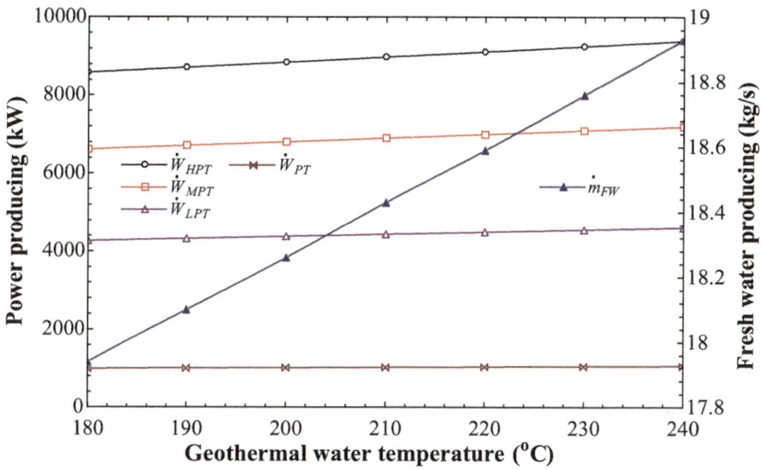

FIGURE 6.42 Effect of the geothermal water temperature on power and freshwater production rates.

Fig. 6.42 shows that power and freshwater production rates are affected by the geothermal water temperature, as expected. Total power and freshwater production rates increase from 20,464 to 22,256 and from 17.94 to 18.93 kg/s, respectively.

6.7.2.1 Reverse double osmosis distillation unit combined with a geothermal energy system

The schematic diagram of a geothermal energy–based power production including freshwater through a double RO desalination plant is given in Fig. 6.43. As seen from the schematic diagram of the quadruple-flash steam geothermal power system, the geothermal fluid is sent to flash chamber 1, where the first flashing process takes place with flow 1. Here, the depressurized fluid is sent to separator 1 as a liquid–vapor mixture. By using separator 1, it is sent to the purifier with flow 3 to increase the steam quality separated from this mixture. Here, the dirty part in the fluid is removed from the system by flow 4. The steam, whose quality is increased, is sent to a very high-pressure turbine in order to produce power with flow 5. The steam here is expanded between flow 5 and flow 6, and power is generated. While the process so far is carried out, the fluid coming out of separator 1 is also transmitted to the flash chamber 2 with flow 14. The fluid, which is converted to the liquid–vapor mixture by decreasing its pressure, is sent to separator 2 with flow 15. The steam coming out of separator 2 is sent to the mixing unit with flow 16 to be mixed with the steam obtained from the first flashing process. Then the steam coming from the very high-pressure turbine

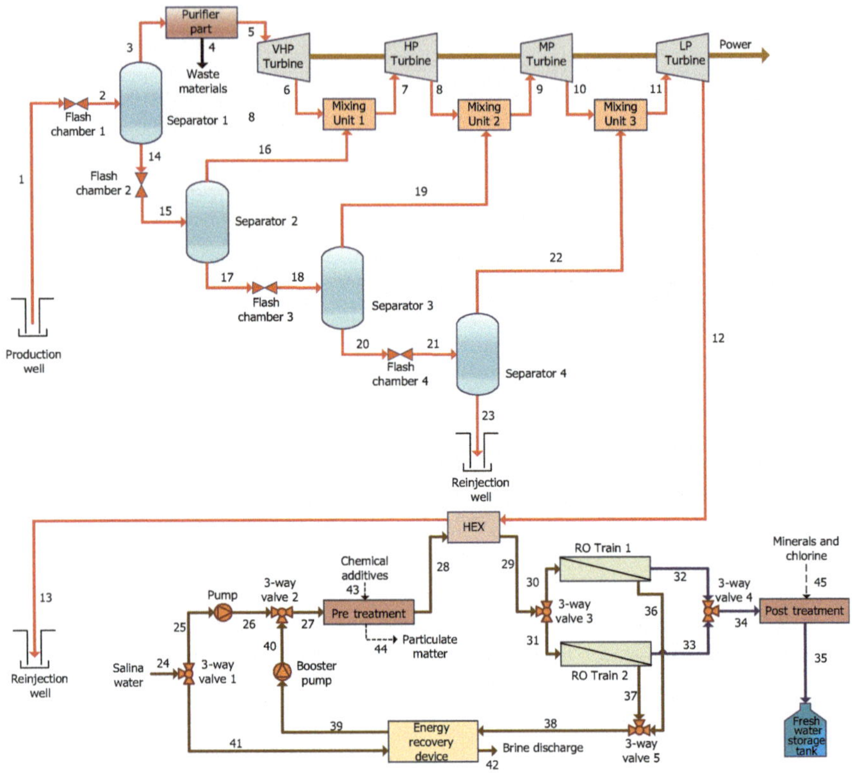

FIGURE 6.43 Simplified flow diagram of geothermal energy–based power production including freshwater through a double reverse osmosis desalination plant.

with flow 6 and the steam coming from separator 2 with flow 16 are mixed in the mixing unit. This steam, which is mixed in the mixing unit, is sent to the high-pressure turbine with flow 7 in order to obtain a power product. The steam here is expanded between flow 7 and flow 8, and power output is produced. Again, while the process up to this point is taking place, the liquid fluid coming out of separator 2 is sent to flash chamber 3 with flow 17 for the third flashing process. Here, the liquid–vapor mixture is obtained by decreasing the pressure of the fluid. Then this mixture is sent to separator 3 with flow 18 to separate the steam part from this mixture. The steam coming out of separator 3 is sent to the mixing unit with flow 19 to be mixed with the steam obtained from the third flashing process. Then the steam coming from the high-pressure turbine with flow 8 and the steam coming from separator 3 with the flow 19 are mixed in the mixing unit. This steam, which is mixed in the mixing unit, is sent to the medium-pressure turbine with flow 9 in order to obtain power. The steam here is expanded between flow 9 and flow 10, and power is generated. Again, while the process up to this point is taking place, the liquid fluid coming out of separator 3 is sent to flash chamber 4 with flow 20 for the fourth flashing process. Here, the liquid–vapor mixture is obtained by decreasing the pressure of the fluid. Then, this mixture is sent to separator 4 with flow 21 to separate the steam part from this mixture. The steam coming out of separator 4 is sent to the mixing unit with flow 22 to be mixed with the steam obtained from the fourth flashing process. Then the steam coming from the medium-pressure turbine with flow 10 and the steam coming from separator 4 with flow 22 are mixed in the mixing unit. This steam, once mixed in the mixing unit, is sent to the low-pressure turbine with flow 11 in order to produce power. The steam here is expanded between flow 11 and flow 12, and power is produced. Then the steam coming out of the turbine is sent to the HEX with flow 12 for energy transfer. The HEX is used for energy transfer between geothermal fluid and fluid in the double RO desalination plant. The fluid coming out of the HEX is sent to the ground with the number 13 flow. A cycle of the geothermal fluid in the geothermal energy–based power production including freshwater through double RO desalination plant is accomplished in this way.

The cycle in which freshwater production is realized in the system is the double RO desalination subsystem. In the double RO desalination subsystem, saline water is first transferred to three-way valve 1 with flow 24. A certain amount of saline water from three-way valve 1 is transferred to the pump with flow 25, and the remaining part of the saline water is transferred to the energy recovery device. With this pump, saline water is transferred to three-way valve 2 with flow 26. Then this saline water and the fluid coming from the booster pump with flow 40 is transferred to the pretreatment component by flow 27. In this component, a chemical process is performed, and then the fluid from this component is transferred to the HEX by flow 28.

By utilizing the HEX, energy transfer is performed between the geothermal fluid and the fluid coming from the pretreatment component. The fluid from the HEX is transferred to three-way valve 3 by flow 29. Here, some amount of the fluid is sent to RO train 1 with flow 30, and the remaining part of the fluid is sent to RO train 1 with flow 31. The freshwater coming to the RO component is passed through semipermeable membranes. All impurities in the water are highly purified. Fluid containing residue remaining from the RO train 1 and RO train 2 is sent to three-way valve 5 with flows 36 and 37. Then this impurities-containing fluid is transferred to the energy recovery device. From this component, the fluid is sent to the booster pump to increase its pressure. Clean freshwater coming from RO train 1 and RO train 2 is transferred to three-way valve 4 by flows 32 and 33. This freshwater is transferred to the pretreatment component by flow 34 and is mixed with minerals and chlorine coming with flow 45 in this component. Then, this freshwater is transferred to the freshwater storage tank with flow 35. Freshwater in the storage tank can be used for various needs. Generally, a cycle of the geothermal energy–based power production including freshwater through a double RO desalination plant is completed in this way.

6.7.3 Case study 6.8

In this case study, the geothermal energy resource–based power production plant including freshwater through double RO desalination plant is investigated by using the energy and exergy analysis viewpoints. The simplified flow diagram for the geothermal energy–based power and freshwater production plant is given in Fig. 6.44. For this case study, the ambient temperature and pressure are taken as 25°C and 101.3 kPa, respectively.

The mass, energy, entropy, and exergy balance equations for the quadruple-flash steam geothermal power system components are previously written in Case Study 6.4. In addition, the mass, energy, entropy, and exergy

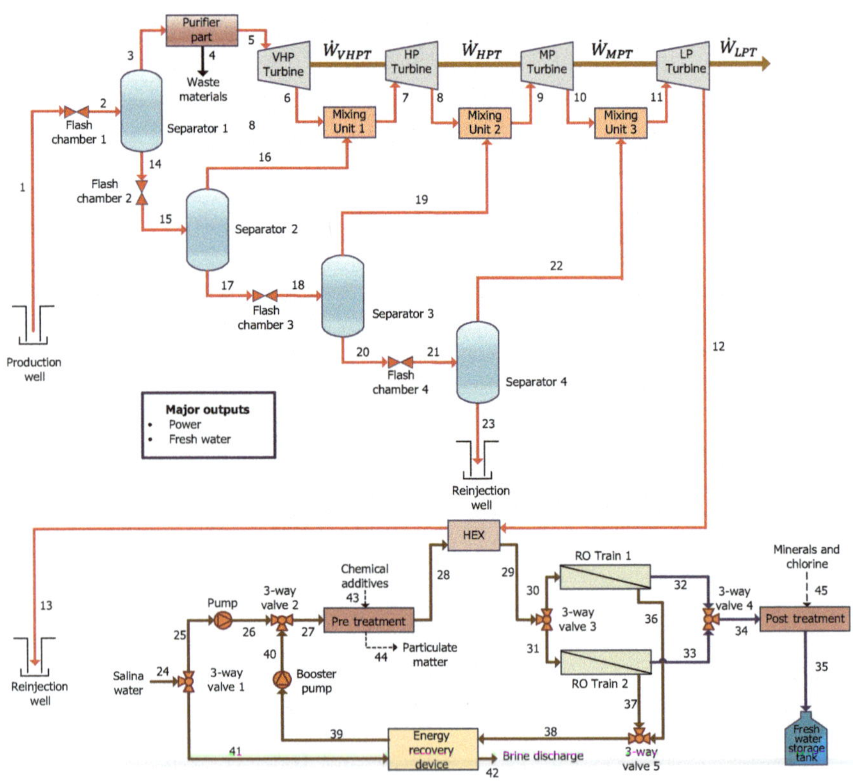

FIGURE 6.44 Schematic diagram of the geothermal energy–based power production including freshwater through a double reverse osmosis desalination plant.

balance equations for the RO distillation unit–based distillation plant components are defined.

The mass, energy, entropy, and exergy balance equations for the three-way valve 1 can be defined:

$$\dot{m}_{24} = \dot{m}_{25} + \dot{m}_{41}$$

$$\dot{m}_{24}h_{24} = \dot{m}_{25}h_{25} + \dot{m}_{41}h_{41}$$

$$\dot{m}_{24}s_{24} + \dot{S}_{g,3wv1} = \dot{m}_{25}s_{25} + \dot{m}_{41}s_{41}$$

$$\dot{m}_{24}ex_{24} = \dot{m}_{25}ex_{25} + \dot{m}_{41}ex_{41} + \dot{E}x_{D,3wv1}$$

The mass, energy, entropy, and exergy balance equations for the pump are:

$$\dot{m}_{25} = \dot{m}_{26}$$

$$\dot{m}_{25}h_{25} + \dot{W}_P = \dot{m}_{26}h_{26}$$

$$\dot{m}_{25}s_{25} + \dot{S}_{g,P} = \dot{m}_{26}s_{26}$$

$$\dot{m}_{25}ex_{25} + \dot{E}x_P^W = \dot{m}_{26}ex_{26} + \dot{E}x_{D,P}$$

The mass, energy, entropy, and exergy balance equations for the three-way valve 2 are:

$$\dot{m}_{26} + \dot{m}_{40} = \dot{m}_{27}$$

$$\dot{m}_{26}h_{26} + \dot{m}_{40}h_{40} = \dot{m}_{27}h_{27}$$

$$\dot{m}_{26}s_{26} + \dot{m}_{40}s_{40} + \dot{S}_{g,3wv2} = \dot{m}_{27}s_{27}$$

$$\dot{m}_{26}ex_{26} + \dot{m}_{40}ex_{40} = \dot{m}_{27}ex_{27} + \dot{E}x_{D,3wv2}$$

The mass, energy, entropy, and exergy balance equations for the pretreatment are defined:

$$\dot{m}_{27} = \dot{m}_{28};\ \dot{m}_{43} = \dot{m}_{44}$$

$$\dot{m}_{27}h_{27} + \dot{m}_{43}h_{43} = \dot{m}_{28}h_{28} + \dot{m}_{44}h_{44}$$

$$\dot{m}_{27}s_{27} + \dot{m}_{43}s_{43} + \dot{S}_{g,PTr} = \dot{m}_{28}s_{28} + \dot{m}_{44}s_{44}$$

$$\dot{m}_{27}ex_{27} + \dot{m}_{43}ex_{43} = \dot{m}_{28}ex_{28} + \dot{m}_{44}ex_{44} + \dot{E}x_{D,PTr}$$

The mass, energy, entropy, and exergy balance equations for the HEX can be described as:

$$\dot{m}_{12} = \dot{m}_{13};\ \dot{m}_{28} = \dot{m}_{29}$$

$$\dot{m}_{12}h_{12} + \dot{m}_{28}h_{28} = \dot{m}_{13}h_{13} + \dot{m}_{29}h_{29}$$

$$\dot{m}_{12}s_{12} + \dot{m}_{28}s_{28} + \dot{S}_{g,HEX} = \dot{m}_{13}s_{13} + \dot{m}_{29}s_{29}$$

$$\dot{m}_{12}ex_{12} + \dot{m}_{28}ex_{28} = \dot{m}_{13}ex_{13} + \dot{m}_{29}ex_{29} + \dot{E}x_{D,HEX}$$

The mass, energy, entropy, and exergy balance equations for the three-way valve 3 are defined:

$$\dot{m}_{29} = \dot{m}_{30} + \dot{m}_{31}$$

$$\dot{m}_{29}h_{29} = \dot{m}_{30}h_{30} + \dot{m}_{31}h_{31}$$

$$\dot{m}_{29}s_{29} + \dot{S}_{g,3wv3} = \dot{m}_{30}s_{30} + \dot{m}_{31}s_{31}$$

$$\dot{m}_{29}ex_{29} = \dot{m}_{30}ex_{30} + \dot{m}_{31}ex_{31} + \dot{E}x_{D,3wv3}$$

The mass, energy, entropy, and exergy balance equations for the RO train 1 can be defined:

$$\dot{m}_{30} = \dot{m}_{32} + \dot{m}_{36}$$

$$\dot{m}_{30}h_{30} = \dot{m}_{32}h_{32} + \dot{m}_{36}h_{36}$$

$$\dot{m}_{30}s_{30} + \dot{S}_{g,ROT1} = \dot{m}_{32}s_{32} + \dot{m}_{36}s_{36}$$

$$\dot{m}_{30}ex_{30} = \dot{m}_{32}ex_{32} + \dot{m}_{36}ex_{36} + \dot{E}x_{D,ROT1}$$

The mass, energy, entropy, and exergy balance equations for the three-way valve 4 can be:

$$\dot{m}_{32} + \dot{m}_{33} = \dot{m}_{34}$$

$$\dot{m}_{32}h_{32} + \dot{m}_{33}h_{33} = \dot{m}_{34}h_{34}$$

$$\dot{m}_{32}s_{32} + \dot{m}_{33}s_{33} + \dot{S}_{g,3wv4} = \dot{m}_{34}s_{34}$$

$$\dot{m}_{32}ex_{32} + \dot{m}_{33}ex_{33} = \dot{m}_{34}ex_{34} + \dot{E}x_{D,3wv4}$$

The mass, energy, entropy, and exergy balance equations for the post-treatment are:

$$\dot{m}_{34} + \dot{m}_{45} = \dot{m}_{35}$$

$$\dot{m}_{34}h_{34} + \dot{m}_{45}h_{45} = \dot{m}_{35}h_{35}$$

$$\dot{m}_{34}s_{34} + \dot{m}_{45}s_{45} + \dot{S}_{g,PoTr} = \dot{m}_{35}s_{35}$$

$$\dot{m}_{34}ex_{34} + \dot{m}_{45}ex_{45} = \dot{m}_{35}ex_{35} + \dot{E}x_{D,PoTr}$$

The mass, energy, entropy, and exergy balance equations for the RO train 2 can be described as:

$$\dot{m}_{31} = \dot{m}_{33} + \dot{m}_{37}$$

$$\dot{m}_{31}h_{31} = \dot{m}_{33}h_{33} + \dot{m}_{37}h_{37}$$

$$\dot{m}_{31}s_{31} + \dot{S}_{g,ROT2} = \dot{m}_{33}s_{33} + \dot{m}_{37}s_{37}$$

$$\dot{m}_{31}ex_{31} = \dot{m}_{33}ex_{33} + \dot{m}_{37}ex_{37} + \dot{E}x_{D,ROT2}$$

The mass, energy, entropy, and exergy balance equations for the three-way valve 5 are defined:

$$\dot{m}_{36} + \dot{m}_{37} = \dot{m}_{38}$$

$$\dot{m}_{36}h_{36} + \dot{m}_{37}h_{37} = \dot{m}_{38}h_{38}$$

$$\dot{m}_{36}s_{36} + \dot{m}_{37}s_{37} + \dot{S}_{g,3wv5} = \dot{m}_{38}s_{38}$$

$$\dot{m}_{36}ex_{36} + \dot{m}_{37}ex_{37} = \dot{m}_{38}ex_{38} + \dot{E}x_{D,3wv5}$$

The mass, energy, entropy, and exergy balance equations for the energy recovery device can be described as:

$$\dot{m}_{38} = \dot{m}_{39};\ \dot{m}_{41} = \dot{m}_{42}$$

$$\dot{m}_{38}h_{38} + \dot{m}_{41}h_{41} = \dot{m}_{39}h_{39} + \dot{m}_{42}h_{42}$$

$$\dot{m}_{38}s_{38} + \dot{m}_{41}s_{41} + \dot{S}_{g,ERD} = \dot{m}_{39}s_{39} + \dot{m}_{42}s_{42}$$

$$\dot{m}_{38}ex_{38} + \dot{m}_{41}ex_{41} = \dot{m}_{39}ex_{39} + \dot{m}_{42}ex_{42} + \dot{E}x_{D,ERD}$$

In order to thermodynamically model the gas turbine–based multigeneration system, some parameters are selected as input data for the simulation. The operating parameters for the quadruple-flash steam geothermal power system integrated with the double RO distillation unit are given in Table 6.19.

The energy and exergy efficiency equations for the quadruple-flash steam geothermal power system can be given as:

$$\eta_{QFGP} = \frac{\dot{W}_{VHPT} + \dot{W}_{HPT} + \dot{W}_{MPT} + \dot{W}_{LPT}}{\dot{m}_1h_1 - \dot{m}_{12}h_{12} - \dot{m}_{23}h_{23}}$$

and

$$\psi_{QFGP} = \frac{\dot{E}x^W_{VHPT} + \dot{E}x^W_{HPT} + \dot{E}x^W_{MPT} + \dot{E}x^W_{LPT}}{\dot{m}_1ex_1 - \dot{m}_{12}ex_{12} - \dot{m}_{23}ex_{23}}$$

TABLE 6.19 Operating parameters for the quadruple-flash steam geothermal power system integrated with the double reverse osmosis distillation unit.

Variables	Values
Geofluid source temperature, T_1	240°C
Geofluid source pressure, P_1	1500 kPa
Geofluid mass flow rate, $\dot{m}_1$	90 kg/s
Separator 1 inlet pressure, P_2	667 kPa
Separator 2 inlet pressure, P_{15}	119 kPa
Separator 3 inlet pressure, P_{18}	62.92 kPa
Separator 4 inlet pressure, P_{21}	45.86 kPa
VHP turbine output pressure, P_6	400 kPa
HP turbine output pressure, P_8	300 kPa
MP turbine output pressure, P_{10}	200 kPa
LP turbine output pressure, P_{12}	100 kPa
Saline water exit temperature from HEX, T_{29}	61.4°C
Freshwater exit temperature from post-treatment, T_{35}	26.8°C
Geofluid reinjection temperature, T_{13}	43.27°C

The energy and exergy efficiency equations for the double RO distillation unit are defined:

$$\eta_{DRODU} = \frac{\dot{m}_{35}h_{35}}{(\dot{m}_{12}h_{12} - \dot{m}_{13}h_{13}) + \dot{m}_{24}h_{24} + \dot{W}_P + \dot{W}_{BP}}$$

and

$$\psi_{DRODU} = \frac{\dot{m}_{35}ex_{35}}{(\dot{m}_{12}ex_{12} - \dot{m}_{13}ex_{13}) + \dot{m}_{24}ex_{24} + \dot{E}x_P^W + \dot{E}x_{BP}^W}$$

The energy and exergy efficiency equations for the geothermal energy—based power production including freshwater through the RO distillation unit are defined:

$$\eta_{QFDP} = \frac{\dot{W}_{net,QFDP} + \dot{m}_{35}h_{35}}{\dot{m}_1h_1 - \dot{m}_{13}h_{13} - \dot{m}_{23}h_{23}}$$

and

$$\psi_{QFDP} = \frac{\dot{E}x^{W}_{net,QFDP} + \dot{m}_{35}ex_{35}}{\dot{m}_1 ex_1 - \dot{m}_{13}ex_{13} - \dot{m}_{23}ex_{23}}$$

where $\dot{W}_{net,QFDP}$ can be defined as:

$$\dot{W}_{net,QFDP} = \dot{W}_{VHPT} + \dot{W}_{HPT} + \dot{W}_{MPT} + \dot{W}_{LPT} - \dot{W}_{P} - \dot{W}_{BP}$$

Based on the thermodynamic analysis balance equations, the results of the energy and exergy analyses for the geothermal energy–based power production plant including freshwater through the RO distillation unit are given in Table 6.20. As given in the table, the energetic and exergetic efficiencies of quadruple-flash combined with RO distillation plant are calculated as 0.4546 and 0.4924, respectively.

The power production rates and mass flow rate of producing freshwater by utilizing the very high-pressure, high-pressure, medium-pressure, low-pressure, and RO desalination subsystems are listed in Table 6.21. The electrical energy consumption of the pump and booster pump are calculated as 65.6 and 34.8 kW, respectively. Based on these data, the net power production rate from the geothermal energy–based power production plant including a freshwater production through RO desalination subsystem is equal to 24,313.6 kW.

In this case study, the base case ambient temperature is taken as 25°C, and the reference temperature is varied between 0°C and 40°C. The effect of the reference temperature on the energy and exergy efficiencies are given in Fig. 6.45. It is observed that the energy and exergy efficiencies of the geothermal energy resource-based power production plant including a freshwater through double RO desalination plant increase from 43.91% to 46.42% and from 47.06% to 50.63%, respectively.

The reference temperature significantly affects the geothermal energy–based plant's power and freshwater production. Fig. 6.46 shows the

TABLE 6.20 Energetic and exergetic efficiencies of the geothermal energy–based power production plant including freshwater through double reverse osmosis distillation unit.

Subsystems/overall system	Energy efficiency (%)	Exergy efficiency (%)
Quadruple-flash steam power plant	42.83	53.91
Double reverse osmosis distillation unit	64.07	61.25
Overall plant	45.46	49.24

TABLE 6.21 Useful output rates from the geothermal energy–based power production including freshwater through reverse osmosis desalination subsystem.

Useful outputs	Value
Power production by using very high-pressure turbine, $\dot{W}_{VHPT}$	7624 kW
Power production by using high-pressure turbine, $\dot{W}_{HPT}$	6853 kW
Power production by using medium-pressure turbine, $\dot{W}_{MPT}$	5916 kW
Power production by using low-pressure turbine, $\dot{W}_{LPT}$	4021 kW
Mass flow rate of producing freshwater, $\dot{m}_{FW}$	18.43 kg/s

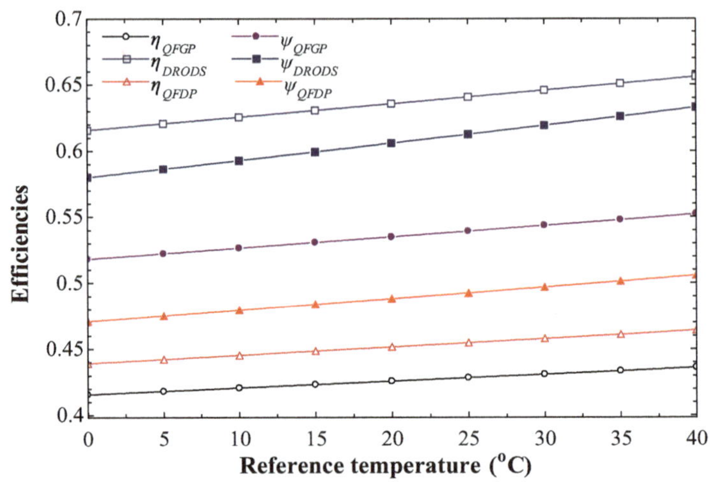

FIGURE 6.45 Impact of the reference temperature on the energy and exergy efficiencies of the combined system and its subsystems.

effect of the reference temperature on the power and freshwater production rates. It is observed that the production rate of useful outputs increases as the reference temperature increases, mainly due to the decrease of heat loss from system components to the surrounding air.

Fig. 6.47 shows the variation with the geothermal water mass flow rate of both energy and exergy efficiencies of the overall plant and its subplants. It is observed that the effectiveness of the plant increases as the geothermal water mass flow rate increases, mainly due to the increase of the working fluid temperature entering the subplants.

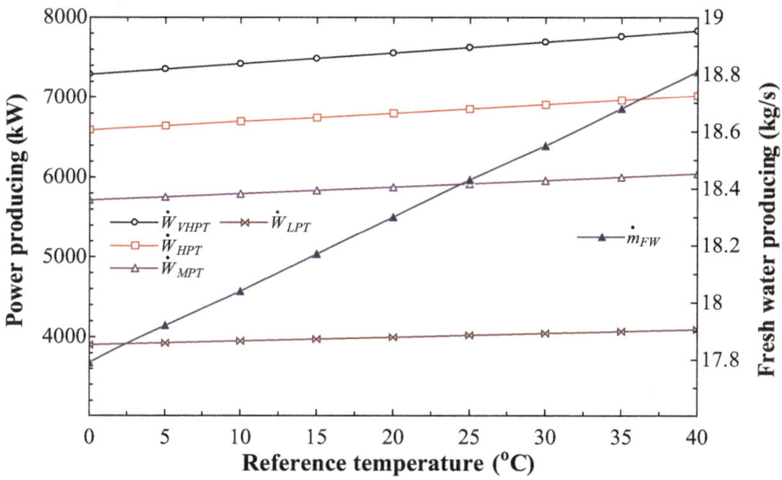

FIGURE 6.46 Impact of the reference temperature on the power and freshwater productions.

In order to better understand more comprehensively the effect of varying geothermal water mass flow rate on the plant performance, the power and freshwater production rates are examined. It is seen in Fig. 6.48 that the total power and freshwater production rates increase from 2383 to 25,598 kW and from 17.63 to 19.25 kg/s, respectively.

Fig. 6.49 shows the variation with geothermal water temperature of both the energetic and exergetic efficiencies for the geothermal energy–based system and its subsystems. It is observed that both the energy and exergy efficiencies of the system and its subsystems increase as geothermal water temperature increases from 200°C to 270°C.

As given in Fig. 6.50, the power production rates for the VHPT, HPT, MPT, and LPT subcomponents increase from 7183 to 7972 kW, from 6482 to 7144 kW, from 5618 to 6149 kW, and from 3833 to 4167 kW, respectively.

6.7.3.1 Multistage flash desalination unit combined with a geothermal energy system

The schematic diagram of an MSF desalination unit combined with a geothermal energy system is illustrated in Fig. 6.51. Explaining the working principle of the system, in general, is useful for understanding how the system works. In order to eliminate particles in the fluid from the geothermal energy source, the fluid is sent to the particle separator by flow 1. Since the pressure in the geothermal energy source is higher than where the particle lifter is located, it is transmitted to the particle lifter after pressure adjustment. After the particles in the fluid are removed, the fluid coming out of the particle extractor is sent to the moisture separator with the number 2 flow to remove the moisture in the fluid. Here, the moisture released from

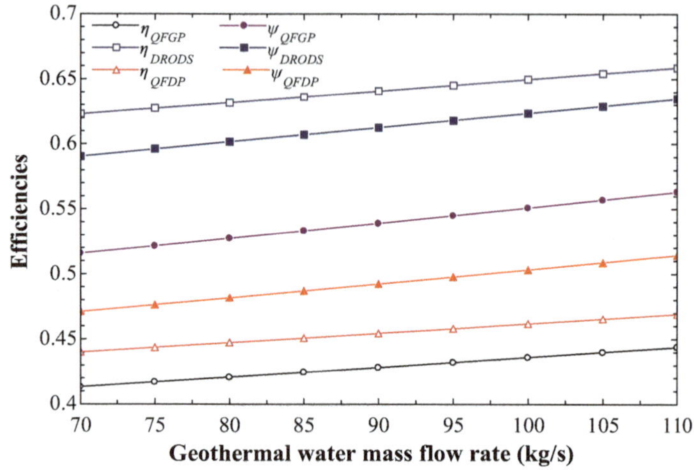

FIGURE 6.47 Impact of the geothermal water mass flow rate on the energy and exergy efficiencies of the combined system and its subsystems.

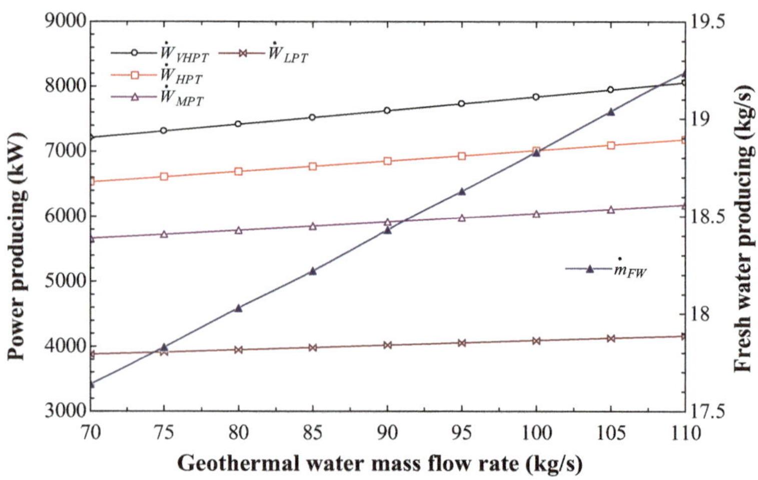

FIGURE 6.48 Impact of the geothermal water mass flow rate on the power and freshwater production rates.

the fluid is sent to the underground with flow 3. Quality fluid coming from the moisture separator is transferred to the valve with flow 4 to adjust its pressure. For this system to serve its essential purpose, the fluid from the valve is sent to turbine 1 to obtain power by flow 5. After the power generation in turbine 1, the expanding steam in turbine 1 is transferred to turbine 2 for power generation with flow 6 in order to benefit from dry steam again. After the power generation in turbine 2 is realized, the expanding steam in

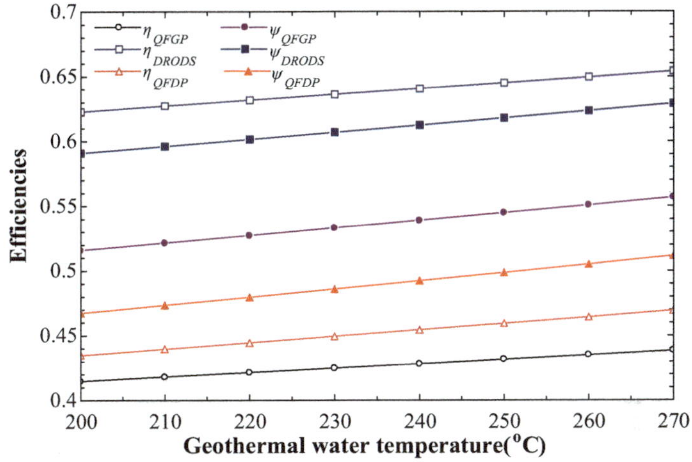

FIGURE 6.49 Impact of the geothermal water temperature on the energy and exergy efficiencies of the combined system and its subsystems.

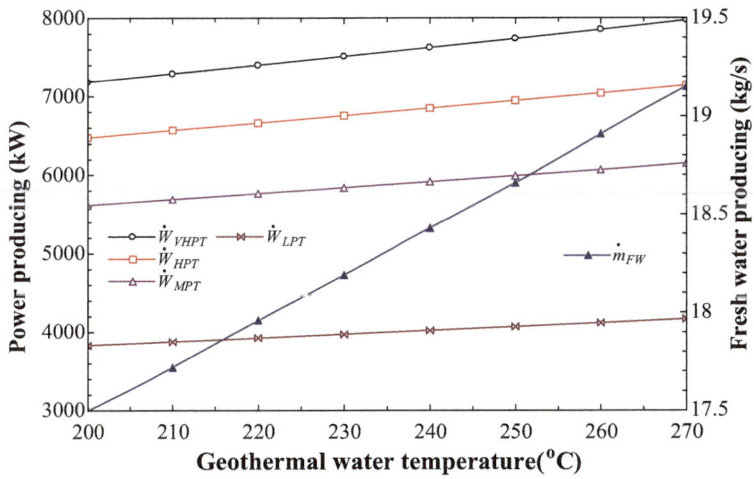

FIGURE 6.50 Impact of the geothermal water temperature on the power and freshwater production rates.

turbine 2 is transferred to turbine 3 with flow 7 in order to realize electricity generation again. In addition, for electricity output, again, the expanding steam in turbine 3 is transferred to turbine 4 with flow 8. After electricity generation by turbine 4, the fluid coming out of turbine 4 is sent to the HEX for energy transfer with flow 9. The HEX is used for energy transfer between the geothermal fluid and the seawater. The fluid coming out of the HEX is sent again to the reinjection well by flow 10. A cycle of the geothermal fluid in the geothermal energy–based power production including freshwater

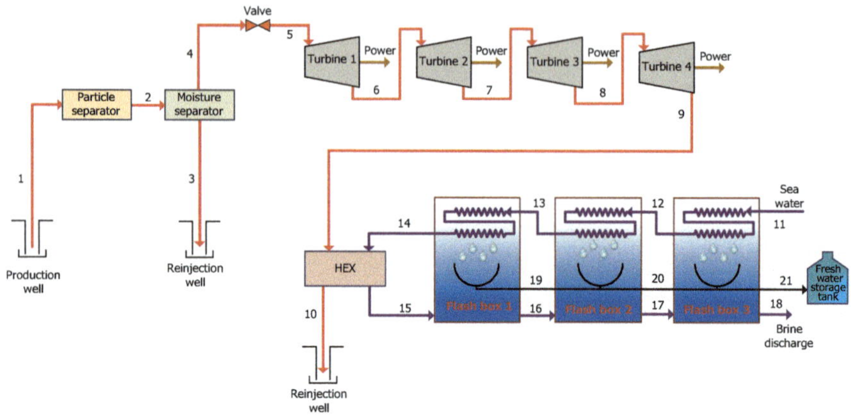

FIGURE 6.51 Simplified flow diagram of geothermal energy–based power production including freshwater through multistage flash desalination unit.

through MSF desalination unit is completed in this way. The cycle in which freshwater production is realized in the system is the MSF desalination subsystem. In this subsystem, seawater is first transferred to flash box 3 with flow 11. MSF desalination cycle takes place between flows 11 and 21. Freshwater coming from flash box 1, flash box 2, and flash box 3 is transferred to the freshwater storage tank with flows 19, 20, and 21, respectively. Freshwater in the storage tank can be used for various needs. Generally, a cycle of the geothermal energy–based power production including freshwater through MSF desalination unit is completed in this way.

6.7.4 Case study 6.9

In this case study, the geothermal energy resource–based power production plant including freshwater through MSF desalination unit is assessed by utilizing the energetic and exergetic analyses. The simplified flow diagram for this integrated plant is shown in Fig. 6.52. For this case study, the ambient temperature and pressure are taken as 25°C and 101.3 kPa, respectively.

The mass, energy, entropy, and exergy balance equations for the multistaged direct system with four turbine components were previously determined in Case Study 6.2. In addition to them, the mass, energy, entropy, and exergy balance equations for the MSF desalination unit–based distillation plant components are defined.

The mass, energy, entropy, and exergy balance equations for the HEX can be described:

$$\dot{m}_9 = \dot{m}_{10};\ \dot{m}_{14} = \dot{m}_{15}$$

$$\dot{m}_9 h_9 + \dot{m}_{14} h_{14} = \dot{m}_{10} h_{10} + \dot{m}_{15} h_{15}$$

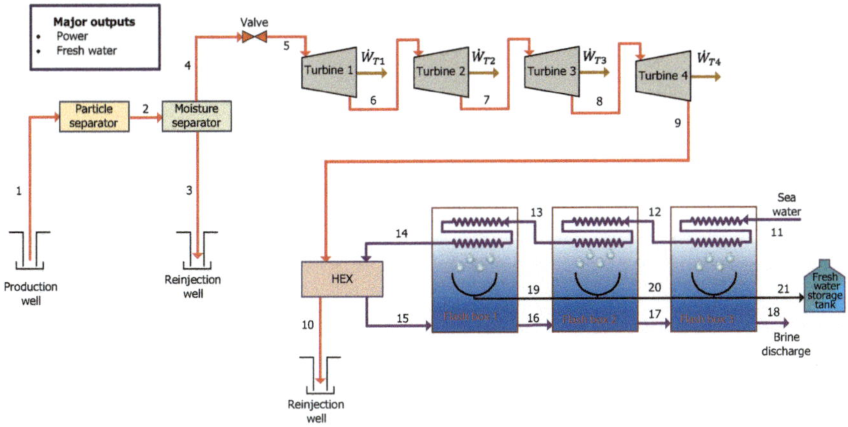

FIGURE 6.52 Schematic diagram of the geothermal energy–based power production including freshwater through an MSF desalination unit.

$$\dot{m}_9 s_9 + \dot{m}_{14} s_{14} + \dot{S}_{g,HEX} = \dot{m}_{10} s_{10} + \dot{m}_{15} s_{15}$$

$$\dot{m}_9 ex_9 + \dot{m}_{14} ex_{14} = \dot{m}_{10} ex_{10} + \dot{m}_{15} ex_{15} + \dot{E}x_{D,HEX}$$

The mass, energy, entropy, and exergy balance equations for the flash box 1 can be obtained:

$$\dot{m}_{13} = \dot{m}_{14} = \dot{m}_{15}; \ \dot{m}_{15} = \dot{m}_{16} + \dot{m}_{19}$$

$$\dot{m}_{13} h_{13} + \dot{m}_{15} h_{15} = \dot{m}_{14} h_{14} + \dot{m}_{16} h_{16} + \dot{m}_{19} h_{19}$$

$$\dot{m}_{13} s_{13} + \dot{m}_{15} s_{15} + \dot{S}_{g,FB1} = \dot{m}_{14} s_{14} + \dot{m}_{16} s_{16} + \dot{m}_{19} s_{19}$$

$$\dot{m}_{13} ex_{13} + \dot{m}_{15} ex_{15} = \dot{m}_{14} ex_{14} + \dot{m}_{16} ex_{16} + \dot{m}_{19} ex_{19} + \dot{E}x_{D,FB1}$$

The mass, energy, entropy, and exergy balance equations for the flash box 2 are given:

$$\dot{m}_{12} = \dot{m}_{13} = \dot{m}_{16}; \ \dot{m}_{16} + \dot{m}_{19} = \dot{m}_{17} + \dot{m}_{20}$$

$$\dot{m}_{12} h_{12} + \dot{m}_{16} h_{16} + \dot{m}_{19} h_{19} = \dot{m}_{13} h_{13} + \dot{m}_{17} h_{17} + \dot{m}_{20} h_{20}$$

$$\dot{m}_{12} s_{12} + \dot{m}_{16} s_{16} + \dot{m}_{19} s_{19} + \dot{S}_{g,FB2} = \dot{m}_{13} s_{13} + \dot{m}_{17} s_{17} + \dot{m}_{20} s_{20}$$

$$\dot{m}_{12} ex_{12} + \dot{m}_{16} ex_{16} + \dot{m}_{19} ex_{19} = \dot{m}_{13} ex_{13} + \dot{m}_{17} ex_{17} + \dot{m}_{20} ex_{20} + \dot{E}x_{D,FB2}$$

The mass, energy, entropy, and exergy balance equations for the flash box 3 can be described as:

$$\dot{m}_{11} = \dot{m}_{12} = \dot{m}_{17}; \ \dot{m}_{17} + \dot{m}_{20} = \dot{m}_{18} + \dot{m}_{21}$$

$$\dot{m}_{11}h_{11} + \dot{m}_{17}h_{17} + \dot{m}_{20}h_{20} = \dot{m}_{12}h_{12} + \dot{m}_{18}h_{18} + \dot{m}_{21}h_{21}$$

$$\dot{m}_{11}s_{11} + \dot{m}_{17}s_{17} + \dot{m}_{20}s_{20} + \dot{S}_{g,FB3} = \dot{m}_{12}s_{12} + \dot{m}_{18}s_{18} + \dot{m}_{21}s_{21}$$

$$\dot{m}_{11}ex_{11} + \dot{m}_{17}ex_{17} + \dot{m}_{20}ex_{20} = \dot{m}_{12}ex_{12} + \dot{m}_{18}ex_{18} + \dot{m}_{21}ex_{21} + \dot{E}x_{D,FB3}$$

The assumptions utilized in the working conditions of the multistaged direct system with four turbines are shown in Table 6.22. The heat and work input/output rate, entropy generation rate, and exergy destruction rates, as well as energy and exergy efficiencies, can be computed by using these balance equalities and chosen conditions.

The energy and exergy efficiency equations for a multistaged direct geothermal energy system with four turbines can be given as:

$$\eta_{MSFT} = \frac{\dot{W}_{T1} + \dot{W}_{T2} + \dot{W}_{T3} + \dot{W}_{T4}}{\dot{m}_1h_1 - \dot{m}_3h_3 - \dot{m}_9h_9}$$

TABLE 6.22 Operating parameters for the multistaged direct geothermal energy system with four turbines.

Variables	Values
Reference temperature, T_0	25°C
Reference pressure, P_0	101.3 kPa
Geofluid source temperature, T_1	280°C
Geofluid source pressure, P_1	3146 kPa
Geofluid mass flow rate, $\dot{m}_1$	90 kg/s
Turbine 1 inlet pressure, P_5	2056 kPa
Turbine 1 inlet temperature, T_5	215.1°C
Turbine 2 inlet pressure, P_6	1452 kPa
Turbine 2 inlet temperature, T_6	196.8°C
Turbine 3 inlet pressure, P_7	825 kPa
Turbine 3 inlet temperature, T_7	171.7°C
Turbine 4 inlet pressure, P_8	540 kPa
Turbine 4 inlet temperature, T_8	154.8°C
HEX inlet temperature for geofluid, T_9	110.3°C
Seawater input temperature, T_{11}	31.5°C
Freshwater input temperature, T_{21}	40.9°C
Geofluid reinjection temperature, T_{10}	48.67°C

and

$$\psi_{MSFT} = \frac{\dot{E}x^{W}_{T1} + \dot{E}x^{W}_{T2} + \dot{E}x^{W}_{T3} + \dot{E}x^{W}_{T4}}{\dot{m}_1 ex_1 - \dot{m}_3 ex_3 - \dot{m}_9 ex_9}$$

The energy and exergy efficiency equations for the MSF desalination unit are defined:

$$\eta_{MSFD} = \frac{\dot{m}_{21} h_{21}}{(\dot{m}_9 h_9 - \dot{m}_{10} h_{10}) + \dot{m}_{11} h_{11}}$$

and

$$\psi_{MSFD} = \frac{\dot{m}_{21} ex_{21}}{(\dot{m}_9 ex_9 - \dot{m}_{10} ex_{10}) + \dot{m}_{11} ex_{11}}$$

The energy and exergy efficiency equations for the geothermal energy—based power production including freshwater through the MSF desalination unit are defined:

$$\eta_{MSDP} = \frac{\dot{W}_{net,MSDP} + \dot{m}_{21} h_{21}}{\dot{m}_1 h_1 - \dot{m}_3 h_3 - \dot{m}_{10} h_{10}}$$

and

$$\psi_{MSDP} = \frac{\dot{E}x^{W}_{net,MSDP} + \dot{m}_{21} ex_{21}}{\dot{m}_1 ex_1 - \dot{m}_3 ex_3 - \dot{m}_{10} ex_{10}}$$

where $\dot{W}_{net,MSDP}$ can be defined as:

$$\dot{W}_{net,MSDP} = \dot{W}_{T1} + \dot{W}_{T2} + \dot{W}_{T3} + \dot{W}_{T4}$$

By using the mass, energy, entropy and exergy balance equations, the results of the energy and exergy analyses for the geothermal energy—based power production plant including freshwater through the MSF desalination unit are given in Table 6.23. As given in the table, the energetic and exergetic efficiencies of the multistaged direct geothermal energy system with four turbines combined with a RO distillation plant are calculated as 0.5316 and 0.5829, respectively.

The power production rates and mass flow rate of producing freshwater by utilizing the steam turbines and the MSF desalination subsystem are written in Table 6.24. The total power production rate from the geothermal energy—based power production plant including freshwater production through an MSF desalination unit subsystem is equal to 24,313.6 kW.

The effects of an increase in reference temperature on the energy and exergy efficiencies of the geothermal energy—based power production including a freshwater through MSF desalination unit are shown in Fig. 6.53. As illustrated in the figure, the energy and exergy efficiencies of the overall plant vary from 51.58% to 54.12% and from 56.01% to 59.7%, respectively,

TABLE 6.23 Energetic and exergetic efficiencies of the geothermal energy–based power production plant including freshwater through a multistage flash desalination unit.

Subsystems/overall system	Energy efficiency (%)	Exergy efficiency (%)
Multistaged direct system with four turbines	62.37	71.82
Multistage flash desalination unit	51.84	46.37
Overall plant	53.16	58.29

TABLE 6.24 Useful output rates from the geothermal energy–based power production including freshwater through multistage flash desalination unit.

Useful outputs	Value
Power production from turbine 1, $\dot{W}_{T1}$	13,800 kW
Power production from turbine 2, $\dot{W}_{T2}$	17,201 kW
Power production from turbine 3, $\dot{W}_{T3}$	19,078 kW
Power production from turbine 4, $\dot{W}_{T4}$	13,082 kW
Mass flow rate of producing freshwater, $\dot{m}_{FW}$	18.43 kg/s

with an increase in reference temperature from 0°C to 40°C. Also, the energy and exergy efficiencies of the multistaged direct system with four turbines increase from 60.83% to 63.31% and from 69.35% to 73.33%, respectively. Finally, the energy and exergy efficiencies of the MSF desalination unit increase from 50.06% to 52.93% and from 44.12% to 47.76%, respectively.

The impact of an increase in reference temperature on the power production and freshwater production rates is shown in Fig. 6.54. As shown in the figure, the total power generation and freshwater production rates are observed to be increasing with the increase in reference temperature. The power production rate increases from 13,261 to 14,133 kW for turbine 1, from 16,611 to 17,564 kW for turbine 2, from 18,515 to 19,422 kW for turbine 3, and from 12,633 to 13,358 kW, and the freshwater production rate varies from 17.19 to 19.21 kg/s.

The increase in geothermal water mass flow rate has a positive impact on the energy and exergy efficiencies of the geothermal energy–based combined plant and its subplants. As shown in Fig. 6.55, the energy and exergy

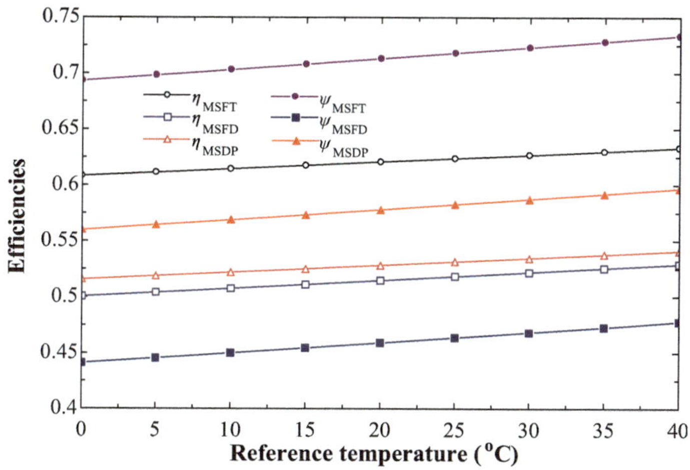

FIGURE 6.53 Effect of the reference temperature on the energy and exergy efficiencies of the combined system and its subsystems.

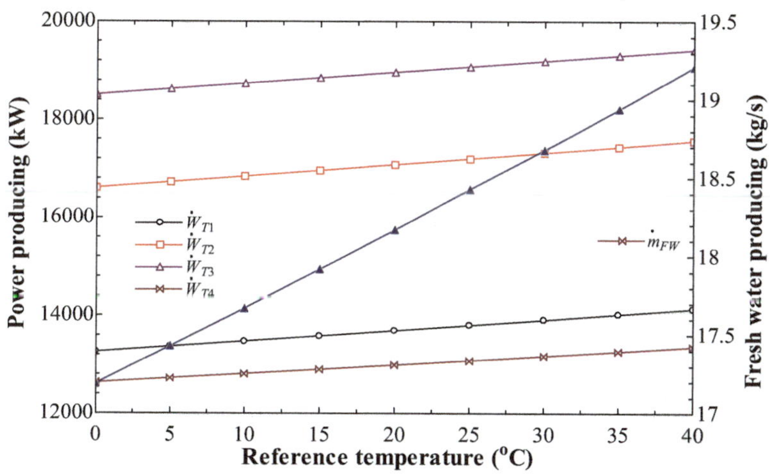

FIGURE 6.54 Effect of the reference temperature on the power and freshwater production rates.

efficiencies of the overall plant increase from 51.69% to 54.66% and from 56.01% to 60.65%, respectively.

The effects of variation in geothermal water mass rate useful outputs are shown in Fig. 6.56. The rate of power production rate is observed to vary from 13,105 to 14,531 kW for turbine 1, from 16,464 to 17,970 kW for turbine 2, from 18,333 to 19,852 kW for turbine 3, and from 12,671 to 13,505 kW, as well as an increase in geothermal water mass flow rate from 70 to 110 kg/s, respectively. On the other hand, the freshwater production

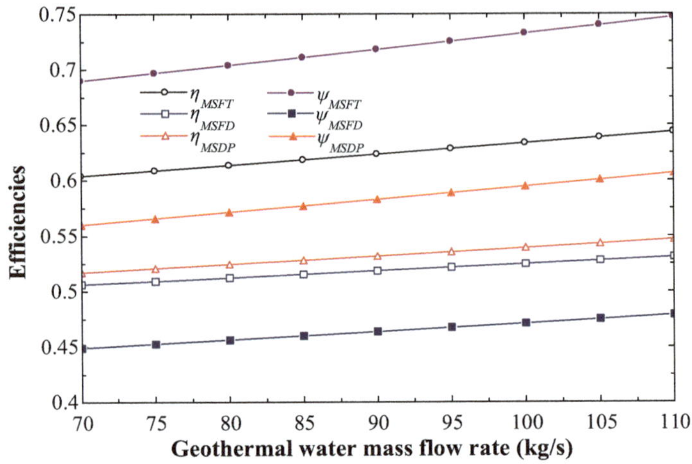

FIGURE 6.55 Effect of the geothermal water mass flow rate on the energy and exergy efficiencies of the combined system and its subsystems.

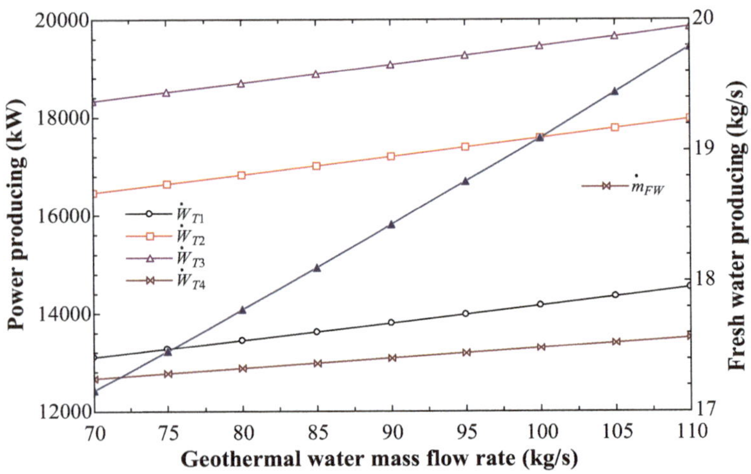

FIGURE 6.56 Effect of the geothermal water mass flow rate on the power and freshwater production rates.

increases from 17.16 to 19.79 kg/s within the same geothermal water mass flow rate interval.

Fig. 6.57 demonstrates the effects of variation in the geothermal water temperature on energy and exergy efficiencies of geothermal energy–based combined plant and its subsystems. The energy and exergy efficiencies of the overall plant are observed to vary from 51.08% to 54.22% and from 55.57% to 59.69%, respectively. Also, the energy and exergy efficiencies of

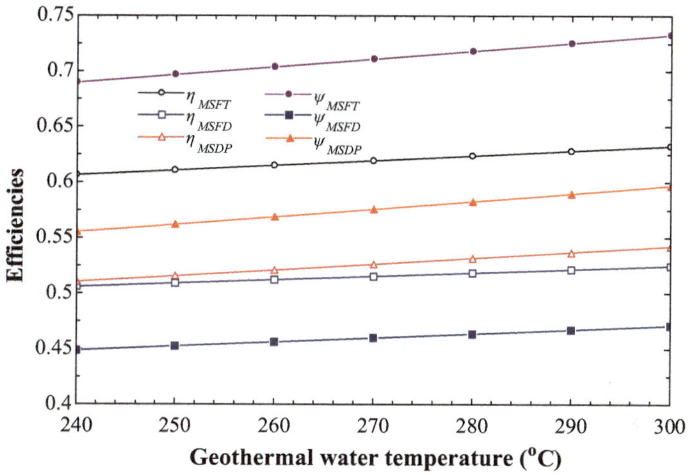

FIGURE 6.57 Effect of the geothermal water temperature on the energy and exergy efficiencies of the combined system and its subsystems.

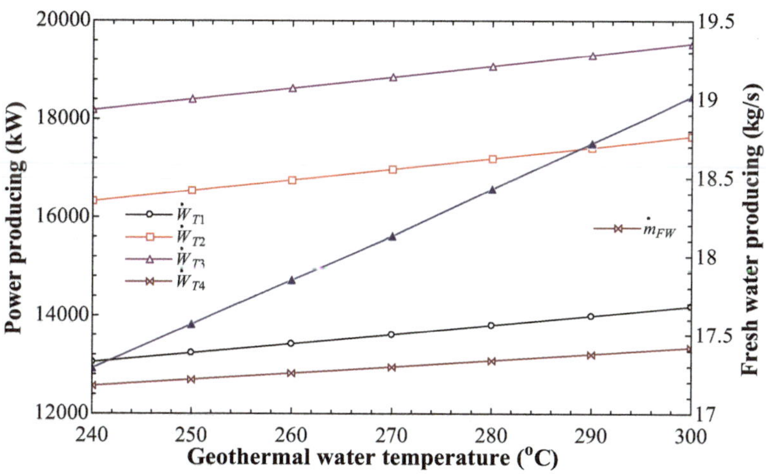

FIGURE 6.58 Effect of the geothermal water temperature on the power and freshwater production rates.

the MSF desalination unit are observed to be varying from 50.61% to 52.46% and from 44.91% to 47.11%, respectively.

Moreover, an increase in the geothermal water temperature has a positive impact on the power and freshwater production rates. As illustrated in Fig. 6.58, the power production rate varies from 13,053 to 14,189 kW for steam turbine 1, from 16,334 to 17,651 kW for steam turbine 2, from 18,189 to 19,538 kW for steam turbine 3, and from 12,571 to 13,344 kW for steam

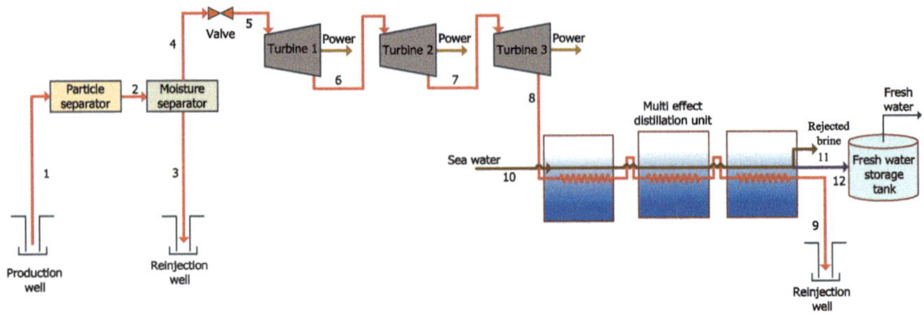

FIGURE 6.59 Simplified flow diagram of geothermal energy–based power production including a freshwater through multieffect distillation unit.

turbine 4, respectively. Also, the freshwater production rate increases from 17.29 to 19.02 kg/s within the same geothermal water temperature interval.

6.7.4.1 Multieffect distillation unit combined with a geothermal energy system

The schematic diagram of geothermal energy–based power production including freshwater through a MED unit is illustrated in Fig. 6.59. Explaining the working principle of the system, in general, is useful for understanding how the system works. To eliminate particles in the fluid from the geothermal energy source, the fluid is sent to the particle separator by flow 1. Since the pressure in the geothermal energy source is higher than where the particle lifter is located, it is transmitted to the particle lifter after a pressure adjustment. After the particles in the fluid are removed, the fluid coming out of the particle extractor is sent to the moisture separator with the number 2 flow to remove the moisture in the fluid. Here, the moisture released from the fluid is sent to the underground with flow 3. Quality fluid coming from the moisture separator is transferred to the valve with flow 4 to adjust its pressure. For this system to serve its essential purpose, the fluid from the valve is sent to turbine 1 to obtain power by flow 5. After the power generation in turbine 1, the expanding steam in turbine 1 is transferred to turbine 2 for electricity generation with flow 6 in order to benefit from dry steam again. After the power generation in turbine 2 is realized, the expanding steam in turbine 2 is transferred to turbine 3 with flow 7 in order to realize electricity generation again. After electricity generation by turbine 3, the fluid coming out of turbine 3 is sent to the MED unit for energy transfer with flow 8. Here, energy transfer is realized between the geothermal fluid and the seawater. The geothermal fluid coming out of the MED unit is sent again to reinjection well by flow 9. A cycle of the geothermal fluid in the geothermal energy–based power production including freshwater through the MED unit is completed in this way.

The cycle in which freshwater production is realized in the system is the MED subsystem. In the MED subsystem, seawater is first transferred to the MED unit with flow 10. The MED cycle takes place between flows 10 and 12. Freshwater coming from the MED unit is transferred to the freshwater storage tank with flow 12. Freshwater in the storage tank can be used to meet various needs. Generally, a cycle of the geothermal energy–based power production including freshwater through a MED unit is completed in this way.

6.7.5 Case study 6.10

In this case study, the geothermal energy resource-based power production plant including freshwater through the MED unit is investigated according to thermodynamic analysis. The simplified flow diagram for this combined power and freshwater production plant is illustrated in Fig. 6.60. For this case study, the ambient temperature and pressure are taken as 25°C and 101.3 kPa, respectively.

The mass, energy, entropy, and exergy balance equations for the MED unit with three turbine components were previously determined in Case Study 6.1. In addition to them, the mass, energy, entropy, and exergy balance equations for the MSF desalination unit–based distillation plant components are defined.

The mass, energy, entropy and exergy balance equations for the MED unit can be defined:

$$\dot{m}_8 = \dot{m}_9; \ \dot{m}_{10} = \dot{m}_{11} + \dot{m}_{12}$$

$$\dot{m}_8 h_8 + \dot{m}_{10} h_{10} = \dot{m}_9 h_9 + \dot{m}_{11} h_{11} + \dot{m}_{12} h_{12}$$

$$\dot{m}_8 s_8 + \dot{m}_{10} s_{10} + \dot{S}_{g,medu} - \dot{m}_9 s_9 + \dot{m}_{11} s_{11} + \dot{m}_{12} s_{12}$$

$$\dot{m}_8 ex_8 + \dot{m}_{10} ex_{10} = \dot{m}_9 ex_9 + \dot{m}_{11} ex_{11} + \dot{m}_{12} ex_{12} + \dot{E}x_{D,medu}$$

The operating indicators utilized of the multistaged direct geothermal energy system with three turbines combined with the MED unit for

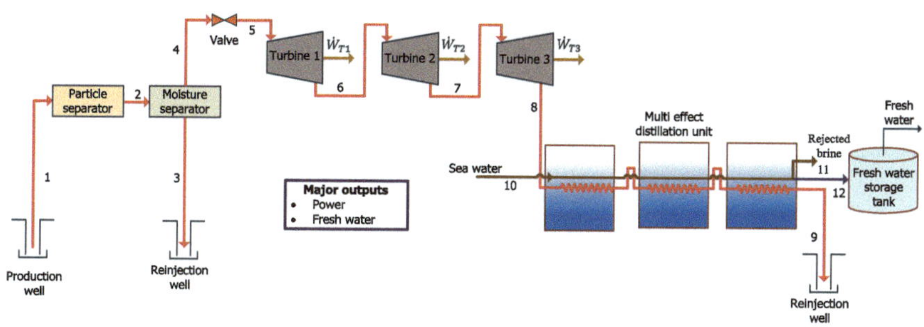

FIGURE 6.60 Schematic diagram of the geothermal energy–based power production including a freshwater through multieffect distillation unit.

TABLE 6.25 Operating parameters for the multistaged direct geothermal energy system with three turbines integrated with the multieffect distillation unit.

Variables	Values
Reference temperature, T_0	25°C
Reference pressure, P_0	101.3 kPa
Geofluid source temperature, T_1	260°C
Geofluid source pressure, P_1	2816 kPa
Geofluid mass flow rate, $\dot{m}_1$	90 kg/s
Turbine 1 inlet pressure, P_5	1580 kPa
Turbine 1 inlet temperature, T_5	204.4°C
Turbine 2 inlet pressure, P_6	972 kPa
Turbine 2 inlet temperature, T_6	178.7°C
Turbine 3 inlet pressure, P_7	525 kPa
Turbine 3 inlet temperature, T_7	153.7°C
Multi effect distillation unit inlet temperature, T_8	80°C
Freshwater exit temperature, T_{12}	41°C
Geofluid reinjection temperature, T_9	48.24°C

producing power and freshwater are listed in Table 6.25. The heat and work input/output rates, entropy generation rate, and exergy destruction rates, as well as energy and exergy effectiveness, are computed by utilizing the balance equalities and chosen assumptions.

The energy and exergy efficiency equations for the multistaged direct system with three turbines can be given as:

$$\eta_{MSDS} = \frac{\dot{W}_{T1} + \dot{W}_{T2} + \dot{W}_{T3}}{\dot{m}_1 h_1 - \dot{m}_3 h_3 - \dot{m}_9 h_9}$$

and

$$\psi_{MSDS} = \frac{\dot{E}x^W_{T1} + \dot{E}x^W_{T2} + \dot{E}x^W_{T3}}{\dot{m}_1 ex_1 - \dot{m}_3 ex_3 - \dot{m}_9 ex_9}$$

The energy and exergy efficiency equations for the MED unit are defined:

$$\eta_{MEDU} = \frac{\dot{m}_{12} h_{12}}{(\dot{m}_8 h_8 - \dot{m}_9 h_9) + \dot{m}_{10} h_{10}}$$

and

$$\psi_{MEDU} = \frac{\dot{m}_{12}ex_{12}}{(\dot{m}_8 ex_8 - \dot{m}_{12}ex_{12}) - \dot{m}_{10}ex_{10}}$$

The energy and exergy efficiency equations for the geothermal energy–based power production including a freshwater through MED unit are defined:

$$\eta_{OVP} = \frac{\dot{W}_{net} + \dot{m}_{12}h_{12}}{\dot{m}_1 h_1 - \dot{m}_3 h_3 - \dot{m}_9 h_9}$$

and

$$\psi_{OVP} = \frac{\dot{E}x^{W}_{net} + \dot{m}_{12}ex_{12}}{\dot{m}_1 ex_1 - \dot{m}_{11}ex_{11} - \dot{m}_{18}ex_{18}}$$

where $\dot{W}_{net,TFDP}$ can be defined as:

$$\dot{W}_{net} = \dot{W}_{T1} + \dot{W}_{T2} + \dot{W}_{T3}$$

Based on the thermodynamic analysis balance equations, the results of the energy and exergy analyses for the geothermal energy–based power production plant including a freshwater through MED unit are given in Table 6.26. As given in Table 6.26, the energetic and exergetic efficiencies of the overall plant are calculated as 0.3819 and 0.6524, respectively.

The power production rates and mass flow rate of producing freshwater by utilizing the turbines and the MED unit are written in Table 6.27. The net power production rate from the geothermal energy–based power production plant including freshwater through a MED unit is equal to 43,206 kW.

Fig. 6.61 presents the effects of reference temperature on the energy and exergy efficiencies of the multistaged direct geothermal energy system with three turbines integrated with a MED unit and its subsystems. As illustrated in Fig. 6.61, the energy and exergy efficiencies of the overall plant vary

TABLE 6.26 Energetic and exergetic efficiencies of the geothermal energy–based power production plant including freshwater through a multieffect distillation unit.

Subsystems/overall system	Energy efficiency (%)	Exergy efficiency (%)
Multistaged direct system with three turbines	31.67	39.29
Multieffect distillation unit	34.95	31.47
Overall plant	42.58	46.13

TABLE 6.27 Useful outputs from the multistaged direct geothermal energy system combined with multieffect distillation unit.

Useful outputs	Value
Power production from turbine 1, $\dot{W}_{T1}$	12,805 kW
Power production from turbine 2, $\dot{W}_{T2}$	16,838 kW
Power production from turbine 3, $\dot{W}_{T3}$	14,018 kW
Mass flow rate of producing freshwater, $\dot{m}_{FW}$	18.43 kg/s

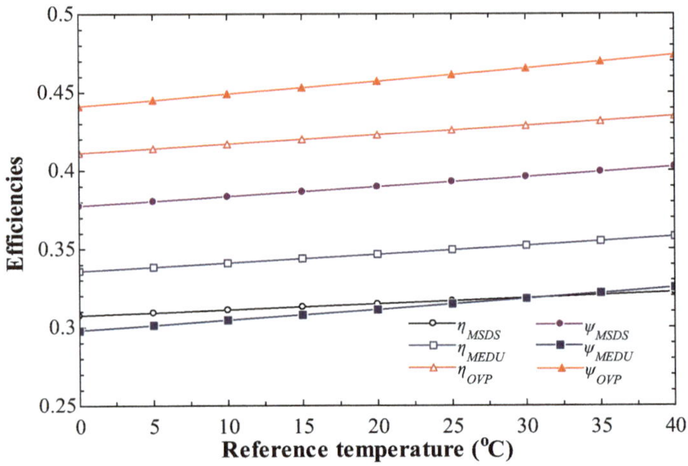

FIGURE 6.61 Impact of the reference temperature on the energy and exergy efficiencies of the whole system and its subsystems.

from 41.12% to 43.48% and from 44.1% to 47.38%, respectively. Also, the energy and exergy efficiencies of the MED unit increase from 33.58% to 35.79% and from 29.79% to 32.51%, respectively.

Fig. 6.62 shows the effect of increasing the reference temperature from 0°C to 40°C on the useful output production rates. The power production rate is observed to vary from 12,244 to 13,153 kW for steam turbine 1, from 16,180 to 17,245 kW for steam turbine 3, from 13,537 to 14,314 kW, respectively. In addition to that, the freshwater production rate of geothermal energy–based combined plant increases from 17.43 to 19.04 kg/s within the same reference temperature interval.

The effects of a rise in geothermal water mass flow rate on the energy and exergy efficiencies of the multistaged direct geothermal energy system with three turbines integrated with a MED unit and its subsystems are shown in Fig. 6.63. The energy and exergy efficiencies of the overall plant are

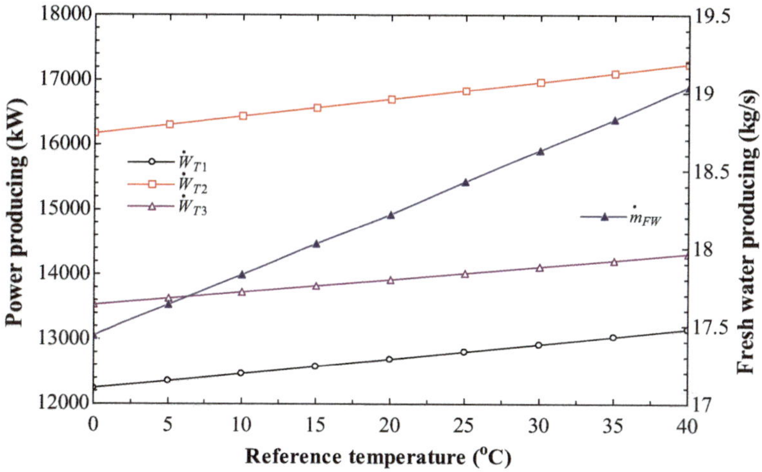

FIGURE 6.62 Impact of the reference temperature on the power and freshwater production rates.

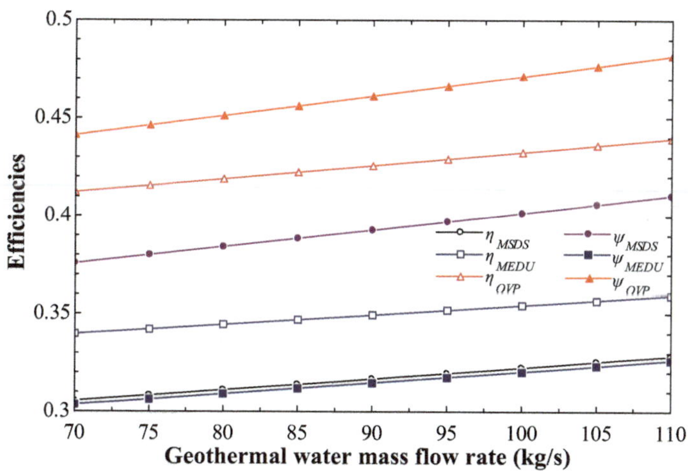

FIGURE 6.63 Impact of the geothermal water mass flow rate on the energy and exergy efficiencies of the whole system and its subsystems.

observed to vary from 41.24% to 43.95 % and from 44.15% to 48.19%, respectively, with a rise in geothermal water mass flow rate from 70 to 110 kg/s. Also, the energy and exergy efficiencies of the MED subsystem increases from 33.98% to 35.93% and from 30.36% to 32.61%, respectively, within the same geothermal water mass flow rate interval.

The power and freshwater production rates versus geothermal water mass flow rate are given in Fig. 6.64. As can be observed from these results, the power production rate increases from 12,112 kW to 13,537 kW for steam

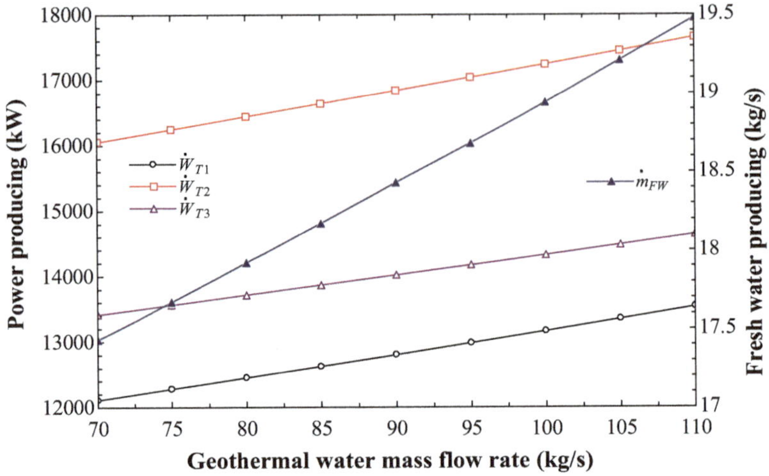

FIGURE 6.64 Impact of the geothermal water mass flow rate on the power and freshwater production rates.

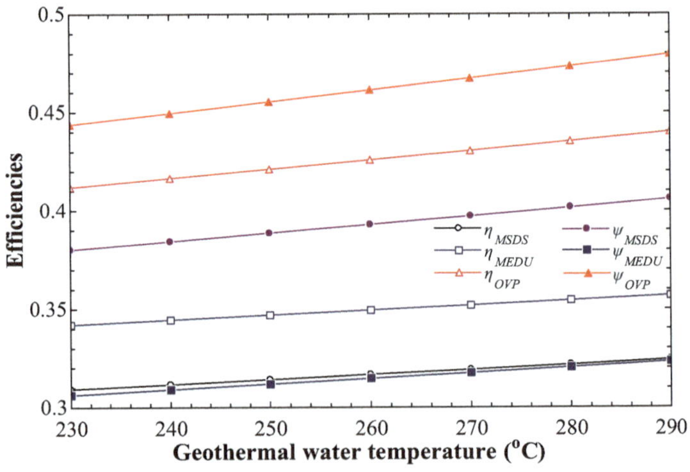

FIGURE 6.65 Impact of the geothermal water temperature on the energy and exergy efficiencies of the whole system and its subsystems.

turbine 1, from 16,053 kW to 17,660 kW for steam turbine 2, and from 13,417 kW to 14,644 kW, respectively. Also, the freshwater production rate rises from 17.43 kg/s to 19.48 kg/s within the same geothermal water mass flow rate interval.

Fig. 6.65 shows that energy and exergy efficiencies of geothermal energy– based combined plant increases with the increasing geothermal water temperature from 230°C to 290°C. It can be seen that, as expected, the

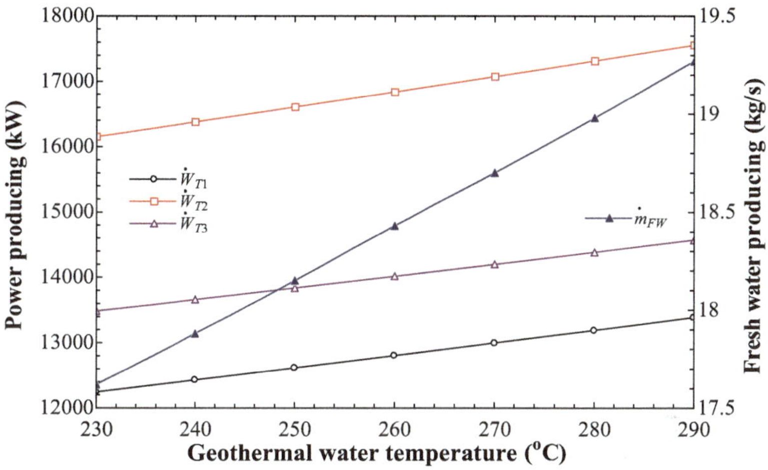

FIGURE 6.66 Impact of the geothermal water temperature on the power and freshwater production rates.

energy and exergy efficiencies of the multistaged direct geothermal energy system with three turbines integrated with a MED unit increase from 41.21% to 44.02% and from 44.37% to 47.95%, respectively. Also, the energy and exergy efficiencies of the MED unit rises from 34.22% to 35.68% and from 30.63% to 32.31%, respectively.

The effect of an increase in geothermal water temperature on the useful outputs from the multistaged direct geothermal energy system with three turbines integrated with a MED unit is illustrated in Fig. 6.66. As shown in the figure, an increase in geothermal water temperature has a positive effect on both useful outputs. The power production rate increases from 12,245 to 13,389 kW for steam turbine 1, from 16,150 to 17,555 kW for steam turbine 2, and from 13,485 to 14,570 kW, respectively. Also, the freshwater production rate rises from 17.62 to 19.27 kg/s within the same geothermal water temperature interval.

6.7.6 Combined/integrated system for power and heating

The most general form of the geothermal energy–based combined power generation system is clearly presented in Fig. 6.67. Looking at the system, it can be seen that the system consists of three subsystems. Explaining the overall operation of this combined power generation system helps to understand the operating principle of the system. Therefore, in this section, what activities this combined power generation system generally contains and which useful products are obtained with these activities are explained.

As can be seen in Fig. 6.67, the combined power generation system starts to operate when the geothermal fluid enters the system with the number 1

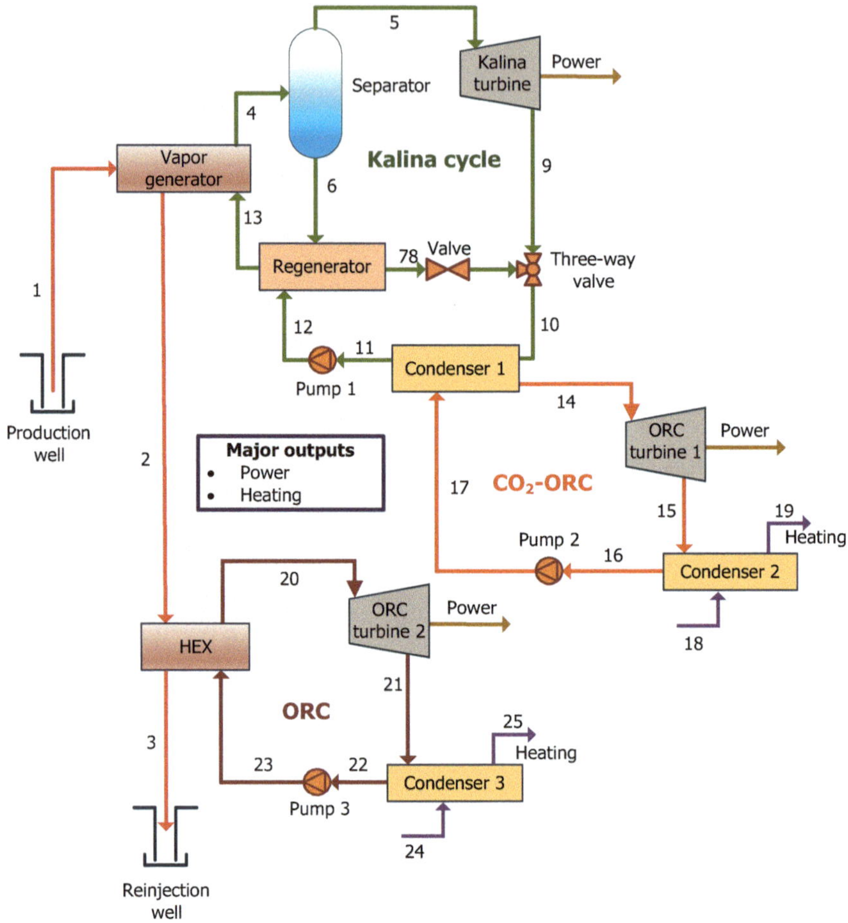

FIGURE 6.67 Schematic flow diagram of the geothermal energy–based combined power production plant.

flow. Heat transfer to the working fluid in the Kalina cycle takes place in the vapor generator with the geothermal fluid loaded into the system with flow 1. So there is a heat transfer between the geothermal fluid and the Kalina cycle working fluid. The working fluid that is energized by taking advantage of the geothermal fluid is sent to the separator with flow 4. A rich mixture is tried to be obtained by increasing the percentage of ammonia in the working fluid in the separator. The working fluid, which is a rich mixture with increased ammonia percentage in the separator, is transferred to the Kalina turbine with flow 5 for electricity production as useful output.

The working fluid with rich ammonia percentage coming to the turbine is expanded between the 5 and 9 numbered flows, and electrical output is obtained. In the meantime, while the working fluid with a high percentage of ammonia is transferred to the turbine, the mixture with a low percentage of ammonia is sent to the regenerator with flow 6. The working fluid with a low ammonia percentage in the regenerator is transferred to the valve by flow 7, where the pressure of the working fluid is reduced. Ammonia-rich working fluid from the turbine with flow 9 and the low percentage of ammonia coming from the valve with flow 8 are combined on the three-way valve and sent to condenser 1 with flow 10. Here, heat transfer takes place between the Kalina cycle working fluid coming to condenser 1 and the ORC working fluid. Some of the energy in the working fluid in the Kalina cycle is transferred to the working fluid in the ORC cycle. To increase the pressure of the Kalina cycle working fluid from condenser 1, this working fluid is sent to pump 1 with flow 11. The working fluid, whose pressure is increased, is sent to the regenerator with flow 12 to increase the temperature. In the regenerator, the temperature of the working fluid is increased by the temperature of the working fluid with a low percentage of ammonia leaving the separator. The working fluid, whose temperature is increased, is transferred to the steam generator with flow 13 to reload it with energy. Thus, the Kalina cycle, a subsystem of the combined power generation system, is completed. However, since the subsystems work integrated with one another, it is useful to talk about the CO_2−ORC subsystem in the continuation of this cycle to understand the working principle of the system.

In the CO_2−ORC cycle, the working fluid is sent to condenser 1 with flow 17, thus increasing the temperature of the working fluid. Working fluid whose temperature is increased is sent to ORC turbine 1 with flow 14 for electricity generation. Here, electricity is produced as a useful product by expanding the working fluid between flows 14 and 15. The working fluid from the ORC turbine 1 is then sent to condenser 2 with flow 15 to obtain the heating output. The CO_2−ORC working fluid condensed in condenser 2 is sent to pump 2 with flow 16 to increase the pressure. The working fluid whose pressure is increased is sent to condenser 1 with flow 17 to increase the temperature, and the CO_2−ORC subsystem, another subsystem of the combined power system, is completed.

The third subsystem in the combined power generation system is the ORC subsystem. The working fluid required for this subsystem is selected as R152a. The geothermal fluid coming out of the steam generator is sent to a heat exchanger (HEX) with flow 2. Here, the geothermal fluid transfers some of its energy to the ORC working fluid. For this, heat transfer takes place between the geothermal and ORC working fluid. The energized ORC working fluid is sent to ORC turbine 2 with flow 20 for electricity generation. In ORC turbine 2, working fluid is expanded between flows 20 and 21, producing electrical output, a useful product. The working fluid from the ORC turbine 2 is transferred to condenser 3 with flow 21 for heating output. The working fluid condensed in condenser 3 is sent to pump 3 with flow 22 to increase the pressure. The working fluid, whose pressure is increased, is transferred to the HEX with flow 23 for energy reload. In this way, the ORC subsystem, the third subsystem of the combined power generation system, is also completed. Also, the geothermal fluid coming out of the HEX is sent underground again to ensure sustainability. The importance of the concept of sustainability is comprehensively described in the relevant sections. In general, the geothermal–based combined power generation system works in this way. The subsystems of this combined power system are described in more detail in the relevant sections. In this section, the working principle of the combined power generation plant, which uses geothermal energy in its most general form, is described.

6.7.7 Case study 6.11

In this case study, the geothermal energy–based combined power plant is analyzed by using thermodynamic analysis. The simplified flow diagram of the combined geothermal power generation system in the first case study is illustrated in Fig. 6.67. To model the geothermal energy–based combined power production plant, some parameters are selected as input data for the thermodynamic analysis. The input data for the energetic and exergetic assessments are listed in Table 6.28.

The energy and exergy efficiencies for the Kalina cycle are given:

$$\eta_{KC} = \frac{\dot{W}_{KT}}{(\dot{m}_4 h_4 - \dot{m}_{13} h_{13}) + \dot{W}_{P1}}$$

and

$$\psi_{KC} = \frac{\dot{E}x^{W}_{KT}}{(\dot{m}_4 ex_4 - \dot{m}_{13} ex_{13}) + \dot{E}x^{W}_{P1}}$$

The energy and exergy efficiencies for the CO_2–ORC cycle can be defined:

$$\eta_{ORC1} = \frac{\dot{W}_{ORCT1}}{(\dot{m}_{14} h_{14} - \dot{m}_{17} h_{17}) + \dot{W}_{P2}}$$

and

$$\psi_{ORC1} = \frac{\dot{E}x^{W}_{ORCT1}}{(\dot{m}_{14} ex_{14} - \dot{m}_{17} ex_{17}) + \dot{E}x^{W}_{P2}}$$

The energy and exergy efficiencies for the ORC cycle are defined as:

$$\eta_{ORC2} = \frac{\dot{W}_{ORCT2}}{(\dot{m}_{20} h_{20} - \dot{m}_{23} h_{23}) + \dot{W}_{P3}}$$

TABLE 6.28 Working parameters of the geothermal energy–based combined power production plant.

Variables	Values
Geofluid source temperature, T_1	210°C
Geofluid source pressure, P_1	1800 kPa
Geofluid mass flow rate, $\dot{m}_1$	140 kg/s
Isentropic efficiency of turbines, η_{tur}	0.75
Separator inlet temperature, T_4	160°C
Separator inlet pressure, P_4	2323 kPa
Kalina turbine inlet temperature, T_5	160°C
Kalina turbine inlet pressure, P_5	2323 kPa
Vapor generator exit temperature, T_2	96°C
Working fluid of ORC 1	CO_2
ORC turbine 1 inlet temperature, T_{14}	35°C
ORC turbine 1 inlet pressure, P_{14}	2000 kPa
Working fluid of ORC 2	R152a
ORC turbine 2 inlet temperature, T_{20}	90°C
ORC turbine 2 inlet pressure, P_{20}	1000 kPa
Geofluid reinjection temperature, T_3	61°C

and

$$\psi_{ORC2} = \frac{\dot{E}x^{W}_{ORCT2}}{(\dot{m}_{20}ex_{20} - \dot{m}_{23}ex_{23}) + \dot{E}x^{W}_{P3}}$$

The energy and exergy efficiencies for the whole system can be described:

$$\eta_{WS} = \frac{\dot{W}_{Net}}{\dot{m}_1(h_1 - h_3)}$$

and

$$\psi_{WS} = \frac{\dot{E}x^{W}_{Net}}{\dot{m}_1(ex_1 - ex_3)}$$

The total energy production rate of the geothermal energy–based combined plant is defined:

$$\dot{W}_{Net} = \dot{W}_{KT} + \dot{W}_{ORCT1} + \dot{W}_{ORCT2} - \dot{W}_{p1} - \dot{W}_{p2} - \dot{W}_{p3}$$

Power productions from the Kalina turbine, CO_2–ORC turbine, and ORC turbine can be calculated by the following equations, and the calculated results are given in Table 6.29.

$$\dot{W}_{KT} = \dot{m}_5(h_5 - h_9),$$

$$\dot{W}_{ORCT1} = \dot{m}_{14}(h_{14} - h_{15}),$$

and

$$\dot{W}_{ORCT2} = \dot{m}_{20}(h_{20} - h_{21})$$

The graph showing the relationship between the geothermal water mass flow rate and the net power generation obtained from the system, the exergy destruction rate of the system and the exergy efficiency of the system are given in Fig. 6.68. As seen in the graphic, the mass flow rate is gradually increased between 100 and 180 kg/s in order to examine the effect of the geothermal water mass flow rate on these three performance measurements. It can be clearly seen that the increase in mass flow rate increases the values of all three parameters. When the mass flow rate is 100, the net power generation, exergy destruction rate, and exergy efficiency values are 3912 kW, 3260 kW, and 0.5496, respectively. When the mass flow rate is 180, the net power generation, exergy destruction rate, and exergy efficiency values are 6671 kW, 4966 kW, and 0.649, respectively. Considering the parameter values in these two flow rates, it can be seen that the amount of energy produced per unit flow rate is higher than the exergy destruction rate per unit flow rate. The exergy efficiency of the system increases from 0.5 to 0.6 with the mass flow rate, increasing from 100 to 180 kg/s, respectively.

TABLE 6.29 Power production and exergy destruction rates, and exergy efficiency for geothermal energy–based combined plant.

Plant outputs	Values
Kalina turbine, $\dot{W}_{KT}$	3299 kW
CO_2-ORC turbine, $\dot{W}_{ORCT1}$	683 kW
ORC turbine, $\dot{W}_{ORCT2}$	1127 kW
Exergy destruction rate, $\dot{E}x_{D,Total}$	4024 kW
Exergy efficiency, ψ_{WS}	59.73%

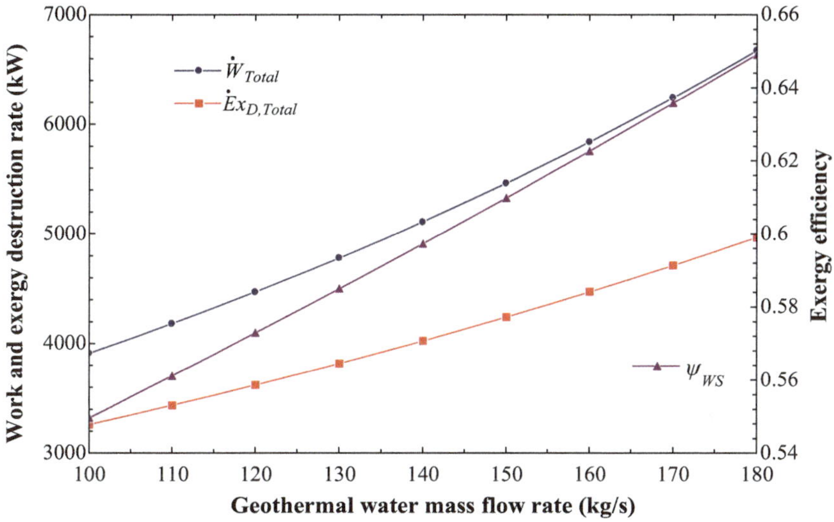

FIGURE 6.68 Effect of geothermal working fluid mass flow rate on net power generation, exergy destruction rates and exergy efficiency for the geothermal energy–based combined plant.

The characteristics of the net power generation, exergy destruction rate, and exergy efficiency of the geothermal energy–based combined plant with the gradual increase of geothermal water temperature are shown in the graph in Fig. 6.69. The geothermal water temperature is gradually increased from 150°C to 230°C in order to perform the analysis of the variation of the characteristics of the variables showing this system performance depending on the geothermal water temperature. With the gradual increase of the geothermal water temperature, the increase in their values in these three variables can be clearly seen from the performance curves in Fig. 6.69. When the geothermal water temperature is 150°C, 200°C, and 230 °C, the net power generation of the combined power plant is 3684, 4838, and 5697 kW, respectively. The exergy destruction rate values of the combined power generation system in these geothermal water temperature values are 3161, 3865, and 4360 kW, respectively. The exergy efficiency values of the geothermal energy–based combined plant at these geothermal temperature values are 0.515, 0.5827, and 0.6275, respectively.

Another parameter that has potential for system performance is the ambient temperature. The changes in the geothermal energy–based combined plant's net power generation, exergy destruction rate, and exergy efficiency due to the gradual increase in the ambient temperature are shown in the graph in Fig. 6.70. To analyze the effect of the reference temperature on these three variables, the reference temperature is gradually increased from 0°C to 40°C. As can be seen from the performance curves in the figure, the gradual increase in the ambient temperature increases the values of net

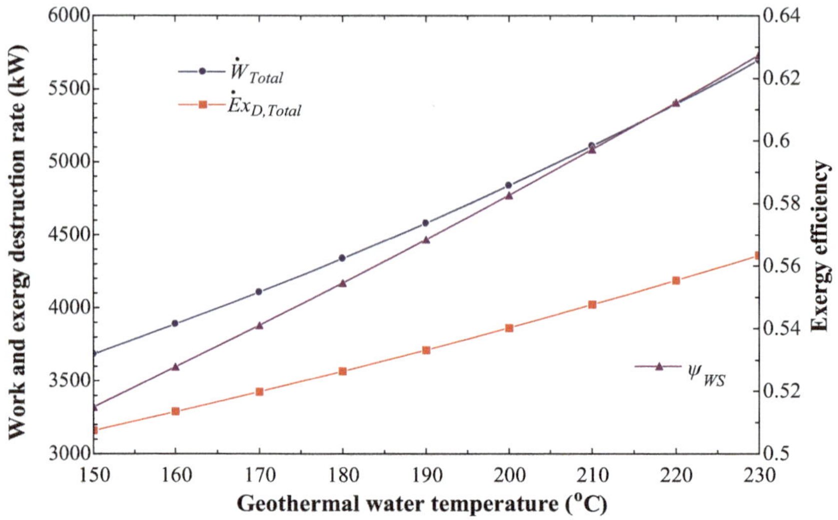

FIGURE 6.69 Effect of geothermal water temperature on net power generation, exergy destruction rate, and exergy efficiency for the geothermal energy–based combined plant.

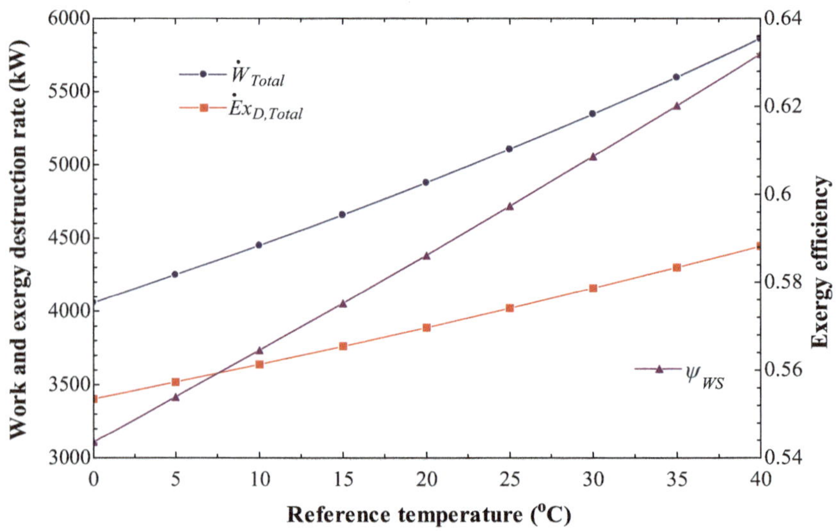

FIGURE 6.70 Effect of reference temperature on net power generation, exergy destruction rate, and exergy efficiency for the geothermal energy–based combined plant.

power generation, exergy destruction rate, and exergy efficiency. When the reference temperature is 10°C, 25°C, and 40 °C, the net power generation of the combined power generation system is 4451, 5109, and 5863 kW, respectively. The exergy destruction rate values of the combined power generation system in these reference temperature values are 3639, 4024, and 4448 kW, respectively. The exergy efficiency values of the geothermal energy–based combined plant at these reference temperature values are 0.5645, 0.5973, and 0.6319, respectively.

6.7.8 Combined/integrated system for cooling production

Absorption cooling systems (ACSs) are one of the most widely used systems in geothermal energy–supported cooling systems [11]. ACSs are generally classified into a refrigerant and combination of an absorber, by type of fluid. Most ACSs use a combination of either lithium bromide–water or ammonia–water as the refrigerant [12].

Usually, the ACSs have been widely used in the geothermal energy–based cooling application because of the low-temperature applications [13]. Using the thermal energy of the working fluid of the power plant at the turbine outlet, it is common to use it in the operation of the cooling plant. Single-, double-, triple-, and quadruple-ACSs are currently used systems.

The single-effect absorption cooling system (SEACS) is one of the first designed kind of ACSs. SEACSs are generally designed with six main components: generator, condenser, absorber, pump, HEX, and evaporator.

The double-effect absorption cooling system (DEACS) is a design development of a single-effect cooling system to achieve more useful output. The design of the DEACS is accomplished by adding a condenser HEX to the system, together with the basic components in the SEACS. In this system, some of the components that are common with SEACS differ in number. These systems are generally designed with a generator, a HEX, and a condenser HEX in addition to the components in SEACS [14].

The triple-effect absorption cooling system (TEACS) is designed by adding a generator and HEX to the dual effect ACS. In other words, this system includes three generators (high-temperature generator, medium-temperature generator, low-temperature generator) and three HEXs (high-temperature HEX, medium-temperature HEX, and low-temperature HEX). The cooling output is obtained with these generators, HEXs, and other basic components in the TEACS.

The quadruple-effect absorption cooling system (QEACS) is designed by adding a generator and a HEX to the components in the TEACS. Since there are more components in the design of this system than in other ACSs, the performance analysis of the system should be made more comprehensively during the design phase. Because of the increase in the number of components, the system has the potential to affect the performance of the system.

The general flowchart of cooling effect generation via geothermal energy is presented in Fig. 6.71. In addition, in this section of the proposed book, geothermal energy–supported power and cooling production systems are described in detail.

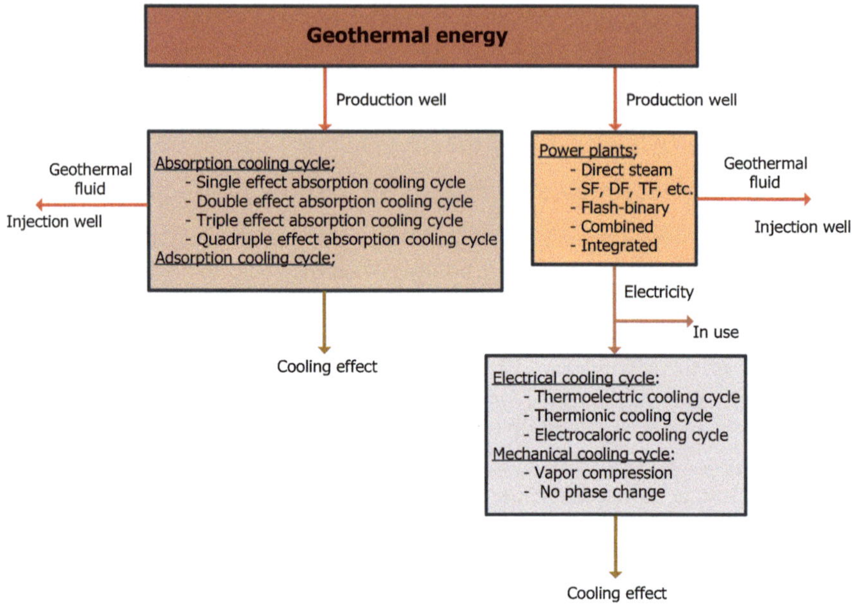

FIGURE 6.71 General flowchart of the cooling effect generation via geothermal energy.

6.7.8.1 Absorption cooling with ejector combined system with geothermal energy

The schematic flow diagram of single-effect absorption cooling (SEAC) with an ejector combined system with geothermal energy is depicted in Fig. 6.72. The main energy source of this proposed system is geothermal water. Basically, the suggested geothermal energy–based cooling production system includes three main subcycles:(1) geothermal water cycle, (2) Kalina cycle with ammonia–water working fluid, and (3) SEAC system with ejector.

First, underground geothermal energy enters the superheater unit of the Kalina cycle at state 1, where the thermal energy required for the working of the Kalina cycle is provided. Subsequently, the geothermal water, which transfers some of its thermal energy, then enters the evaporator at state 2 and enters the economizer at state 3, respectively. The geothermal water coming out of the economizer unit enters the generator of the SEAC subcycle at state 4 and is then reinjected at the lower temperature again at state 5. Therewithal, geothermal fluid transfers its thermal energy to the Kalina cycle and SEAC subsystems the at superheater, evaporator, economizer, and generator components, respectively, and after that generally power, as well as cooling, generation is performed in these systems.

Another subcycle of the proposed system is the Kalina cycle, which is working ammonia–water working fluid. Since the Kalina cycle operates at

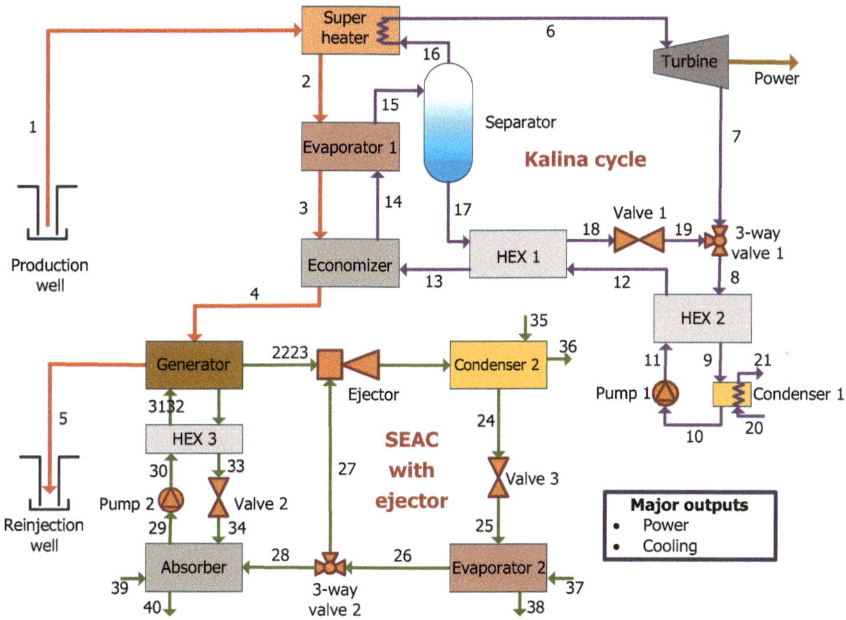

FIGURE 6.72 Schematic flow diagram of geothermal energy combined with SEAC with ejector cooling plant. *SEAC*, Single-effect absorption cooling.

low temperatures, it is widely used especially in geothermal energy applications. The Kalina cycle mainly consists of a separator, turbine, HEXs, condenser, economizer, evaporator, and superheater. The ammonia–water mixture working fluid from state 16 enters the turbine in the vapor phase by taking the heat of the geothermal fluid in the superheater section, and then electrical power is produced in the turbine part. The rich ammonia–water solution from state 7 and the weak ammonia–water solution from state 19 enters the three-way valve, and then it condenses in the condenser subcomponent between the states 9 and 10, respectively. Later, the Kalina cycle continues with the increasing temperature between states from 11 to 15, respectively, and the system operates in this way continuously.

The last other subsystem shown in Fig. 6.72 is the SEAC system with an ejector. The SEAC system includes a generator, condenser, absorber, evaporator, three-way valve, HEX, and ejector. By transferring the geothermal fluid heat to the SEAC subsystem in the generator, the thermal energy required for the cooling system is provided. The SEAC system can work with LiBr-H_2O or NH_3-H_2O working fluids. Steam coming out of point 22 enters the ejector, condenser throttle valve, and evaporator, respectively. Cooling occurs by pulling heat from the external environment in the evaporator, and the working fluid leaves here as steam and then enters the three-way valve at state 26. The steam coming from state 28 and the weak solution from state 34 mix in the absorber and then pass through the 29, 30, and 31 states and come to the generator. In the SEAC system, while cooling can be done in the evaporator, it is also heated in the condenser, and it is preferred in geothermal-supported systems because it is suitable for low temperatures. Consequently, as summarized in Fig. 6.72, power and cooling can be produced in the Kalina and SEAC subcycles with geothermal water thermal energy.

6.7.9 Case study 6.12

In the second case study, the schematic flow diagram of geothermal energy combined with the SEAC cooling plant is investigated by utilizing the energetic and exergetic viewpoints. The simplified flow diagram of the combined geothermal cooling plant for this case study is shown in Fig. 6.72. Also, the working parameters of this cooling plant are given in Table 6.30.

The ejector subsystem, which is utilized in the single-effect cooling process, has three sections: the nozzle, mixing chamber, and diffuser. For the nozzle section, the energy equation for the adiabatic main current based on the steady-state condition should be:

$$h_{n1} = h_{n2} + \left(v_{n2}^2\right)/2$$

The nozzle performance needs to be completed:

$$\eta_{nz} = \frac{(h_{n1} - h_{n2})}{\left(h_{n1} - h_{n2,s}\right)}$$

Based on these equations, the outlet velocity of the main flow is described as:

$$v_{n2} = \sqrt{2\eta_{nz}\left(h_{n1} - h_{n2,s}\right)}$$

To make a performance assessment, the energetic and exergetic performance equations for subplants and the whole plant are given. The energy and exergy efficiency equations for the Kalina cycle are:

$$\eta_{KC} = \frac{\dot{W}_{KT} + \dot{m}_6(h_6 - h_7)}{\dot{m}_{13}(h_{14} - h_{13}) + \dot{m}_{14}(h_{15} - h_{14}) + \dot{m}_{16}(h_6 - h_{16})}$$

and

$$\psi_{KC} = \frac{\dot{W}_{KT} + \dot{m}_6(ex_6 - ex_7)}{\dot{m}_{13}(ex_{14} - ex_{13}) + \dot{m}_{14}(ex_{15} - ex_{14}) + \dot{m}_{16}(ex_6 - ex_{16})}$$

TABLE 6.30 Working parameters of the geothermal energy combined with a SEAC plant.

Parameters	Values
Reference temperature, T_o	20°C
Reference pressure, P_o	101.3 kPa
Geofluid source temperature, T_1	210°C
Geofluid source pressure, P_1	1800 kPa
Geofluid mass flow rate, $\dot{m}_1$	140 kg/s
Isentropic efficiency of turbines, η_{tur}	0.75
Effectiveness of superheater, evaporator and economizer	0.78
Working fluid of Kalina cycle	Ammonia–water
Separator inlet temperature, T_{15}	115°C
Turbine inlet temperature, T_6	160°C
Turbine inlet pressure, P_6	2323 kPa
Turbine exit pressure, P_7	459 kPa
Working fluid of SEAC	Ammonia–water
Generator inlet temperature, T_4	96°C
Evaporator temperature, T_{eva}	8°C
Energetic coefficient of performance, COP_{en}	1.052
Exergetic coefficient of performance, COP_{ex}	0.284
Ejector entrainment ratio of the ejector, ϕ_{ejec}	0.4460
Geofluid reinjection temperature, T_5	61°C

SEAC, Single-effect absorption cooling.

The energy and exergy efficiency equations for the single-effect absorption cooling with the ejector subsystem can be defined as:

$$\eta_{SEACE} = \frac{\dot{Q}_{Cooling}}{(\dot{m}_4 h_4 - \dot{m}_5 h_5) + \dot{W}_{P2}}$$

and

$$\psi_{SEACE} = \frac{\dot{E}x^{Q}_{Cooling}}{(\dot{m}_4 ex_4 - \dot{m}_5 ex_5) + \dot{W}_{P2}}$$

The energy and exergy efficiency equations for the whole system can be written:

$$\eta_{WS} = \frac{\dot{W}_{Net} + \dot{Q}_{Cooling}}{\dot{m}_1(h_1 - h_5)}$$

and

$$\psi_{WS} = \frac{\dot{E}x^{W}_{Net} + \dot{E}x^{Q}_{Cooling}}{\dot{m}_1(ex_1 - ex_5)}$$

Total energy production of the geothermal energy based combined system can be calculated:

$$\dot{W}_{Net} = \dot{W}_{KT} - \dot{W}_{p1} - \dot{W}_{p2}$$

Also, the calculated results for this parametric study are given in Table 6.31. To ensure the accuracy and validity of the ESS simulation code, the PEM electrolyzer simulation is validated with the experimental data from the open literature.

Some parameters guide the characteristics of the system performance in a geothermal energy–based system. One of these parameters is the geothermal water mass flow rate. Due to the gradual increase of the geothermal water mass flow rate, the change in useful outputs obtained from the system, exergy destruction of the system, and exergy efficiency of the system are shown in the graph in Fig. 6.73. In order to analyze this change, the geothermal water mass flow rate is gradually increased from 100 to 180 kg/s. Considering the performance curves in the figure in general, the gradual increase in mass flow rate increases the values of useful outputs, exergy destruction, and exergy efficiency. In cases where the geothermal water mass flow rate is 100, 150, and 180 kg/s, the net power generations of the system are 3355, 4684, and 5722 kW, respectively. For these geothermal water flow rate values, the cooling outputs obtained from the system are

TABLE 6.31 Calculated results for the geothermal energy combined with SEAC plant.

Plant outputs	Values
Obtained work rate by Kalina turbine, $\dot{W}_{KT}$	4382 kW
Cooling producing rate, $\dot{Q}_{Cooling}$	3676 kW
Total exergy destruction rate, $\dot{E}x_{D,total}$	6391 kW
Whole system exergy efficiency, η_{WS}	0.5608

SEAC, Single-effect absorption cooling.

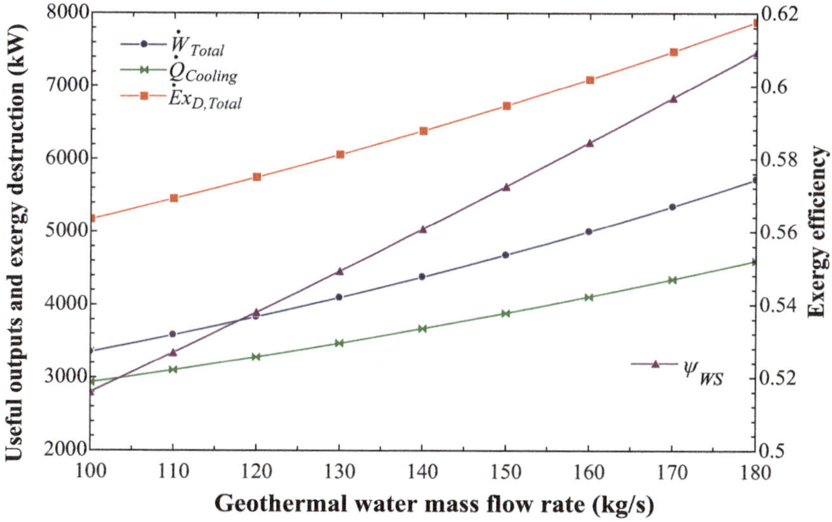

FIGURE 6.73 Effect of geothermal water mass flow rate on net power generation, cooling effect production, and exergy destruction rate, as well as exergy efficiency, for the geothermal energy combined with the SEAC plant. *SEAC*, Single-effect absorption cooling.

2933, 3889, and 4605 kW, respectively. In addition, the exergy destruction rates of the system for these mass flow rate values are 5178, 6736, and 7887 kW, respectively. An important performance criterion directed by the geothermal water mass flow rate is the exergy efficiency of the system. When the geothermal water flow rates are 100, 140, 160, and 180 kg/s, the combined system's exergy efficiencies are 0.516, 0.5608, 0.5846, and 0.6094, respectively.

Another parameter that has an important potential for system performance is geothermal water temperature. Depending on the gradual increase of the geothermal water temperature, the change in beneficial outputs produced in the system, the exergy destruction of the system, and the exergy efficiency of the system are clearly shown in the graph in Fig. 6.74. Here, the gradual increase of geothermal water temperature increases beneficial outcomes, exergy destruction, and exergy efficiency. Looking at the graph in the figure, the change in performance curves can be seen generally due to the gradual increase of the geothermal water temperature from 150°C to 230°C. When the geothermal water temperatures are 150°C, 200°C, and 230°C, the combined system's net power generations are 3160, 4149, and 4886 kW, respectively. For these geothermal water temperature values, the cooling outputs obtained from the system are 2758, 3504, and 4045 kW, respectively. Also, the exergy destruction rates of the system for these geothermal water temperature values are 5021, 6139, and 6925 kW, respectively. In addition to these, in cases where the geothermal water temperatures are

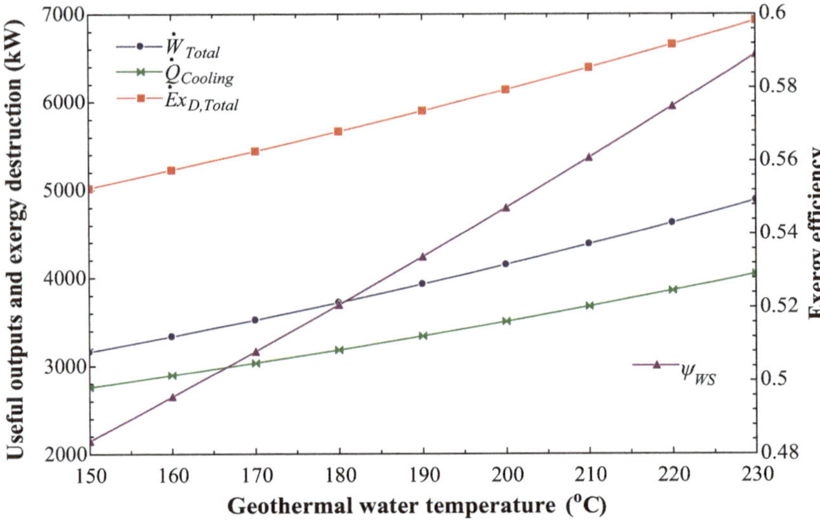

FIGURE 6.74 Effect of geothermal water temperature on net power generation, cooling effect production, and exergy destruction rate, as well as exergy efficiency, for the geothermal energy combined with SEAC plant. *SEAC*, Single-effect absorption cooling.

150°C, 180°C, 210°C, and 230°C, the combined system's exergy efficiencies are 0.4835, 0.5207, 0.5608, and 0.5891, respectively.

Another parameter that guides the system performance is the ambient temperature. When the ambient temperature is increased gradually, changes occur in the useful outputs obtained from the system, in the exergy destruction and in the exergy efficiency of the system. These changes depending on the ambient temperature are also shown in the graph in Fig. 6.75. In order to see the effect of the ambient temperature on the system, the ambient temperature is gradually increased from 0°C to 40°C. As can be clearly seen from the graph in Fig. 6.75, the gradual increase in the ambient temperature increases the net power generation, exergy destruction, and exergy efficiency of the combined system, while reducing the cooling output from the combined system. When the reference temperature values are 10°C, 25°C, and 40°C, the combined system's net power generations are 3817, 4382, and 5029 kW, respectively. For these reference temperature values, the cooling outputs obtained from the system are 3970, 3676, and 3403 kW, respectively. As can be seen from the values of the cooling output, as the ambient temperature increases, the amount of cooling output obtained from the combined system decreases. Also, the exergy destruction rates of the system for these reference temperature values are 5781, 6391, and 7065 kW, respectively. In addition to these, in cases where the reference temperatures are 10°C, 25°C, 30°C, and 40°C, the combined system's exergy efficiencies are 0.53, 0.5608, 0.5714, and 0.5933, respectively.

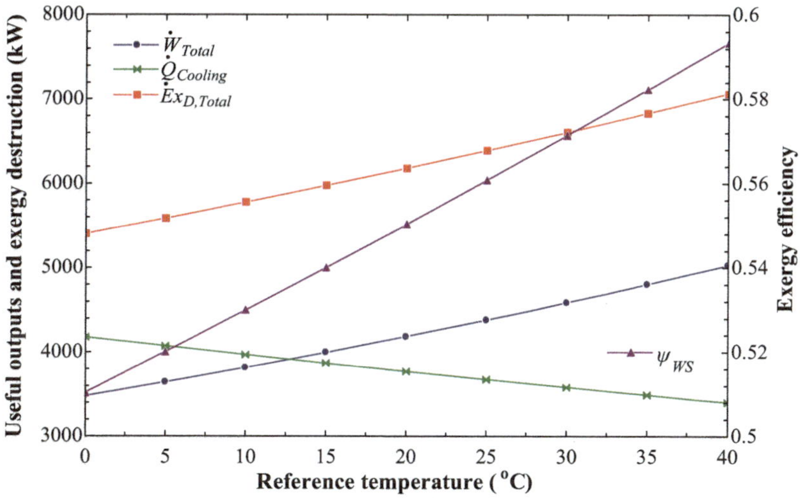

FIGURE 6.75 Effect of reference temperature on net power generation, cooling effect production, and exergy destruction rate, as well as exergy efficiency, for the geothermal energy combined with SEAC plant. *SEAC*, Single-effect absorption cooling.

6.7.9.1 Ejector cooling combined with geothermal energy

The second proposed system in this section of the chapter is the ejector cooling combined with the geothermal power plant, as shown in Fig. 6.76. The proposed geothermal energy–based power plant integrated with the ejector cooling subsystem shown in Fig. 6.10 mainly consists of a geothermal fluid cycle, an ORC, and an ejector cooling system. The main difference between the ejector cooling system from traditional vapor compression cooling systems is that the compressor is replaced by the HEX, circulation pump, and ejector. In general, an ejector consists of a nozzle, a suction chamber, a mixing chamber, and a diffuser.

First, geothermal fluid enters the ORC generator at state 1 where the thermal energy transfers from the geothermal fluid to the generator, and the geothermal fluid is reinjected at state 2. ORC working fluid comes as a saturated liquid from state 6; after that, it enters the turbine as superheated steam at state 3 with the thermal energy received from the geothermal fluid in the generator. The ORC fluid in the liquid–vapor mixture exiting from point 4 transfers its heat to the HEX 1 in the ejector cooling system working fluid and then enters the pump as a saturated liquid at point 5. The ORC system continues to operate continuously in this way.

The ejector cooling system, driven by thermal energy, has been widely used in cooling applications. The system has some advantages, such as being simple and without moving parts. The biggest advantage is that they can produce cooling using waste heat or solar energy. The ejector cooling system includes a

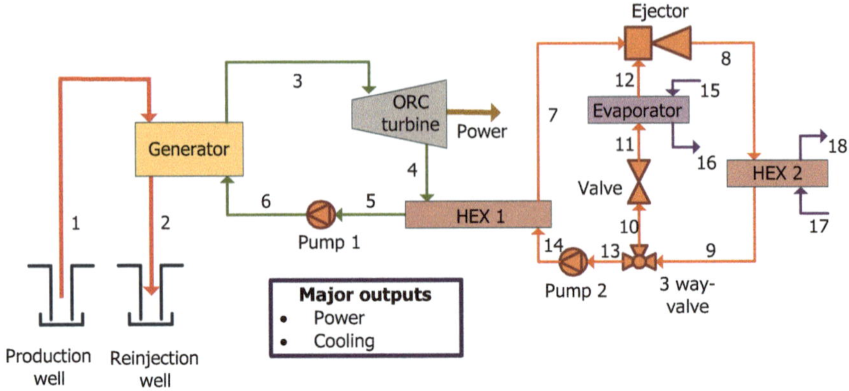

FIGURE 6.76 Schematic flow diagram of the ejector cooling combined with a geothermal power plant.

pump, an evaporator, an ejector, a three-way valve, an expansion valve, and two HEXs, as seen between states 7 and 18 in Fig. 6.76. In HEX 1, the waste heat from the ORC turbine is transferred to the cooling system with an ejector, and the thermal energy required for the system operation is provided. In this integrated process, the ORC working fluid enters the HEX, which then supplies high-temperature working fluid to the ejector subcomponent where the pressure rate of the high-temperature working fluid drops. The exiting low-pressure working fluid from the ejector passes through the evaporator where the water–vapor mixture extracts heat from the working fluid flowing through the evaporator. Finally, the cooling process occurs in the evaporator section.

6.7.10 Case study 6.13

In this case study, the schematic flow diagram of ejector cooling combined with the geothermal power plant is assessed by using energy and exergy analyses. The simplified flow diagram of the ejector cooling combined with a geothermal power plant for this case study is shown in Fig. 6.76. In addition, the working parameters of the ejector cooling combined with geothermal power are given in Table 6.32.

The energy and exergy efficiencies for the ORC subsystem are given:

$$\eta_{ORC} = \frac{\dot{W}_{ORC} + \dot{m}_{14}(h_7 - h_{14})}{\dot{m}_6(h_3 - h_6)}$$

and

$$\psi_{ORC} = \frac{\dot{E}x^W_{ORC} + \dot{m}_{14}(ex_7 - ex_{14})}{\dot{m}_6(ex_3 - ex_6)}$$

TABLE 6.32 Working parameters of the ejector cooling system combined with geothermal power.

Parameters	Values
Reference temperature, T_o	20°C
Reference pressure, P_o	101.3 kPa
Geofluid source temperature, T_1	210°C
Geofluid source pressure, P_1	1800 kPa
Geofluid mass flow rate, $\dot{m}_1$	140 kg/s
Working fluid of ORC	Isobutane
ORC turbine inlet temperature, T_3	150°C
ORC turbine inlet pressure, P_3	3000 kPa
ORC turbine exit pressure, P_4	300 kPa
Working fluid of ejector cooling system	Isobutane
Ejector inlet temperature, T_7	96°C
Ejector inlet pressure, P_7	1821 kPa
Pump 2 inlet temperature, T_{13}	36.9°C
Ejector entrainment ratio, ϕ_{ejec}	0.4468
Evaporator temperature, T_{eva}	8°C
Nozzle efficiency of the ejector, η_{noz}	0.85
Mixing efficiency of the ejector, η_{mix}	0.90
Diffuser efficiency of the ejector, η_{ejc}	0.85
Energetic coefficient of performance, COP_{en}	1.728
Exergetic coefficient of performance, COP_{ex}	0.436
Geofluid reinjection temperature, T_2	74°C

The energy and exergy efficiencies for the ejector cooling subsystem can be defined:

$$\eta_{EC} = \frac{\dot{Q}_{Cooling}}{(\dot{m}_7 h_7 - \dot{m}_{14} h_{14}) + \dot{W}_{P2}}$$

and

$$\psi_{EC} = \frac{\dot{E}x^{Q}_{Cooling}}{(\dot{m}_7 ex_7 - \dot{m}_{14} ex_{14}) + \dot{E}x^{W}_{P2}}$$

The energy and exergy efficiencies for the whole system are:

$$\eta_{WS} = \frac{\dot{W}_{Net} + \dot{Q}_{Cooling}}{\dot{m}_1(h_1 - h_2)}$$

and

$$\psi_{WS} = \frac{\dot{E}x^{W}_{Net} + \dot{E}x^{Q}_{Cooling}}{\dot{m}_1(ex_1 - ex_2)}$$

The total energy production of the geothermal energy–based ejector cooling plant can be calculated:

$$\dot{W}_{Net} = \dot{W}_{ORC} - \dot{W}_{p1} - \dot{W}_{p2}$$

The results calculated for the geothermal energy–based ejector cooling plant are written in Table 6.33, along with some other parametric studies, which are given by using these results.

The geothermal water mass flow rate is gradually increased from 100 to 180 kg/s in order to examine how the geothermal water mass flow rate affects system outputs, the exergy destruction rate, and the exergy efficiency of the ejector cooling system. This effect of geothermal water mass flow rate on the ejector cooling system is given in the graph in Fig. 6.77. Looking at the graph in the figure, the increase in mass flow rate in general increases the values of net power generation, cooling output, exergy destruction rate, and exergy efficiency. In cases where the geothermal water mass flow rate is 100, 140, and 180 kg/s, the net power generations of the system are 4807, 6278, and 8198 kW, respectively. For these geothermal water flow rate values, the cooling outputs obtained from the system are 2237, 2804, and 3513 kW, respectively. Also, the exergy destruction rates of the system for these mass flow rate values are 6225, 7683, and 9481 kW, respectively. An important performance measurement directed by the geothermal water mass flow rate is the exergy efficiency of the combined system. When the

TABLE 6.33 Calculated results for the ejector cooling system combined with geothermal power.

Plant outputs	Values
Work rate generated by ORC turbine, $\dot{W}_{ORC}$	6278 kW
Cooling producing rate, $\dot{Q}_{Cooling}$	2804 kW
Total exergy destruction rate, $\dot{E}x_{D,total}$	7683 kW
Whole system exergy efficiency, η_{WS}	0.5491

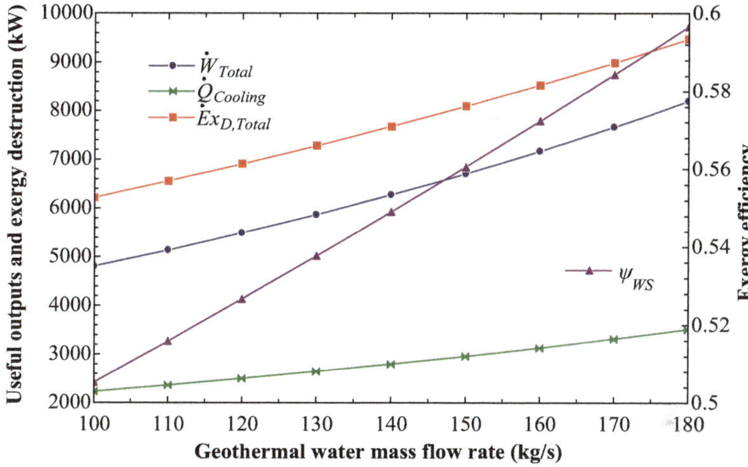

FIGURE 6.77 Effect of geothermal water mass flow rate on net power generation, cooling effect production and exergy destruction rates, and exergy efficiency of the ejector cooling system combined with geothermal power.

geothermal water flow rates are 100, 150, and 180 kg/s, the ejector cooling system's exergy efficiencies are 0.5053, 0.5606, and 0.5966, respectively.

Another parameter that has an important effect on system performance is geothermal water temperature. Depending on the gradual increase of the geothermal water temperature, the change in useful outputs generated in the system, the exergy destruction of the system, and the exergy efficiency of the system are clearly shown in the graph in Fig. 6.78. Here, the gradual increase of geothermal water temperature increases beneficial outcomes, exergy destruction, and exergy efficiency. Looking at the graph in the figure, the change in performance curves can be seen generally due to the gradual increase of the geothermal water temperature from 150°C to 230°C. When the geothermal water temperatures are 150°C, 180°C, and 230°C, the ejector cooling system's net power generations are 4527, 5331, and 7000 kW, respectively. For these geothermal water temperature values, the cooling outputs obtained from the system are 2104, 2429, and 3085 kW, respectively. Also, the exergy destruction rates of the system for these geothermal water temperature values are 6037, 6810, and 8325 kW, respectively. In addition, in cases where the geothermal water temperatures are 150°C, 170°C, 200°C, and 230°C, the ejector cooling system's exergy efficiencies are 0.4734, 0.4974, 0.5357, and 0.5768, respectively.

Another parameter that directs system performance is the reference temperature. When the reference temperature is increased gradually, changes occur in the beneficial products obtained from the system, in the exergy destruction of the system, and in the exergy efficiency of the system. These changes, depending on the reference temperature, are also shown in the

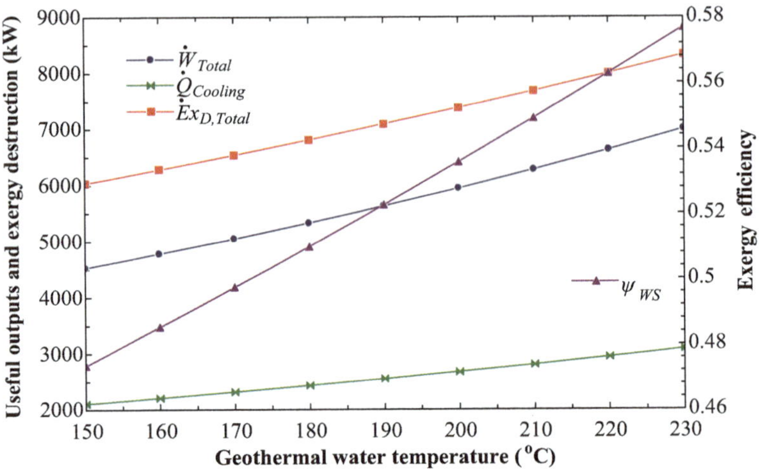

FIGURE 6.78 Effect of geothermal water temperature on net power generation, cooling effect production and exergy destruction rates, and exergy efficiency for the ejector cooling system combined with geothermal power.

graph in Fig. 6.79. In order to see this effect of the reference temperature on the ejector cooling system, the reference temperature is gradually increased from 0°C to 40°C. As can be clearly seen from the graph in Fig. 6.13, the gradual increase in the reference temperature increases the net power generation, exergy destruction, and exergy efficiency of the ejector cooling system, while reducing the cooling output from the system. When the reference temperature values are 10°C, 20°C, 25°C, 30°C, and 40°C, the ejector cooling system's net power generations are 5469, 5996, 6278, 6573, and 7205 kW, respectively. For these reference temperature values, the cooling outputs obtained from the system are 3028, 2876, 2804, 2732, and 2596 kW, respectively. As can be seen from the values of the cooling output here, as the ambient temperature increases, the amount of cooling output obtained from the ejector cooling system decreases. Also, the exergy destruction rates of the system for these reference temperature values are 6949, 7430, 7683, 7944, and 8493 kW, respectively. In addition, in cases where the reference temperatures are 10°C, 25°C, 35°C, and 40°C, the ejector cooling system's exergy efficiencies are 0.5189, 0.5491, 0.5701, and 0.5809, respectively.

6.7.10.1 Cascaded refrigeration combined with geothermal energy

Refrigeration systems are needed in some industrial applications. The systems in which cooling output can be obtained vary depending on the amount, level, and cost of the cooling requirement. Especially in applications requiring moderate low temperatures, systems that perform a cooling operation in two or more stages

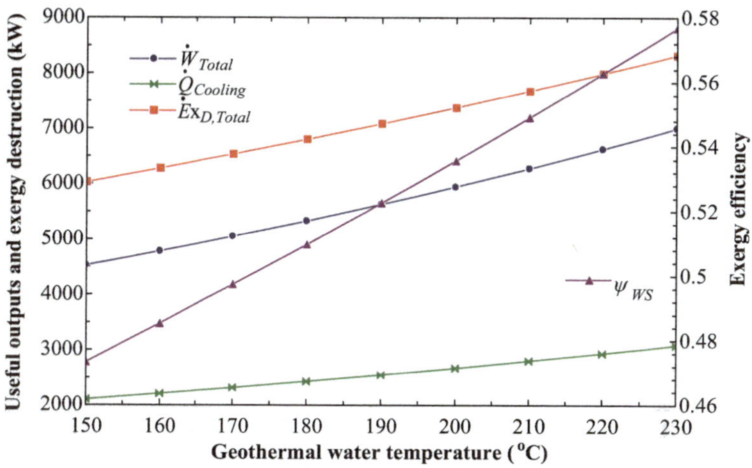

FIGURE 6.79 Effects of geothermal water temperature on net power generation, cooling effect production and exergy destruction rates, and exergy efficiency for the ejector cooling system combined with geothermal power.

in series are more useful than single steam cycle compression systems. Systems that perform this cooling process in two or more stages in series are called cascade refrigeration systems. These systems can work for a wide temperature range to create high-temperature differences between a heat source and a heat sink. Cascade systems have application areas for temperatures from −70°C to 100°C. For temperature values below −70, it is more convenient to use a cascade steam compression cycle instead of multistage steam compression cycles. Because in general, freezing temperatures of refrigerants may occur at these and lower temperatures. Therefore, cascade systems have a large working area for liquefying natural gas and some other gases.

The two most important features of these systems that they are potentially the refrigerants of choice depending on the desired properties and they do not require large system components. More than one evaporator can be used for compression in cascade systems. Another important feature of these systems is that the refrigerants used for each stage may differ. Therefore, the refrigerants may be selected to obtain the performance closest to the desired performance from the system.

In general, cascade cooling systems can be used to reach very low temperatures. Two-stage cascade systems can be used effectively to reach temperatures around 85°C. These two-stage cascade systems can reach this temperature value by connecting two cooling systems in series. There are also systems with a single compressor that can reach temperatures lower than this temperature value. However, these systems are not widely used. These systems can also be called autocascading systems. The most important disadvantage of these systems is that they operate with a special refrigerant mixture.

Depending on the special refrigerant mixture, various problems may occur while the system is operating. During operation, a leakage in the system can change the ratio of the different types of refrigerant in the mixture. Depending on this, the desired efficiency may not be obtained from the system. To overcome this, the refrigerant mixture in the system needs to be replaced with a new one. Since this refrigerant mixture is a special mixture containing different types of refrigerants, this change is costly for the system, and it is not easy to obtain it from the usual sources; therefore this special mixture can be difficult and costly to supply. These systems are not used in a wide range since it is difficult to reach qualified personnel for repair and maintenance. Considering the operation of this type of cascade systems in general, this type of cascade systems can create disadvantages in terms of cost and time.

The last advanced geothermal energy–based cooling system in the part of this chapter is cascaded refrigeration combined with a geothermal power plant, as seen in Fig. 6.80. As observed from this figure, the suggested system consists of the three main subcycles: the geothermal fluid subcycle, an ORC, and a cascaded absorption cooling subcycle. The main difference of the study is that the ACS is integrated with a vapor compression cooling system. These two cooling systems are combined with the cascade condenser subcomponent.

First, the geothermal fluid enters HEX 1 at point 1 and transfers its thermal energy to the ORC working fluid. After that, the geothermal fluid enters the generator and transfers its thermal energy to the cooling working fluid. ORC working fluid, which is a saturated liquid at state 7, enters the ORC turbine at state 4 in the saturated vapor phase with thermal energy coming from the geothermal fluid in HEX 1, where power generation occurs. Then the ORC working fluid entering condenser 1 at state 5 condenses here and then enters the pump at state 6 as the saturated liquid.

The cooling subsystem includes the generator, condenser 2, cascade condenser, absorber, HEX 3, valves, pump, compressor, and evaporator. At the outlet of the generator, the refrigerant in the vapor phase enters condenser 2 and passes through the throttle valve and then into the cascade condenser. Here, the cascade condenser acts as both an evaporator and a condenser. Refrigerant, which is a saturated liquid at point 13, passes through the throttle valve; then its pressure and temperature decrease and enters the evaporator where cooling occurs. The steam coming from state 17 and the weak solution from state 23 mix in the absorber and pass through states 18, 19, and 20, respectively, to the generator. Briefly, by using the thermal energy of geothermal fluid, both the power production from the ORC subsystem and the cooling production from the refrigeration subsystem are realized.

6.7.11 Case study 6.14

In this case study, the schematic flow diagram of cascaded refrigeration combined with the geothermal power plant is analyzed by using thermodynamic

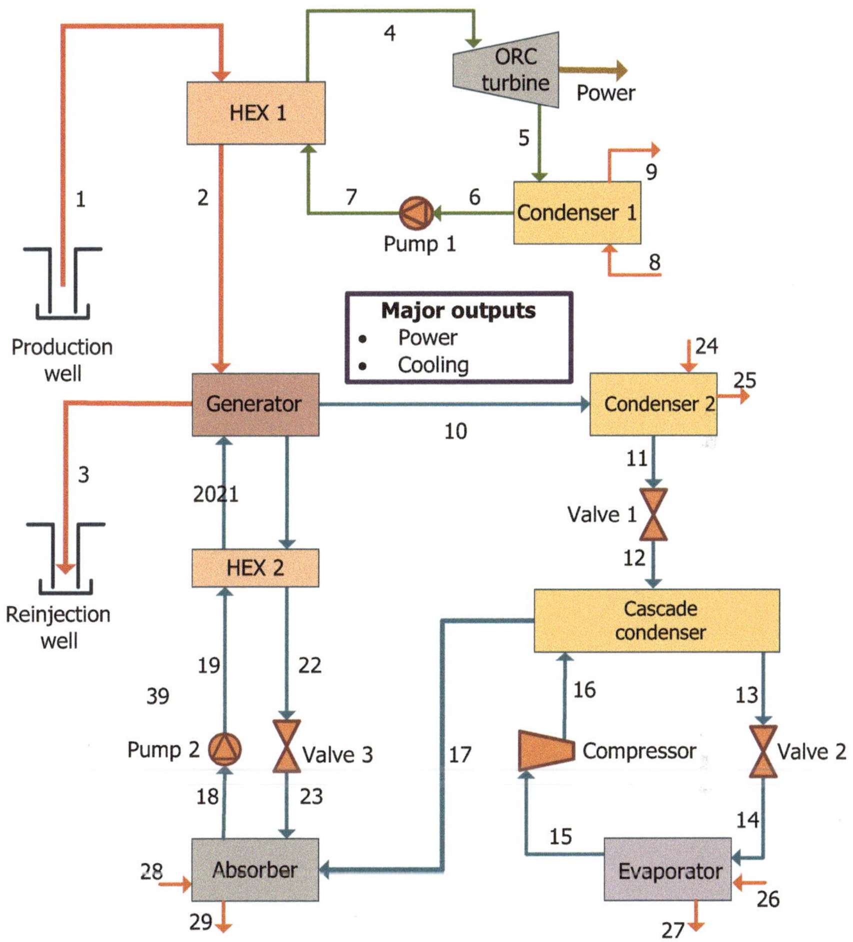

FIGURE 6.80 Schematic flow diagram of the cascaded refrigeration combined with a geothermal power plant.

performance assessment. The simplified flow diagram of the cascaded refrigeration combined with the geothermal power plant for this case study is shown in Fig. 6.80. In addition, the working parameters of the cascaded refrigeration combined with geothermal power are shown in Table 6.34.

To make a thermodynamic performance assessment, the energetic and exergetic efficiency equations for the whole plant and its subplants must be given. The thermodynamic efficiency equations for the ORC subsystem are given as:

$$\eta_{ORC} = \frac{\dot{W}_{ORC} + \dot{m}_8(h_9 - h_8)}{\dot{m}_7(h_4 - h_7)}$$

and

$$\psi_{ORC} = \frac{\dot{E}x_{ORC}^{W} + \dot{m}_8(ex_9 - ex_8)}{\dot{m}_7(ex_4 - ex_7)}$$

The energy and exergy efficiency equations for the cascade refrigeration subsystem can be defined:

$$\eta_{CRS} = \frac{\dot{Q}_{Cooling}}{\dot{m}_2(h_2 - h_3) + \dot{W}_{P2} + \dot{W}_{Comp}}$$

and

$$\psi_{CRS} = \frac{\dot{E}x_{Cooling}^{Q}}{\dot{m}_2(ex_2 - ex_3) + \dot{E}x_{P2}^{W} + \dot{E}x_{Comp}^{W}}$$

The energy and exergy efficiency equations for the whole system are given as:

$$\eta_{WS} = \frac{\dot{W}_{Net} + \dot{Q}_{Cooling}}{\dot{m}_1(h_1 - h_3)}$$

and

$$\psi_{WS} = \frac{\dot{E}x_{Net}^{W} + \dot{E}x_{Cooling}^{Q}}{\dot{m}_1(ex_1 - ex_3)}$$

The total energy production of the cascaded refrigeration system combined with geothermal energy can be defined:

$$\dot{W}_{Net} = \dot{W}_{ORC} - \dot{W}_{p1} - \dot{W}_{p2} - \dot{W}_{Comp}$$

The computed outputs for the cascaded refrigeration system combined with geothermal energy are given in Table 6.35, and some parametric works are also done utilizing these outputs.

The geothermal water mass flow rate is gradually increased from 100 to 180 kg/s in order to examine how the geothermal water mass flow rate affects useful outputs, the exergy destruction rate, and the exergy efficiency of the system. This effect of geothermal water mass flow rate on the system is given in the graph in Fig. 6.81. Looking at the graph in the figure, the increase in mass flow rate increases the values of net power generation, cooling output, exergy destruction rate, and exergy efficiency. When the geothermal water mass flow rate is 100, 140, and 180 kg/s, the net power generations of the system are 4676, 5926, and 7509 kW, respectively. For these geothermal water flow rate values, the cooling outputs obtained from the system are 2198, 2683, and 3273 kW, respectively. The exergy destruction rates of the system for these mass flow rate values are 6227, 7426, and 8855 kW, respectively. A significant performance measurement directed by

TABLE 6.34 Working parameters of the cascaded refrigeration combined with a geothermal power plant.

Parameters	Values
Reference temperature, T_o	20°C
Reference pressure, P_o	101.3 kPa
Geofluid source temperature, T_1	210°C
Geofluid source pressure, P_1	1800 kPa
Geofluid mass flow rate, $\dot{m}_1$	140 kg/s
Working fluid of ORC	Isobutane
ORC turbine inlet temperature, T_4	150°C
ORC turbine inlet pressure, P_4	3000 kPa
ORC turbine exit pressure, P_5	300 kPa
Energetic COP, COP_{en}	0.6124
Exergetic COP, COP_{ex}	0.3258
Working fluid of ejector cooling system	H_2O-LiBr
Effectiveness of cascade condenser, ϵ_{CC}	0.72
Isentropic efficiency of refrigeration compressor, η_{Comp}	0.66
Electrical efficiency of refrigeration pump, η_{P2}	0.91
Compressor inlet pressure, P_{15}	434.5 kPa
Compressor inlet temperature, T_{15}	−4.1°C
Compressor outlet pressure, P_{16}	860.5 kPa
Compressor outlet temperature, T_{16}	38.3°C
Pump 2 inlet pressure, P_{18}	1.228 kPa
Pump 2 inlet temperature, T_{18}	40°C
Pump 2 outlet pressure, P_{19}	8.931 kPa
Pump 2 outlet temperature, T_{19}	40.2°C
Generator outlet temperature, T_{10}	90°C
Condenser 2 outlet temperature, T_{11}	43.6°C
Geofluid reinjection temperature, T_3	68 °C

the geothermal water mass flow rate is the exergy efficiency of the cascaded refrigeration system combined with geothermal energy. When the geothermal water flow rates are 100, 150, and 180 kg/s, the system's exergy efficiencies are 0.5022, 0.5491, and 0.5792, respectively.

TABLE 6.35 Output commodities for the cascaded refrigeration system combined with geothermal energy.

Plant outputs	Values
Work rate generated by ORC turbine, $\dot{W}_{ORC}$	5926 kW
Cooling producing rate, $\dot{Q}_{Cooling}$	2683 kW
Total exergy destruction rate, $\dot{E}x_{D,total}$	7426 kW
Whole system exergy efficiency, η_{WS}	0.5394

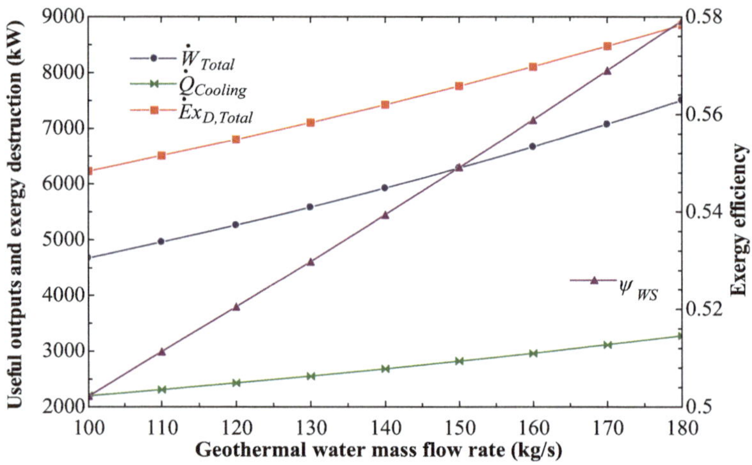

FIGURE 6.81 Effect of the geothermal water mass flow rate on net power generation, cooling effect production and exergy destruction rates, and exergy efficiency for the cascaded refrigeration system combined with geothermal energy.

Another indicator that has a potential effect on system performance is geothermal water temperature. Depending on the gradual increase of the geothermal water temperature, the change in useful outputs generated in the system, the exergy destruction of the system, and the exergy efficiency of the system are clearly shown in the graph in Fig. 6.82. Here, the gradual increase of geothermal water temperature increases beneficial outcomes, exergy destruction, and exergy efficiency. Looking at the graph in the figure, the change in performance curves can be seen to be generally due to the gradual increase of the geothermal water temperature from 150°C to 230°C. When the geothermal water temperatures are 150°C, 190°C, and 230°C, the cascaded refrigeration system's net power generations are 4447, 5385, and 6520 kW, respectively. For these geothermal water temperature values, the cooling outputs obtained from the cascaded refrigeration system are 2145,

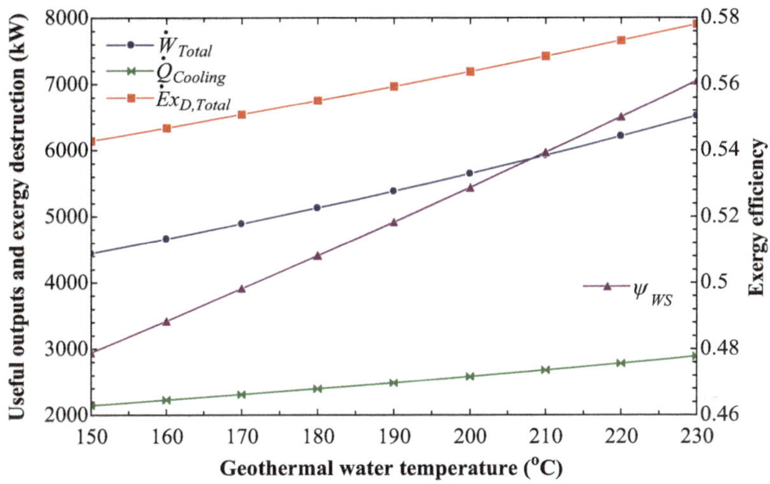

FIGURE 6.82 Effect of the geothermal water temperature on net power generation, cooling effect production and exergy destruction rates, and exergy efficiency for the cascaded refrigeration system combined with geothermal energy.

2490, and 2890 kW, respectively. Also, the exergy destruction rates of the system for these geothermal water temperature values are 6147, 6972, and 7908 kW, respectively. In addition to these, in cases where the geothermal water temperatures are 150°C, 180°C, 200°C, and 230°C, the cascaded refrigeration system's exergy efficiencies are 0.4789, 0.5082, 0.5288, and 0.5611, respectively.

Another parameter that affects system performance is the reference temperature. When the reference temperature is increased gradually, changes occur in the useful products obtained from the system, in the exergy destruction of the system, and in the exergy efficiency of the system. These changes in response to the reference temperature are also shown in the graph in Fig. 6.83. In order to see the effect of the reference temperature on the cascaded refrigeration system, the reference temperature is gradually increased from 0°C to 40°C. As can be clearly seen from the graph, the gradual increase in the reference temperature increases the net power generation, exergy destruction, and exergy efficiency of the cascaded refrigeration system, while reducing the cooling output from the system. When the reference temperature values are 10°C, 25°C, and 40°C, the cascaded refrigeration system's net power generations are 5253, 5926, and 6685 kW, respectively. For these reference temperature values, the cooling outputs obtained from the system are 2855, 2683, and 2520 kW, respectively. As can be seen from the values of the cooling output here, as the ambient temperature increases, the amount of cooling output obtained from the combined system decreases. The exergy destruction rates of the system for these reference temperature

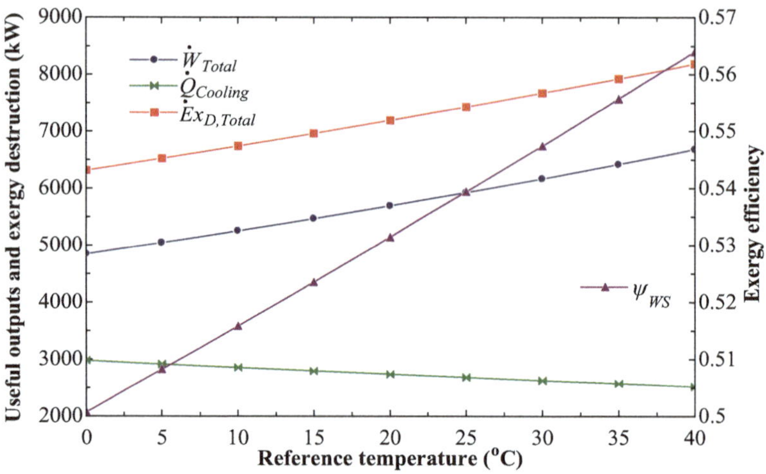

FIGURE 6.83 Effect of the reference temperature on net power generation, cooling effect production and exergy destruction rates, and exergy efficiency for the cascaded refrigeration system combined with geothermal energy.

values are 6736, 7426, and 8185 kW, respectively. In addition, where the reference temperatures are 10°C, 20°C, 25°C, 30°C, and 40°C, the cascaded refrigeration system's exergy efficiencies are 0.5158, 0.5314, 0.5394, 0.5474 and 0.564, respectively.

6.7.12 Combined/integrated system for hydrogen production

One of the biggest challenges in the world is the increase in energy demands with increasing population and living standards, and, accordingly, this need is met in a sustainable and environmentally friendly manner [15]. Environmental problems such as global warming, acid rain, ozone depletion, etc. are on the rise, as a large part of the energy need is still met by combusting fossil fuels. In this context, for a sustainable environment, the importance of renewable energy sources is increasing, and their usage is inevitable. Renewable energy sources are considered sustainable alternatives to fossil fuels. However, since most renewable energy sources have intermittent and fluctuating structures, storage will make the systems more efficient. Renewable energies can be stored in the form of electricity or chemical energy (such as hydrogen).

Due to its advantages, hydrogen, more specifically renewable hydrogen, has become synonymous with sustainability in both energy supply and storage. In particular, the use of renewable energy–assisted hydrogen generation systems is very important from an environmental point of view. Renewable-based hydrogen has the potential to provide economically viable, financially promising, and energy-efficient solutions to problems related to ever increasing global energy demand, including global warming [16]. In this context, geothermal energy–supported hydrogen production, which has low-temperature applications, has been among the topics of interest in recent years. Power generation technology such as ORC derived from geothermal energy, which comes from the Earth's crust, has developed considerably across the globe. It is possible to produce hydrogen in PEM electrolysis with some of the surplus electricity produced in the power cycle [17]. In addition, the production of the electrical energy required for the PEM electrolyzer with geothermal energy has made these systems more environmentally friendly [18]. Hydrogen production of electrical power without any environmental detriments has an important role in achieving a sustainable future [4]. Fig. 6.84 shows the general flowchart of hydrogen generation via geothermal energy.

Hydrogen generally functions as an energy carrier. So it can be used as a fuel for the production of other commodities [19]. The potential of hydrogen among energy sources shows an increasing profile. In the studies conducted, it can be seen that basic energy carriers tend toward electricity and hydrogen due to both increasing global warming and environmental policies. There are different systems for hydrogen production. However, it is becoming increasingly important to meet energy needs by producing hydrogen in a way that supports environmental policies [5]. As mentioned, benefiting from

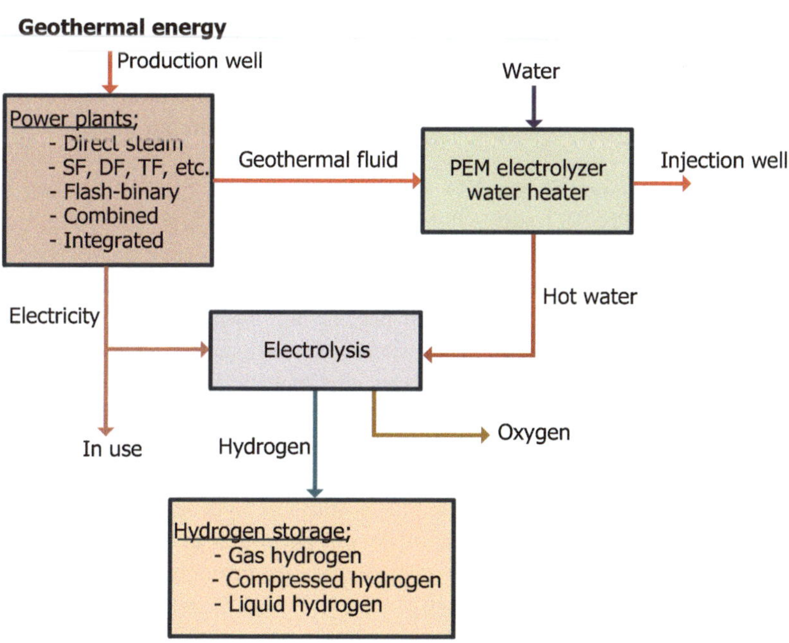

FIGURE 6.84 General flowchart of hydrogen generation via geothermal energy.

renewable energy sources is an effective way to ensure sustainability in energy resources and to support environmental policies. Geothermal energy sources with high sustainability indexes are also among the options that can be used among renewable energy sources. Especially if regions rich in geothermal energy sources can fully utilize the potential of these energy sources in hydrogen production, they can achieve meaningful results in terms of both regional development and supporting environmental policies. The storage of produced hydrogen also means that an energy source may have the potential for use in different applications. The use of hydrogen technologies generally depends on an advanced technology level. In studies carried out at the advanced technology level, hydrogen production and storage can be accomplished using geothermal energy, which is a renewable energy source, and the produced hydrogen can be used as an energy source for various applications.

The schematic flow diagram of the hydrogen production and compression process combined with a geothermal power plant is seen in Fig. 6.84. The proposed system consists of four main subsystems: (1) geothermal fluid cycle, (2) ORC, (3) seawater desalination unit, and (4) hydrogen production and compression systems. In this system, geothermal fluid's thermal energy is first transferred by using the generator to the working fluid of ORC. Then geothermal fluid enters underground again at point 2.

Another subsystem, the power generation cycle (ORC) fluid, enters the turbine in the superheated vapor phase with thermal energy from the geothermal fluid at state 3, where electricity is produced. As is clearly seen in Fig. 6.84, the ORC fluid entering the condenser at state 4 transfers its heat to clean water coming from state 20 and then enters the pump at state 5 in the saturated liquid phase. The ORC cycle continues to operate continuously in this way.

Another subsystem is the RO subsystem, which is the clean water production system. The RO subsystem includes low- and high-pressure pumps, chemical processing, a Pelton turbine, RO, and a three-way valve. The seawater enters at point 7 and then passes through several subcomponents: a low-pressure pump, a filter, chemical processing, and a high-pressure pump. While seawater exits the RO unit as clean water at state 17, it leaves the system as brine at state 13. The clean water entering the ORC condenser at state 20 reaches the approximate PEM electrolyzer working temperature with the thermal energy of the ORC fluid and enters PEM electrolysis at state 21. At point 23, the hydrogen produces gas phases with the PEM electrolyzer. Since storage is a problem due to the specific volume of hydrogen produced in gas form, it is possible to store it as compressed hydrogen in the hydrogen compression subunit. The gas form hydrogen passing through points 24, 25, 26, 27, and 28, that is, hydrogen compressors 1, 2, and 3, respectively, is stored under high pressure in the hydrogen compression tank. Briefly, in this section, as seen in Fig. 6.84, it is aimed at generating power by using the thermal energy of geothermal fluid and then to produce clean water and hydrogen from the waste energy of ORC fluid and seawater.

6.7.13 Case study 6.15

In this case study, the schematic flow diagram of the hydrogen production and compression process combined with a geothermal power plant is investigated by using energy and exergy analyses. The simplified flow diagram of the hydrogen production and compression process combined with a geothermal power plant for this case study is shown in Fig. 6.85. In addition, the working parameters of the hydrogen production and compression process combined with a geothermal power plant are listed in Table 6.36.

To make a detailed thermodynamic analysis, the overall chemical reaction equality of H_2O decomposition for the PEM electrolyzer is defined:

$$H_2O(l) \rightarrow H_2(g) + 1/2O_2(g)$$

Here, l and g show the liquid and gaseous forms. The following chemical reactions take place in the PEM anode and cathode components.

$$H_2O(l) \rightarrow 1/2O_2(g) + H^+(aq) + 2e^-$$

and

$$H^+(aq) + 2e^- \rightarrow 2H_2$$

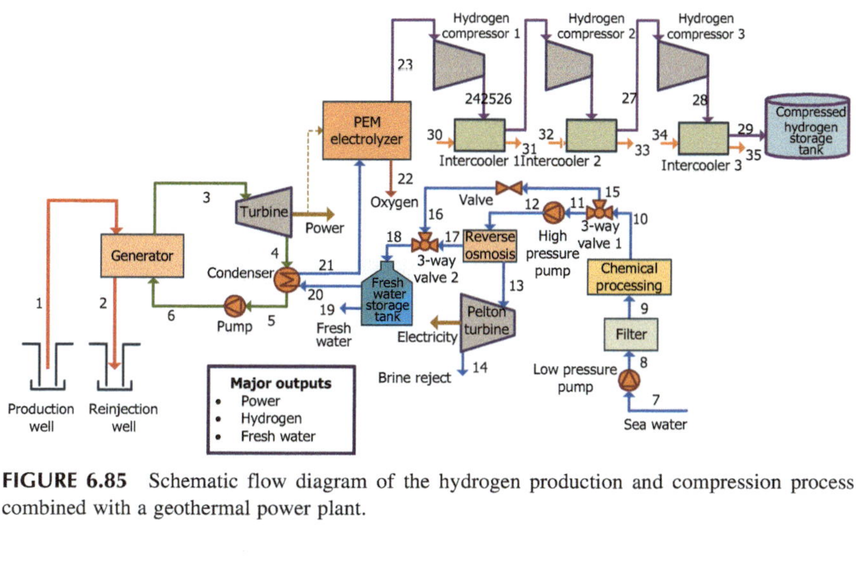

FIGURE 6.85 Schematic flow diagram of the hydrogen production and compression process combined with a geothermal power plant.

TABLE 6.36 Working parameters of the hydrogen production and compression process combined with a geothermal power plant.

Parameters	Values
Reference temperature, T_o	20°C
Reference pressure, P_o	101.3 kPa
Geofluid source temperature, T_1	210°C
Geofluid source pressure, P_1	1800 kPa
Geofluid mass flow rate, $\dot{m}_1$	140 kg/s
Pump isentropic efficiency, η_p	0.80
ORC turbine isentropic efficiency, η_{Tur}	0.80
Pelton turbine isentropic efficiency, η_{PT}	0.75
Isentropic efficiency of hydrogen compressor, η_{HC}	78%
Working fluid of ORC	Isobutane
ORC turbine inlet temperature, T_3	150°C
ORC turbine inlet pressure, P_3	3000 kPa
ORC turbine exit pressure, P_4	300 kPa
Hydrogen compressor 1 outlet pressure, P_{24}	1378 kPa
Hydrogen compressor 2 outlet pressure, P_{25}	19,257 kPa
Compressed hydrogen temperature, T_{29}	25°C
Compressed hydrogen pressure, P_{29}	42,000 kPa
Seawater salinity	35,000 ppm
Product-water salinity	450 ppm
Seawater feeding temperature, T_7	20°C
Membrane recovery ratio for filter	0.6
Pelton turbine inlet pressure, P_{13}	5100 kPa
Pelton turbine inlet temperature, T_{13}	41.7°C
Pelton turbine outlet pressure, P_{14}	101.3 kPa
Pelton turbine outlet temperature, T_{14}	40.62°C
Freshwater temperature, T_{18}	40.62°C
The pressure of compressed hydrogen storage tank, P_{chst}	2500 kPa
Geofluid reinjection temperature, T_2	56°C

The generated hydrogen and oxygen flow rate from the anode and cathode components can be written as:

$$\dot{N}_{H_{2,out}} = J/2F = \dot{N}_{H_2O}$$

and

$$\dot{N}_{O_{2,out}} = J/4F$$

Here, J shows the current density, F is the Faraday constant, and $\dot{N}_{H_2O}$ is the water consumed rate. In order to generate hydrogen, the electric energy must be used for the PEM electrolyzer, and this case should be defined as:

$$\dot{E}_{elec} = \dot{E}x_{elec} = JV$$

Here, V shows the cell potential and should be defined as:

$$V = V_o + \eta_{act,a} + \eta_{act,c} + \eta_{ohm}$$

where V_o shows the reversible potential and is related to the difference in free energy between reactants and products. Reversible potential should be defined by utilizing the Nernst equality. $\eta_{act,a}$, $\eta_{act,c}$, and η_{ohm} are the anode activation overpotential, cathode activation overpotential, and electrolyte ohmic overpotential. Also, η_{ohm} is to be attached to the resistance of the membrane to hydrogen ions crossing over the electrolyzer and should be described as:

$$\eta_{ohm} = JR$$

Here, R is the overall ohmic resistance.

$$R = \int_0^D \frac{dx}{\sigma_{PEM}[\lambda(x)]}$$

where D shows the PEM thickness, and $\sigma_{PEM}[\lambda(x)]$ is the membrane's local ionic conductivity and should be computed as:

$$\sigma_{PEM}[\lambda(x)] = [0.5139\lambda(x) - 0.326]exp\left[1268\left(\frac{1}{303} - \frac{1}{T}\right)\right]$$

Here, x and $\lambda(x)$ are the distance in the PEM evaluated from the cathode membrane interface and water content at a location x in the PEM and can be computed as [20]:

$$\lambda(x) = \frac{\lambda_a - \lambda_c}{D}x + \lambda_c$$

where λ_a and λ_c show the water content for the anode and cathode membrane interface, respectively. Also, η_{act} should be defined as [21,22]:

$$\eta_{act,i} = \frac{RT}{F}\sinh^{-1}\left(\frac{J}{2J_{o,i}}\right), i = a, c$$

and

$$J = J_{o,i}\left[\exp\left(\frac{\alpha z F \eta_{act,i}}{RT}\right) - \exp\left(\frac{(1-\alpha) z F \eta_{act,i}}{RT}\right)\right], i = a, c$$

as well as

$$J_{o,i} = J_i^{ref}\exp\left(-\frac{E_{act,i}}{RT}\right), i = a, c$$

Here, subscripts a and c show the anode and cathode, J_o gives the exchange current density, and α shows the anode and cathode reactions charge transfer coefficient and usually equals ½. z shows the number of electrons involved per reaction and is equal to 2. J_i^{ref} shows the preexponential factor, and $E_{act,i}$ gives the activation energy for the anode and cathode parts. Some significant design indicators for the PEM electrolyzer are shown in Table 6.37.

The energy and exergy efficiencies for the ORC subsystem can be defined as:

$$\eta_{ORC} = \frac{\dot{W}_{ORC} + \dot{m}_{20}(h_{21} - h_{20})}{\dot{m}_6(h_3 - h_6)}$$

and

$$\psi_{ORC} = \frac{\dot{E}x^W_{ORC} + \dot{m}_{20}(ex_{21} - ex_{20})}{\dot{m}_6(ex_3 - ex_6)}$$

The energy and exergy efficiencies for the distillation subsystem are given as:

$$\eta_{DP} = \frac{\dot{m}_{18}h_{18} + \dot{W}_{PT}}{\dot{m}_7h_7 + \dot{W}_{LPP} + \dot{W}_{HPP}}$$

TABLE 6.37 Some important design parameters of PEM electrolyzer.

Parameter	Value
Operating current density	5000 A/m^2
Anode activation energy, $E_{act,a}$	76 kJ/mol
Cathode activation energy, $E_{act,c}$	18 kJ/mol
Water content at the anode membrane, λ_a	14
Water content at the cathode membrane, λ_c	10
Exchange current density of anode, $J_{ref,a}$	1×10^5 A/m^2
Exchange current density of cathode, $J_{ref,c}$	10 A/m^2

Data from [23].

and

$$\psi_{DP} = \frac{\dot{m}_{18}ex_{18} + \dot{E}x_{PT}^{W}}{\dot{m}_{7}ex_{7} + \dot{E}x_{LPP}^{W} + \dot{E}x_{HPP}^{W}}$$

The energy and exergy efficiencies for the hydrogen production and compression subplant can be written as:

$$\eta_{HPC} = \frac{\dot{m}_{22}h_{22} + \dot{m}_{29}h_{29}}{\dot{m}_{21}h_{21} + \dot{W}_{PEM} + \dot{W}_{HyC1} + \dot{W}_{HyC2} + \dot{W}_{HyC3}}$$

and

$$\psi_{HPC} = \frac{\dot{m}_{22}ex_{22} + \dot{m}_{29}ex_{29}}{\dot{m}_{21}ex_{21} + \dot{E}x_{PEM}^{W} + \dot{E}x_{HyC1}^{W} + \dot{E}x_{HyC2}^{W} + \dot{E}x_{HyC3}^{W}}$$

The energy and exergy efficiencies for the whole system are written:

$$\eta_{WS} = \frac{\dot{W}_{Net} + \dot{m}_{18}h_{18} + \dot{m}_{22}h_{22} + \dot{m}_{29}h_{29}}{\dot{m}_{1}(h_{1} - h_{2})}$$

and

$$\psi_{WS} = \frac{\dot{E}x_{Net}^{W} + \dot{m}_{18}ex_{18} + \dot{m}_{22}ex_{22} + \dot{m}_{29}ex_{29}}{\dot{m}_{1}(ex_{1} - ex_{2})}$$

Total energy production of the hydrogen production and compression process combined with geothermal power plant can be defined:

$$\dot{W}_{Net} = \dot{W}_{ORC} - \dot{W}_{p} - \dot{W}_{LPP} - \dot{W}_{HPP} - \dot{W}_{HyC1} - \dot{W}_{HyC2} - \dot{W}_{HyC3}$$

The calculated results for the hydrogen production and compression process combined with geothermal power plants are included in Table 6.38; in addition, several parametric studies are given by using these results.

TABLE 6.38 Calculated results for the hydrogen production and compression process combined with a geothermal power plant.

Plant outputs	Values
Work rate generated by ORC turbine, $\dot{W}_{ORC}$	3793 kW
Generated hydrogen mass flow rate, $\dot{m}_{hydrogen}$	0.0725 kg/s
Freshwater production rate, $\dot{m}_{fw}$	6.4 kg/s
Total exergy destruction rate, $\dot{E}x_{D,total}$	8076 kW
Overall system exergy efficiency, η_{WS}	0.6108

The graph showing the relationship between the geothermal water mass flow rate and the net power generation, the exergy destruction rate, freshwater production rate, hydrogen production rate, and the exergy efficiency of the system are given in Fig. 6.86A and B.

As seen in the figure, the mass flow rate is gradually increased between 100 and 180 kg/s in order to examine the effect of the geothermal water mass flow rate on these five performance measurements. It can be clearly seen that the increase in mass flow rate increases the values of all five

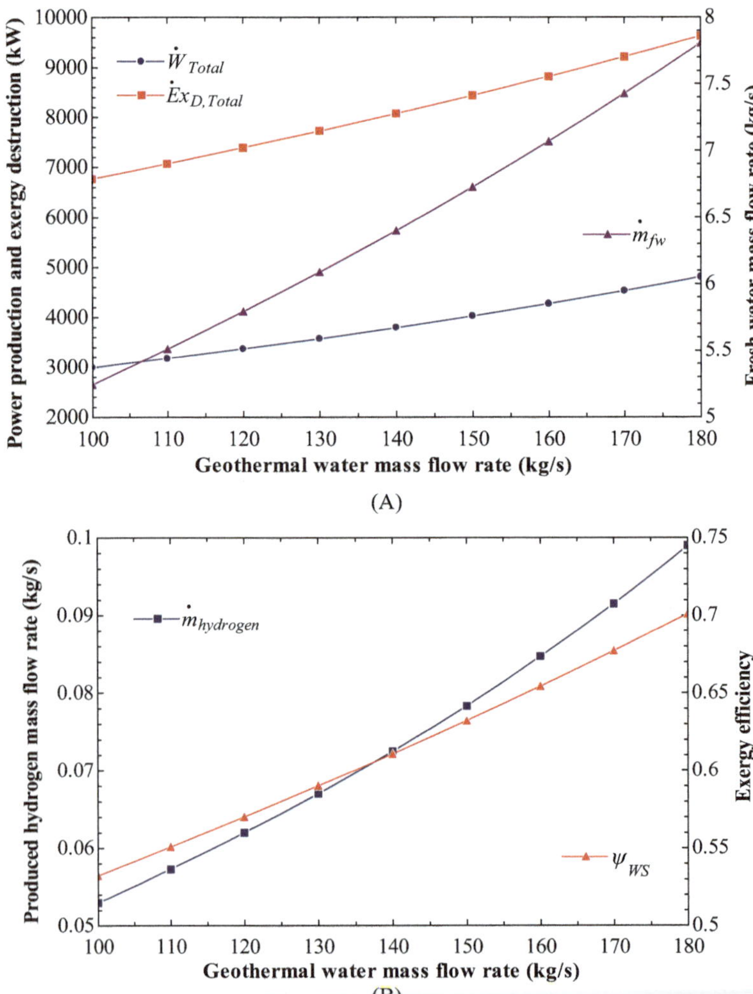

FIGURE 6.86 Effect of geothermal water mass flow rate on (A) net power generation, exergy destruction and freshwater production rates, (B) hydrogen production rate and exergy efficiency for the geothermal energy–based integrated plant.

performance measurements. In cases where the geothermal water mass flow rates are 100, 150, and 180 kg/s, the net power generations of the system are 2993, 4024, and 4806 kW, respectively. In addition, the exergy destruction rates of the system for these mass flow rate values are 6772, 8439, and 9630 kW, respectively. For these geothermal water flow rate values, the freshwater production rates of the system are 5.245, 6.726, and 7.808 kg/s, respectively. Additionally, for these geothermal water flow rate values, the hydrogen production rates of the system are 0.053, 0.0783, and 0.099 kg/s, respectively. An important performance criterion directed by the geothermal water mass flow rate is the exergy efficiency of the system. When the geothermal water flow rates are 100, 140, 160, and 180 kg/s, the geothermal power plant's exergy efficiencies are 0.5322, 0.6108, 0.6543, and 0.7009, respectively.

Another parameter that has an important potential for system performance is the geothermal water temperature. Depending on the gradual increase of the geothermal water temperature, the change in net power generation produced in the system, the exergy destruction rate of the system, freshwater production rate, hydrogen production rate, and the exergy efficiency of the system are clearly shown in Fig. 6.87A and B. Here, the gradual increase of geothermal water temperature increases net power generation, exergy destruction rate, freshwater production rate, hydrogen production rate, and exergy efficiency.

Looking at the figure, the change in performance curves can be seen to be generally due to the gradual increase of the geothermal water temperature from 150°C to 230°C. When the geothermal water temperatures are 150°C, 200°C, and 230°C, the geothermal power plant's net power generations are 2846, 3615, and 4173, respectively. Also, the exergy destruction rates of the system for these geothermal water temperature values are 6685, 7825, and 8601 kW, respectively. For these geothermal water temperature values, the freshwater production rates of the system are 5.245, 6.726, and 7.808 kg/s, respectively. Additionally, for these geothermal water temperature values, the hydrogen production rates of the system are 0.0477, 0.0676, and 0.0833 kg/s, respectively. In addition, in cases where the geothermal water temperatures are 150°C, 180°C, 210°C, and 230°C, the combined system's exergy efficiencies are 0.5175, 0.5622, 0.6108, and 0.6454, respectively.

Another indicator that guides the system performance is the reference temperature. When the reference temperature is increased gradually, changes occur in net power generation produced from the system, the exergy destruction rate, freshwater production rate, hydrogen production rate, and exergy efficiency of the system. These changes in response to the reference temperature are also shown in the graph in Fig. 6.88A and B. In order to see this effect of the reference temperature on the system, the reference temperature is gradually increased from 0°C to 40°C. As can be clearly seen from the graph in Fig. 6.88A and B, the gradual increase in the reference temperature

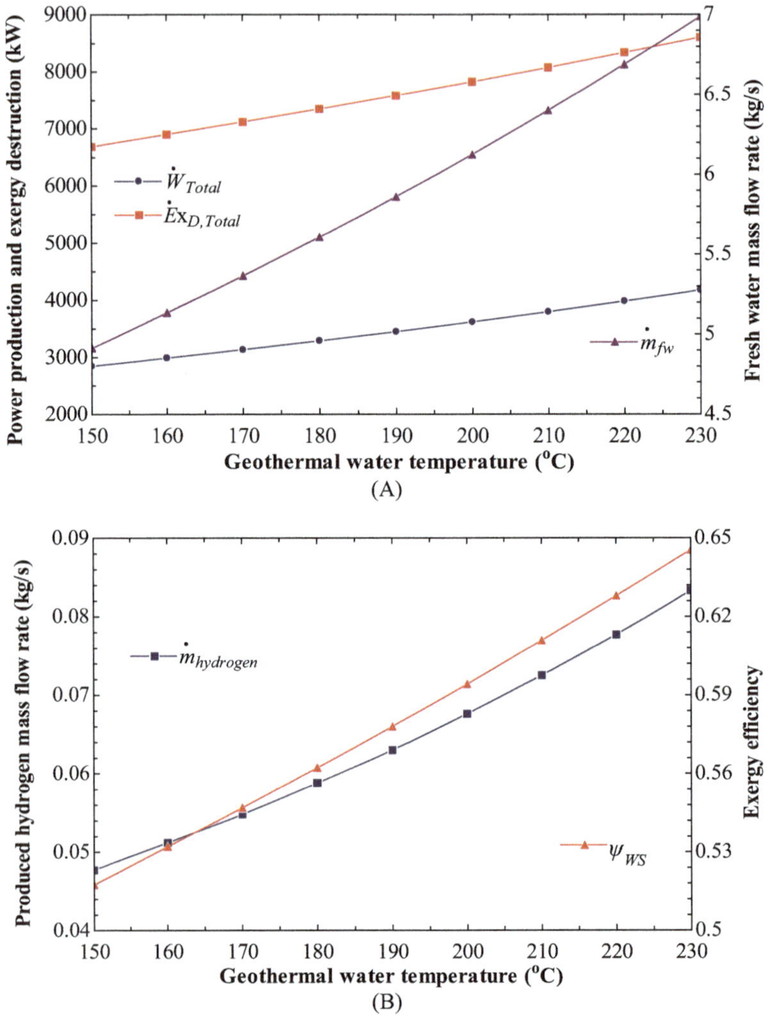

FIGURE 6.87 Effect of geothermal water temperature on (A) net power generation, exergy destruction and freshwater production rate, (B) hydrogen production rate and exergy efficiency for the geothermal energy–based integrated plant.

increases net power generation, the exergy destruction rate, freshwater production rate, hydrogen production rate, and the exergy efficiency of the geothermal power plant. When the reference temperature values are 10°C, 25°C, and 40°C, the geothermal power plant's net power generations are 3491, 3793, and 4120 kW, respectively. The exergy destruction rates of the system for these reference temperature values are 7632, 8076, and 8545 kW, respectively. For these reference temperature values, the freshwater production rates of the system are 5.806, 6.4, and 7.054 kg/s, respectively. Also, for these reference temperature values, the hydrogen production rates of the system are 0.061, 0.0725, and 0.0861 kg/s, respectively. In addition to these, in cases where the reference temperatures are 10°C, 25°C, 30°C, and 40°C, the geothermal power plant's exergy efficiencies are 0.5823, 0.6108, 0.6205, and 0.6406, respectively.

6.7.14 Combined/integrated system for ammonia production

Considering the comprehensive analysis studies carried out in general, obtaining beneficial outputs by utilizing carbon-free fuels such as hydrogen and ammonia from energy generation systems has great potential in terms of both supporting environmental policies and sustainability [24,25]. Ammonia, which is one of the carbon-free fuels, has an important place among energy sources with its feature of being a hydrogen carrier. Looking at the structure of ammonia, it is seen that it contains one nitrogen atom and three hydrogen atoms. As can be seen from its structure, ammonia does not contain any carbon atoms in its structure. At the same time, ammonia is considered the fuel that provides clean energy since it has a high hydrogen content in its structure. In a comparative analysis that was conducted, it was seen that ammonia reduces carbon dioxide emissions. No carbon dioxide is formed during the combustion reaction since water and nitrogen production is realized. Therefore, the use of ammonia as a mixture of conventional fuels has great potential to reduce carbon dioxide emissions.

Interest in the energy storage phenomenon is gradually increasing [17]. Expectations for hydrogen and hydrogen carriers are also increasing to realize this storage. Since hydrogen is in the gas form under normal environmental conditions, the storage and transportation of hydrogen become important. Hydrogen carriers are defined in the literature as mediums that enable the conversion of hydrogen into chemicals with high hydrogen content. In this context, ammonia is a hydrogen carrier and transport fuel containing high hydrogen atoms. Ammonia has many specific prominent properties. Ammonia can perform its functions in the solid and liquid states for various purposes. Another highlight is its ability to be stored and transported at low pressures. Since it can also be used directly in fuel cells, ammonia has potential in this area compared to other types of fuel [26]. Ammonia can also be used as a refrigerant in various situations. Many existing resources have the potential to be used, including renewable resources for the production of ammonia.

As a hydrogen carrier, ammonia production and storage are becoming increasingly important, considering the potential of ammonia to support both environmental policies and transfer between energy sources. Realizing ammonia production with renewable energy–based systems is beneficial in terms of supporting both sustainability and environmental policies. In this section, a geothermal energy–based energy generation system is used, where

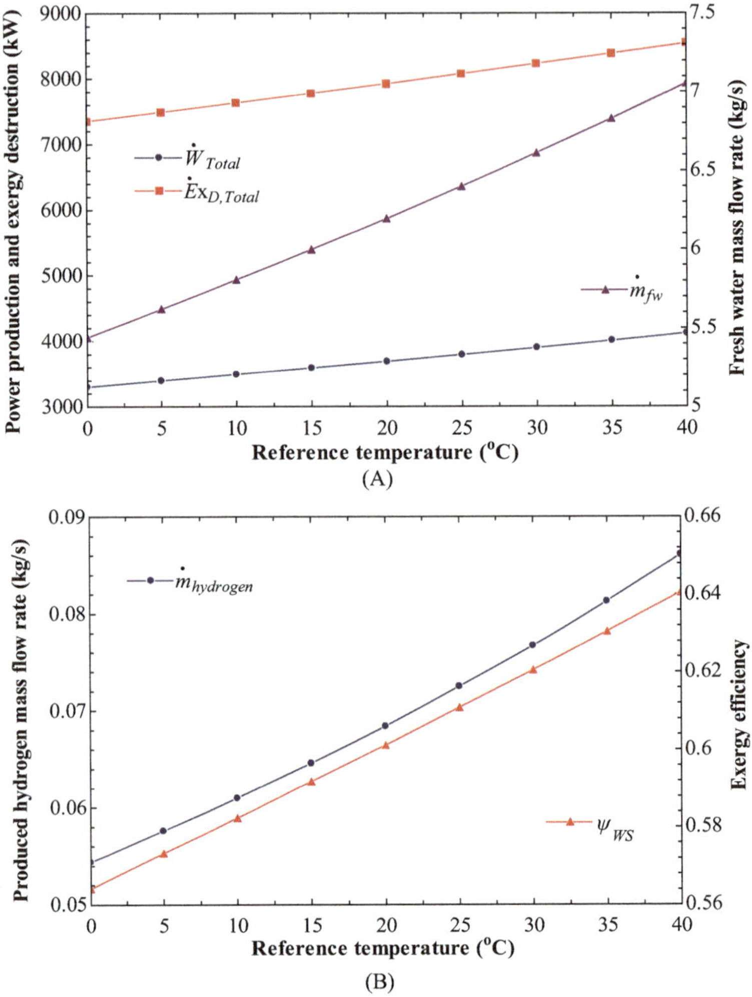

FIGURE 6.88 Effect of reference temperature on (A) net power generation, exergy destruction, and freshwater production rate, (B) hydrogen production rate and exergy efficiency for the geothermal energy–based integrated plant.

geothermal energy is used as the renewable energy source. The production of ammonia from this geothermal energy–based energy system also enables the utilization of the existing potential of geothermal energy by producing different useful outputs.

By integrating geothermal energy–supported power generation systems with multiple generation systems, many useful outputs can be obtained [26]. These useful outcomes can be expressed as power, heating, cooling, hydrogen, and ammonia production. The electrical power obtained from geothermal energy can be used in hydrogen production in electrolysis. In addition to the hydrogen gas produced, air-separated nitrogen gas is sent to the ammonia synthesis reactor, where ammonia is produced as a result of the following endothermic reaction:

$$N_2 + 3H_2 = 2NH_3$$

In the ammonia reactor, only heat is required to start the reactions, and the reaction itself is exothermic. The ammonia reactor is used to produce ammonia from hydrogen and nitrogen. The hydrogen is generated by electrolysis and the nitrogen is generated by the cryogenic air separation process for ammonia production. The general flowchart for ammonia generation via geothermal energy is given in Fig. 6.89.

The schematic representation of the geothermal energy supported hydrogen and ammonia production system is illustrated in Fig. 6.90. Considering the system in general, the suggested power plant consists of a double separator and double turbine power generation subsystem, clean water generation subsystem that is RO, PEM electrolyzer for hydrogen production, and ammonia reactor subsystem for ammonia production. The geothermal fluid enters the separator at state 2, where the vapor phase geothermal fluid is separated, and then enters the purifier part at point 3 and then the HP turbine at state 5. At state 10, high-temperature geothermal fluid enters separator 2 at state 11

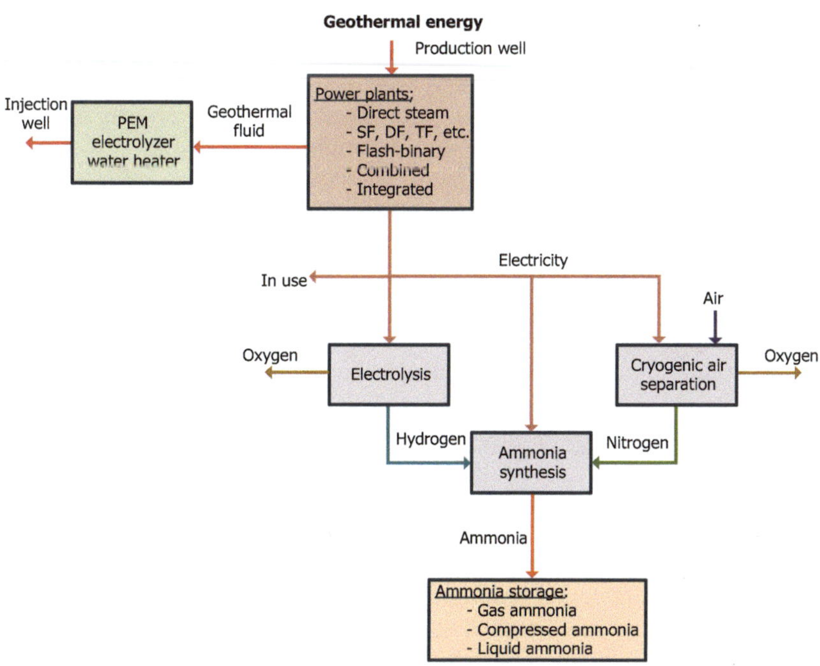

FIGURE 6.89 General flowchart of ammonia generation via geothermal energy.

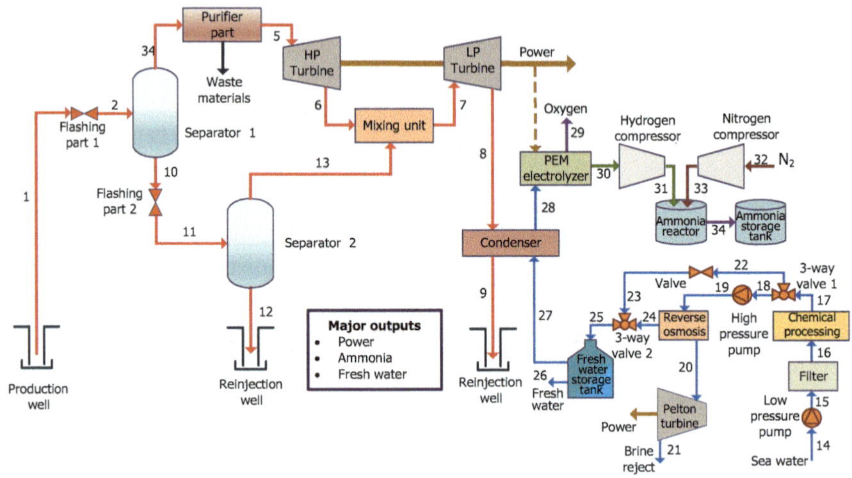

FIGURE 6.90 Schematic flow diagram of the ammonia production process combined with double-flash geothermal power plant.

and passes into the mixing chamber at state 13. The heat of the geothermal fluid coming from state 6 is increased and then enters the LP turbine at state 7. Then it is passed through the condenser and sent underground at state 9.

Seawater entering the RO unit at state 14 is passed through various steps and sent to the freshwater storage tank as clean water at state 25. Clean water at point 27 is heated by the thermal energy discharged in the condenser. After that, it enters the PEM electrolysis at point 28 at about 80°C. In PEM electrolysis, it combines with electrical power coming from the power plant, and hydrogen production in the gas phase occurs.

The gas-phase hydrogen at state 30 enters the hydrogen compressor, and then it is pressured to the ammonia reactor pressure. Nitrogen gas, which is separated from air, enters the nitrogen compressor at 32 points, and it is pressured at the ammonia reactor pressure. With pressurized hydrogen from state 31, nitrogen from state 33 enters the ammonia synthesis reaction, where ammonia is produced. This is how power, hydrogen, clean water, and ammonia are produced with geothermal energy–based plants.

6.7.15 Case study 6.16

In this case study, the schematic flow diagram of the ammonia production process combined with a double-flash geothermal power plant is analyzed by using the energetic and exergetic assessments. The simplified flow diagram of the ammonia production process combined with a double-flash geothermal power plant for this case study is shown in Fig. 6.90. In addition, the working parameters of the ammonia production process combined with a double-flash geothermal power plant are given in Table 6.39.

TABLE 6.39 **Working parameters of the ammonia production process combined with double-flash geothermal power plant.**

Parameters	Values
Reference temperature, T_o	20°C
Reference pressure, P_o	101.3 kPa
Geofluid source temperature, T_1	210°C
Geofluid source pressure, P_1	2795 kPa
Geofluid mass flow rate, $\dot{m}_1$	140 kg/s
HP turbine inlet temperature, T_5	163°C
HP turbine inlet pressure, P_5	666.5 kPa
LP turbine inlet temperature, T_7	98.58°C
LP turbine inlet pressure, P_7	96.4 kPa
The fraction of vapor at flash part 1 outlet	0.1454
The fraction of vapor at flash part 2 outlet	0.1042
Pump isentropic efficiency, η_p	0.80
Pelton turbine isentropic efficiency, η_{PT}	0.75
Isentropic efficiency of hydrogen compressor, η_{HC}	78%
Isentropic efficiency of nitrogen compressor, η_{NC}	80%
Hydrogen compressor outlet pressure, P_{31}	10,000 kPa
Nitrogen compressor outlet pressure, P_{33}	10,000 kPa
Hydrogen compressor outlet temperature, T_{31}	250°C
Nitrogen compressor outlet temperature, T_{33}	250°C
Seawater salinity	35,000 ppm
Product-water salinity	450 ppm
Seawater feeding temperature, T_{14}	20°C
Membrane recovery ratio	0.6
Pelton turbine inlet pressure, P_{20}	5100 kPa
Pelton turbine inlet temperature, T_{20}	41.7°C
Pelton turbine outlet pressure, P_{21}	101.3 kPa
Pelton turbine outlet temperature, T_{21}	40.62°C
Freshwater temperature, T_{25}	40.62°C
The pressure of compressed hydrogen storage tank, P_{chst}	2500 kPa
Geofluid reinjection temperature, T_9	42°C
Geofluid reinjection temperature, T_{12}	98.38°C

The energy and exergy efficiencies for the double-flash geothermal power system are written:

$$\eta_{DFGP} = \frac{\dot{W}_{HPT} + \dot{W}_{LPT}}{\dot{m}_1 h_1 - \dot{m}_9 h_9 - \dot{m}_{12} h_{12}}$$

and

$$\psi_{DFGP} = \frac{\dot{E}x^W_{HPT} + \dot{E}x^W_{LPT}}{\dot{m}_1 ex_1 - \dot{m}_9 ex_9 - \dot{m}_{12} ex_{12}}$$

The energy and exergy efficiencies for the distillation subsystem can be defined:

$$\eta_{DP} = \frac{\dot{m}_{25} h_{25} + \dot{W}_{PT}}{\dot{m}_{14} h_{14} + \dot{W}_{LPP} + \dot{W}_{HPP}}$$

and

$$\psi_{DP} = \frac{\dot{m}_{25} ex_{25} + \dot{E}x^W_{PT}}{\dot{m}_{14} ex_{14} + \dot{E}x^W_{LPP} + \dot{E}x^W_{HPP}}$$

The energy and exergy efficiencies for the hydrogen production and compression subsystem are given as:

$$\eta_{HPC} = \frac{\dot{m}_{29} h_{29} + \dot{m}_{31} h_{31}}{\dot{m}_{28} h_{28} + \dot{W}_{PEM} + \dot{W}_{HC}}$$

and

$$\psi_{HPC} = \frac{\dot{m}_{29} ex_{29} + \dot{m}_{31} ex_{31}}{\dot{m}_{28} ex_{28} + \dot{E}x^W_{PEM} + \dot{E}x^W_{HC}}$$

The energy and exergy efficiencies for the ammonia production and compression subsystem are defined:

$$\eta_{APC} = \frac{\dot{m}_{34} h_{34}}{\dot{m}_{31} h_{31} + \dot{m}_{32} h_{32} + \dot{W}_{NC}}$$

and

$$\psi_{APC} = \frac{\dot{m}_{34} ex_{34}}{\dot{m}_{31} ex_{31} + \dot{m}_{32} ex_{32} + \dot{E}x^W_{NC}}$$

The energy and exergy efficiencies for the whole system can be defined:

$$\eta_{WS} = \frac{\dot{W}_{Net} + \dot{m}_{25} h_{25} + \dot{m}_{29} h_{29} + \dot{m}_{34} h_{34}}{(\dot{m}_1 h_1 - \dot{m}_9 h_9 - \dot{m}_{12} h_{12}) + \dot{m}_{14} h_{14}}$$

and

$$\psi_{WS} = \frac{\dot{E}x^W_{Net} + \dot{m}_{25} ex_{25} + \dot{m}_{29} ex_{29} + \dot{m}_{34} ex_{34}}{(\dot{m}_1 ex_1 - \dot{m}_9 ex_9 - \dot{m}_{12} ex_{12}) + \dot{m}_{14} ex_{14}}$$

Total energy production of the ammonia production process combined with double-flash geothermal power plant can be defined:

$$\dot{W}_{Net} = \dot{W}_{HPT} + \dot{W}_{LPT} - \dot{W}_{LPP} - \dot{W}_{HPP} - \dot{W}_{HC} - \dot{W}_{NC} - \dot{W}_{PEM}$$

The computed outputs for the ammonia production process combined with the double-flash geothermal power plant are written in Table 6.40; in addition, several parametric works are done by utilizing these outputs.

The graph showing the relationship between the geothermal water mass flow rate and the net power generation, the exergy destruction rate, freshwater production rate, hydrogen production rate, ammonia production rate, and the exergy efficiency of the system are given in Fig. 6.91A and B. As seen in the graphics, the mass flow rate is gradually increased between 100 and 180 kg/s in order to assess the effect of the geothermal water mass flow rate on these six performance measurements. It can be clearly seen that the increase in the mass flow rate increases the values of these six performance measurements. In cases where the geothermal water mass flow rates are 100, 140, and 180 kg/s, the net power generations of the system are 5156, 6486, and 8157 kW, respectively.

In addition, the exergy destruction rates of the system for these mass flow rate values are 6706, 7936, and 9391 kW, respectively. For these geothermal water flow rate values, the freshwater production rates of the system are 4.847, 5.87, and 7.107 kg/s, respectively. Additionally, for these geothermal water flow rate values, the hydrogen production rates of the system are 0.0399, 0.0526, and 0.0692 kg/s, respectively. The ammonia production rates for these geothermal water flow rate values are 0.1619, 0.2052, and 0.26 kg/s, respectively. A significant performance measurement directed by the geothermal water mass flow rate is the exergy efficiency of the system. When the geothermal water flow rates are 100, 140, 150, and 180 kg/s, the geothermal

TABLE 6.40 Computed outputs for the ammonia production process combined with double-flash geothermal power plant.

Plant outputs	Values
Work rate generated by HP turbine, $\dot{W}_{HP}$	2639 kW
Work rate generated by LP turbine, $\dot{W}_{LP}$	3847 kW
Generated hydrogen mass flow rate, $\dot{m}_{hydrogen}$	0.0526 kg/s
Generated ammonia mass flow rate, $\dot{m}_{ammonia}$	0.2052 kg/s
Freshwater production rate, $\dot{m}_{fw}$	5.87 kg/s
Total exergy destruction rate, $\dot{E}x_{D,total}$	7936 kW
Overall system exergy efficiency, η_{WS}	0.6428

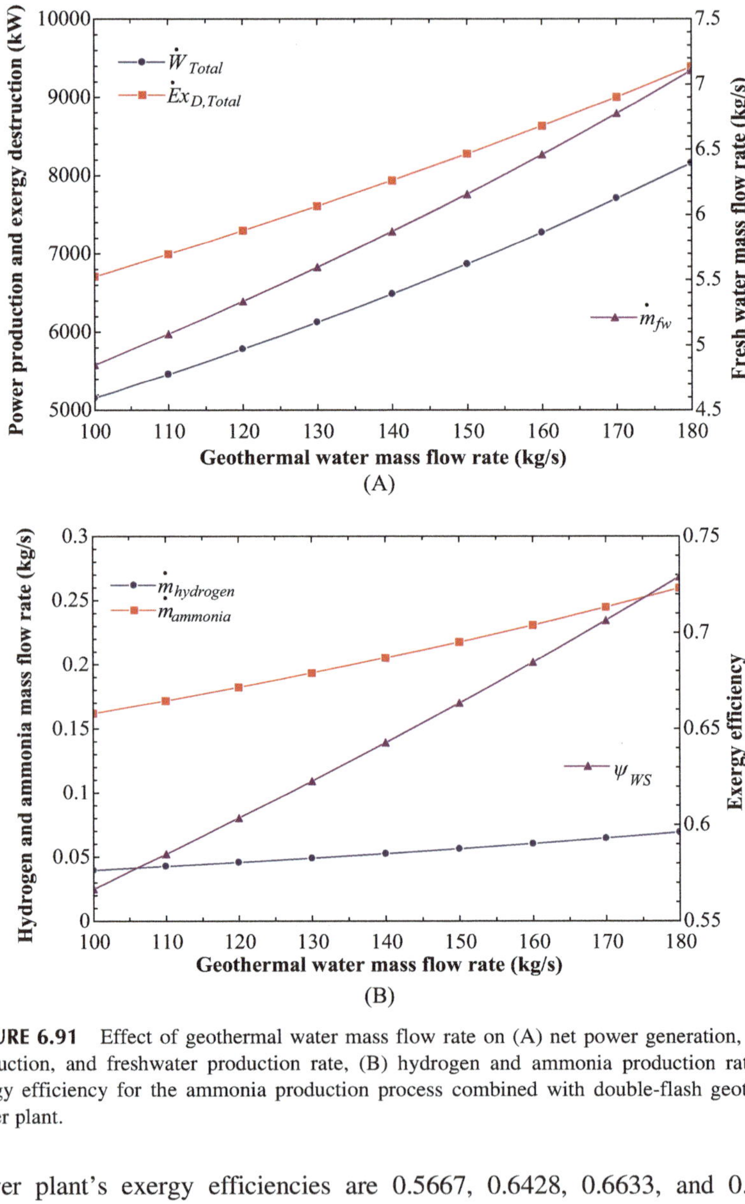

FIGURE 6.91 Effect of geothermal water mass flow rate on (A) net power generation, exergy destruction, and freshwater production rate, (B) hydrogen and ammonia production rates and exergy efficiency for the ammonia production process combined with double-flash geothermal power plant.

power plant's exergy efficiencies are 0.5667, 0.6428, 0.6633, and 0.7291, respectively.

Another indicator that has significant potential for system performance is geothermal water temperature. Depending on the gradual increase of the geothermal water temperature, the change in net power generation, the exergy destruction rate, freshwater production rate, hydrogen production rate, ammonia production rate, and exergy efficiency of the double-flash geothermal power plant are clearly shown in the graph in Fig. 6.92A and B. Here, the gradual increase of geothermal water temperature increases net power generation, exergy destruction rate, freshwater production rate, hydrogen production rate, ammonia production rate, and exergy efficiency. Looking at the graph in the figure, the change in performance curves can be seen generally due to the gradual increase of the geothermal water temperature from 150°C to 230°C. When the geothermal water temperatures are

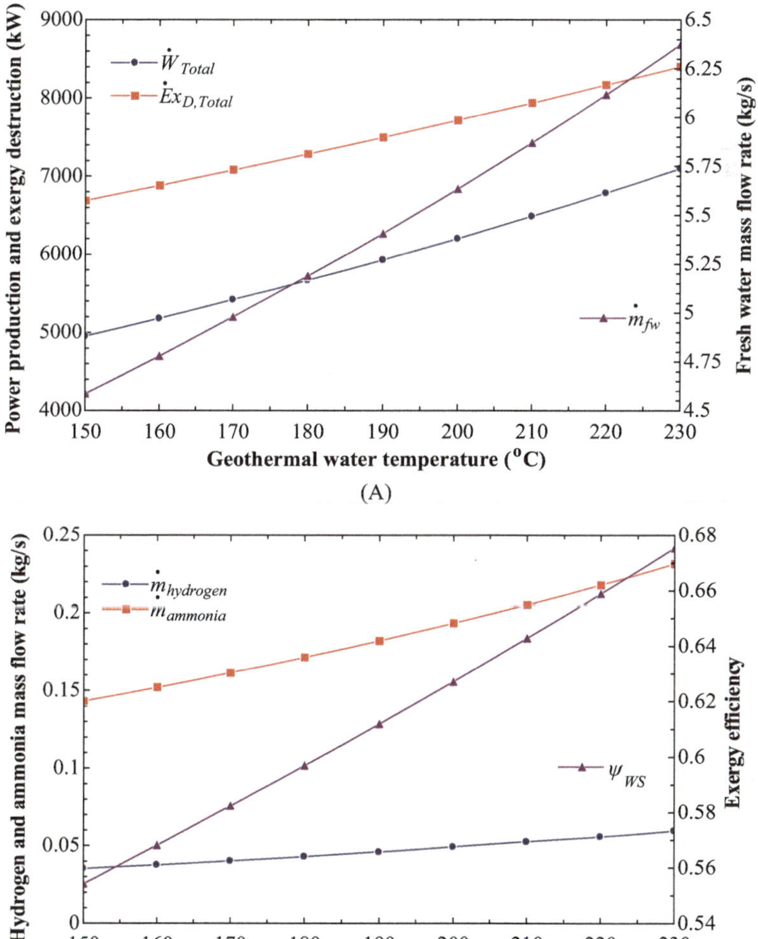

FIGURE 6.92 Effect of geothermal water temperature on (A) net power generation, exergy destruction and freshwater production rates, (B) hydrogen and ammonia production rate and exergy efficiency for the ammonia production process combined with a double-flash geothermal power plant.

150°C, 200°C, and 230°C, the double-flash geothermal power plant's net power generations are 4952, 6200, and 7096 kW, respectively. The exergy destruction rates of the system for these geothermal water temperature values are 6685, 7712, and 8402 kW, respectively. For these geothermal water temperature values, the freshwater production rates of the system are 4.585, 5.633, and 6.373 kg/s, respectively. Additionally, for these geothermal water temperature values, the hydrogen production rates of the system are 0.0352, 0.0492, and 0.0596 kg/s, respectively, and the ammonia production rates of the system are 0.143, 0.1932, and 0.2314 kg/s, respectively. In addition, in cases where the geothermal water temperatures are 150°C, 180°C, 210°C, and 230°C, the double-flash geothermal power plant's exergy efficiencies are 0.5542, 0.5969, 0.6428, and 0.6753, respectively.

Another parameter that guides the system performance is the reference temperature. When the reference temperature is increased gradually, changes occur in net power generation, the exergy destruction rate, freshwater production rate, hydrogen production rate, ammonia production rate, and the exergy efficiency of the double-flash geothermal power plant. These changes in response to the reference temperature are shown in the graph in Fig. 6.93A and B. In order to see this effect of the reference temperature on the plant, the reference temperature is gradually increased from 0°C to 40°C. As can be clearly seen from the graph in Fig. 6.93A and B, the gradual increase in the reference temperature increases net power generation, the exergy destruction rate, freshwater production rate, hydrogen production rate, ammonia production rate, and exergy efficiency of the double-flash geothermal power plant. When the reference temperature values are 10°C, 25°C, and 40°C, the double-flash geothermal power plant's net power generations are 6022, 6486, and 6984 kW, respectively. The exergy destruction rates of the system for these reference temperature values are 7566, 7936, and 8323 kW, respectively. At these reference temperature values, the freshwater production rates of the system are 5.371, 5.87, and 6.414 kg/s, respectively. Also, for these reference temperature values, the hydrogen production rates of the system are 0.0446, 0.0526, and 0.0619 kg/s, respectively, and the ammonia production rates of the system are 0.1793, 0.2052, and 0.2348 kg/s, respectively. In addition, in cases where the reference temperatures are 10°C, 25°C, 30°C, and 40°C, the geothermal power plant's exergy efficiencies are 0.6183, 0.6428, 0.6511, and 0.6682, respectively.

6.8 Closing remarks

In this chapter, we have dealt with the advanced versions of geothermal energy systems and their energy and exergy analyses. These systems are treated in five categories: multistaging, multiflashing, multistaging with binary, multiflashing with binary, and combining/integrating. In this regard, any combination of these basic methods can be used to design more complex

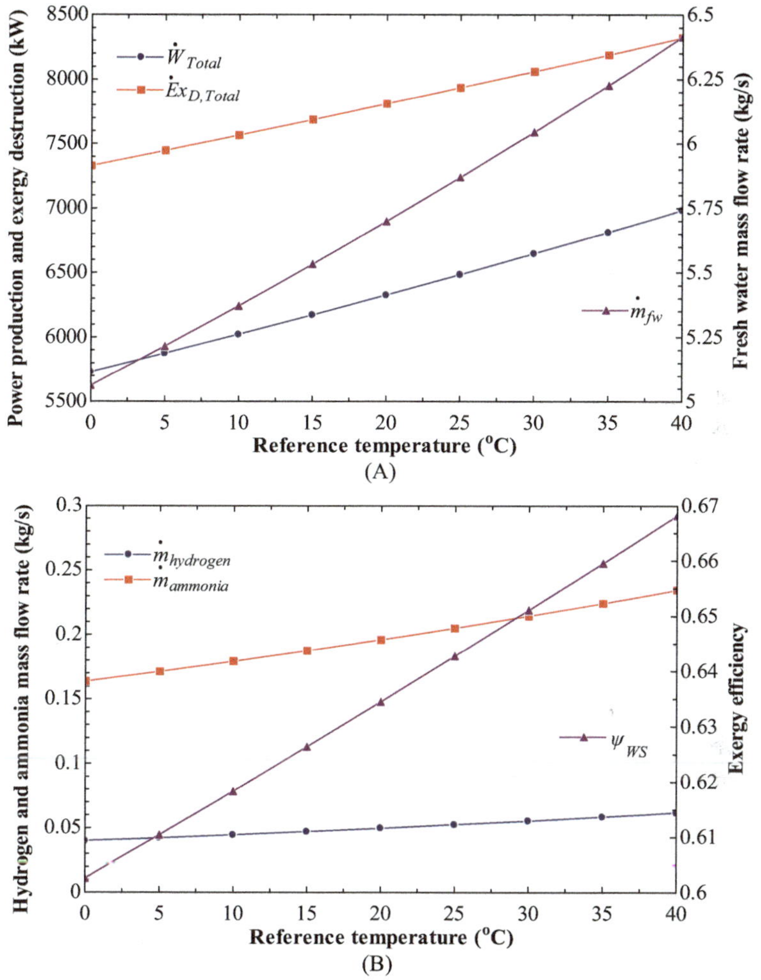

FIGURE 6.93 Effect of reference temperature on (A) net power generation, exergy destruction and freshwater production rates, (B) hydrogen and ammonia production rate and exergy efficiency for the ammonia production process combined with double-flash geothermal power plant.

and more robust systems capable of achieving higher performance. Several criteria are also introduced for the assessment of the advanced geothermal energy systems for various outputs. The energy and exergy efficiencies are presented for each type of advanced geothermal energy–based combined systems, including power and heating, power and cooling, power and freshwater, power and hydrogen, and power and ammonia production options. Exergy efficiency is one of the most important parameters for assessment and for comparing one advanced geothermal energy system with another, especially with conventional systems. Illustrative examples and some practical case studies are also provided to better understand the technical details of the advanced geothermal energy systems and to highlight the importance of geothermal energy–based power systems' design, analysis, and assessment. In addition to that, the effects of environmental conditions and working parameters on the effectiveness of geothermal energy–based combined plants and their subplants are also investigated. The benefit of the advanced geothermal energy system is emphasized more when the system is compared with independent systems that generate the same outputs. Due to their ability to regenerate energy internally, integrated geothermal energy systems are always better than independent single-generation systems.

Nomenclature

A	Area (m^2)
E	Energy (kJ)
e	Specific energy (kJ/kg)
$\dot{E}$	Energy rate (kW)
ex	Specific exergy (kJ/kg)
$\dot{E}x$	Exergy rate (kW)
$\dot{E}x_d$	Exergy destruction rate (kW)
$\dot{E}x^Q$	Exergy transfer rate associated with heat transfer (kW)
$\dot{E}x^W$	Exergy transfer rate associated with work (kW)
F	Faraday constant (C/mol)
G	Gibbs free energy (kJ)
h	Specific enthalpy (kJ/kg)
H	Enthalpy (kJ)
J	Current density (A/m^2)
J_o	Exchange current density (A/m^2)
J_i^{ref}	Preexponential factor (A/m^2)
L	Length (m)
m	Mass (kg)
$\dot{m}$	Mass flow rate (kg/s)
P	Pressure (kPa)
q	Specific heat transfer (kJ/kg)
Q	Heat (kJ)
$\dot{q}$	Specific heat transfer rate (kW/kg)
$\dot{Q}$	Heat rate (kW)
R_{PEM}	Proton exchange membrane resistance (Ω)
R_u	Universal gas constant (kJ/mol K)
s	Specific entropy (kJ/kg K)
S	Entropy (kJ/K)
$\dot{S}$	Entropy rate (kW/K)
t	Time (s)
T	Temperature (°C, K)
u	Internal energy (kJ/kg)
V	Volume (m^3)
V_0	Reversible potential (V)

V_{act}	Activation overpotential (V)
$V_{act,a}$	Anode activation overpotential (V)
$V_{act,c}$	Cathode activation overpotential (V)
W	Work (kJ)
$\dot{w}$	Specific work rate (kW/kg)
$\dot{W}$	Work Rate (kW)

Greek letters

Δ	Change in variable
λ_a	Water content at anode-membrane interface $[\Omega^{-1}]$
λ_c	Water content at cathode-membrane interface (Ω^{-1})
$\lambda(x)$	Water content at location x in the membrane Ω^{-1}
σ_{PEM}	Proton conductivity in PEM (s/m)
$\sigma(x)$	Local ionic PEM conductivity (s/m)
$\in_{HEX}$	HEX efficiency factor
η	Energy efficiency
ψ	Exergy efficiency

Subscript

a	Air
abs	Absorber
AC	Air compressor
Cmp	Compressor
cooling	Cooling load
Con	Condenser
D	Destruction
e	Exit condition
Ej	Ejector
en	Energy
Erd	Energy recovery device
Eva	Evaporator
Ev	Expansion valve
ex	Exergy
f	Fuel
fls	Flashing
g	Generation
Gen	Generator
heating	Heating load
HP	High pressure
l	Liquid
LP	Low pressure
i	Inlet condition
mdm	Membrane distillation module
MP	Medium pressure
MR	Moisture remover
Mx	Mixer
ohm	Ohmic

p	Pump
pr	Particulate remover
RO	Reserve osmosis
Sep	Separator
ST	Steam turbine
tot	Total
Tur	Turbine
Vl	Valve
wf	Working fluid
1...74	State numbers
0	Ambient or reference condition

Superscripts

.	Rate
Ch	Chemical

Acronyms

ACS	Absorption cooling system
CHP	Combined heat and power
COP	Coefficient of performance
DEACS	Double effect absorption cooling system
DFGP	Double-flash geothermal plant
DRODS	Double reverse osmosis desalination system
DS	Double-stage
EES	Engineering Equation Solver
GPS	Geothermal power system
HEX	Heat exchanger
HTSE	High-temperature steam electrolysis
KC	Kalina cycle
KCGP	Kalina cycle geothermal plant
MSBS	Multistaged with binary system
MSFT	Multistaged direct steam with four turbines
MSTT	Multistaged direct steam with three turbines
ORC	Organic Rankine cycle
PEM	Proton exchange membrane
QFGP	Quadruple-flash geothermal plant
RODS	Reverse osmosis desalination system
SEACS	Single effect absorption cooling system
SS	Single-stage
TFGP	Triple-flash geothermal plant

References

[1] I. Dincer, C. Zamfirescu, Advanced Power Generation Systems, Elsevier, New York, NY, 2014.

[2] Y.E. Yuksel, M. Ozturk, I. Dincer, Development of a geothermal based integrated plant for generating clean hydrogen and other useful commodities, J. Energy Resour. Technol. 142 (9) (2020) 1–13.

[3] S. Seyam, I. Dincer, M. Agelin-Chaab, Thermodynamic analysis of a hybrid energy system using geothermal and solar energy sources with thermal storage in a residential building, Energy Storage 2 (1) (2020) 1–22.
[4] Y.E. Yuksel, M. Ozturk, I. Dincer, Analysis and performance assessment of a combined geothermal power-based hydrogen production and liquefaction system, Int. J. Hydrog. Energy 43 (22) (2018) 10268–10280.
[5] Y.E. Yuksel, M. Ozturk, Thermodynamic and thermoeconomic analyses of a geothermal energy based integrated system for hydrogen production, Int. J. Hydrog. Energy 42 (4) (2017) 2530–2546.
[6] N. Tukenmez, Y.E. Yuksel, M. Ozturk, Thermodynamic assessment and optimisation of a biomass energy-based combined system for multigeneration, Int. J. Exergy 30 (3) (2019) 201–238.
[7] I. Dincer, H. Ozcan, Geothermal Energy, Comprehensive Energy Systems, Elsevier, 2018, pp. 702–732.
[8] T.A.H. Ratlamwala, I. Dincer, Comparative efficiency assessment of novel multi-flash integrated geothermal systems for power and hydrogen production, Appl. Therm. Eng. 48 (2012) 359–366.
[9] T.A.H. Ratlamwala, I. Dincer, Energetic and exergetic investigation of novel multi-flash geothermal systems integrated with electrolyzers, J. Power Sources 254 (2014) 306–315.
[10] World Bank, The Role of Desalination in an Increasingly Water-Scarce World; Water Papers, World Bank, Washington, DC, USA, 2019.
[11] I. Dincer, Refrigeration Systems and Applications, third ed., John Wiley & Sons Ltd, 2017.
[12] F. Suleman, I. Dincer, M. Agelin-Chaab, Development of an integrated renewable energy system for multigeneration, Energy 78 (2014) 196–204.
[13] M. Ozturk, Energy and exergy analysis of a combined ground source heat pump system, Appl. Therm. Eng. 73 (1) (2014) 362–370.
[14] I. Dincer, M.A. Rosen, Exergy: Energy, Environment and Sustainable Development, Elsevier, Oxford, UK, 2012.
[15] I. Dincer, C. Acar, A review on clean energy solutions for better sustainability, Int. J. Energy Res. 39 (5) (2015) 585–606.
[16] C. Acar, I. Dincer, Review and evaluation of hydrogen production options for better environment, J. Clean. Prod. 218 (2019) 835–849.
[17] Y.E. Yuksel, M. Ozturk, I. Dincer, Performance assessment of a solar tower based multigeneration system with thermal energy storage, Energy Storage 1 (4) (2019) 1–19.
[18] M. Yari, Exergetic analysis of various types of geothermal power plants, Renew. Energy 35 (2010) 112–121.
[19] A.A. AlZahrani, I. Dincer, G.F. Naterer, Performance evaluation of a geothermal based integrated system for power, hydrogen and heat generation, Int. J. Hydrog. Energy 38 (34) (2013) 14505–14511.
[20] V. Gurau, F. Barbir, H. Liu, An analytical solution of a half-cell model for PEM fuel cells, J. Electrochem. Soc. 147 (7) (2000) 2468–2477.
[21] C.H. Hamann, A. Hamnett, W. Vielstich, Electrochemistry, Wiley-VCH, Weinheim, 2007.
[22] T. Thampan, S. Malhotra, J. Zhang, R. Datta, PEM fuel cell as a membrane reactor, Catal. Today 67 (1–3) (2001) 15–32.
[23] M. Ni, M.K.H. Leung, D.Y.C. Leung, Energy and exergy analysis of hydrogen production by a proton exchange membrane (PEM) electrolyzer plant, Energy Convers. Manag. 49 (2008) 2748–2756.

[24] M. Ozturk, I. Dincer, Thermodynamic assessment of an integrated solar power tower and coal gasification system for multi-generation purposes, Energy Convers. Manag. 76 (2013) 1061–1072.
[25] M. Ozturk, I. Dincer, Thermodynamic analysis of a solar-based multi-generation system with hydrogen production, Appl. Therm. Eng. 51 (1–2) (2013) 1235–1244.
[26] M. Al-Zareer, I. Dincer, M.A. Rosen, Transient thermodynamic analysis of a novel integrated ammonia production, storage and hydrogen production system, Int. J. Hydrog. Energy 44 (33) (2019) 18214–18224.

Study questions and problems

6.1. What indicators affect the sustainability of geothermal energy sources? What working policies can be used to improve the sustainability of this renewable source?

6.2. What are the benefits of an advanced geothermal energy system?

6.3. What are the critical steps that are generally needed to successfully improve a combined geothermal energy plant?

6.4. The energy efficiency of a cogeneration plant involving power and process heat outputs may be expressed as the sum of the power and process heat divided by the heat input. This is also sometimes known as the utilization efficiency. Is it true that adding heat and power enhances efficiency analysis? Can you get around this awkward situation by using exergy efficiency?

6.5. Explain the Rankine cycle, organic Rankine cycle, and Kalina cycle. Why are they important for the utilization of geothermal energy sources for power production?

6.6. What is the estimated minimum temperature for a beneficial geothermal power production plant that operates binary technology? What determines this estimated temperature?

6.7. What is the integrated system with a geothermal source? What role can it play in geothermal power and other useful output generation systems? What would be the benefits of such an integrated system?

6.8. The most popular application of geothermal resource is power production. Some other uses are heating, cooling, clean and hot water, some chemical production, greenhouse heating, and heating for fish farming. From a thermodynamic point of view and considering the quality of the energy, explain which of these integrated with power production uses you recommend most.

6.9. What would be the advantages of running the geothermal plant for power and cooling, power and hydrogen, or power and ammonia production?

6.10. Is exergy analysis more useful for geothermal energy–based cogeneration plants? Explain.

6.11. Is the statement "the exergy of work is equal to work" always correct? Explain.

6.12. In the geothermal energy–based advanced systems, such as (1) multistaged direct systems, (2) multiflashing systems, (3) multistaged binary systems, (4) multiflashing binary systems, and (5) combined/integrated systems, exergy destructions occur in various components. What are the causes of exergy destructions in each of these plants?

6.13. Describe the geothermal energy–based absorption, ejector, and cascaded refrigeration plants.

6.14. Compare reverse osmosis, MSF, and MED systems from an exergetic point of view.

6.15. Describe the single-, double-, triple-, and quadruple-effect absorption cooling systems. How do you connect these absorption cooling plants with geothermal energy sources?

6.16. Consider the system shown in Fig. 6.67. Conduct a detailed exergy analysis and create a bar chart showing the exergy destruction rate in each component of the geothermal energy–based integrated system. Consider the working parameters given in Table 6.28. Vary the isentropic efficiency of turbines from 0.55 to 0.85 and see its impacts on the energy and exergy efficiencies and on the exergy destruction rate of the overall system.

6.17. Consider the system shown in Fig. 6.72. Conduct detailed exergy analysis and create a bar chart showing the exergy destruction rate in each component of the geothermal energy–based power and cooling production system. Consider the working parameters given in Table 6.30. Vary the effectiveness of superheater, evaporator, and economizer from 0.65 to 0.90 and see its impacts on the energy and exergy efficiencies and on the exergy destruction rate of the overall integrated system.

6.18. Consider the system shown in Fig. 6.85. Conduct detailed exergy analysis and create a bar chart showing the exergy destruction rate in each component of the geothermal energy–based hydrogen production and compression system. Consider the working parameters given in Table 6.36. Vary the effectiveness of the ORC turbine inlet temperature from 110°C to 170°C and see its impacts on the energy and exergy efficiencies and on exergy destruction rate of the overall hydrogen production and compression system.

6.19. Consider the system shown in Fig. 6.90. Conduct a detailed exergy analysis and create a bar chart showing the exergy destruction rate in each component of the geothermal energy–based ammonia production system. Consider the working parameters given in Table 6.39. Vary the effectiveness of the HP turbine inlet pressure from 450 to 850 kPa and see its impacts on the energy and exergy efficiencies and on the exergy destruction rate of the overall ammonia production integrated system.

6.20. Rework the case studies provided in this chapter using the given input data and try to duplicate the results. If your results differ from those given in the example, discuss why. Propose methods for improving the performance of the system based on reducing or minimizing exergy destruction.

Chapter 7

Multigenerational geothermal energy systems

7.1 Introduction

Multigenerational energy systems are recognized as energy systems where different subsystems are designed to work together to generate multiple useful outputs in harmony with one another. The potential of multigeneration energy systems vary according to the useful outputs obtained from the system. Also, multigeneration energy systems that include different subsystems can produce many useful outputs with a single energy source [1]. Useful outputs such as heating, cooling, power, fresh-/hot water, and hydrogen privilege integrated energy generation systems. Multigeneration energy systems can benefit from renewable energy sources as an energy source and provide even more benefits in terms of both efficiency and environmental impact [2] because renewable energy sources and energy production will both reduce gas emissions and contribute to the sustainability of energy sources. As a renewable energy source, geothermal energy sources have sustainable energy generation potential for multigeneration energy systems [3].

In this chapter, geothermal energy–based power generation systems for multigeneration aims are defined in general. In these proposed systems, geothermal water is used as the main energy source [4]. By looking at the geothermal water output properties (temperature, steam level, liquid level, or enthalpy level), power generation systems are designed as binary or in different types of flash. Depending on the geothermal water temperature, as a result of the water coming out under pressure and vapor phase, the geothermal source is sent directly to the energy conversion system, and electricity generation takes place; for example, the basic diagram of a single-generation system is given in Fig. 7.1A. In these systems, the geothermal fluid generally goes above ground on the basis of steam or liquid–vapor phase and is passed through a separator directly, and then it enters the turbine where it generates electricity. Subsequently, the geothermal water coming out of the turbine is sent underground again without using the energy.

On the other hand, high-temperature geothermal water is used in the energy conversion plant to produce electricity and then enters the heat recovery subsystem for a water heating application, seen in Fig. 7.1B. In these

Geothermal Energy Systems. DOI: https://doi.org/10.1016/B978-0-12-820775-8.00008-8

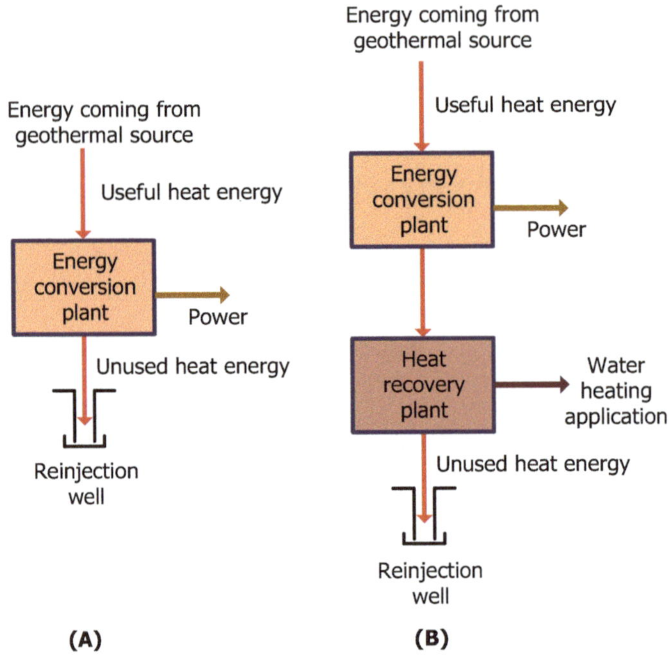

FIGURE 7.1 Basic diagram of (A) single-generation plant, (B) cogeneration plant supported by geothermal energy.

systems, it is possible to cogenerate by using the thermal heat of geothermal water. The high-temperature geothermal water coming out of the energy conversion plant enters the heat recovery subsystem, where it transfers its heat to the clean water coming with ambient temperature and pressure and then enters the reinjection well. Geothermal energy–supported cogeneration plants are more efficient than single-generation plants, as seen in Fig. 7.1B, which produces secondary outputs such as power generation and hot water production.

In general, single-generation systems (A) and cogeneration systems (B), shown in Fig. 7.1, are systems that generate power using the heat of geothermal water and a second product, such as hot water. Finally, these systems can be depicted that are more environmentally sustainable because they use renewable energy sources such as geothermal energy [5].

7.2 Geothermal energy–based multigeneration

Geothermal water energy can be used in different configurations such as triple- and quadruple-generation systems depending on underground exit temperature and pressure. In these systems, the system design generally changes depending on the thermal energy of the geothermal water. Since the geothermal water coming out of the energy conversion systems has high thermal energy, the system efficiency may be drastically increased by using it in different subsystems for various purposes such as heating, cooling, water heating, etc.

The schematic diagram of the trigeneration plant and quadruple plant based on simple modification systems are given separately in Fig. 7.2A and B. Generally, as shown in Fig. 7.2A, trigeneration systems can be defined that provide three different energy outputs with single energy input. Briefly, the schematic flowchart of the geothermal energy–assisted trigeneration system is presented in Fig. 7.2A. The geothermal water coming from the production well transfers its thermal energy to the energy conversion systems—the flash power system, binary system, or organic Rankine cycle (ORC)—where electrical power generation occurs. By using the thermal energy of geothermal water, electricity production takes place in these systems and subsequently enters the heat recovery system without entering the geothermal water reinjection well [6]. Heat recovery systems can be designed differently depending on the temperature of the geothermal water. Here, it is possible to

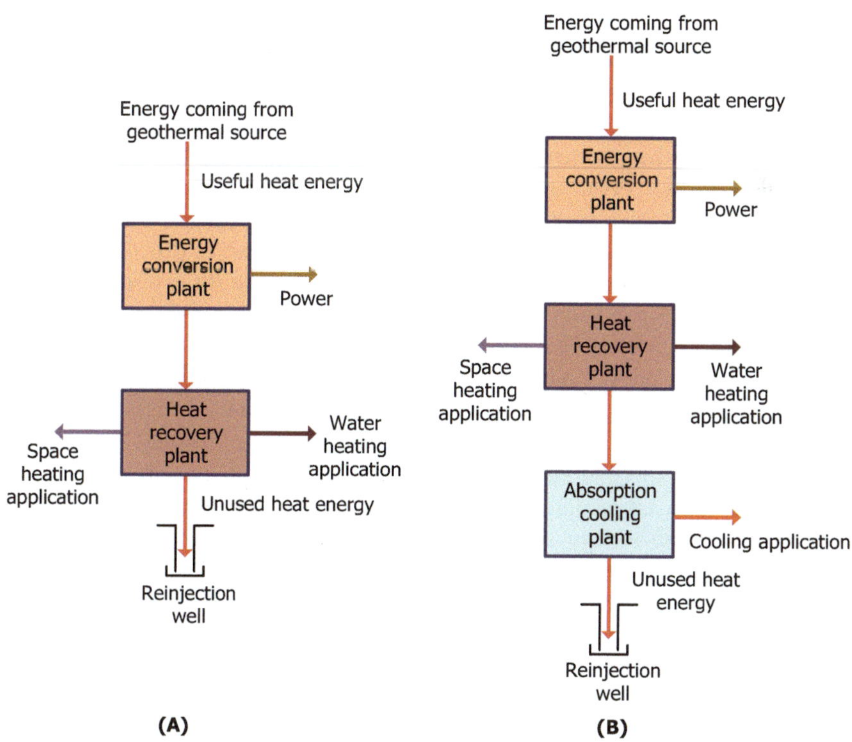

FIGURE 7.2 Schematic diagram of (A) trigeneration plant and (B) quadruple plant based on the simple modification supported by geothermal energy.

use the thermal energy of geothermal water for the purpose of water heating and space heating.

On the other hand, the quadruple plant based on simple modification is presented in Fig. 7.2B. In this system, unlike the trigeneration system, another subsystem, an absorption cooling plant, is added, and a cooling application is applied. The geothermal water transfers its thermal energy, respectively, to the energy conversion system where electricity is produced, to the heat recovery system where heating applications are carried out, and finally to the absorption cooling system (ACS) where the cooling application is done. Finally, the geothermal fluid returns to the reinjection well. Quadruple power plant efficiency is greater than that of the trigeneration system in that production is occurring with the same energy input [7].

Briefly, geothermal energy—supported triple and quadruple systems offer many advantages in terms of both renewable energy and performance [8]. Another advantage of these systems is to help recover some of the heat rejected from the energy conversion systems when combined in stages.

Multigeneration systems can make significant contributions due to their potential for high efficiency as well as low operating costs and pollution emissions per energy output. In the present chapter the design parameters of multigenerational geothermal energy systems are considered and analyzed. As given in Chapter 8, Geothermal District Energy Systems, another important application area for geothermal energy sources is the geothermal energy-based district systems. Also, recent works on the future trends and emerging opportunities of the geothermal source–based energy production plants with multigeneration are given in Chapter 9, Future Directions. In a way, it is the use of waste heat for a specific purpose and efficiently. Many useful products can be obtained with a single energy input [9]. The general flowcharts of the quintuple plant and the sextuple plant based on a simple modification system are presented in Fig. 7.3. As shown in Fig. 7.3A, the quintuple plant consists of an energy conversion plant, three heat recovery plants, a heat engine, a proton exchange membrane (PEM) electrolyzer, and an absorption cooling subsystem. Useful outputs from this proposed study are electricity, hydrogen, oxygen, hot water, heating, space heating, and cooling. First, the geothermal fluid transfers its thermal energy to the energy conversion system, where electricity generation takes place. Afterward, the waste heat or geothermal fluid coming out of the energy conversion system transfers its energy to the heat recovery system and from there to the cooling system.

The thermal energy required for the operation of the heat engine is provided from the energy conversion system, and the electricity produced in the heat engine is sent to PEM electrolysis. In PEM electrolysis, pure water decomposes into hydrogen and oxygen atoms by using electrical energy. Pure water, which is not used in PEM, then enters the heat recovery system and exits as hot water [10]. The thermal energy coming from the heat machine and the cooling system enters the last stage of the heat recovery

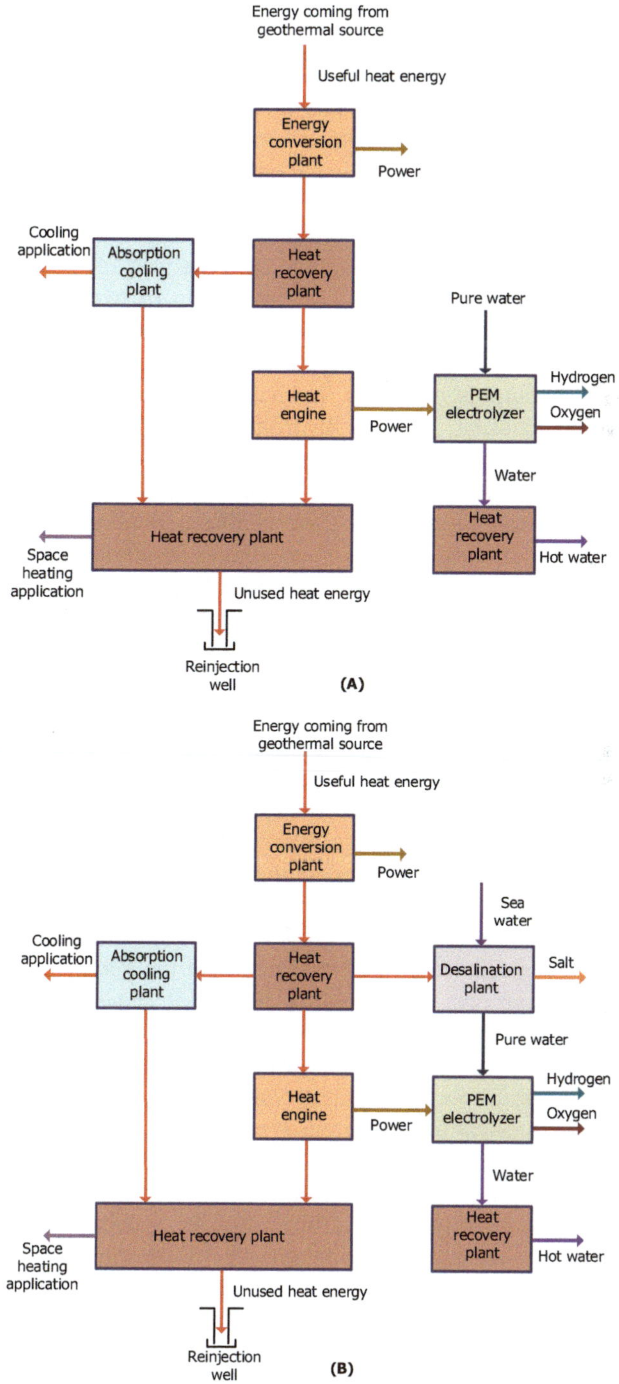

FIGURE 7.3 Schematic diagram of (A) quintuple plant and (B) sextuple plant based on the simple modification supported by geothermal energy.

system and is used for space heating. The geothermal water then enters the reinjection well again.

The main difference of the sextuple plant proposed in Fig. 7.3B from the quintuple plant shown in Fig. 7.3A is the distillation unit. The geothermal fluid coming out of the energy conversion system enters the heat recovery plant, and from there it provides thermal energy in order to meet the working of the required energy for the cooling and distillation system. The seawater entering the distillation unit at approximately environmental conditions is distilled with the thermal energy coming from the heat recovery system, and pure water production is realized. The production of hydrogen and oxygen takes place in PEM electrolysis with pure water produced there [11]. In the schematic representation of the sextuple plant shown in Fig. 7.3B, the production of many outputs, with the thermal energy of the geothermal fluid, such as heating, cooling, oxygen, hydrogen, and electricity are presented clearly. In this system, the cooling application is provided by the ACS. The geothermal fluid, at a temperature approximately above 90°C, enters the generator of the ACS and provides the thermal energy required for the operation of the ACS. These cooling systems have a great advantage because they can work with low-energy sources such as waste heat or geothermal and solar energy. The cooling load takes place in the evaporator subsection of the ACS, or heat exchanger (HEX). The geothermal fluid eventually leaves the heat recovery system at low temperatures and then enters the reinjection well.

Generally, a multigeneration geothermal energy–based integrated system is defined as a system that produces many useful outputs with single energy input, such as heating, cooling, power, oxygen, hydrogen, hot water, etc. The energy input in these systems is the thermal energy of the geothermal fluid. Depending on the pressure and temperature of the injection well outlet of the geothermal fluid, different types of power production systems such as single-, double-, triple-, or more flashed systems and the binary system can be designed. The main feature of their design is that the geothermal water coming from the underground well is in different phase states such as steam, liquid, or liquid-steam. The skeleton of the system can be created according to the characteristics of the geothermal water coming from underground and according to the useful outputs desired. Since geothermal fluids are generally low-temperature applications, different types of ORC and Kalina cycles can be used in power generation in binary systems. From another point of view, power generation can be obtained after the geothermal fluid released in the vapor phase is sent directly to the turbine after being dehumidified.

As seen in Fig. 7.4, the geothermal energy–based multigeneration system can be used for many useful product applications separately or simultaneously. As is clearly seen in the figure, the first and most important is electricity production from geothermal-supported systems. Electrical power generation practices are considered to be important in countries that are in

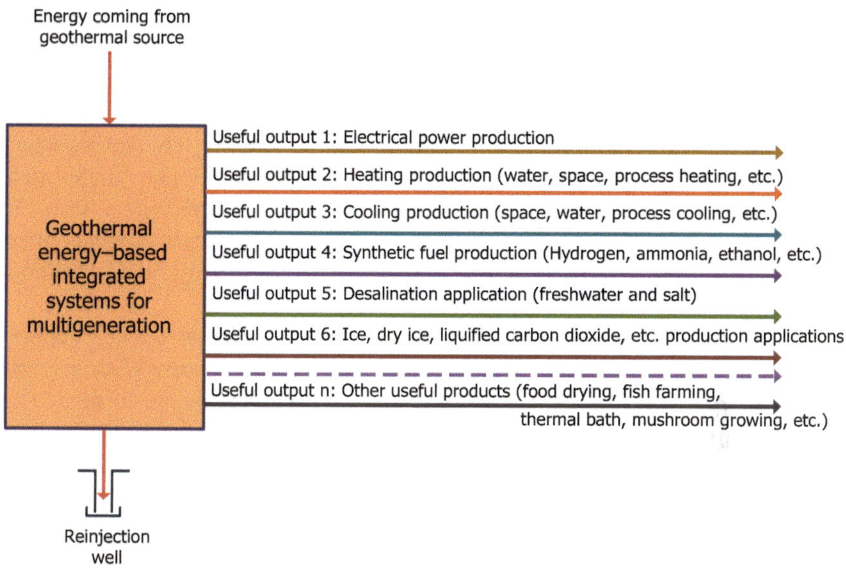

FIGURE 7.4 Useful production outputs based on geothermal energy.

the development and development stages and are often an important part of the sustainable future. In terms of environmental detriments, electricity generation from geothermal energy–supported systems has an important place since they do not use fossil fuels.

Second, another useful output from these systems is in heating applications. By using the thermal energy of geothermal energy, heat can be obtained by using heat recovery units for the heating of spaces and water. By means of a HEX, the geothermal fluid can transfer its thermal energy to domestic water, and the heating application can take place.

Third, geothermal energy–supported cooling applications can be realized with ACS s operating at about 90°C. The geothermal fluid transfers its heat to the other fluid in the generator part of the ACS, where the thermal energy required for the operation of the ACS is provided.

Fourth is the production of synthetic fuels as a result of using electrical power or waste heat from geothermal energy–powered power generation systems. Here, some or all of the electrical energy produced in the turbines of the power systems are delivered to PEM electrolysis to separate the water into oxygen and hydrogen. It is possible to produce ammonia by sending the nitrogen in the air with the hydrogen produced in PEM to the ammonia synthesis unit.

Another sub output, which is the fifth useful output, is the production of clean water and salt using the distillation unit. Many distillation systems are used in clean water production. One of these is the thermal distillation system, for which the thermal energy needed can be obtained with the steam–liquid mixture coming out of the turbine. In short, the seawater that usually enters under reference conditions in the desalination unit is met by the thermal energy required by this unit, that is, the thermal energy of the geothermal fluid or a different fluid coming out of the power cycle.

The sixth useful output of these systems can be ice, dry ice, and liquefaction of carbon dioxide. Finally, other than the ones previously mentioned, these outputs can be obtained from geothermal energy–supported systems in different areas such as product drying/dehumidification, fish farming, thermal bath, etc.—especially at temperatures between 30°C and 50°C—without sending geothermal fluid underground.

Briefly, multigeneration systems assisted by geothermal energy support can be used in the production of many different useful outputs. With a single energy input, it is possible to increase the thermodynamic performance of these systems by obtaining many useful outputs such as heating, cooling, hydrogen, oxygen, ammonia, clean water, salt, and electricity. In the design of the multigeneration system, how to obtain the thermal energy needed by the subsystems should be taken into consideration.

7.2.1 Case study 7.1

The geothermal energy-based integrated energy system consisting of nine main subsystems is given in Fig. 7.5 in its most general form. These subsystems are (1) geothermal power cycle (GPC), (2) isobutene power cycle (IPC), (3) single-effect absorption cooling (SEAC) system, (4) hydrogen production system (HPS), (5) hydrogen liquefaction system (HLS), (6) ammonia production system (λ), (7) seawater distillation system, (8) drying system (DS), and (9) hot water production system (HWPS). As can be seen from these subsystems, many useful outputs are obtained from this integrated system. In particular, the production of hydrogen and ammonia from this integrated system provides a privileged position among the systems integrated into this system. In addition to hydrogen and ammonia production from this integrated system, other useful outputs, such as electricity, heating, cooling, freshwater, and hot water, can be obtained. Explaining the working principle of the integrated system, in general, provides a better understanding of the system.

The integrated system starts working by including the geothermal fluid in the system with flow 1. The geothermal fluid included in the system is a mixture of vapor and liquid phases. In order for the geothermal fluid to be used efficiently in the production of electricity with the turbine, the higher amount must be vapor phase. For this reason, the geothermal fluid entering the system with the number 1 flow is first reduced with the flash chamber. In order to increase the efficiency of geothermal fluid in electricity generation, the geothermal fluid, whose pressure decreases, is transferred to separator 1 with flow 2. In geothermal fluid separator 1, in which there is a

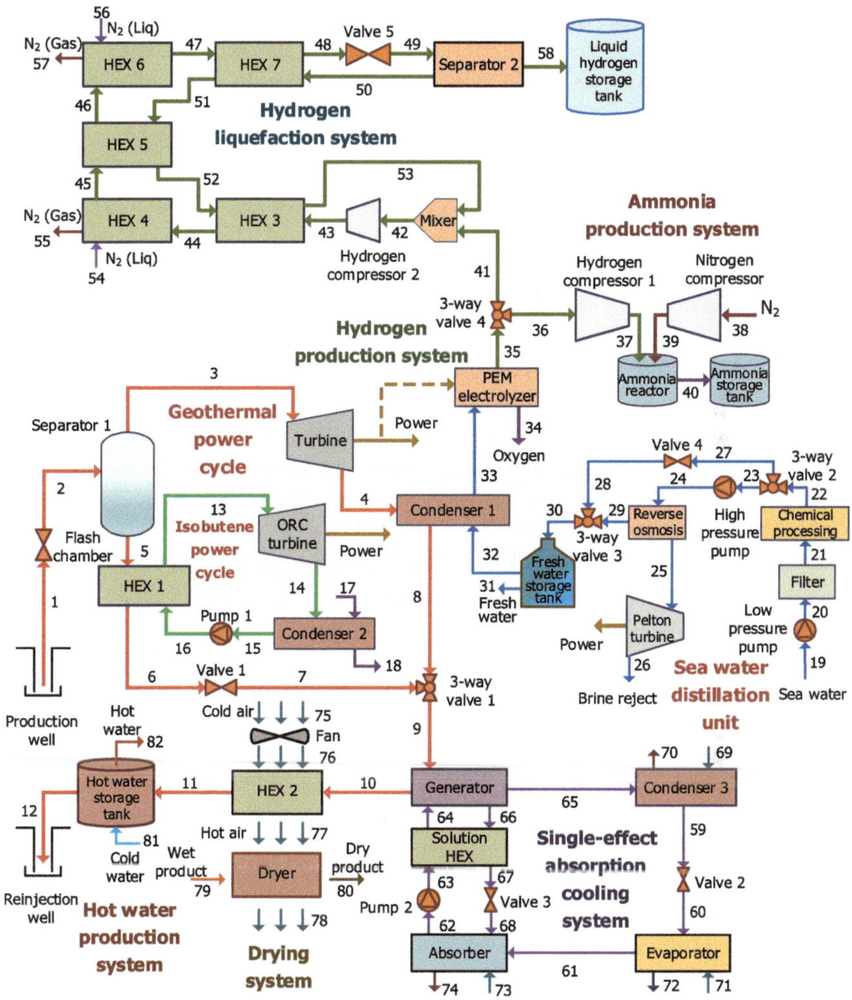

FIGURE 7.5 Geothermal energy–based integrated system for multigeneration.

mixture of liquid and vapor phase, the mixture is divided into two parts as liquid and vapor phases. The part having a high vapor density is delivered to the turbine with flow 3 in order to realize the electricity production. The production of electricity is provided by expanding the geothermal fluid with high vapor density between flows 3 and 4. The geothermal fluid with reduced pressure from the turbine is then sent to condenser 1 for condensation with flow 4. While the GPC gets to this point, the geothermal fluid with a high density of liquid phase is transferred from separator 1 to HEX 1 with flow 5. By using HEX 1, the heat energy present in the highly liquid phase geothermal fluid supplies energy to the other power generation cycle, the isobutane power cycle. The geothermal fluid from HEX 1 is sent to valve 1 with flow 6 to reduce its pressure. The pressure-reduced geothermal fluid is transferred to three-way valve 1 with flow 7. Fluid with reduced temperature and pressure from condenser 1 is sent to three-way valve 1 with flow 8. In this way, the GPC is completed. Considering the useful outputs produced until the GPC process is completed, the electricity generated by the turbine can be clearly seen first. In addition to the generated electricity, energy transfers to different subsystems of the integrated system take place in this cycle. By utilizing condenser 1 from the geothermal fluid in the GPC, an energy transfer occurs to the freshwater sent to the PEM electrolyzer. Also, the energy transfer from the geothermal fluid in the GPC to the working fluid in the isobutane power cycle is realized by HEX 1. In addition, the heat energy required for a SEAC system, DS, and HWPS is also supplied from the geothermal fluid coming out of the GPC. These energy transfers take place with the geothermal fluid in the GPC.

Another subsystem where electricity production in the integrated system takes place is the isobutane power cycle. This cycle is an ORC. Since isobutane was chosen as the working fluid used in the ORC, this subsystem was called the isobutane power cycle. Heat energy is transferred to the working fluid in the isobutane power cycle by the geothermal fluid in the GPC. The working fluid, which is charged with HEX 1, is transferred to the ORC turbine with the number 13 flow for electricity generation. Electricity is produced in the ORC turbine based on the expanding working fluid between flows 13 and 14. In order to condense the working fluid with a certain amount of heat energy from the ORC turbine and reduced pressure, it is sent to condenser 2 with flow 14. In condenser 2, some of the heat energy of the working fluid is transferred to a different fluid between flows 17 and 18. Working fluid from condenser 2, whose pressure and temperature have decreased, is first transferred to pump 1 with flow 15 to increase its pressure. The working fluid whose pressure is increased is then transferred to HEX 1 with flow 16 to increase its temperature. In this way, a cycle of the isobutane power system is completed. Looking at the useful outputs from the isobutane power cycle, the electricity produced in the ORC turbine can be clearly seen first. In addition, during the condensation of the working fluid in condenser 2, heat energy transferred to another fluid can be seen.

The geothermal fluid coming from three-way valve 1 with flow 9 comes first to the generator in a SEAC cycle, providing heat energy to this subsystem. After a certain part of the heat energy of the geothermal fluid is can be transferred to the working fluid in the SEAC system by the generator, the geothermal fluid is transferred to HEX 2 with flow 10 to supply energy to the DS. The geothermal fluid from HEX 2 is also transferred to the HWPS with flow 11 to store hot water. The geothermal fluid coming from the HWPS is transferred back to the ground with flow 12 to ensure the sustainability of the energy source.

The subsystem in the integrated system to obtain freshwater from seawater is the seawater distillation unit. First, seawater is transferred to this subsystem with flow 19. By using the low-pressure pump, the seawater coming to this subsystem is sent to the filter component where the seawater is filtered with flow 20. Various particles that can bring a load to the components in this subsystem are separated from the seawater by filtration. Seawater, which passes through the filtering process, is sent to the chemical process with flow 21. Freshwater from this chemical treatment component is sent to three-way valve 2 with flow 22. A certain amount of freshwater from chemical processing is transferred to the high-pressure pump with flow 23. Here, the freshwater with increased pressure is sent to the reverse osmosis component with flow 24. The seawater that came to this subsystem first entered a filtering process. In this way, some particles present in seawater were separated from it. However, the quality of the incoming fluid can be further improved. The high-pressure freshwater coming to the reverse osmosis component is passed through semipermeable membranes. Almost all of the impurities in the water are removed completely. To take advantage of the residue remaining from the treatment, this impurity-containing fluid is sent to the Pelton turbine with flow 25 for electrical output. Electricity is produced by expanding this fluid between flows 25 and 26 in the Pelton turbine. In this way, the fluid containing impurities coming from the purification process is taken from the turbine to outside the system. While the process from chemical processing to this point is taking place, the remaining part of the freshwater from chemical processing is sent to valve 4 with flow 27. By utilizing this valve, the pressure of the water is reduced. Freshwater, with reduced pressure, is sent to three-way valve 3 with flow 28. This freshwater is mixed with the freshwater coming from the reverse osmosis component with flow 29 and sent to the freshwater storage tank with flow 30. Freshwater in this tank can be used for various needs with flow 31. A certain amount of freshwater in the freshwater storage tank is sent to condenser 1 with flow 32 to be used for hydrogen and ammonia production. In condenser 1, the temperature of the freshwater is increased by using the heat energy of the geothermal fluid.

Another subsystem that provides useful output in the integrated system is the hydrogen production subsystem. For the production of hydrogen in the integrated system, freshly adjusted water from condenser 1 is transferred to the PEM electrolyzer with flow 33. The electrical energy required for the operation of the PEM electrolyzer is met by the electricity produced from the turbine in the GPC. In the PEM electrolyzer, sweet hydrogen and oxygen are divided into two parts. The produced oxygen can be used with flow 34 as required. The produced hydrogen is transferred to three-way valve 4 with flow 35, for the hydrogen liquefaction process and ammonia production.

A certain amount of hydrogen coming into three-way valve 4 is sent to the hydrogen compressor 1 with flow 36 to produce ammonia. The nitrogen gas obtained by separation from the air is sent to the ammonia compressor with flow 38. Hydrogen compressed with hydrogen compressor 1 and nitrogen compressed with the nitrogen compressor are transferred to the ammonia reactor with flows 37 and 39, respectively. Ammonia production is carried out by reacting hydrogen and nitrogen in the ammonia reactor. The produced ammonia is transferred to the ammonia storage tank with flow 40.

The other part of the hydrogen coming to three-way valve 4 is transferred to the hydrogen liquefaction subsystem by flow 41. The Linde-Hampson liquefaction cycle is used to liquefy the produced hydrogen. The hydrogen pressure coming to the mixer with flow 41 is sent to hydrogen compressor 2 with flow 42 to increase the pressure. The pressurized hydrogen is sent to HEX 3 by flow 43. Here, the temperature of the hydrogen is slightly reduced. Hydrogen is then transferred to HEX 3 by flow 44. With flow 54, the temperature of the hydrogen is further reduced by making use of the nitrogen liquid entering the HEX 4. Nitrogen liquid entering the HEX 4 exits here as a gas with flow 55. The hydrogen, whose temperature decreases, is transferred to HEX 5 with flow 45 to further reduce its temperature. From here, with flow 46, hydrogen is sent to HEX 6, where nitrogen fluid is used to reduce the temperature. With flow 56, nitrogen liquid enters HEX 6. Then the nitrogen, whose temperature rises, exits HEX 6 as a gas with flow 57. The hydrogen, whose temperature decreases significantly, is transferred to HEX 7 with flow 47, to reduce the temperature again. In the process so far, a certain part of the hydrogen coming to the liquefaction process is liquefied. Hydrogen from HEX 7 is transferred to valve 5 with flow 48 to adjust the pressure. This liquid–gas mixture is then transferred to separator 2 with flow 49. The liquid part of the hydrogen in the liquid–gas mixture is transferred from separator 2 to the liquid hydrogen storage tank by flow 58. The hydrogen in the gas phase is transferred to the mixer again through the flows 50, 51, 52, and 53, respectively. While the hydrogen directed to the mixer passes through these flows, the hydrogen performs heat transfer with the hydrogen entering the liquefaction process. Since it comes to the mixer, hydrogen is mixed with the hydrogen coming to the liquefaction subsystem. This is how the hydrogen liquefaction subsystem works.

The cooling output in the integrated system is achieved with a SEAC subsystem. In these systems, cooling can be performed with different working fluids. Ammonia–water mixture is used in the subsystem in this integrated system. The rich working fluid from the absorber is transferred to pump 2 with flow 62 to increase the pressure. The pressurized working fluid is sent to the solution HEX with flow 63, and then this working fluid is transferred to the generator with flow 64. Ammonia vapor is obtained from this fluid by making use of the temperature in the geothermal fluid in the generator. The weak solution in the generator is sent to the HEX first with flow 66 to be transferred back to the absorber. It is then sent to valve 3 with flow 67 to reduce its pressure. The low-pressure weak solution is also transferred to the absorber with flow 68. Ammonia vapor in the generator is sent to condenser 3 with flow 65 to be condensed. After the ammonia vapor is condensed in condenser 3, it is sent to valve 2 with flow 59. With valve 2, the pressure of the working fluid is reduced and, with its pressure decreased, is transferred to the evaporator with flow 60. In this subsystem, the pressure of the working fluid is adjusted by the valve, since the evaporator performs its function at low pressure. By utilizing the evaporator, the cooling output is obtained. Here, the working fluid is evaporated by taking heat from the environment. The vapor working fluid is sent to the absorber with flow 61. The weak solution coming from flow 68 forms a mixture in the absorber along with the steam coming from the evaporator. In this way, the SEAC subsystem performs its function.

In the integrated system, the drying subsystem can meet its energy needs based on its geothermal fluid. The cold air coming to the fan with flow 75 is sent to HEX 2 with flow 76. By using the energy in the geothermal fluid in the HEX, the heated air is transferred to the dryer with flow 77. The wet products coming to the dryer with flow 79 are dried and output from the system with flow 80. In this way, the need for drying is met. The need for hot water is met from the hot water storage tank. By utilizing the energy in the geothermal fluid, the cold water coming to the hot water storage tank with flow 81 can be used as hot water with flow 82. Hot water is also obtained in this way.

In general, the integrated system works this way. Electricity, ammonia, liquefied hydrogen, heating, cooling, freshwater, hot water, and drying can be obtained from this integrated system consisting of nine subsystems. In this case study, the geothermal energy–based combined power plant is analyzed by using the thermodynamic analysis. To make a detailed thermodynamic analysis, the working parameters of the geothermal energy–based integrated system for multigeneration are given in Table 7.1.

To conduct a thermodynamic analysis, the mass, energy, entropy, and exergy balance equations for geothermal energy–based integrated system components need to be written. For this reason, the mass, energy, entropy and exergy balance equations for the flash chamber are defined:

$$\dot{m}_1 = \dot{m}_2$$

$$\dot{m}_1 h_1 = \dot{m}_2 h_2$$

$$\dot{m}_1 s_1 + \dot{S}_g^{fc} = \dot{m}_2 s_2$$

$$\dot{m}_1 ex_1 = \dot{m}_2 ex_2 + \dot{E}x_d^{fc}$$

TABLE 7.1 Parameters selected for the geothermal energy–based integrated plant for multigenerational purposes.

Parameters	Values
Reference temperature and pressure, T_o and P_o	25°C and 101.3 kPa
Geofluid source temperature, T_1	210°C
Geofluid source pressure, P_1	3975 kPa
Geofluid mass flow rate, $\dot{m}_1$	140 kg/s
Isentropic efficiency of the turbine, η_{Tur}	0.80
Isentropic efficiency of ORC turbine, η_{ORCT}	0.75
Separator 1 inlet temperature, T_2	170°C
Separator 1 inlet pressure, P_2	1250 kPa
Fraction of vapor at separator 1 outlet	0.16
Turbine inlet temperature, T_3	170°C
Turbine inlet pressure, P_3	1000 kPa
Turbine outlet temperature, T_4	98.51°C
Turbine outlet pressure, P_4	80 kPa
Working fluid of ORC	Isobutene
ORC turbine inlet temperature, T_{13}	109.7°C
ORC turbine inlet pressure, P_{13}	1500 kPa
ORC turbine exit pressure, P_{14}	200 kPa
Seawater salinity	35,000 ppm
Product water salinity	450 ppm
Seawater feeding temperature, T_{19}	20°C
Membrane recovery ratio	0.6
Pelton turbine inlet pressure, P_{25}	5100 kPa
Pelton turbine inlet temperature, T_{25}	41.7°C
Pelton turbine outlet pressure, P_{26}	101.3 kPa
Pelton turbine outlet temperature, T_{26}	40.62°C
PEM electrolyzer temperature, T_{PEM}	81°C
Isentropic efficiency of hydrogen compressor, η_{HC}	78%
Isentropic efficiency of nitrogen compressor, η_{NC}	80%
Hydrogen compressor 1 outlet pressure, P_{37}	10,000 kPa

(Continued)

TABLE 7.1 (Continued)

Parameters	Values
Nitrogen compressor outlet pressure, P_{39}	10,000 kPa
Hydrogen compressor 1 outlet temperature, T_{37}	250°C
Nitrogen compressor outlet temperature, T_{39}	250°C
Hydrogen compressor 2 outlet pressure, P_{43}	8080 kPa
Hydrogen compressor 2 outlet temperature, T_{43}	52.8°C
Liquid hydrogen temperature, T_{58}	−252.8°C
Energetic coefficient of performance, COP_{en}	1.052
Exergetic coefficient of performance, COP_{ex}	0.284
Working fluid of SEAC	Ammonia–water
Generator temperature, T_{Gen}	92°C
Condenser 2 temperature, T_{Con2}	38.9°C
Evaporator temperature, T_{Eva}	−10°C
Absorber temperature, T_{Ab}	38.9°C
Geofluid reinjection temperature, T_{12}	38°C

The mass, energy, entropy and exergy balance equations for the separator 1 can be expressible:

$$\dot{m}_2 = \dot{m}_3 + \dot{m}_5$$

$$\dot{m}_2 h_2 = \dot{m}_3 h_3 + \dot{m}_5 h_5$$

$$\dot{m}_2 s_2 + \dot{S}_g^{Sep1} = \dot{m}_3 s_3 + \dot{m}_5 s_5$$

$$\dot{m}_2 ex_2 = \dot{m}_3 ex_3 + \dot{m}_5 ex_5 + \dot{E}x_d^{Sep1}$$

The mass, energy, entropy, and exergy balance equations for the turbine can be expressed:

$$\dot{m}_3 = \dot{m}_4$$

$$\dot{m}_3 h_3 = \dot{m}_4 h_4 + \dot{W}_T$$

$$\dot{m}_3 s_3 + \dot{S}_g^T = \dot{m}_4 s_4$$

$$\dot{m}_3 ex_3 = \dot{m}_4 ex_4 + \dot{E}x_T^W + \dot{E}x_d^T$$

The mass, energy, entropy, and exergy balance equations for the HEX 1 can be expressed:

$$\dot{m}_5 = \dot{m}_6; \dot{m}_{16} = \dot{m}_{13}$$

$$\dot{m}_5 h_5 + \dot{m}_{16} h_{16} = \dot{m}_6 h_6 + \dot{m}_{13} h_{13}$$

$$\dot{m}_5 s_5 + \dot{m}_{16} s_{16} + \dot{S}_g^{HEX1} = \dot{m}_6 s_6 + \dot{m}_{13} s_{13}$$

$$\dot{m}_5 ex_5 + \dot{m}_{16} ex_{16} = \dot{m}_6 ex_6 + \dot{m}_{13} ex_{13} + \dot{E}x_d^{HEX1}$$

The mass, energy, entropy, and exergy balance equations for the pump 1 are defined:

$$\dot{m}_{15} = \dot{m}_{16}$$

$$\dot{m}_{15} h_{15} + \dot{W}_{P1} = \dot{m}_{16} h_{16}$$

$$\dot{m}_{15} s_{15} + \dot{S}_g^{P1} = \dot{m}_{16} s_{16}$$

$$\dot{m}_{15} ex_{15} + \dot{E}x_{P1}^{W} = \dot{m}_{16} ex_{16} + \dot{E}x_d^{P1}$$

The mass, energy, entropy and exergy balance equations for the condenser 1 can be expressed:

$$\dot{m}_4 = \dot{m}_8; \dot{m}_{32} = \dot{m}_{33}$$

$$\dot{m}_4 h_4 + \dot{m}_{32} h_{32} = \dot{m}_8 h_8 + \dot{m}_{33} h_{33}$$

$$\dot{m}_4 s_4 + \dot{m}_{32} s_{32} + \dot{S}_g^{con1} = \dot{m}_8 s_8 + \dot{m}_{33} s_{33}$$

$$\dot{m}_4 ex_4 + \dot{m}_{32} ex_{32} = \dot{m}_8 ex_8 + \dot{m}_{33} ex_{33} + \dot{E}x_d^{Con1}$$

The mass, energy, entropy and exergy balance equations for the valve 1 can be expressed:

$$\dot{m}_6 = \dot{m}_7$$

$$\dot{m}_6 h_6 = \dot{m}_7 h_7$$

$$\dot{m}_6 s_6 + \dot{S}_g^{vl1} = \dot{m}_7 s_7$$

$$\dot{m}_6 ex_6 = \dot{m}_7 ex_7 + \dot{E}x_d^{vl1}$$

The mass, energy, entropy, and exergy balance equations for the ORC turbine are expressible:

$$\dot{m}_{13} = \dot{m}_{14}$$

$$\dot{m}_{13} h_{13} = \dot{m}_{14} h_{14} + \dot{W}_{ORC,T}$$

$$\dot{m}_{13}s_{13} + \dot{S}_g^{ORC,T} = \dot{m}_{14}s_{14}$$

$$\dot{m}_{13}ex_{13} = \dot{m}_{14}ex_{14} + \dot{E}x_{ORC,T}^{W} + \dot{E}x_d^{ORC,T}$$

The mass, energy, entropy, and exergy balance equations for the condenser 2 can be written:

$$\dot{m}_{14} = \dot{m}_{15}; \dot{m}_{17} = \dot{m}_{18}$$

$$\dot{m}_{14}h_{14} + \dot{m}_{17}h_{17} = \dot{m}_{15}h_{15} + \dot{m}_{18}h_{18}$$

$$\dot{m}_{14}s_{14} + \dot{m}_{17}s_{17} + \dot{S}_g^{con2} = \dot{m}_{15}s_{15} + \dot{m}_{18}s_{18}$$

$$\dot{m}_{14}ex_{14} + \dot{m}_{17}ex_{17} = \dot{m}_{15}ex_{15} + \dot{m}_{18}ex_{18} + \dot{E}x_d^{Con2}$$

The mass, energy, entropy, and exergy balance equations for the three-way valve 1 can be written:

$$\dot{m}_7 + \dot{m}_8 = \dot{m}_9$$

$$\dot{m}_7h_7 + \dot{m}_8h_8 = \dot{m}_9h_9$$

$$\dot{m}_7s_7 + \dot{m}_8s_8 + \dot{S}_g^{3wvl1} = \dot{m}_9s_9$$

$$\dot{m}_7ex_7 + \dot{m}_8ex_8 = \dot{m}_9ex_9 + \dot{E}x_d^{3wvl1}$$

The mass, energy, entropy, and exergy balance equations for the low-pressure pump are expressed:

$$\dot{m}_{19} = \dot{m}_{20}$$

$$\dot{m}_{19}h_{19} + \dot{W}_{LPP} = \dot{m}_{20}h_{20}$$

$$\dot{m}_{19}s_{19} + \dot{S}_g^{LPP} = \dot{m}_{20}s_{20}$$

$$\dot{m}_{19}ex_{19} + \dot{E}x_{LPP}^{W} = \dot{m}_{20}ex_{20} + \dot{E}x_d^{LPP}$$

The mass, energy, entropy, and exergy balance equations for the filter are defined:

$$\dot{m}_{20} = \dot{m}_{21}$$

$$\dot{m}_{20}h_{20} = \dot{m}_{21}h_{21}$$

$$\dot{m}_{20}s_{20} + \dot{S}_g^{fl} = \dot{m}_{21}s_{21}$$

$$\dot{m}_{20}ex_{20} = \dot{m}_{21}ex_{21} + \dot{E}x_d^{fl}$$

The mass, energy, entropy, and exergy balance equations for the chemical filter can be given:

$$\dot{m}_{21} = \dot{m}_{22}$$

$$\dot{m}_{21}h_{21} = \dot{m}_{22}h_{22}$$

$$\dot{m}_{21}s_{21} + \dot{S}_g^{cf} = \dot{m}_{22}s_{22}$$

$$\dot{m}_{21}ex_{21} = \dot{m}_{22}ex_{22} + \dot{E}x_d^{cf}$$

The mass, energy, entropy, and exergy balance equations for the three-way valve 2 can be defined:

$$\dot{m}_{22} = \dot{m}_{23} + \dot{m}_{27}$$

$$\dot{m}_{22}h_{22} = \dot{m}_{23}h_{23} + \dot{m}_{27}h_{27}$$

$$\dot{m}_{22}s_{22} + \dot{S}_g^{3wv2} = \dot{m}_{23}s_{23} + \dot{m}_{27}s_{27}$$

$$\dot{m}_{22}ex_{22} = \dot{m}_{23}ex_{23} + \dot{m}_{27}ex_{27} + \dot{E}x_d^{3wv2}$$

The mass, energy, entropy, and exergy balance equations for the valve 4 can be written:

$$\dot{m}_{27} = \dot{m}_{28}$$

$$\dot{m}_{27}h_{27} = \dot{m}_{28}h_{28}$$

$$\dot{m}_{27}s_{27} + \dot{S}_g^{val4} = \dot{m}_{28}s_{28}$$

$$\dot{m}_{27}ex_{27} = \dot{m}_{28}ex_{28} + \dot{E}x_d^{val4}$$

The mass, energy, entropy, and exergy balance equations for the high-pressure pump are defined:

$$\dot{m}_{23} = \dot{m}_{24}$$

$$\dot{m}_{23}h_{23} + \dot{W}_{HPP} = \dot{m}_{24}h_{24}$$

$$\dot{m}_{23}s_{23} + \dot{S}_g^{HPP} = \dot{m}_{24}s_{24}$$

$$\dot{m}_{23}ex_{23} + \dot{E}x_{HPP}^{W} = \dot{m}_{24}ex_{24} + \dot{E}x_d^{HPP}$$

The mass, energy, entropy, and exergy balance equations for the reverse osmosis are written:

$$\dot{m}_{24} = \dot{m}_{25} + \dot{m}_{29}$$

$$\dot{m}_{24}h_{24} = \dot{m}_{25}h_{25} + \dot{m}_{29}h_{29}$$

$$\dot{m}_{24}s_{24} + \dot{S}_g^{ro} = \dot{m}_{25}s_{25} + \dot{m}_{29}s_{29}$$

$$\dot{m}_{24}ex_{24} = \dot{m}_{25}ex_{25} + \dot{m}_{29}ex_{29} + \dot{E}x_d^{ro}$$

The mass, energy, entropy, and exergy balance equations for the Pelton turbine can be defined:

$$\dot{m}_{25} = \dot{m}_{26}$$

$$\dot{m}_{25}h_{25} = \dot{m}_{26}h_{26} + \dot{W}_{PT}$$

$$\dot{m}_{25}s_{25} + \dot{S}_g^{PT} = \dot{m}_{26}s_{26}$$

$$\dot{m}_{25}ex_{25} = \dot{m}_{26}ex_{26} + \dot{E}x_{PT}^{W} + \dot{E}x_d^{PT}$$

The mass, energy, entropy, and exergy balance equations for the three-way valve 3 are written:

$$\dot{m}_{28} + \dot{m}_{29} = \dot{m}_{30}$$

$$\dot{m}_{28}h_{28} + \dot{m}_{29}h_{29} = \dot{m}_{30}h_{30}$$

$$\dot{m}_{28}s_{28} + \dot{m}_{29}s_{29} + \dot{S}_g^{3wv3} = \dot{m}_{30}s_{30}$$

$$\dot{m}_{28}ex_{28} + \dot{m}_{29}ex_{29} = \dot{m}_{30}ex_{30} + \dot{E}x_d^{3wv3}$$

The mass, energy, entropy, and exergy balance equations for the freshwater storage tank can be written:

$$\dot{m}_{30} = \dot{m}_{31} + \dot{m}_{32}$$

$$\dot{m}_{30}h_{30} = \dot{m}_{31}h_{31} + \dot{m}_{32}h_{32}$$

$$\dot{m}_{30}s_{30} + \dot{S}_g^{fwst} = \dot{m}_{31}s_{31} + \dot{m}_{32}s_{32}$$

$$\dot{m}_{30}ex_{30} = \dot{m}_{31}ex_{31} + \dot{m}_{32}ex_{32} + \dot{E}x_d^{fwst}$$

The mass, energy, entropy, and exergy balance equations for the PEM electrolyzer can be defined:

$$\dot{m}_{33} = \dot{m}_{34} + \dot{m}_{35}$$

$$\dot{m}_{33}h_{33} + \dot{W}_{PEM} = \dot{m}_{34}h_{34} + \dot{m}_{35}h_{35}$$

$$\dot{m}_{33}s_{33} + \dot{S}_g^{PEM} = \dot{m}_{34}s_{34} + \dot{m}_{35}s_{35}$$

$$\dot{m}_{33}ex_{33} + \dot{E}x_{PEM}^{W} = \dot{m}_{34}ex_{34} + \dot{m}_{35}ex_{35} + \dot{E}x_d^{PEM}$$

The mass, energy, entropy, and exergy balance equations for the three-way valve 4 can be written:

$$\dot{m}_{35} = \dot{m}_{36} + \dot{m}_{41}$$

$$\dot{m}_{35}h_{35} = \dot{m}_{36}h_{36} + \dot{m}_{41}h_{41}$$

$$\dot{m}_{35}s_{35} + \dot{S}_g^{3wv4} = \dot{m}_{36}s_{36} + \dot{m}_{41}s_{41}$$

$$\dot{m}_{35}ex_{35} = \dot{m}_{36}ex_{36} + \dot{m}_{41}ex_{41} + \dot{E}x_d^{3wv4}$$

The mass, energy, entropy, and exergy balance equations for the hydrogen compressor 1 can be defined:

$$\dot{m}_{36} = \dot{m}_{37}$$

$$\dot{m}_{36}h_{36} + \dot{W}_{HyC1} = \dot{m}_{37}h_{37}$$

$$\dot{m}_{36}s_{36} + \dot{S}_g^{HyC1} = \dot{m}_{37}s_{37}$$

$$\dot{m}_{36}ex_{36} + \dot{E}x_{HyC1}^{W} = \dot{m}_{37}ex_{37} + \dot{E}x_d^{HyC1}$$

The mass, energy, entropy, and exergy balance equations for the nitrogen compressor are written:

$$\dot{m}_{38} = \dot{m}_{39}$$

$$\dot{m}_{38}h_{38} + \dot{W}_{NC} = \dot{m}_{39}h_{39}$$

$$\dot{m}_{38}s_{38} + \dot{S}_g^{NC} = \dot{m}_{39}s_{39}$$

$$\dot{m}_{38}ex_{38} + \dot{E}x_{NC}^{W} = \dot{m}_{39}ex_{39} + \dot{E}x_d^{NC}$$

The mass, energy, entropy, and exergy balance equations for the ammonia reactor can be written:

$$\dot{m}_{37} + \dot{m}_{39} = \dot{m}_{40}$$

$$\dot{m}_{37}h_{37} + \dot{m}_{39}h_{39} = \dot{m}_{40}h_{40}$$

$$\dot{m}_{37}s_{37} + \dot{m}_{39}s_{39} + \dot{S}_g^{AR} = \dot{m}_{40}s_{40}$$

$$\dot{m}_{37}ex_{37} + \dot{m}_{39}ex_{39} = \dot{m}_{40}ex_{40} + \dot{E}x_d^{AR}$$

The mass, energy, entropy, and exergy balance equations for the mixer can be defined:

$$\dot{m}_{41} + \dot{m}_{53} = \dot{m}_{42}$$

$$\dot{m}_{41}h_{41} + \dot{m}_{53}h_{53} = \dot{m}_{42}h_{42}$$

$$\dot{m}_{41}s_{41} + \dot{m}_{53}s_{53} + \dot{S}_g^{Mix} = \dot{m}_{42}s_{42}$$

$$\dot{m}_{41}ex_{41} + \dot{m}_{53}ex_{53} = \dot{m}_{42}ex_{42} + \dot{E}x_d^{Mix}$$

The mass, energy, entropy, and exergy balance equations for the hydrogen compressor 2 are defined:

$$\dot{m}_{42} = \dot{m}_{43}$$

$$\dot{m}_{42}h_{42} + \dot{W}_{HyC2} = \dot{m}_{43}h_{43}$$

$$\dot{m}_{42}s_{42} + \dot{S}_g^{HyC2} = \dot{m}_{43}s_{43}$$

$$\dot{m}_{42}ex_{42} + \dot{E}x_{HyC2}^{W} = \dot{m}_{43}ex_{43} + \dot{E}x_d^{HyC2}$$

The mass, energy, entropy, and exergy balance equations for the HEX 3 can be defined:

$$\dot{m}_{43} = \dot{m}_{44};\dot{m}_{52} = \dot{m}_{53}$$

$$\dot{m}_{43}h_{43} + \dot{m}_{52}h_{52} = \dot{m}_{44}h_{44} + \dot{m}_{53}h_{53}$$

$$\dot{m}_{43}s_{43} + \dot{m}_{52}s_{52} + \dot{S}_g^{HEX3} = \dot{m}_{44}s_{44} + \dot{m}_{53}s_{53}$$

$$\dot{m}_{43}ex_{43} + \dot{m}_{52}ex_{52} = \dot{m}_{44}ex_{44} + \dot{m}_{53}ex_{53} + \dot{E}x_d^{HEX3}$$

The mass, energy, entropy, and exergy balance equations for the HEX 4 can be written:

$$\dot{m}_{44} = \dot{m}_{45};\dot{m}_{54} = \dot{m}_{55}$$

$$\dot{m}_{44}h_{44} + \dot{m}_{54}h_{54} = \dot{m}_{45}h_{45} + \dot{m}_{55}h_{55}$$

$$\dot{m}_{44}s_{44} + \dot{m}_{54}s_{54} + \dot{S}_g^{HEX4} = \dot{m}_{45}s_{45} + \dot{m}_{55}s_{55}$$

$$\dot{m}_{44}ex_{44} + \dot{m}_{54}ex_{54} = \dot{m}_{45}ex_{45} + \dot{m}_{55}ex_{55} + \dot{E}x_d^{HEX4}$$

The mass, energy, entropy, and exergy balance equations for the HEX 5 can be defined:

$$\dot{m}_{45} = \dot{m}_{46};\dot{m}_{51} = \dot{m}_{52}$$

$$\dot{m}_{45}h_{45} + \dot{m}_{51}h_{51} = \dot{m}_{46}h_{46} + \dot{m}_{52}h_{52}$$

$$\dot{m}_{45}s_{45} + \dot{m}_{51}s_{51} + \dot{S}_g^{HEX5} = \dot{m}_{46}s_{46} + \dot{m}_{52}s_{52}$$

$$\dot{m}_{45}ex_{45} + \dot{m}_{51}ex_{51} = \dot{m}_{46}ex_{46} + \dot{m}_{52}ex_{52} + \dot{E}x_d^{HEX5}$$

The mass, energy, entropy, and exergy balance equations for the HEX 6 are defined:

$$\dot{m}_{46} = \dot{m}_{47}; \dot{m}_{56} = \dot{m}_{57}$$

$$\dot{m}_{46}h_{46} + \dot{m}_{56}h_{56} = \dot{m}_{47}h_{47} + \dot{m}_{57}h_{57}$$

$$\dot{m}_{46}s_{46} + \dot{m}_{56}s_{56} + \dot{S}_g^{HEX6} = \dot{m}_{47}s_{47} + \dot{m}_{57}s_{57}$$

$$\dot{m}_{46}ex_{46} + \dot{m}_{56}ex_{56} = \dot{m}_{47}ex_{47} + \dot{m}_{57}ex_{57} + \dot{E}x_d^{HEX6}$$

The mass, energy, entropy, and exergy balance equations for the HEX 7 are written:

$$\dot{m}_{47} = \dot{m}_{48}; \dot{m}_{50} = \dot{m}_{51}$$

$$\dot{m}_{47}h_{47} + \dot{m}_{50}h_{50} = \dot{m}_{48}h_{48} + \dot{m}_{51}h_{51}$$

$$\dot{m}_{47}s_{47} + \dot{m}_{50}s_{50} + \dot{S}_g^{HEX7} = \dot{m}_{48}s_{48} + \dot{m}_{51}s_{51}$$

$$\dot{m}_{47}ex_{47} + \dot{m}_{50}ex_{50} = \dot{m}_{48}ex_{48} + \dot{m}_{51}ex_{51} + \dot{E}x_d^{HEX7}$$

The mass, energy, entropy, and exergy balance equations for the valve 5 can be written:

$$\dot{m}_{48} = \dot{m}_{49}$$

$$\dot{m}_{48}h_{48} = \dot{m}_{49}h_{49}$$

$$\dot{m}_{48}s_{48} + \dot{S}_g^{vl5} = \dot{m}_{49}s_{49}$$

$$\dot{m}_{48}ex_{48} = \dot{m}_{49}ex_{49} + \dot{E}x_d^{vl5}$$

The mass, energy, entropy, and exergy balance equations for the separator 2 are defined:

$$\dot{m}_{49} = \dot{m}_{50} + \dot{m}_{58}$$

$$\dot{m}_{49}h_{49} = \dot{m}_{50}h_{50} + \dot{m}_{58}h_{58}$$

$$\dot{m}_{49}s_{49} + \dot{S}_g^{Sep2} = \dot{m}_{50}s_{50} + \dot{m}_{58}s_{58}$$

$$\dot{m}_{49}ex_{49} = \dot{m}_{50}ex_{50} + \dot{m}_{58}ex_{58} + \dot{E}x_d^{Sep2}$$

The mass, energy, entropy, and exergy balance equations for the generator can be defined:

$$\dot{m}_9 = \dot{m}_{10}; \dot{m}_{64} = \dot{m}_{65} + \dot{m}_{66}$$

$$\dot{m}_9h_9 + \dot{m}_{64}h_{64} = \dot{m}_{10}h_{10} + \dot{m}_{65}h_{65} + \dot{m}_{66}h_{66}$$

$$\dot{m}_9 s_9+\dot{m}_{64}s_{64}+\dot{S}_g^{Gen}=\dot{m}_{10}s_{10}+\dot{m}_{65}s_{65}+\dot{m}_{66}s_{66}$$

$$\dot{m}_9 ex_9+\dot{m}_{64}ex_{64}=\dot{m}_{10}ex_{10}+\dot{m}_{65}ex_{65}+\dot{m}_{66}ex_{66}+\dot{E}x_d^{Gen}$$

The mass, energy, entropy, and exergy balance equations for the condenser 3 are written:

$$\dot{m}_{65}=\dot{m}_{59};\dot{m}_{69}=\dot{m}_{70}$$

$$\dot{m}_{65}h_{65}+\dot{m}_{69}h_{69}=\dot{m}_{59}h_{59}+\dot{m}_{70}h_{70}$$

$$\dot{m}_{65}s_{65}+\dot{m}_{69}s_{69}+\dot{S}_g^{Con3}=\dot{m}_{59}s_{59}+\dot{m}_{70}s_{70}$$

$$\dot{m}_{65}ex_{65}+\dot{m}_{69}ex_{69}=\dot{m}_{59}ex_{59}+\dot{m}_{70}ex_{70}+\dot{E}x_d^{Con3}$$

The mass, energy, entropy, and exergy balance equations for the evaporator can be written:

$$\dot{m}_{60}=\dot{m}_{61};\dot{m}_{71}=\dot{m}_{72}$$

$$\dot{m}_{60}h_{60}+\dot{m}_{71}h_{71}=\dot{m}_{61}h_{61}+\dot{m}_{72}h_{72}$$

$$\dot{m}_{60}s_{60}+\dot{m}_{71}s_{71}+\dot{S}_g^{Eva}=\dot{m}_{61}s_{61}+\dot{m}_{72}s_{72}$$

$$\dot{m}_{60}ex_{60}+\dot{m}_{71}ex_{71}=\dot{m}_{61}ex_{61}+\dot{m}_{72}ex_{72}+\dot{E}x_d^{Eva}$$

The mass, energy, entropy, and exergy balance equations for the absorber are defined:

$$\dot{m}_{73}=\dot{m}_{74};\dot{m}_{61}+\dot{m}_{68}=\dot{m}_{62}$$

$$\dot{m}_{61}h_{61}+\dot{m}_{68}h_{68}+\dot{m}_{73}h_{73}=\dot{m}_{62}h_{62}+\dot{m}_{74}h_{74}$$

$$\dot{m}_{61}s_{61}+\dot{m}_{68}s_{68}+\dot{m}_{73}s_{73}+\dot{S}_g^{Abs}=\dot{m}_{62}s_{62}+\dot{m}_{74}s_{74}$$

$$\dot{m}_{61}ex_{61}+\dot{m}_{68}ex_{68}+\dot{m}_{73}ex_{73}=\dot{m}_{62}ex_{62}+\dot{m}_{74}ex_{74}+\dot{E}x_d^{Abs}$$

The mass, energy, entropy, and exergy balance equations for the pump 2 can be defined:

$$\dot{m}_{62}=\dot{m}_{63}$$

$$\dot{m}_{62}h_{62}+\dot{W}_{P2}=\dot{m}_{63}h_{63}$$

$$\dot{m}_{62}s_{62}+\dot{S}_g^{P2}=\dot{m}_{63}s_{63}$$

$$\dot{m}_{62}ex_{62}+\dot{E}x_{P2}^{W}=\dot{m}_{63}ex_{63}+\dot{E}x_d^{P2}$$

The mass, energy, entropy, and exergy balance equations for the valve 3 can be written:

$$\dot{m}_{67} = \dot{m}_{68}$$

$$\dot{m}_{67}h_{67} = \dot{m}_{68}h_{68}$$

$$\dot{m}_{67}s_{67} + \dot{S}_g^{vl3} = \dot{m}_{68}s_{68}$$

$$\dot{m}_{67}ex_{67} = \dot{m}_{68}ex_{68} + \dot{E}x_d^{vl3}$$

The mass, energy, entropy, and exergy balance equations for the solution HEX are written:

$$\dot{m}_{63} = \dot{m}_{64};\dot{m}_{66} = \dot{m}_{67}$$

$$\dot{m}_{63}h_{63} + \dot{m}_{66}h_{66} = \dot{m}_{64}h_{64} + \dot{m}_{67}h_{67}$$

$$\dot{m}_{63}s_{63} + \dot{m}_{66}s_{66} + \dot{S}_g^{SHEX} = \dot{m}_{64}s_{64} + \dot{m}_{67}s_{67}$$

$$\dot{m}_{63}ex_{63} + \dot{m}_{66}ex_{66} = \dot{m}_{64}ex_{64} + \dot{m}_{67}ex_{67} + \dot{E}x_d^{SHEX}$$

The mass, energy, entropy, and exergy balance equations for the fan can be defined:

$$\dot{m}_{75} = \dot{m}_{76}$$

$$\dot{m}_{75}h_{75} + \dot{W}_{Fn} = \dot{m}_{76}h_{76}$$

$$\dot{m}_{75}s_{75} + \dot{S}_g^{Fn} = \dot{m}_{76}s_{76}$$

$$\dot{m}_{75}ex_{75} + \dot{E}x_{Fn}^{W} = \dot{m}_{76}ex_{76} + \dot{E}x_d^{Fn}$$

The mass, energy, entropy, and exergy balance equations for the HEX 2 are defined:

$$\dot{m}_{10} = \dot{m}_{11};\dot{m}_{76} = \dot{m}_{77}$$

$$\dot{m}_{10}h_{10} + \dot{m}_{76}h_{76} = \dot{m}_{11}h_{11} + \dot{m}_{77}h_{77}$$

$$\dot{m}_{10}s_{10} + \dot{m}_{76}s_{76} + \dot{S}_g^{HEX2} = \dot{m}_{11}s_{11} + \dot{m}_{77}s_{77}$$

$$\dot{m}_{10}ex_{10} + \dot{m}_{76}ex_{76} = \dot{m}_{11}ex_{11} + \dot{m}_{77}ex_{77} + \dot{E}x_d^{HEX2}$$

The mass, energy, entropy, and exergy balance equations for the dryer can be written:

$$\dot{m}_{77} = \dot{m}_{78};\dot{m}_{79} = \dot{m}_{80}$$

$$\dot{m}_{77}h_{77} + \dot{m}_{79}h_{79} = \dot{m}_{78}h_{78} + \dot{m}_{80}h_{80}$$

$$\dot{m}_{77}s_{77} + \dot{m}_{79}s_{79} + \dot{S}_g^{Dr} = \dot{m}_{78}s_{78} + \dot{m}_{80}s_{80}$$

$$\dot{m}_{77}ex_{77} + \dot{m}_{79}ex_{79} = \dot{m}_{78}ex_{78} + \dot{m}_{80}ex_{80} + \dot{E}x_d^{Dr}$$

The mass, energy, entropy, and exergy balance equations for the hot water storage tank can be defined:

$$\dot{m}_{11} = \dot{m}_{12}; \dot{m}_{81} = \dot{m}_{82}$$

$$\dot{m}_{11}h_{11} + \dot{m}_{81}h_{81} = \dot{m}_{12}h_{12} + \dot{m}_{82}h_{82}$$

$$\dot{m}_{11}s_{11} + \dot{m}_{81}s_{81} + \dot{S}_g^{hwst} = \dot{m}_{12}s_{12} + \dot{m}_{82}s_{82}$$

$$\dot{m}_{11}ex_{11} + \dot{m}_{81}ex_{81} = \dot{m}_{12}ex_{12} + \dot{m}_{82}ex_{82} + \dot{E}x_d^{hwst}$$

As given in Fig. 7.5, the generated hydrogen can be converted into the liquid phase by using any hydrogen liquefaction plant. The hydrogen liquefaction technique considered in this case study is the Linde-Hampson hydrogen liquefaction process. It should be accepted that the liquefied hydrogen mass flow is kept constant. By utilizing this assumption, the liquid hydrogen yield can be described as [12]:

$$y = (h_{51} - h_{47})/(h_{51} - h_{58})$$

$$h_{44} = h_{43} - (1 - y)(h_{53} - h_{52})$$

$$h_{53} = h_{52} + \in_{HEX}(h_{53'} - h_{52})$$

$$h_{46} = h_{45} - (1 - y)(h_{52} - h_{51})$$

$$h_{52} - h_{51} + \in_{HEX}(h_{52'} - h_{51})$$

$$h_{51} = h_g + \in_{HEX}\left(h_{51'} - h_g\right)$$

$$h_{48} = h_{47} - (1 - y)\left(h_{50} - h_g\right)$$

Here, $\in_{HEX}$ shows the HEX efficiency factor and varies between 0.85 and 1. Based on the computed cycle performance the flow enthalpies, which figure in the computation of compression electrical power needed for the working fluids' mass flow rates in the hydrogen liquefaction process, can be determined. Also, the electrical energy needed for nitrogen flows should be considered 7760 kJ/kgN_2.

To assess system components' performance, the energy and exergy efficiency equations for geothermal energy–based integrated system components must be investigated. For this reason, the energy and exergy efficiency equations for the flash chamber can be written:

$$\eta_{fc} = \frac{\dot{m}_2 h_2}{\dot{m}_1 h_1}$$

and

$$\psi_{fc} = \frac{\dot{m}_2 ex_2}{\dot{m}_1 ex_1}$$

The energy and exergy efficiency equations for the separator 1 can be defined:

$$\eta_{Sep1} = \frac{\dot{m}_3 h_3 + \dot{m}_5 h_5}{\dot{m}_2 h_2}$$

and

$$\psi_{Sep1} = \frac{\dot{m}_3 ex_3 + \dot{m}_5 ex_5}{\dot{m}_2 ex_2}$$

The energy and exergy efficiency equations for the turbine can be written:

$$\eta_{Tur} = \frac{\dot{W}_{Tur}}{\dot{m}_3 (h_3 - h_4)}$$

and

$$\psi_{Tur} = \frac{\dot{E}x^{W}_{Tur}}{\dot{m}_3 (ex_3 - ex_4)}$$

The energy and exergy efficiency equations for the HEX 1 can be defined:

$$\eta_{HEX1} = \frac{\dot{m}_{13} h_{13} - \dot{m}_{16} h_{16}}{\dot{m}_5 h_5 - \dot{m}_6 h_6}$$

and

$$\psi_{HEX1} = \frac{\dot{m}_{13} ex_{13} - \dot{m}_{16} ex_{16}}{\dot{m}_5 ex_5 - \dot{m}_6 ex_6}$$

The energy and exergy efficiency equations for the pump 1 can be written:

$$\eta_{P1} = \frac{\dot{m}_{15} (h_{16} - h_{15})}{\dot{W}_{P1}}$$

and

$$\psi_{P1} = \frac{\dot{m}_{15} (ex_{16} - ex_{15})}{\dot{E}x^{W}_{P1}}$$

The energy and exergy efficiency equations for the condenser 1 can be written:

$$\eta_{Con1} = \frac{\dot{m}_{33} h_{33} - \dot{m}_{32} h_{32}}{\dot{m}_4 h_4 - \dot{m}_8 h_8}$$

and

$$\psi_{Con1} = \frac{\dot{m}_{33}ex_{33} - \dot{m}_{32}ex_{32}}{\dot{m}_4 ex_4 - \dot{m}_8 ex_8}$$

The energy and exergy efficiency equations for the valve 1 can be defined:

$$\eta_{vl1} = \frac{\dot{m}_7 h_7}{\dot{m}_6 h_6}$$

and

$$\psi_{vl1} = \frac{\dot{m}_7 ex_7}{\dot{m}_6 ex_6}$$

The energy and exergy efficiency equations for the ORC turbine can be written:

$$\eta_{ORC,T} = \frac{\dot{W}_{ORC,T}}{\dot{m}_{13}(h_{13} - h_{14})}$$

and

$$\psi_{ORC,T} = \frac{\dot{E}x^{W}_{ORC,T}}{\dot{m}_{13}(ex_{13} - ex_{14})}$$

The energy and exergy efficiency equations for the condenser 2 can be described:

$$\eta_{Con2} = \frac{\dot{m}_{18}h_{18} - \dot{m}_{17}h_{17}}{\dot{m}_{14}h_{14} - \dot{m}_{15}h_{15}}$$

and

$$\psi_{Con2} = \frac{\dot{m}_{18}ex_{18} - \dot{m}_{17}ex_{17}}{\dot{m}_{14}ex_{14} - \dot{m}_{15}ex_{15}}$$

The energy and exergy efficiency equations for the three-way valve 1 can be defined:

$$\eta_{3wv1} = \frac{\dot{m}_9 h_9}{\dot{m}_7 h_7 + \dot{m}_8 h_8}$$

and

$$\psi_{3wv1} = \frac{\dot{m}_9 ex_9}{\dot{m}_7 ex_7 + \dot{m}_8 ex_8}$$

The energy and exergy efficiency equations for the low-pressure pump can be described:

$$\eta_{LPP} = \frac{\dot{m}_{19}(h_{20} - h_{19})}{\dot{W}_{LPP}}$$

and

$$\psi_{LPP} = \frac{\dot{m}_{19}(ex_{20} - ex_{19})}{\dot{E}x^{W}_{LPP}}$$

The energy and exergy efficiency equations for the filter can be defined:

$$\eta_{flt} = \frac{\dot{m}_{21}h_{21}}{\dot{m}_{20}h_{20}}$$

and

$$\psi_{flt} = \frac{\dot{m}_{21}ex_{21}}{\dot{m}_{20}ex_{20}}$$

The energy and exergy efficiency equations for the chemical processing can be defined:

$$\eta_{cp} = \frac{\dot{m}_{22}h_{22}}{\dot{m}_{21}h_{21}}$$

and

$$\psi_{cp} = \frac{\dot{m}_{22}ex_{22}}{\dot{m}_{21}ex_{21}}$$

The energy and exergy efficiency equations for the three-way valve 2 can be written:

$$\eta_{3wv2} = \frac{\dot{m}_{23}h_{23} + \dot{m}_{27}h_{27}}{\dot{m}_{22}h_{22}}$$

and

$$\psi_{3wv2} = \frac{\dot{m}_{23}ex_{23} + \dot{m}_{27}ex_{27}}{\dot{m}_{22}ex_{22}}$$

The energy and exergy efficiency equations for the high-pressure pump can be described:

$$\eta_{HPP} = \frac{\dot{m}_{23}(h_{24} - h_{23})}{\dot{W}_{HPP}}$$

and

$$\psi_{HPP} = \frac{\dot{m}_{23}(ex_{24} - ex_{23})}{\dot{E}x^{W}_{HPP}}$$

The energy and exergy efficiency equations for the valve 4 can be defined:

$$\eta_{vl4} = \frac{\dot{m}_{28}h_{28}}{\dot{m}_{27}h_{27}}$$

and

$$\psi_{vl4} = \frac{\dot{m}_{28} ex_{28}}{\dot{m}_{27} ex_{27}}$$

The energy and exergy efficiency equations for the reverse osmosis can be written:

$$\eta_{ro} = \frac{\dot{m}_{25} h_{25} + \dot{m}_{29} h_{29}}{\dot{m}_{24} h_{24}}$$

and

$$\psi_{ro} = \frac{\dot{m}_{25} ex_{25} + \dot{m}_{29} ex_{29}}{\dot{m}_{24} ex_{24}}$$

The energy and exergy efficiency equations for the Pelton turbine can be defined:

$$\eta_{PT} = \frac{\dot{W}_{PT}}{\dot{m}_{25}(h_{25} - h_{26})}$$

and

$$\psi_{PT} = \frac{\dot{E}x_{PT}^{W}}{\dot{m}_{25}(ex_{25} - ex_{26})}$$

The energy and exergy efficiency equations for the three-way valve 3 can be written:

$$\eta_{3wv3} = \frac{\dot{m}_{30} h_{30}}{\dot{m}_{28} h_{28} + \dot{m}_{29} h_{29}}$$

and

$$\psi_{3wv3} = \frac{\dot{m}_{30} ex_{30}}{\dot{m}_{28} ex_{28} + \dot{m}_{29} ex_{29}}$$

The energy and exergy efficiency equations for the freshwater storage tank can be defined:

$$\eta_{fwst} = \frac{\dot{m}_{31} h_{31} + \dot{m}_{32} h_{32}}{\dot{m}_{30} h_{30}}$$

and

$$\psi_{fwst} = \frac{\dot{m}_{31} ex_{31} + \dot{m}_{32} ex_{32}}{\dot{m}_{30} ex_{30}}$$

The energy and exergy efficiency equations for the PEM electrolyzer can be written:

$$\eta_{PEM} = \frac{\dot{m}_{34} h_{34} + \dot{m}_{35} h_{35}}{\left(\dot{m}_{33} h_{33} + \dot{W}_{PEM}\right)}$$

and

$$\psi_{PEM} = \frac{\dot{m}_{34}ex_{34} + \dot{m}_{35}ex_{35}}{\left(\dot{m}_{33}ex_{33} + \dot{E}x_{PEM}^{W}\right)}$$

The energy and exergy efficiency equations for the three-way valve 4 can be written:

$$\eta_{3wv4} = \frac{\dot{m}_{36}h_{36} + \dot{m}_{41}h_{41}}{\dot{m}_{35}h_{35}}$$

and

$$\psi_{3wv4} = \frac{\dot{m}_{36}ex_{36} + \dot{m}_{41}ex_{41}}{\dot{m}_{35}ex_{35}}$$

The energy and exergy efficiency equations for the hydrogen compressor 1 can be written:

$$\eta_{HyC1} = \frac{\dot{m}_{36}(h_{37} - h_{36})}{\dot{W}_{HyC1}}$$

and

$$\psi_{HyC1} = \frac{\dot{m}_{36}(ex_{37} - ex_{36})}{\dot{E}x_{HyC1}^{W}}$$

The energy and exergy efficiency equations for the nitrogen compressor can be defined:

$$\eta_{NC} = \frac{\dot{m}_{38}(h_{39} - h_{38})}{\dot{W}_{NC}}$$

and

$$\psi_{NC} = \frac{\dot{m}_{38}(ex_{39} - ex_{38})}{\dot{E}x_{NC}^{W}}$$

The energy and exergy efficiency equations for the ammonia reactor can be written:

$$\eta_{AR} = \frac{\dot{m}_{40}h_{40}}{\dot{m}_{37}h_{37} + \dot{m}_{39}h_{39}}$$

and

$$\psi_{AR} = \frac{\dot{m}_{40}ex_{40}}{\dot{m}_{37}ex_{37} + \dot{m}_{39}ex_{39}}$$

The energy and exergy efficiency equations for the mixer can be defined:

$$\eta_{Mix} = \frac{\dot{m}_{42}h_{42}}{\dot{m}_{41}h_{41} + \dot{m}_{53}h_{53}}$$

and

$$\psi_{Mix} = \frac{\dot{m}_{42}ex_{42}}{\dot{m}_{41}ex_{41} + \dot{m}_{53}ex_{53}}$$

The energy and exergy efficiency equations for the hydrogen compressor 2 can be written:

$$\eta_{HyC2} = \frac{\dot{m}_{42}(h_{43} - h_{42})}{\dot{W}_{HyC2}}$$

and

$$\psi_{HyC2} = \frac{\dot{m}_{42}(ex_{43} - ex_{42})}{\dot{E}x^{W}_{HyC2}}$$

The energy and exergy efficiency equations for the HEX 3 can be defined:

$$\eta_{HEX3} = \frac{\dot{m}_{44}h_{44} - \dot{m}_{43}h_{43}}{\dot{m}_{52}h_{52} - \dot{m}_{53}h_{53}}$$

and

$$\psi_{HEX3} = \frac{\dot{m}_{44}ex_{44} - \dot{m}_{43}ex_{43}}{\dot{m}_{52}ex_{52} - \dot{m}_{53}ex_{53}}$$

The energy and exergy efficiency equations for the HEX 4 can be described:

$$\eta_{HEX4} = \frac{\dot{m}_{45}h_{45} - \dot{m}_{44}h_{44}}{\dot{m}_{54}h_{54} - \dot{m}_{55}h_{55}}$$

and

$$\psi_{HEX4} = \frac{\dot{m}_{45}ex_{45} - \dot{m}_{44}ex_{44}}{\dot{m}_{54}ex_{54} - \dot{m}_{55}ex_{55}}$$

The energy and exergy efficiency equations for the HEX 5 can be written:

$$\eta_{HEX5} = \frac{\dot{m}_{46}h_{46} - \dot{m}_{45}h_{45}}{\dot{m}_{51}h_{51} - \dot{m}_{52}h_{52}}$$

and

$$\psi_{HEX5} = \frac{\dot{m}_{46}ex_{46} - \dot{m}_{45}ex_{45}}{\dot{m}_{51}ex_{51} - \dot{m}_{52}ex_{52}}$$

The energy and exergy efficiency equations for the HEX 6 can be defined:

$$\eta_{HEX6} = \frac{\dot{m}_{47}h_{47} - \dot{m}_{46}h_{46}}{\dot{m}_{56}h_{56} - \dot{m}_{57}h_{57}}$$

and

$$\psi_{HEX6} = \frac{\dot{m}_{47}ex_{47} - \dot{m}_{46}ex_{46}}{\dot{m}_{56}ex_{56} - \dot{m}_{57}ex_{57}}$$

The energy and exergy efficiency equations for the HEX 7 can be described:

$$\eta_{HEX7} = \frac{\dot{m}_{48}h_{48} - \dot{m}_{47}h_{47}}{\dot{m}_{50}h_{50} - \dot{m}_{51}h_{51}}$$

and

$$\psi_{HEX7} = \frac{\dot{m}_{48}ex_{48} - \dot{m}_{47}ex_{47}}{\dot{m}_{50}ex_{50} - \dot{m}_{51}ex_{51}}$$

The energy and exergy efficiency equations for the valve 5 can be defined:

$$\eta_{vl5} = \frac{\dot{m}_{49}h_{49}}{\dot{m}_{48}h_{48}}$$

and

$$\psi_{vl5} = \frac{\dot{m}_{49}ex_{49}}{\dot{m}_{48}ex_{48}}$$

The energy and exergy efficiency equations for the separator 2 can be defined:

$$\eta_{Sep2} = \frac{\dot{m}_{50}h_{50} + \dot{m}_{58}h_{58}}{\dot{m}_{49}h_{49}}$$

and

$$\psi_{Sep2} = \frac{\dot{m}_{50}ex_{50} + \dot{m}_{58}ex_{58}}{\dot{m}_{49}ex_{49}}$$

The energy and exergy efficiency equations for the generator can be defined:

$$\eta_{Gen} = \frac{(\dot{m}_{65}h_{65} + \dot{m}_{66}h_{66} - \dot{m}_{64}h_{64})}{(\dot{m}_{9}h_{9} - \dot{m}_{10}h_{10})}$$

and

$$\psi_{Gen} = \frac{(\dot{m}_{65}ex_{65} + \dot{m}_{66}ex_{66} - \dot{m}_{64}ex_{64})}{(\dot{m}_{9}ex_{9} - \dot{m}_{10}ex_{10})}$$

The energy and exergy efficiency equations for the condenser 3 can be written:

$$\eta_{Con3} = \frac{\dot{m}_{70}h_{70} - \dot{m}_{69}h_{69}}{\dot{m}_{65}h_{65} - \dot{m}_{59}h_{59}}$$

and

$$\psi_{Con3} = \frac{\dot{m}_{70}ex_{70} - \dot{m}_{69}ex_{69}}{\dot{m}_{65}ex_{65} - \dot{m}_{59}ex_{59}}$$

The energy and exergy efficiency equations for the valve 2 can be defined:

$$\eta_{vl2} = \frac{\dot{m}_{60}h_{60}}{\dot{m}_{59}h_{59}}$$

and

$$\psi_{vl2} = \frac{\dot{m}_{60}ex_{60}}{\dot{m}_{59}ex_{59}}$$

The energy and exergy efficiency equations for the evaporator can be defined:

$$\eta_{Eva} = \frac{\dot{m}_{72}h_{72} - \dot{m}_{71}h_{71}}{\dot{m}_{60}h_{60} - \dot{m}_{61}h_{61}}$$

and

$$\psi_{Eva} = \frac{\dot{m}_{72}ex_{72} - \dot{m}_{71}ex_{71}}{\dot{m}_{60}ex_{60} - \dot{m}_{61}ex_{61}}$$

The energy and exergy efficiency equations for the absorber can be written:

$$\eta_{Abs} = \frac{(\dot{m}_{74}h_{74} - \dot{m}_{73}h_{73})}{(\dot{m}_{61}h_{61} + \dot{m}_{68}h_{68} - \dot{m}_{62}h_{62})}$$

and

$$\psi_{Abs} = \frac{(\dot{m}_{74}ex_{74} - \dot{m}_{73}ex_{73})}{(\dot{m}_{61}ex_{61} + \dot{m}_{68}ex_{68} - \dot{m}_{62}ex_{62})}$$

The energy and exergy efficiency equations for the pump 2 can be written:

$$\eta_{P2} = \frac{\dot{m}_{62}(h_{63} - h_{62})}{\dot{W}_{P2}}$$

and

$$\psi_{P2} = \frac{\dot{m}_{62}(ex_{63} - ex_{62})}{\dot{E}x^{W}_{P2}}$$

The energy and exergy efficiency equations for the valve 3 can be defined:

$$\eta_{vl3} = \frac{\dot{m}_{68}h_{68}}{\dot{m}_{67}h_{67}}$$

and

$$\psi_{vl3} = \frac{\dot{m}_{68}ex_{68}}{\dot{m}_{67}ex_{67}}$$

The energy and exergy efficiency equations for the solution HEX can be written:

$$\eta_{SHEX} = \frac{\dot{m}_{64}h_{64} - \dot{m}_{63}h_{63}}{\dot{m}_{66}h_{66} - \dot{m}_{67}h_{67}}$$

and

$$\psi_{SHEX} = \frac{\dot{m}_{64}ex_{64} - \dot{m}_{63}ex_{63}}{\dot{m}_{66}ex_{66} - \dot{m}_{67}ex_{67}}$$

The energy and exergy efficiency equations for the fan can be defined:

$$\eta_{Fn} = \frac{\dot{m}_{75}(h_{76} - h_{75})}{\dot{W}_{fn}}$$

and

$$\psi_{Fn} = \frac{\dot{m}_{75}(ex_{76} - ex_{75})}{\dot{E}x_{fn}^{W}}$$

The energy and exergy efficiency equations for the HEX 2 can be defined:

$$\eta_{HEX2} = \frac{\dot{m}_{77}h_{77} - \dot{m}_{76}h_{76}}{\dot{m}_{10}h_{10} - \dot{m}_{11}h_{11}}$$

and

$$\psi_{HEX2} = \frac{\dot{m}_{77}ex_{77} - \dot{m}_{76}ex_{76}}{\dot{m}_{10}ex_{10} - \dot{m}_{11}ex_{11}}$$

The energy and exergy efficiency equations for the dryer can be written:

$$\eta_{Dr} = \frac{\dot{m}_{80}h_{80} - \dot{m}_{79}h_{79}}{\dot{m}_{77}h_{77} - \dot{m}_{78}h_{78}}$$

and

$$\psi_{Dr} = \frac{\dot{m}_{80}ex_{80} - \dot{m}_{79}ex_{79}}{\dot{m}_{77}ex_{77} - \dot{m}_{78}ex_{78}}$$

The energy and exergy efficiency equations for the hot water storage tank can be written:

$$\eta_{hwst} = \frac{\dot{m}_{82}h_{82} - \dot{m}_{81}h_{81}}{\dot{m}_{11}h_{11} - \dot{m}_{12}h_{12}}$$

and

$$\psi_{hwst} = \frac{\dot{m}_{82}ex_{82} - \dot{m}_{81}ex_{81}}{\dot{m}_{11}ex_{11} - \dot{m}_{12}ex_{12}}$$

To investigate the effect of some design conditions on the effectiveness of geothermal energy–based integrated system for multigeneration, the thermodynamic efficiency equations must be given. For this aim, the energy and exergy efficiency equations for the GPC can be defined:

$$\eta_{GPC} = \frac{\dot{W}_{Tur} + \dot{Q}_{HEX1} + \dot{Q}_{Con1}}{\dot{m}_1(h_1 - h_9)}$$

and

$$\psi_{GPC} = \frac{\dot{E}x^W_{Tur} + \dot{E}x^Q_{HEX1} + \dot{E}x^Q_{Con1}}{\dot{m}_1(ex_1 - ex_9)}$$

The energy and exergy efficiency equations for the IPC are written:

$$\eta_{IPC} = \frac{\dot{W}_{ORCT} + \dot{Q}_{Con2}}{\dot{m}_{16}(h_{13} - h_{16})}$$

and

$$\psi_{IPC} = \frac{\dot{E}x^W_{ORCT} + \dot{E}x^Q_{Con2}}{\dot{m}_{16}(ex_{13} - ex_{16})}$$

The energy and exergy efficiencies for the seawater distillation unit can be given:

$$\eta_{SWDU} = \frac{\dot{m}_{30}h_{30} + \dot{W}_{PT}}{\dot{m}_{19}h_{19} + \dot{W}_{LPP} + \dot{W}_{HPP}}$$

and

$$\psi_{SWDU} = \frac{\dot{m}_{30}ex_{30} + \dot{E}x^W_{PT}}{\dot{m}_{19}ex_{19} + \dot{E}x^W_{LPP} + \dot{E}x^W_{HPP}}$$

The energy and exergy efficiencies for the HPS are defined:

$$\eta_{HPS} = \frac{\dot{m}_{34}h_{34} + \dot{m}_{35}h_{35}}{\dot{m}_{33}h_{33} + \dot{W}_{PEM}}$$

and

$$\psi_{HPS} = \frac{\dot{m}_{34}ex_{34} + \dot{m}_{35}ex_{35}}{\dot{m}_{33}ex_{33} + \dot{E}x^W_{PEM}}$$

The energy and exergy efficiency equations for the APS can be written:

$$\eta_{APS} = \frac{\dot{m}_{40}h_{40}}{\dot{m}_{36}h_{36} + \dot{m}_{38}h_{38} + \dot{W}_{HyC1} + \dot{W}_{NC}}$$

and

$$\psi_{APS} = \frac{\dot{m}_{40}ex_{40}}{\dot{m}_{36}ex_{36} + \dot{m}_{38}ex_{38} + \dot{E}x^{W}_{HyC1} + \dot{E}x^{W}_{NC}}$$

The energy and exergy efficiencies for the HLS are described:

$$\eta_{HLS} = \frac{\dot{m}_{58}h_{58} + \dot{Q}_{HEX4} + \dot{Q}_{HEX6}}{\dot{m}_{41}h_{41} + \dot{W}_{HyC2}}$$

and

$$\psi_{HLS} = \frac{\dot{m}_{58}ex_{58} + \dot{E}x^{Q}_{HEX4} + \dot{E}x^{Q}_{HEX6}}{\dot{m}_{41}ex_{41} + \dot{E}x^{W}_{HyC2}}$$

The energy and exergy efficiencies for the SEAC system can be defined:

$$\eta_{SEACS} = \frac{\dot{Q}_{Cooling}}{(\dot{m}_{9}h_{9} - \dot{m}_{10}h_{10}) + \dot{W}_{P2}}$$

and

$$\psi_{SEACS} = \frac{\dot{E}x^{Q}_{Cooling}}{(\dot{m}_{9}ex_{9} - \dot{m}_{10}ex_{10}) + \dot{E}x^{W}_{P2}}$$

The energy and exergy efficiency equations for the DS are written:

$$\eta_{DS} = \frac{\dot{m}_{80}h_{80} - \dot{m}_{79}h_{79}}{\dot{m}_{75}h_{75} - \dot{m}_{78}h_{78} + \dot{W}_{fan}}$$

and

$$\psi_{DS} = \frac{\dot{m}_{80}ex_{80} - \dot{m}_{79}ex_{79}}{\dot{m}_{75}ex_{75} - \dot{m}_{78}ex_{78} + \dot{E}x^{W}_{fan}}$$

The energy and exergy efficiencies for the HWPS can be expressed:

$$\eta_{HWPS} = \frac{\dot{m}_{82}h_{82} - \dot{m}_{81}h_{81}}{\dot{m}_{11}h_{11} - \dot{m}_{12}h_{12}}$$

and

$$\psi_{HWPS} = \frac{\dot{m}_{82}ex_{82} - \dot{m}_{81}ex_{81}}{\dot{m}_{11}ex_{11} - \dot{m}_{12}ex_{12}}$$

The energy and exergy efficiency equations for the overall system are expressed:

$$\eta_{OS} = \frac{\dot{W}_{Net} + \dot{m}_{31}h_{31} + \dot{m}_{34}h_{34} + \dot{m}_{40}h_{40} + \dot{m}_{58}h_{58} + \dot{Q}_{Cooling} + \dot{Q}_{Heating} + \dot{Q}_{Drying} + \dot{Q}_{Hot\_water}}{\dot{m}_1(h_1 - h_{12})}$$

and

$$\psi_{OS} = \frac{\dot{E}x^{W}_{Net} + \dot{m}_{31}ex_{31} + \dot{m}_{34}ex_{34} + \dot{m}_{40}ex_{40} + \dot{m}_{58}ex_{58} + \dot{E}x^{Q}_{Cooling} + \dot{E}x^{Q}_{Heating} + \dot{E}x^{Q}_{Drying} + \dot{E}x^{Q}_{Hot\_water}}{\dot{m}_1(ex_1 - ex_{12})}$$

The net energy production of the geothermal energy–based integrated system for multigeneration can be calculated using the following formula:

$$\dot{W}_{Net} = \dot{W}_{Tur} + \dot{W}_{ORCT} + \dot{W}_{PT} - \sum \dot{W}_p - \sum \dot{W}_{HyC} - \dot{W}_{NC} - \dot{W}_{fan} - \dot{W}_{LPP} - \dot{W}_{HPP}$$

In addition, the energetic and exergetic performance coefficients for a SEAC system can be expressed:

$$COP_{en} = \frac{\dot{Q}_{Eva}}{\dot{Q}_{Gen} + \dot{W}_{P2}}$$

and

$$COP_{ex} = \frac{\dot{E}x^{Q}_{Eva}}{\dot{E}x^{Q}_{Gen} + \dot{E}x^{W}_{P2}}$$

For the proposed geothermal power–based integrated system for multigeneration as a case study, the energetic and exergetic assessments are comparatively performed. The exergetic assessment is more useful than the energetic assessment as it gives more meaningful outputs than energy analysis. In addition, the energy assessment also does not illustrate the plant components' losses. By means of exergetic assessment, where and how much exergy destruction occurs can be understood. Table 7.2 demonstrates the energetic and exergetic assessments with exergy destruction rates of the integrated plant and its subsystems. Based on the table, the lowest exergetic performance occurs in the SEAC system. The energy and exergy efficiencies of the whole system have been computed as 62.21% and 58.95%, respectively.

The major exergy destruction rates of geothermal energy–based combined plant components are shown in Fig. 7.6. The separator, generator, turbine, PEM electrolyzer, condenser 1, HEX 1, ammonia reactor, hot water storage tank, and drying subcomponents exhibit higher a exergy destruction rate than other system components. Also, the generator has the next largest

TABLE 7.2 Thermodynamic assessment results of the geothermal energy–based integrated system for multigeneration.

Subplants/ whole plant	Energetic efficiency (%)	Exergetic efficiency (%)	Exergy destruction rate (kW)	Exergy destruction ratio (%)
Geothermal power cycle	65.57	68.29	5637	32.45
Isobutene power cycle	24.93	21.26	2816	16.21
Sea water distillation unit	74.26	71.18	993	5.72
Hydrogen production system	71.63	67.34	1792	10.32
Ammonia production system	72.94	68.79	1583	9.11
Hydrogen liquefaction system	64.28	60.93	874	5.03
Single-effect absorption cooling system	20.74	16.28	1673	9.63
Drying system	67.83	63.92	1136	6.54
Hot water production system	69.51	66.37	866	4.99
Overall system	62.21	58.95	17270	100

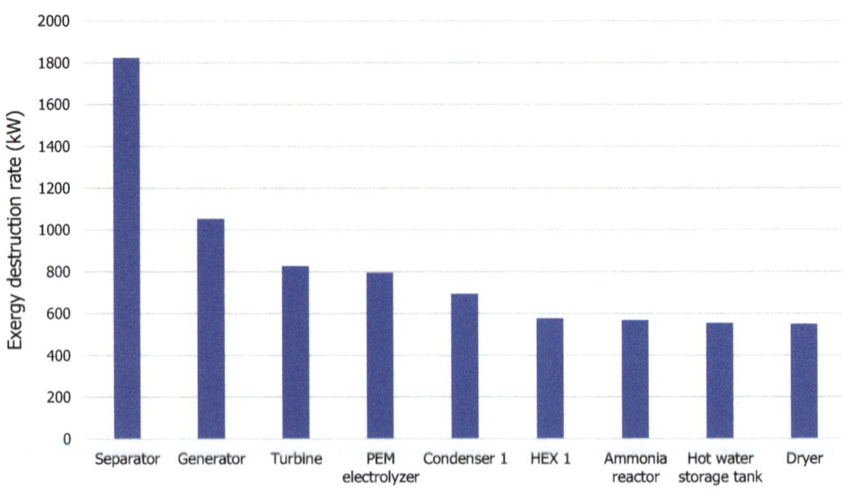

FIGURE 7.6 Comparison of the exergy destruction rates of major components of the geothermal energy–based integrated system for multigeneration.

TABLE 7.3 Geothermal energy–based integrated system outputs.

Plant outputs	Values
Power production of the geothermal power cycle, $\dot{W}_{Tur}$	8283 kW
Power production rate of the isobutene power cycle, $\dot{W}_{ORCT}$	4073 kW
Produced hydrogen mass flow rate, $\dot{m}_{Hydrogen}$	0.0413 kg/s
Produced ammonia mass flow rate, $\dot{m}_{Ammonia}$	0.1612 kg/s
Cooling producing rate, $\dot{Q}_{Cooling}$	2693 kW
Heating producing rate, $\dot{Q}_{Heating}$	1837 kW
Hot water production capacity, $\dot{Q}_{Hot\_water}$	3128 kW
Drying production capacity, $\dot{Q}_{Drying}$	2436 kW

exergy destruction rate, mainly due to the temperature difference between geothermal fluid and SEAC system working fluid passing through the generator but also due to the pressure drop across the subcomponent.

The outputs of geothermal energy–based integrated system for multigeneration are calculated based on the balance and efficiency equations and are given in Table 7.3. Total power productions by GPC and IPC have been found to be 12,356 kW, and the cooling producing rate has been calculated as 2693 kW. With selected key indicators, hydrogen and ammonia production mass flow rates are 0.0413 kg/s and 0.1612 kg/s, respectively.

7.2.1.1 Effect of the reference temperature

The graphs given in Fig. 7.7 show how the reference temperature has an effect on the energy efficiency of the integrated system and its subsystems. To consider the effect of the reference temperature, the reference temperature is gradually increased from 0°C to 40°C. In general, considering the performance characteristics of the integrated system and its subsystems, when the reference temperature is increased gradually, the energy efficiency of the integrated system and of all its subsystems increases, except for the SEAC subsystem. When the reference temperature values are 5°C, 25°C, and 40°C, the GPC's energy efficiencies are 0.6401, 0.6557, and 0.6675, respectively. As can be seen from these values, the energy efficiency of GPC increases based on the gradual increase in reference temperature. At these reference temperature values, the energy efficiencies of the IPC are 0.2463, 0.2493, and 0.6675, respectively.

For these reference temperature values, the energy efficiencies of the sea water distillation unit (SWDU) are 0.7308, 0.7426, and 0.7515, respectively. HPS's energy efficiencies for these reference temperature values are 0.7021,

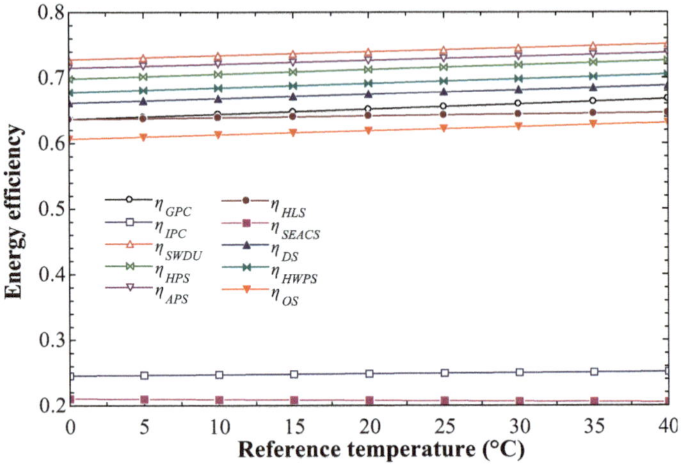

FIGURE 7.7 Effect of reference temperature on the energy efficiency of the integrated system and its subsystems.

0.7163, and 0.727, respectively. The energy efficiency of the HPS also decreases according to a gradual increase in reference temperature values. At these reference temperature values, the energy efficiencies of the APS are 0.7178, 0.7294, and 0.7381, respectively. HLS's energy efficiencies for these reference temperature values are 0.6376, 0.6428, and 0.6466, respectively. The energy efficiency of the SEACS decreases with increasing reference temperature. SEACS's energy efficiencies for these reference temperature values are 0.2099, 0.2074, and 0.2055, respectively. For these reference temperature values, the energy efficiencies of the DS are 0.6649, 0.6783, and 0.6885, respectively. HWPS's energy efficiencies for these reference temperature values are 0.6813, 0.6951, and 0.7055, respectively. In addition, in cases where the reference temperatures are 10°C, 25°C, and 40°C, the whole system's energy efficiencies are 0.6128, 0.6221, and 0.6314, respectively. As can be seen from these values, the energy efficiency of the integrated plant increases based on the gradual increase in reference temperature.

The performance characteristics given in Fig. 7.8 show the effect of the reference temperature on the exergy efficiency of the integrated energy system and its subplants. In general, considering the exergy efficiency performance curves of the integrated system and its subsystems, when the reference temperature is increased gradually, exergy efficiencies of the integrated system and all its subsystems increase, except for the SEAC subsystem. When the reference temperature value is 10°C, exergy efficiencies of the GPC, IPC, SWDU, HPS, APS, HLS, SEACS, DS, and HWPS are 0.6667, 0.2094, 0.6991, 0.6594, 0.6756, 0.602, 0.1652, 0.6259, and 0.6499, respectively. Also, when the reference temperature value is 25°C, exergy efficiencies of the GPC, IPC, SWDU, HPS, APS, HLS, SEACS, DS, and

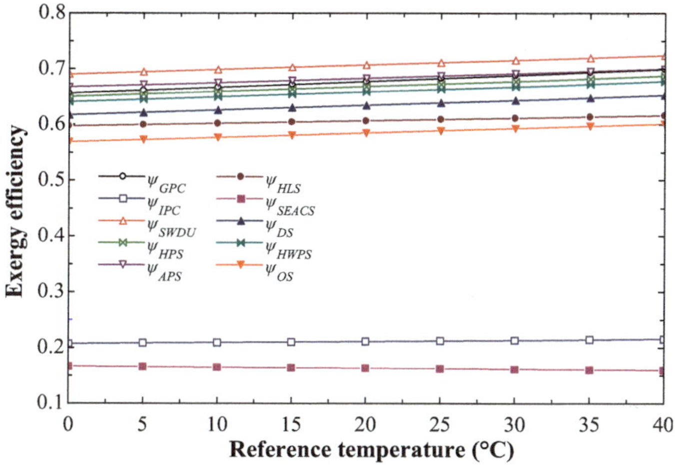

FIGURE 7.8 Effect of reference temperature on the exergy efficiency of the integrated system and its subsystems.

HWPS are 0.6829, 0.2126, 0.7118, 0.6734, 0.6879, 0.6093, 0.1628, 0.6392, and 0.6637, respectively. At the maximum reference temperature value, these exergy efficiencies are 0.6994, 0.2158, 0.7246, 0.6876, 0.7003, 0.6166, 0.1603, 0.6527, and 0.6777, respectively.

Note that, as can be understood from these values, when the reference temperature is increased gradually, exergy efficiencies of all subsystems of the integrated plant increase, except for the SEAC subsystem. In cases where the reference temperatures are 10°C, 20°C, 25°C, and 40°C, the whole system's exergy efficiencies are 0.5772, 0.5854, 0.5895, and 0.6019, respectively.

The performance curves given in Fig. 7.9 show the effect of the reference temperature on the exergy destruction rate of the integrated system's subplants. In general, considering the performance curves of the subsystems of the integrated system, when the reference temperature is increased gradually, exergy destruction rates of all subsystems decrease, except for the SEAC subsystem. When the reference temperature value is 10°C, the exergy destruction rates of the GPC, IPC, SWDU, HPS, APS, HLS, SEACS, DS, and HWPS are 5807 kW, 2875 kW, 1017 kW, 1840 kW, 1621 kW, 889 kW, 1638 kW, 1166 kW, and 889 kW, respectively. Also, when the reference temperature value is 25°C, the exergy destruction rates of the GPC, IPC, SWDU, HPS, APS, HLS, SEACS, DS, and HWPS are 5637 kW, 2816 kW, 993 kW, 1792 kW, 1583 kW, 874 kW, 1673 kW, 1136 kW, and 866 kW, respectively. At the maximum reference temperature value, these exergy destruction rates are 5471 kW, 2756 kW, 969 kW, 1744 kW, 1545 kW, 858 kW, 1708 kW, 1105 kW, and 843 kW, respectively. As can be seen from these values of exergy destruction rate, when the reference temperature is

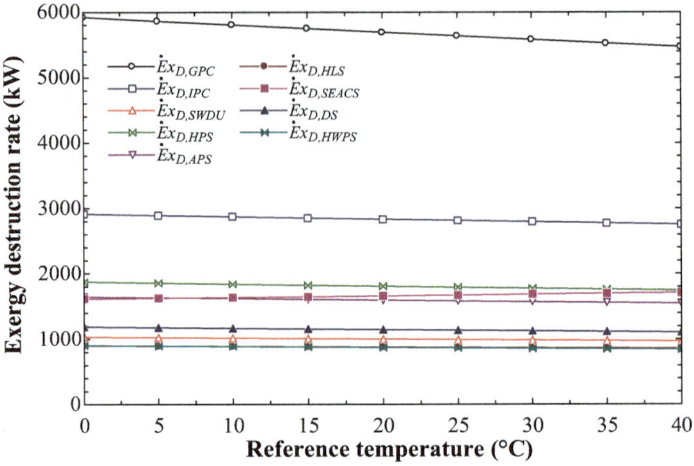

FIGURE 7.9 Effect of reference temperature on the exergy destruction rate of the subsystems.

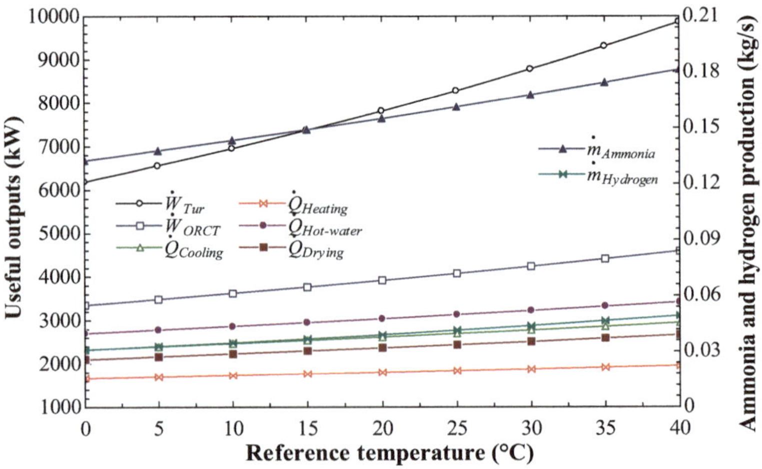

FIGURE 7.10 Effect of reference temperature on the power, cooling, heating, hot water, drying, and ammonia and hydrogen production rates.

increased gradually, exergy destruction rates of all subsystems of the integrated plant decrease, except for the SEAC subsystem.

The graphs given in Fig. 7.10 show the effect of the reference temperature on the power generation, heating–cooling production, hot water production, drying performance, and ammonia and hydrogen production of the integrated system. In general, considering the performance curves of the integrated plant, when the reference temperature is increased gradually, the production of these useful outputs increases. When the reference temperature value is 10°C, power generations of the GPC and IPC are 6954 kW and 3620 kW, respectively. Also, at this reference temperature value, cooling, heating, hot water, and drying outputs of the integrated system are 2464 kW, 1731 kW, 2862 kW, and 2229 kW, respectively. For this reference, temperature value, ammonia and hydrogen outputs of the integrated system are 0.1433 kg/s and 0.0346 kg/s, respectively. When the reference temperature value is 25°C, power generations of the GPC and IPC are 8283 kW and 4073 kW, respectively. Also, at this reference temperature value, cooling, heating, hot water, and drying outputs of the integrated system are 2693 kW, 1837 kW, 3128 kW, and 2436 kW, respectively. For this reference, temperature value, ammonia and hydrogen outputs of the integrated system are 0.1612 kg/s and 0.0413 kg/s, respectively. At the maximum reference temperature value, power generations of the GPC and IPC are 9865 kW and 4581 kW, respectively. Also, at this reference temperature value, cooling, heating, hot water, and drying outputs of the integrated system are 2942 kW, 1949 kW, 3418 kW, and 2661 kW, respectively. For this reference temperature value, the ammonia and hydrogen outputs of the integrated system are 0.1813 kg/s and 0.0491 kg/s, respectively. As can be seen from these values of the integrated system outputs, when the reference temperature is increased gradually, all outputs of the integrated plant increase.

7.2.1.2 Effect of geothermal water temperature

The performance curves given in Fig. 7.11 show the effect of geothermal water temperature on the energy efficiency of the integrated energy system and its subsystems. To assess the effect of geothermal water temperature,

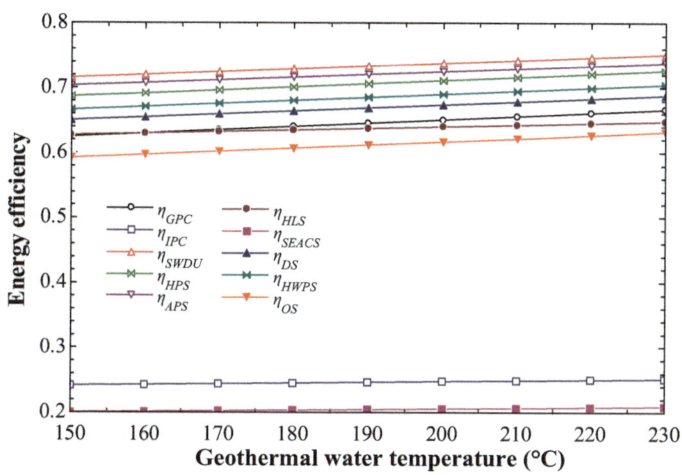

FIGURE 7.11 Impact of geothermal water temperature on the energy efficiency of the integrated system and its subsystems.

geothermal water temperature is gradually increased from 150°C to 230°C. In general, considering the energy efficiencies of the integrated system and its subsystems, when the geothermal water temperature is increased gradually, energy efficiencies of the integrated system and all its subsystems increase. When the geothermal water temperature value is 160°C, energy efficiencies of the GPC, IPC, SWDU, HPS, APS, HLS, SEACS, DS, and HWPS are 0.63, 0.2431, 0.7207, 0.6917, 0.7079, 0.63, 0.2022, 0.655, and 0.6712, respectively. When the geothermal water temperature value is 200°C, the energy efficiencies of the GPC, IPC, SWDU, HPS, APS, HLS, SEACS, DS, and HWPS are 0.6504, 0.248, 0.7381, 0.7113, 0.725, 0.6402, 0.2063, 0.6735, and 0.6902, respectively. At the maximum geothermal water temperature value, these exergy efficiencies are 0.6662, 0.2517, 0.7515, 0.7263, 0.7381, 0.6479, 0.2094, 0.6878, and 0.7048, respectively. As can be understood from these values, when the geothermal water temperature is increased gradually, energy efficiencies of all subsystems of the integrated plant increase. In cases where geothermal water temperatures are 150°C, 180°C, 200°C, and 230°C, the whole system's energy efficiencies are 0.5929, 0.6074, 0.6171 and 0.632, respectively.

The performance graphs in Fig. 7.12 show the effect of geothermal water temperature on the exergy efficiency of the integrated system and its subplants. Considering the exergy efficiency performance characteristics of the integrated system and its subsystems, when the geothermal water temperature is increased gradually, exergy efficiencies of the integrated system and of all its subsystems increase.

When the geothermal water temperature value is 160°C, the exergy efficiencies of the GPC, IPC, SWDU, HPS, APS, HLS, SEACS, DS, and HWPS

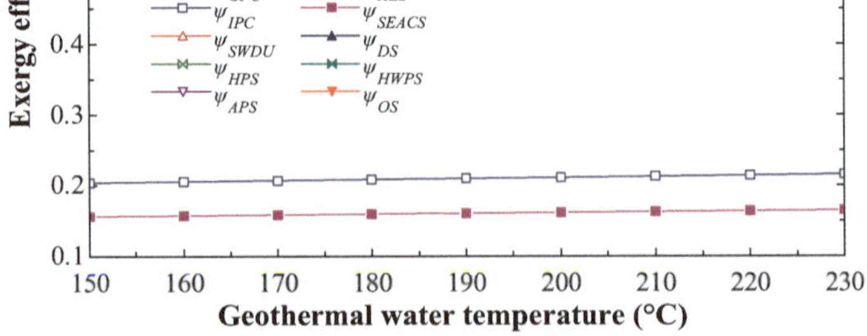

FIGURE 7.12 Impact of geothermal water temperature on the exergy efficiency of the integrated system and its subsystems.

are 0.6497, 0.2053, 0.6839, 0.6438, 0.661, 0.5913, 0.1572, 0.6111, and 0.6346, respectively. When the geothermal water temperature value is 200°C, the exergy efficiencies of the GPC, IPC, SWDU, HPS, APS, HLS, SEACS, DS, and HWPS are 0.6761, 0.2111, 0.7061, 0.6673, 0.6824, 0.6056, 0.1616, 0.6334, and 0.6577, respectively. At the maximum geothermal water temperature, these exergy efficiencies are 0.6966, 0.2155, 0.7232, 0.6855, 0.6989, 0.6166, 0.165, 0.6507, and 0.6757, respectively. As can be seen from these exergy efficiency values, when the geothermal water temperature is increased gradually, exergy efficiencies of all the integrated plant's subsystems increase. In cases where geothermal water temperatures are 150°C, 180°C, 200°C, and 230°C, the whole system's exergy efficiencies are 0.552, 0.5704, 0.583, and 0.6025, respectively.

The performance curves given in Fig. 7.13 show the effect of geothermal water temperature on the exergy destruction rate of the integrated system's subplants. In general, considering the performance curves of the integrated system' subsystems, when the geothermal water temperature is increased gradually, exergy destruction rates of all subsystems decrease. When geothermal water temperature value is 150°C, exergy destruction rates of the GPC, IPC, SWDU, HPS, APS, HLS, SEACS, DS, and HWPS are 6055 kW, 2971 kW, 1054 kW, 1913 kW, 1680 kW, 916 kW, 1765 kW, 1205 kW, and 919 kW, respectively. When the geothermal water temperature value is 200°C, the exergy destruction rates of the GPC, IPC, SWDU, HPS, APS, HLS, SEACS, DS, and HWPS are 5704 kW, 2841 kW, 1002 kW, 1811 kW, 1598 kW, 880 kW, 1688 kW, 1147 kW, and 874 kW, respectively. At the maximum geothermal water temperature value, these exergy destruction rates

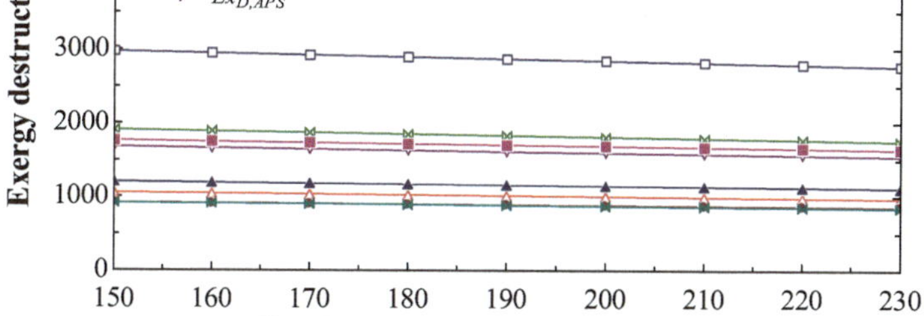

FIGURE 7.13 Impact of geothermal water temperature on the exergy destruction rate of the subsystems.

are 5504 kW, 2765 kW, 973 kW, 1753 kW, 1551 kW, 860 kW, 1643 kW, 1113 kW, and 848 kW, respectively. As can be seen from these values of exergy destruction rate, when the geothermal water temperature is increased gradually, the exergy destruction rates of all subsystems of the integrated plant decrease.

The graphs in Fig. 7.14 show the effect of geothermal water temperature on power generation, heating–cooling production, hot water production, drying performance, ammonia, and hydrogen production of the integrated system. In general, considering the performance curves of the integrated plant, when the geothermal water temperature is increased gradually, the production of these useful outputs increases. When the geothermal water temperature value is 160°C, power generations of the GPC and IPC are 5905 kW and 3191 kW, respectively. Also, at this geothermal water temperature value, the cooling, heating, hot water, and drying outputs of the integrated system are 2213 kW, 1584 kW, 2570 kW, and 2002 kW, respectively. For this geothermal water temperature value, the ammonia and hydrogen outputs of the integrated system are 0.1263 kg/s and 0.0294 kg/s, respectively. When geothermal water temperature value is 200°C, power generations of the GPC and IPC are 7741 kWand 3879 kW, respectively. At this geothermal water temperature value, the cooling, heating, hot water, and drying outputs of the integrated system are 2589 kW, 1783 kW, 3007 kW, and 2342 kW, respectively. For this geothermal water temperature value, ammonia and hydrogen outputs of the integrated system are 0.1535 kg/s and 0.0385 kg/s, respectively. At the maximum geothermal water temperature value, power generations of the GPC and IPC are 9483 kW and 4490 kW, respectively. Also, at this geothermal water temperature value, the cooling, heating, hot water, and

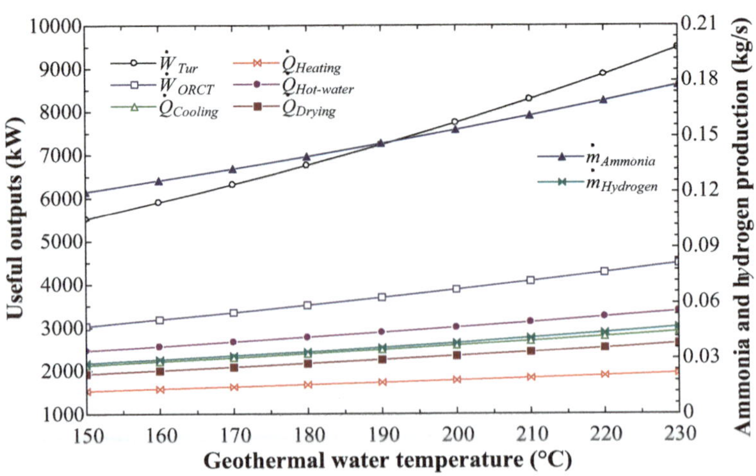

FIGURE 7.14 Impact of geothermal water temperature on the power, cooling, heating, hot water, drying, and ammonia and hydrogen production rates.

drying outputs of the integrated system are 2912 kW, 1948 kW, 3383 kW, and 2634 kW, respectively. For this geothermal water temperature value, the ammonia and hydrogen outputs of the integrated system are 0.1777 kg/s and 0.0472 kg/s, respectively. As can be seen from these values of the integrated system's outputs, when the geothermal water temperature is increased gradually, all outputs of the integrated plant increase.

7.2.1.3 *Effect of geothermal water mass flow rate*

The graphs given in Fig. 7.15 show the effect of geothermal water mass flow rate on the energy efficiency of the integrated energy system and its subsystems. To assess the effect of geothermal water mass flow rate, the geothermal water mass flow rate is gradually increased from 100 to 180 kg/s. In general, considering the energy efficiencies of the integrated system and its subsystems, when geothermal water mass flow rate is increased gradually, energy efficiencies of the integrated system and of all its subsystems increase. When the geothermal water mass flow rate value is 100 kg/s, energy efficiencies of the GPC, IPC, SWDU, HPS, APS, HLS, SEACS, DS, and HWPS are 0.6301, 0.2424, 0.7193, 0.691, 0.7065, 0.6276, 0.2016, 0.6544, and 0.6706, respectively. When geothermal water mass flow rate value is 140 kg/s, the energy efficiencies of the GPC, IPC, SWDU, HPS, APS, HLS, SEACS, DS, and HWPS are 0.6557, 0.2493, 0.7426, 0.7163, 0.7294, 0.6428, 0.2074, 0.6783, and 0.6951, respectively. At the maximum geothermal water mass flow rate value, these exergy efficiencies are 0.6823, 0.2563, 0.7666, 0.7424, 0.753, 0.6583, 0.2132, 0.703, and 0.7204, respectively. As can be understood from these values, when geothermal water mass

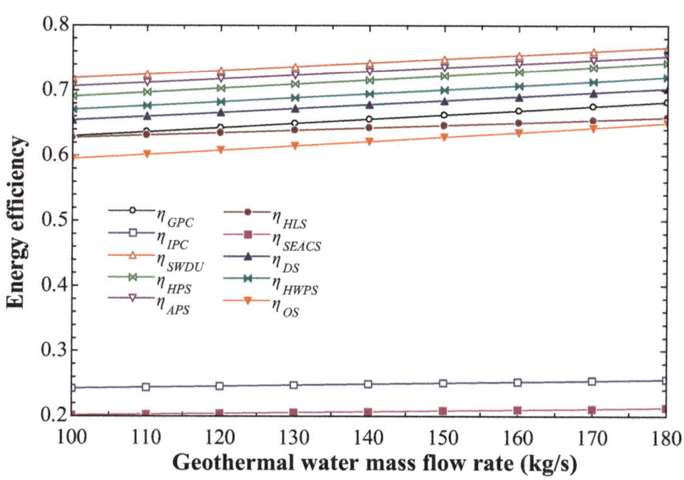

FIGURE 7.15 Effect of geothermal water mass flow rate on the energy efficiency of the integrated system and its subsystems.

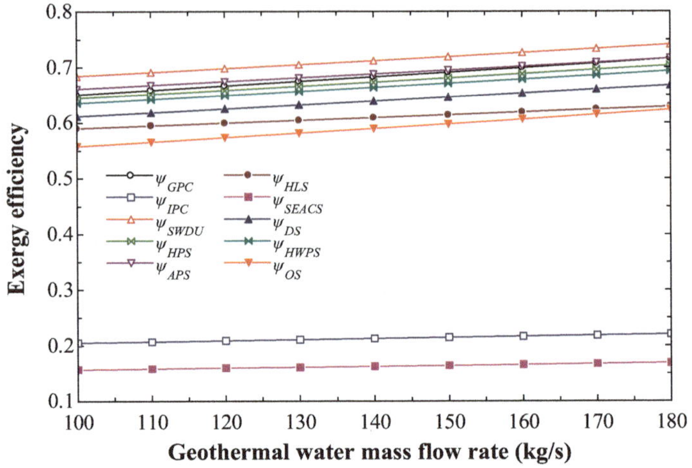

FIGURE 7.16 Effect of geothermal water mass flow rate on the exergy efficiency of the integrated system and its subsystems.

flow rate is increased gradually, the energy efficiencies of all subsystems of the integrated plant increase. In cases where geothermal water mass flow rates are 100 kg/s, 140 kg/s, 150 kg/s, and 180 kg/s, the whole system's energy efficiencies are 0.5954, 0.6221, 0.6289, and 0.6499, respectively.

The performance curves given in Fig. 7.16 show the effect of geothermal water mass flow rate on the exergy efficiency of the integrated system and its subplants. Considering the exergy efficiency performance characteristics of the integrated system and its subsystems, when the geothermal water mass flow rate is increased gradually, exergy efficiencies of the integrated system and all its subsystems increase. When geothermal water mass flow rate value is 110 kg/s, exergy efficiencies of the GPC, IPC, SWDU, HPS, APS, HLS, SEACS, DS, and HWPS are 0.6588, 0.2069, 0.6908, 0.6516, 0.6676, 0.5949, 0.1584, 0.6185, and 0.6422, respectively. Also, when the geothermal water mass flow rate value is 160 kg/s, exergy efficiencies of the GPC, IPC, SWDU, HPS, APS, HLS, SEACS, DS, and HWPS are 0.6993, 0.2164, 0.7261, 0.6882, 0.7017, 0.619, 0.1657, 0.6533, and 0.6783, respectively. At the maximum geothermal water mass flow rate, the exergy efficiencies are 0.7162, 0.2203, 0.7407, 0.7035, 0.7158, 0.629, 0.1687, 0.6677, and 0.6933, respectively. As can be seen from these exergy efficiency values, when the geothermal water mass flow rate is increased gradually, the exergy efficiencies of all the integrated plant's subsystems increase slightly. In cases where geothermal water mass flow rates are 100 kg/s, 140 kg/s, 160 kg/s, and 180 kg/s, the whole the system's exergy efficiencies are 0.5576, 0.5895, 0.6061, and 0.6232, respectively.

The performance characteristics given in Fig. 7.17 show the effect of the geothermal water mass flow rate on the exergy destruction rate of the

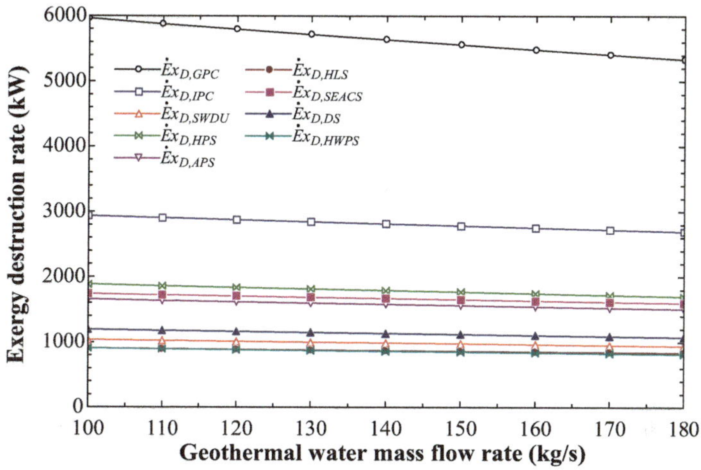

FIGURE 7.17 Effect of geothermal water mass flow rate on the exergy destruction rate of the subsystems.

integrated system's subplants. In general, considering the performance curves of the integrated system's subsystems, when geothermal water mass flow rate is increased gradually, the exergy destruction rates of all subsystems decrease. When the geothermal water mass flow rate value is 120 kg/s, exergy destruction rates of the GPC, IPC, SWDU, HPS, APS, HLS, SEACS, DS, and HWPS are 5795 kW, 2878 kW, 1016 kW, 1838 kW, 1621 kW, 891 kW, 1710 kW, 1165 kW, and 888 kW, respectively. When the geothermal water mass flow rate value is 160 kg/s, the exergy destruction rates of the GPC, IPC, SWDU, HPS, APS, HLS, SEACS, DS, and HWPS are 5482 kW, 2755 kW, 969 kW, 1746 kW, 1545 kW, 856 kW, 1636 kW, 1107 kW, and 843 kW, respectively. At the maximum geothermal water mass flow rate value, these exergy destruction rates are 5332 kW, 2695 kW, 946 kW, 1701 kW, 1509 kW, 839 kW, 1601 kW, 1078 kW, and 822 kW, respectively. As can be seen from these values of the exergy destruction rate, when the geothermal water mass flow rate is increased gradually, the exergy destruction rates of all the integrated plant's subsystems decrease.

The curves given in Fig. 7.18 show the effects of geothermal water mass flow rate on power generation, heating–cooling production, hot water production, drying performance, ammonia, and hydrogen production of the integrated system. In general, considering the performance curves of the integrated plant, when the geothermal water mass flow rate is increased gradually, the production of these useful outputs increases. When geothermal water mass flow rate value is 100 kg/s, the power generations of the GPC and IPC are 5867 kW and 3107 kW, respectively. Also, at this geothermal water mass flow rate value, cooling, heating, hot water, and drying outputs of the integrated system are 2133 kW, 1570 kW, 2477 kW, and 1929 kW,

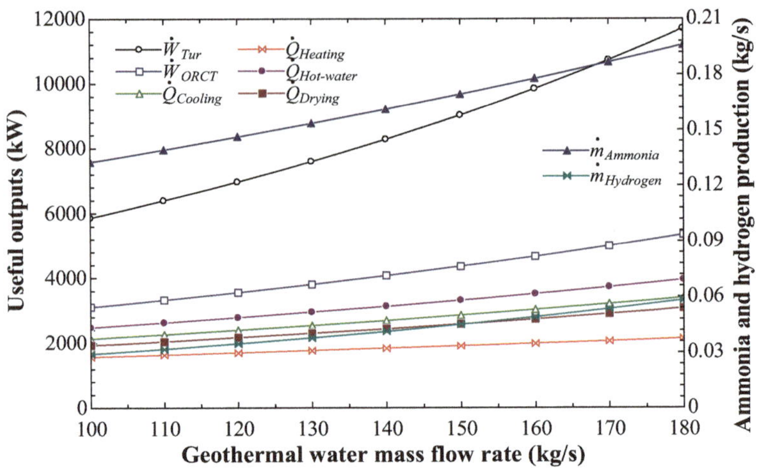

FIGURE 7.18 Effect of geothermal water mass flow rate on the power, cooling, heating, hot water, drying, ammonia, and hydrogen production rates.

respectively. For this geothermal water mass flow rate value, the ammonia and hydrogen outputs of the integrated system are 0.1326 kg/s and 0.0292 kg/s, respectively. When the geothermal water mass flow rate value is 140 kg/s, the power generations of the GPC and IPC are 8283 kW and 4073 kW, respectively. Also, at this geothermal water mass flow rate value, cooling, heating, hot water, and drying outputs of the integrated system are 2693 kW, 1837 kW, 3128 kW, and 2436 kW, respectively. For this geothermal water mass flow rate value, the ammonia and hydrogen outputs of the integrated system are 0.1612 kg/s and 0.0413 kg/s, respectively. At the maximum geothermal water mass flow rate value, the power generations of the GPC and IPC are 11,692 kW and 5338 kW, respectively. Also, at this geothermal water mass flow rate value, cooling, heating, hot water, and drying outputs of the integrated system are 3399 kW, 2148 kW, 3949 kW, and 3075 kW, respectively. For this geothermal water mass flow rate value, the ammonia and hydrogen outputs of the integrated system are 0.1959 kg/s and 0.0582 kg/s, respectively. As can be seen from these values of the integrated system' outputs, when the geothermal water mass flow rate is increased gradually, all outputs of the integrated plant increase.

7.2.1.4 Effect of geothermal power cycle turbine inlet temperature

The performance curves given in Fig. 7.19 show the effect of turbine inlet temperature on the energy efficiency of the integrated energy system and its subsystems. To examine this effect of turbine inlet temperature, the temperature is gradually increased from 150°C to 190°C. In general, considering the

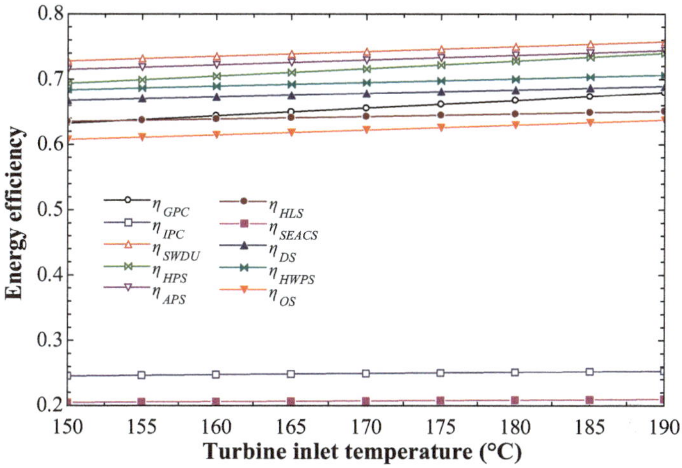

FIGURE 7.19 Impact of turbine inlet temperature on the energy efficiency of the integrated system and its subsystems.

energy efficiencies of the integrated system and its subsystems, when the turbine inlet temperature is increased gradually, the energy efficiencies of the integrated system and of all its subsystems increase. When the turbine inlet temperature value is 150°C, the energy efficiencies of the GPC, IPC, SWDU, HPS, APS, HLS, SEACS, DS, and HWPS are 0.6326, 0.2453, 0.7279, 0.6938, 0.7149, 0.6351, 0.2049, 0.6675, and 0.684, respectively. When the turbine inlet temperature value is 165°C, the energy efficiencies of the GPC, IPC, SWDU, HPS, APS, HLS, SEACS, DS, and HWPS are 0.6498, 0.2483, 0.7389, 0.7106, 0.7257, 0.6408, 0.2067, 0.6755, and 0.6923, respectively. At the maximum turbine inlet temperature value, the exergy efficiencies are 0.6796, 0.2533, 0.7575, 0.7394, 0.744, 0.6505, 0.2099, 0.6892, and 0.7062, respectively. As can be seen from these values, when the turbine inlet temperature is increased gradually, energy efficiencies of all subsystems of the integrated plant increase. In cases where the turbine inlet temperatures are 150°C, 160°C, 175°C, and 190°C, the whole system's energy efficiencies are 0.6073, 0.6147, 0.6258 and 0.6371, respectively.

The performance curves given in Fig. 7.20 show the effects of turbine inlet temperature on the exergy efficiency of the integrated system and its subplants. Considering the exergy efficiency performance characteristics of the integrated system and its subsystems, when turbine inlet temperature is increased gradually, exergy efficiencies of the integrated system and all its subsystems increase. When turbine inlet temperature value is 160°C, exergy efficiencies of the GPC, IPC, SWDU, HPS, APS, HLS, SEACS, DS, and HWPS are 0.6681, 0.21, 0.7019, 0.6614, 0.6736, 0.6032, 0.1611, 0.6315, and 0.6558, respectively. Also, when turbine inlet temperature value is 170°C, exergy efficiencies of the GPC, IPC, SWDU, HPS, APS, HLS, SEACS, DS,

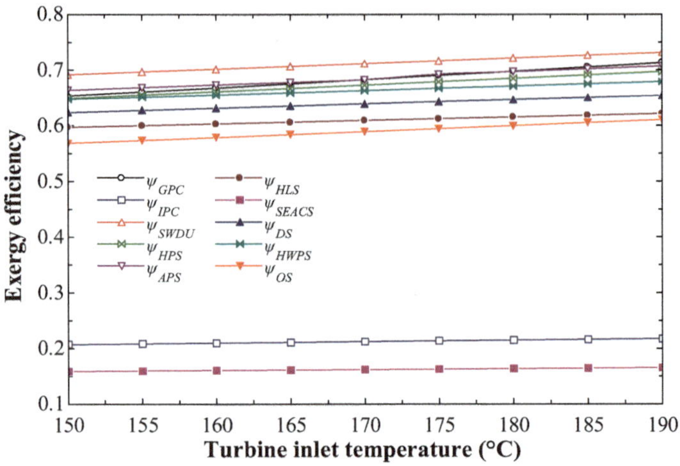

FIGURE 7.20 Impact of turbine inlet temperature on the exergy efficiency of the integrated system and its subsystems.

and HWPS are 0.6829, 0.2126, 0.7118, 0.6734, 0.6831, 0.6093, 0.1628, 0.6392, and 0.6637, respectively. At the maximum turbine inlet temperature, the exergy efficiencies are 0.7134, 0.2177, 0.7319, 0.6979, 0.7073, 0.6215, 0.166, 0.6546, and 0.6797, respectively. As can be seen from these exergy efficiency values, when the turbine inlet temperature is increased gradually, the exergy efficiencies of all the integrated plant's subsystems increase. In cases where turbine inlet temperatures are 150°C, 165°C, 170°C, and 230°C, the whole system's exergy efficiencies are 0.5687, 0.5842, 0.5895, and 0.611, respectively.

The performance graphs given in Fig. 7.21 show the effects of the turbine inlet temperature on the exergy destruction rate of the integrated system's subplants. In general, considering performance curves of the integrated system' subsystems, when turbine inlet temperature is increased gradually, the exergy destruction rates of all subsystems decrease. When the turbine inlet temperature value is 150°C, the exergy destruction rates of the GPC, IPC, SWDU, HPS, APS, HLS, SEACS, DS, and HWPS are 5935 kW, 2907 kW, 1029 kW, 1872 kW, 1640 kW, 898 kW, 1720 kW, 1172 kW, and 894 kW, respectively. When turbine inlet temperature value is 180°C, exergy destruction rates of the GPC, IPC, SWDU, HPS, APS, HLS, SEACS, DS, and HWPS are 5493 kW, 2771 kW, 975 kW, 1753 kW, 1554 kW, 861 kW, 1649 kW, 1118 kW, and 852 kW, respectively. At the maximum turbine inlet temperature value, these exergy destruction rates are 5353 kW, 2727 kW, 958 kW, 1715 kW, 1527 kW, 849 kW, 1626 kW, 1100 kW, and 838 kW, respectively. As can be seen from these values of exergy destruction rate, when the turbine inlet temperature is increased gradually, exergy destruction rates of all subsystems of the integrated plant decrease.

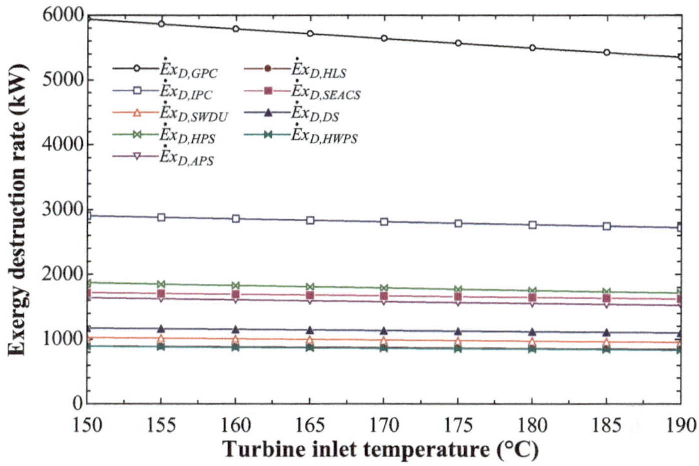

FIGURE 7.21 Impact of turbine inlet temperature on the exergy destruction rate of the subsystems.

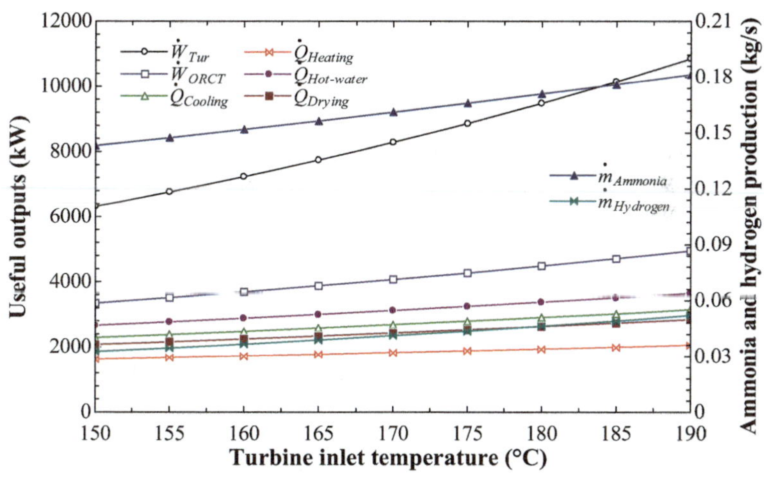

FIGURE 7.22 Impact of turbine inlet temperature on the power, cooling, heating, hot water, drying, ammonia and hydrogen production rates.

The curves given in Fig. 7.22 show the effect of turbine inlet temperature on power generation, heating–cooling production, hot water production, drying performance, ammonia, and hydrogen production of the integrated system. In general, considering the performance curves of the integrated plant, when the turbine inlet temperature is increased gradually, the production of these useful outputs increases. When the turbine inlet temperature value is 165°C, power generations of the GPC and IPC are 7741 kW, 3879 kW, respectively. Also, at this turbine inlet temperature value, cooling, heating, hot water, and drying outputs of the integrated system are 2589 kW, 1783 kW, 3007 kW, and 2342 kW, respectively. For this turbine inlet temperature value, the ammonia and hydrogen outputs of the integrated system are 0.1565 kg/s and 0.0367 kg/s, respectively. When the turbine inlet temperature value is 170°C, the power generations of the GPC and IPC are 8283 kW and 4073 kW, respectively. Also, at this turbine inlet temperature value, cooling, heating, hot water, and drying outputs of the integrated system are 2693 kW, 1837 kW, 3128 kW, and 2436 kW, respectively. For this turbine inlet temperature value, the ammonia and hydrogen outputs of the integrated system are 0.1612 kg/s and 0.0413 kg/s, respectively. In addition to these, when the turbine inlet temperature value is 175°C, the power generations of the GPC and IPC are 8862 kW and 4276 kW, respectively. Also, at this turbine inlet temperature value, cooling, heating, hot water, and drying outputs of the integrated system are 2800 kW, 1892 kW, 3253 kW, and 2533 kW, respectively. For this turbine inlet temperature value, the ammonia and hydrogen outputs of the integrated system are 0.166 kg/s and 0.0437 kg/s, respectively. As can be understood from these values of the integrated system's outputs, when the turbine inlet temperature is increased gradually, all outputs of the integrated plant increase.

7.2.1.5 Effect of geothermal power cycle turbine inlet pressure

The performance characteristics given in Fig. 7.23 show the effect of turbine inlet pressure on the energy efficiency of the integrated energy system and its subsystems. To assess this effect of the turbine inlet pressure, it is gradually increased from 850 to 1650 kPa. In general, considering the energy

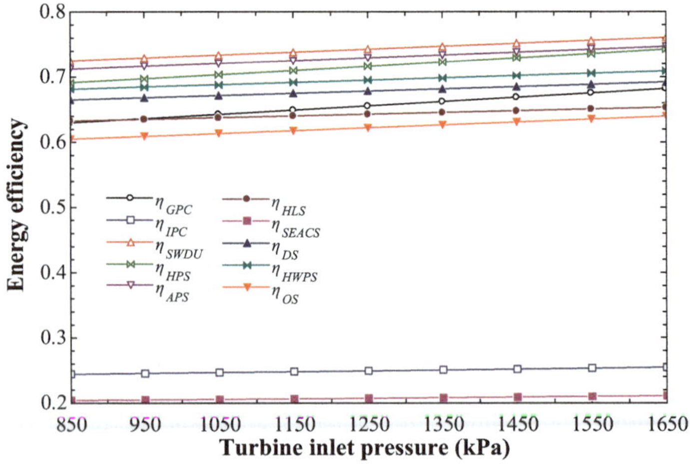

FIGURE 7.23 Effect of turbine inlet pressure on the energy efficiency of the integrated system and its subsystems.

efficiencies of the integrated system and its subsystems, when turbine inlet pressure is increased gradually, the energy efficiencies of the integrated system and all its subsystems increase. When the turbine inlet pressure value is 1050 kPa, the energy efficiencies of the GPC, IPC, SWDU, HPS, APS, HLS, SEACS, DS, and HWPS are 0.6427, 0.2468, 0.7337, 0.7035, 0.7207, 0.6376, 0.2057, 0.6715, and 0.6882, respectively. Also, when the turbine inlet pressure value is 1450 kPa, the energy efficiencies of the GPC, IPC, SWDU, HPS, APS, HLS, SEACS, DS, and HWPS are 0.6688, 0.2517, 0.7515, 0.7292, 0.7381, 0.6479, 0.209, 0.685, and 0.702, respectively. At the maximum turbine inlet pressure value, the exergy efficiencies are 0.6823, 0.2543, 0.7605, 0.7424, 0.747, 0.6531, 0.2107, 0.6919, and 0.7091, respectively. As can be understood from these values, when the turbine inlet pressure is increased gradually, the energy efficiencies of all subsystems of the integrated plant increase. In cases where the turbine inlet pressure values are 850 kPa, 1150 kPa, 1450 kPa, and 1650 kPa, the whole system's energy efficiencies are 0.6049, 0.6177, 0.6308, and 0.6397, respectively.

The performance graphs given in Fig. 7.24 show the effects of turbine inlet pressure on the exergy efficiency of the integrated system and its subplants. Considering the exergy efficiency performance characteristics of the integrated system and its subsystems, when the turbine inlet pressure is increased gradually, the exergy efficiencies of the integrated system and all its subsystems increase. When the turbine inlet pressure value is 850 kPa, the exergy efficiencies of the GPC, IPC, SWDU, HPS, APS, HLS, SEACS, DS, and HWPS are 0.651, 0.2067, 0.6894, 0.6445, 0.6663, 0.5948, 0.1589, 0.6216, and 0.6454, respectively. When the turbine inlet pressure value is 1150 kPa, the exergy efficiencies of the GPC, IPC, SWDU, HPS, APS, HLS,

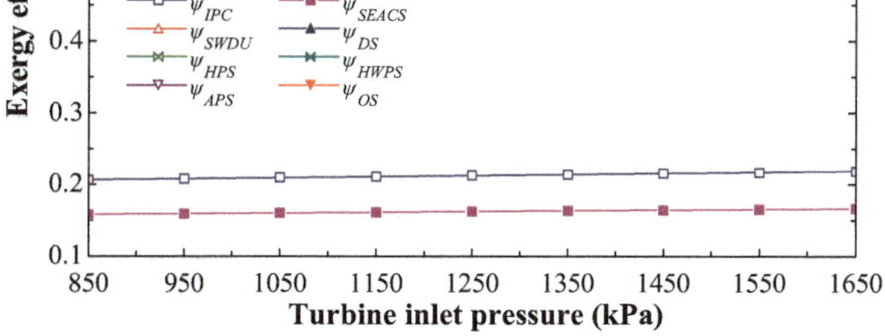

FIGURE 7.24 Effect of turbine inlet pressure on the exergy efficiency of the integrated system and its subsystems.

SEACS, DS, and HWPS are 0.6748, 0.2111, 0.7061, 0.666, 0.6824, 0.6056, 0.1618, 0.6347, and 0.659, respectively. At the maximum turbine inlet pressure, the exergy efficiencies are 0.7162, 0.2186, 0.7348, 0.7035, 0.7101, 0.624, 0.1667, 0.6572, and 0.6824, respectively. As can be seen from these exergy efficiency values, when turbine inlet pressure is increased gradually, exergy efficiencies of all the integrated plant's subplants increase. In cases where the geothermal water mass flow rates are 850 kPa, 1250 kPa, 1450 kPa, and 1650 kPa, the whole system's exergy efficiencies are 0.5687, 0.5895, 0.6001 and 0.611, respectively.

The performance characteristics given in Fig. 7.25 show the effect of turbine inlet pressure on the exergy destruction rates of the integrated system's subplants. In general, considering performance curves of an integrated system' subsystems, when the turbine inlet pressure is increased gradually, the exergy destruction rates of all subsystems decrease. When the turbine inlet pressure value is 850 kPa, the exergy destruction rates of the GPC, IPC, SWDU, HPS, APS, HLS, SEACS, DS, and HWPS are 5982 kW, 2918 kW, 1033 kW, 1887 kW, 1647 kW, 902 kW, 1727 kW, 1182 kW, and 901 kW, respectively. When the turbine inlet pressure value is 1250 kPa, the exergy destruction rates of the GPC, IPC, SWDU, HPS, APS, HLS, SEACS, DS, and HWPS are 5637 kW, 2816 kW, 993 kW, 1792 kW, 1583 kW, 874 kW, 1673 kW, 1136 kW, and 866 kW, respectively. At the maximum turbine inlet pressure value, the exergy destruction rates are 5311 kW, 2716 kW, 954 kW, 1701 kW, 1521 kW, 846 kW, 1620 kW, 1091 kW, and 831 kW, respectively. As can be understood from these values of exergy destruction rate, when the turbine inlet pressure is increased gradually, the exergy destruction rates of all the integrated system's subsystems decrease.

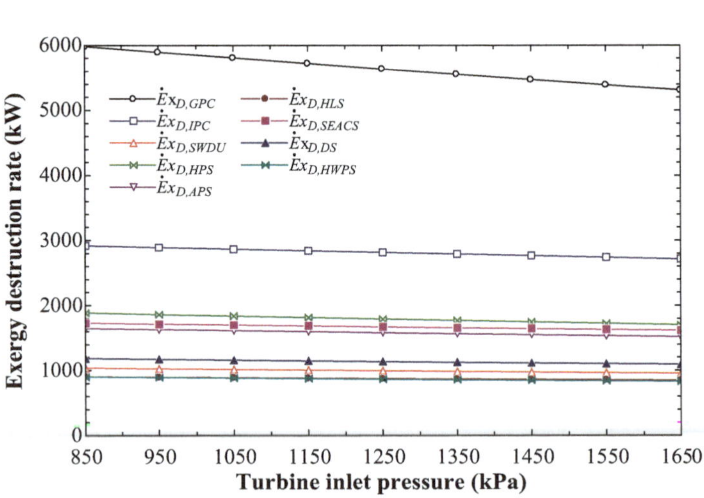

FIGURE 7.25 Effect of turbine inlet pressure on the exergy destruction rate of the subsystems.

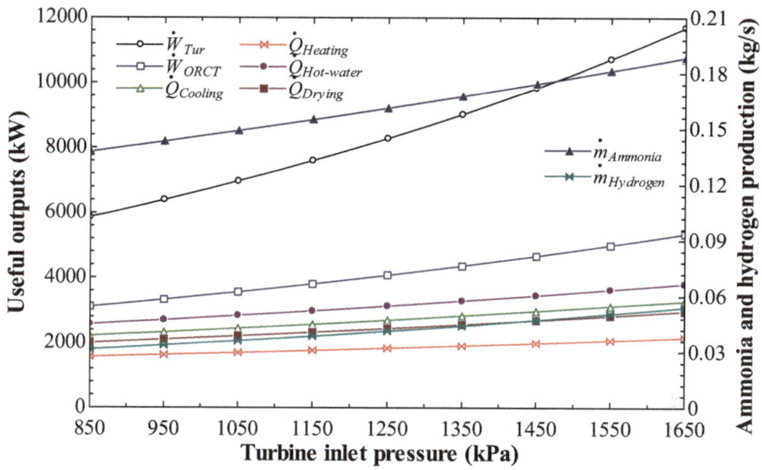

FIGURE 7.26 Effect of turbine inlet pressure on the power, cooling, heating, hot water, drying, ammonia and hydrogen production rates.

The performance graphs given in Fig. 7.26 show the effect of turbine inlet pressure on power generation, heating–cooling production, hot water production, drying performance, ammonia, and hydrogen production of the integrated system. In general, considering the performance curves of the integrated plant, when the turbine inlet pressure is increased gradually, the production of these useful outputs increases. When the turbine inlet pressure value is 950 kPa, the power generations of the GPC and IPC are 6395 kW and 3324 kW, respectively. At this turbine inlet pressure value, cooling, heating, hot water, and drying outputs of the integrated system are 2326 kW, 1633 kW, 2702 kW, and 2104 kW, respectively. For this turbine inlet pressure value, the ammonia and hydrogen outputs of the integrated system are 0.1433 kg/s and 0.1433 kg/s, respectively. When the turbine inlet pressure value is 1050 kPa, the power generations of the GPC and IPC are 6971 kWand 3557 kW, respectively. This turbine inlet pressure value, the cooling, heating, hot water, and drying outputs of the integrated system are 2442 kW, 1698 kW, 2837 kW, and 2209 kW, respectively. For this turbine inlet pressure value, the ammonia and hydrogen outputs of the integrated system are 0.149 kg/s and 0.036 kg/s, respectively. When the turbine inlet pressure value is 1450 kPa, the power generations of the GPC and IPC are 9841 kW and 4663 kW, respectively. At this turbine inlet pressure value, the cooling, heating, hot water, and drying outputs of the integrated system are 2968 kW, 1986 kW, 3448 kW, and 2685 kW, respectively. For this turbine inlet pressure value, the ammonia and hydrogen outputs of the integrated system are 0.1743 kg/s, and 0.0472 kg/s, respectively. As can be seen from these values of the integrated system's outputs, when turbine inlet pressure is increased gradually, all outputs of the integrated plant increase.

7.3 Closing remarks

In this section, geothermally driven integrated systems are introduced and discussed for multigenerational applications where multiple useful outputs are produced to meet the needs of the local communities in an environmentally friendly and sustainable manner. In these integrated plants, the plant design generally changes depending on the thermal energy of the geothermal water. Since the geothermal water coming out of the energy conversion systems has high thermal energy, the system efficiency can be increased by using it in different subplants for various purposes such as heating, cooling, water heating, etc. It further explains in detail the energy generation system configurations that can provide different products using geothermal energy. These configurations are single, cogeneration, trigeneration, quadruple plant, etc. Also in this section, with a case study, a geothermal energy–based energy generation system is modeled, and thermodynamic analysis was performed. The effects of different system indicators on the modeled system are evaluated comparatively by the parametric studies undertaken. The useful products obtained from the modeled system were shown comprehensively in the tables given. The effects of the different system indicators on the performance characteristics of the system and its subsystems were presented comparatively in the graphs.

Nomenclature

A	area (m^2)
E	energy (kJ)
e	specific energy (kJ/kg)
$\dot{E}$	energy rate (kW)
ex	specific exergy (kJ/kg)
$\dot{E}x$	exergy rate (kW)
$\dot{E}x_d$	exergy destruction rate (kW)
$\dot{E}x^Q$	exergy transfer rate associated with heat transfer (kW)
$\dot{E}x^W$	exergy transfer rate associated with work (kW)
F	Faraday constant (C/mol)
G	Gibbs free energy (kJ)
h	specific enthalpy (kJ/kg)
H	enthalpy (kJ)
J	current density (A/m^2)
J_o	exchange current density (A/m^2)
J_i^{ref}	preexponential factor (A/m^2)
L	length (m)
m	mass (kg)
$\dot{m}$	mass flow rate (kg/s)
P	presuure (kPa)
q	specific heat transfer (kJ/kg)
Q	heat (kJ)

$\dot{q}$	specific heat transfer rate (kW/kg)
$\dot{Q}$	heat rate (kW)
R_{PEM}	proton exchange membrane resistance (Ω)
R_u	universal gas constant (kJ/mol K)
s	specific entropy (kJ/kg K)
S	entropy (kJ/K)
$\dot{S}$	entropy rate (kW/K)
t	time (s)
T	temperature (°C, K)
u	internal energy (kJ/kg)
V	volume (m^3)
ν	velocity (m/s)
V_0	reversible potential (V)
V_{act}	activation overpotential (V)
$V_{act,a}$	anode activation overpotential (V)
$V_{act,c}$	cathode activation overpotential (V)
w	weight (N)
W	work (kJ)
$\dot{w}$	specific work rate (kW/kg)
$\dot{W}$	work rate (kW)

Greek letters

Δ	change in variable
λ_a	water content at anode–membrane interface (Ω^{-1})
λ_c	water content at cathode–membrane interface (Ω^{-1})
$\lambda(x)$	water content at location x in the membrane (Ω^{-1})
σ_{PEM}	proton conductivity in PEM (s/m)
$\sigma(x)$	local ionic PEM conductivity (s/m)
ϵ_{HEX}	HEX efficiency factor
η	energy efficiency
ψ	exergy efficiency

Subscript

a	air
abs	absorber
AC	air compressor
Cmp	compressor
cooling	cooling load
Con	condenser
CV	control volume
D	destruction
e	exit condition
Ej	ejector
en	energy
Erd	energy recovery device
Eva	evaporator
Ev	expansion valve

ex	exergy
f	fuel
fls	flashing
g	generation
Gen	generator
heating	heating load
HP	high pressure
l	liquid
LP	low pressure
i	inlet condition
mdm	membrane distillation module
mine	mineralizer
MP	medium pressure
MR	moisture remover
Mx	mixer
p	pump
pr	particulate remover
pst	posttreatment
pt	pretreatment
RO	reserve osmosis
Sep	separator
ST	steam turbine
tot	total
Tur	turbine
Vl	valve
wf	working fluid
1...74	state numbers
0	ambient or reference condition

Superscripts

.	rate
Ch	chemical

Acronyms

ACS	absorption cooling system
APS	ammonia production system
CHP	combined heat and power
COP	coefficient of performance
DS	drying system
GPC	geothermal power cycle
HEX	heat exchanger
HLS	hydrogen liquefaction system
HPS	hydrogen production system
HWPS	hot water production system
IPC	isobutene power cycle
ORC	organic Rankine cycle
OS	overall system

PEM	proton exchange membrane
SEACS	single-effect absorption cooling system
SS	single-stage
SWDU	sea water distillation unit

References

[1] Y.E. Yuksel, M. Ozturk, Thermodynamic analysis and assessment of a novel integrated geothermal energy-based system for hydrogen production and storage, Int. J. Hydrog. Energy 43 (9) (2018) 4233–4243.

[2] I. Dincer, M.A. Rosen, Exergy: Energy, Environment and Sustainable Development, Elsevier, Oxford, UK, 2012.

[3] Y.E. Yuksel, M. Ozturk, I. Dincer, Development of a geothermal based integrated plant for generating clean hydrogen and other useful commodities, J. Energy Resour. Technol. 142 (9) (2020) 1–13.

[4] M. Ozturk, I. Dincer, Geothermal energy conversion, Comprehensive Energy Systems, Elsevier, 2018, pp. 474–544.

[5] Y.E. Yuksel, M. Ozturk, I. Dincer, Analysis and performance assessment of a combined geothermal power-based hydrogen production and liquefaction system, Int. J. Hydrog. Energy 43 (22) (2018) 10268–10280.

[6] S. Salehi, S. Mohammad, S. Mahmoudi, M. Yari, M.A. Rosen, Multi-objective optimization of two double-flash geothermal power plants integrated with absorption heat transformation and water desalination, J. Clean. Prod. 195 (2018) 796–809.

[7] Y.E. Yuksel, M. Ozturk, Energetic and exergetic performance evaluations of a geothermal power plant based integrated system for hydrogen production, Int. J. Hydrog. Energy 43 (1) (2018) 78–90.

[8] T.A.H. Ratlamwala, I. Dincer, Comparative efficiency assessment of novel multi-flash integrated geothermal systems for power and hydrogen production, Appl. Therm. Eng. 48 (2012) 359–366.

[9] I. Dincer, C. Zamfirescu, USA, Elsevier Advanced Power Generation Systems, Elsevier, New York, NY, 2014.

[10] V. Zare, A comparative thermodynamic analysis of two tri-generation systems utilizing low-grade geothermal energy, Energy Convers. Manag. 118 (2016) 264–274.

[11] S. Islam, I. Dincer, Development, analysis and performance assessment of a combined solar and geothermal energy-based integrated system for multigeneration, Sol. Energy 147 (2017) 328–343.

[12] T.K. Nandi, S. Sarangi, Performance and optimization of hydrogen liquefaction cycles, Int. J. Hydrog. Energy 18 (2) (1993) 131–139.

Study questions and problems

7.1. Conduct a thorough investigation and determine how many multigenerational plants have been commissioned in (a) your community and (b) your country.

7.2. Propose methods of increasing the exergetic performance of geothermal energy–based integrated systems.

7.3. What is the geothermal energy–based integrated system for multigeneration? Describe some of the main advantages of multigeneration plants.
7.4. Identify several methods for reducing or minimizing the exergy destructions in a geothermal energy–based integrated system for multigeneration.
7.5. Identify and describe several methods for increasing the exergetic performance of the multigeneration plants described in this chapter.
7.6. How are the overall energy and exergy utilization indicators of the trigeneration and multigeneration plants investigated in this chapter defined?
7.7. Why does an increase in the outputs of a geothermal energy–based integrated system lead to a reduction in impact on the environment?
7.8. Describe the usefulness of exergy analysis in evaluating hydrogen generation processes using fossil fuels and geothermal energy sources.
7.9. Compare the results of the exergy analysis of hydrogen production using fossil fuels and geothermal energy sources.
7.10. Identify a geothermal energy–based integrated system for power, heating, cooling, hot and freshwater, drying, and hydrogen and ammonia production, and perform energy and exergy analyses of the integrated system.
7.11. How are the overall energy and exergy utilization indicators of the geothermal energy–based integrated system for the multigeneration plant examined in this chapter defined?
7.12. What are the important operating parameters in geothermal energy–based integrated systems for multigeneration? What are the effects on geothermal energy–based integrated systems energy and exergy efficiencies of varying these design parameters?
7.13. Identify the sources of exergy losses in the geothermal energy–based integrated plant for multigeneration examined in this book chapter, and propose methods for reducing or minimizing them.
7.14. Calculate the exergy destruction rate in each component of the geothermal energy–based integrated system for multigeneration presented in Case study 7.1. Based on the exergy destruction rate, determine which component performs worst in an integrated system from an exergy perspective, and suggest ways for improving the exergy destruction rate.
7.15. For the geothermal energy–based integrated system for multigeneration examined in this book chapter, what will be the effect on the overall system performance if the quadruple-flash steam geothermal power system is replaced by a flash binary geothermal power system?

7.16. For the geothermal energy–based integrated system for multigeneration examined in this chapter, what is the effect on the overall system performance if the reverse osmosis desalination plant is replaced by a double reverse osmosis desalination plant?
7.17. For the multigeneration system investigated in this chapter, propose an alternative method for achieving the same product outputs. For this alternative method, conduct detailed energy and exergy analyses, and compare the results with those for the multigeneration system examined in this chapter.
7.18. Obtain a published article on a geothermal energy–based integrated system. Write general mass, energy, entropy, exergy balance equations, as well as energy and exergy efficiency equations, and explain their differences.
7.19. Obtain a published article on a geothermal energy–based integrated system for multigeneration. Using the data provided in the article, try to duplicate the results. Compare your results to those in the original article.
7.20. Obtain actual operating data for the geothermal energy system, and perform an exergy analysis. Discuss the results.

Chapter 8

Geothermal district energy systems

8.1 Introduction

For a long time, heating applications of spaces have been obtained by providing energy sources to the places and burning those energy sources in different methods. The energy sources have changed from biomass through charcoal to coal, coke, oil, natural gas, and renewable energy. From the end of the 19th up to the middle of the 20th century, most of the energy sources were burned in open fires and enclosed stoves. In most developed countries, the technique became progressively popular of burning energy sources to heat a working fluid or to generate steam, which was then circulated throughout the building to supply heating systems, thus acting as a central heating application. The progressive popularity began with large plants and was maintained through domestic plants [1]. After that, steam distribution became the most known district heating technique (it can be also defined as a district energy plant, community energy plant, or neighborhood energy plant). On the other hand, steam could be delivered to the distribution plant at high temperatures, which was a very common method, and this had a significant effect on the design indicators for building heating processes [2]. We should not forget that the important growth of district heating systems (DHS) in the world started during the 1970s, when the primary heating energy source, oil, rapidly rose in cost as an effect of the decisions of the Organization of Petroleum Exporting Countries (OPEC). After that energy crisis, except for oil-exporting countries, a change from oil to coal or natural gas in the energy generation sectors for most government administrators became a partially compulsory choice. Also, numerous countries integrated this with special progress in district heating plants to utilize the thermal energy recovered from electricity generation plants. In other words, some countries improved district heating plants in response to the need to increase primary energy performance and hence decrease the importation of fossil energy sources. Several countries were able to take advantage of local power sources through district heating plants. Geothermal power is well-known as an effective technique for hot water preparation and space heating in residential buildings. For example,

Geothermal Energy Systems. DOI: https://doi.org/10.1016/B978-0-12-820775-8.00011-8

Iceland, the United States, Turkey, Chile, and other countries have suggested that its abundant geothermal power sources could be utilized efficiently for heating buildings by using district heating plants. The invitation to increase the energetic performance of power generation plants and, in numerous countries, a noticeable shift from oil to coal and natural gas in the energy market have performed combined heat and power generation (CHP) attractive, and district heating plants' temperature levels have become very important to the whole performance of cogeneration plants. With CHP generation supplying a rising quantity of thermal energy for district heating aims, it became more important to calculate the correct balance between the production well and injection well temperature differences, distribution design and operation, investment in generation and circulation plants, and working costs. The relations of geothermal energy–based district plants with production, distribution, and plant design options are illustrated in Fig. 8.1, [3].

Nowadays, novel district heating plants can supply cities or small communities. The most significant property of such plants is the utilization of a power resource that offers an important cost differential in producing thermal energy compared with classic heating systems utilizing boilers or

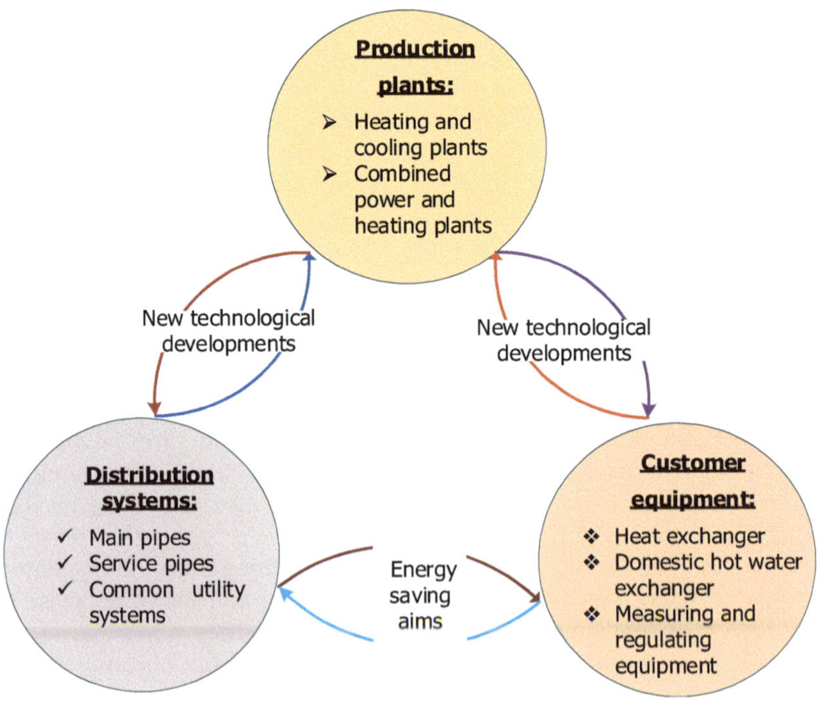

FIGURE 8.1 Relations of district energy plant parts. *Modified from [3].*

direct electrical energy heating. Also, studies on the connection of district heating and cooling processes with CHP plants become very important to increase the performance of integrated systems by simultaneously producing power, heating, and cooling. Finally, some theoretical studies for geothermal district energy systems for the multigeneration of power, hot/freshwater, heating/cooling, hydrogen, ammonia, etc. have environmental system design aims.

The application studies of geothermal district energy systems with multigeneration capability show that the introduction of these plants contributed remarkably to decreasing greenhouse gas emissions. World total energy consumption based on the energy carrier for (1) commercial buildings and (2) residential buildings by end-use in 2010 is illustrated in Fig. 8.2, [4]. Also, residential sector energy consumption in the world for (1) commercial buildings and (2) residential buildings by end-use in 2018 is shown in Fig. 8.3.

8.2 Classification of district energy systems

District energy systems are efficient energy systems in which thermal energy conversion occurs in a centralized plant and the thermal energy is distributed as hot water or steam to the end user. District energy systems have been known and used in Europe since the 14th century, and they are still being operated in France. The first district system in the United States was established in 1853 and in Canada in 1924.

As seen in Fig. 8.4, district energy systems can be grouped under five main headings: district heating, district cooling, combined district systems (heating and cooling), cogeneration district systems (power and heating), and integrated district systems (multiple products). District energy systems have the technology to use renewable energy sources or waste heat as a source of energy. Numerous studies show that district energy systems have the ability to utilize low-temperature heat from the sources, and these types of systems have proved themselves. Industrial waste heat and waste heat from power generation may be used as low-temperature heat sources. Yuksel et al. [5] also have suggested that incorporating thermal energy storage into the district energy systems may improve system performance.

District heating has been popular for about 40 years due to its high efficiency and low cost. Another advantage of the DHS is to utilize the waste heat that otherwise would be useless. Due to increasing environmental concerns, the design works of DHS have become more important than in the past. The need for heating and the type of source available should be determined. In this manner, the type of production unit should be decided, and then optimization of the configuration should be performed.

In district cooling systems, the cooling effect is produced in a centralized plant like the district heating plant. Then, by means of the distribution

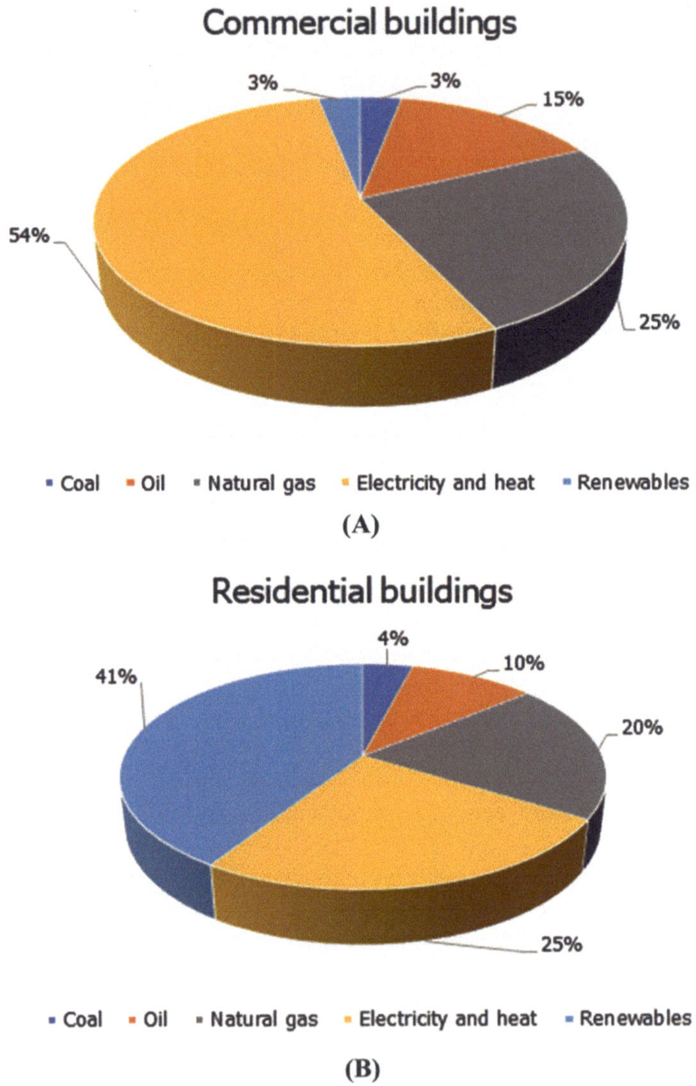

FIGURE 8.2 World total energy consumption based on the energy carrier for (A) commercial buildings and (B) residential buildings by end use in 2018. *Based on data taken from IEA [4].*

network, coolant is circulated to users in multiple buildings. District cooling systems are more efficient systems than traditional cooling methods. Similar to DHS, optimization is necessary to improve the performance of district cooling systems.

The third option, which is district heating and cooling systems, has the opportunity to reduce energy input and then to increase the efficiency by

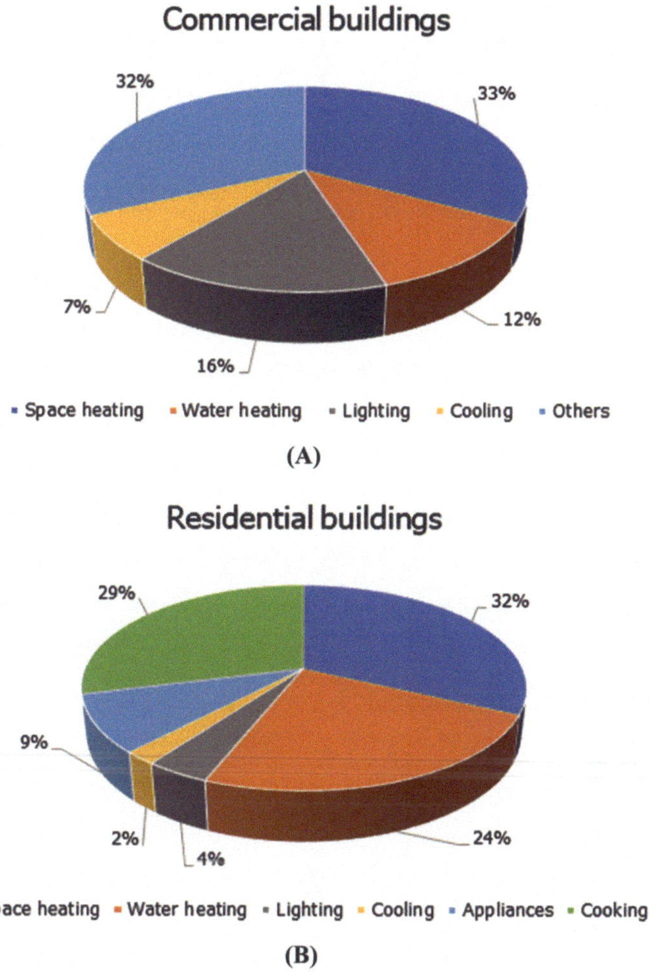

FIGURE 8.3 Residential sector energy consumption in the world for (A) commercial buildings and (B) residential buildings by end use in 2018. *Based on data taken from IEA [4].*

combining two systems. In district heating and cooling systems, hot and cold fluids are produced and distributed to users via network pipelines. The utilization of energy sources in such efficient systems may reduce fuel consumption, cost of service, and emissions.

On the other hand, some district energy systems that can use the waste heat of a power generation plant can be said to be able to produce power and heat simultaneously; that is called cogeneration. In this type of system, heat is distributed by circulating hot water or low-pressure steam via underground pipelines. Due to the proven attributes of cogeneration systems,

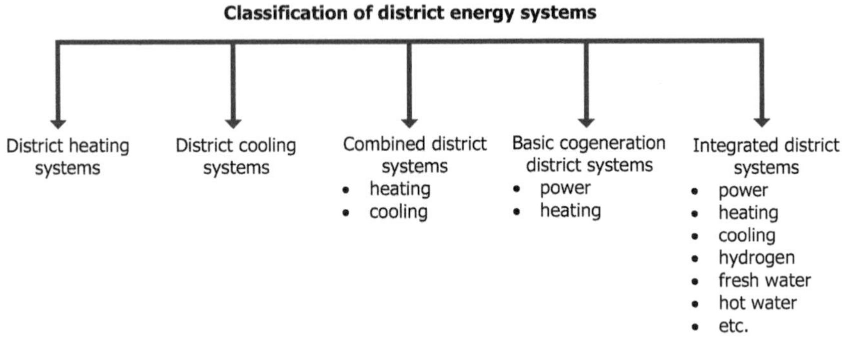

FIGURE 8.4 Schematic diagram of district energy systems.

these cogeneration district energy systems are also highly efficient, clean, and cost-effective.

After cogeneration and trigeneration systems gained popularity due to their efficiencies and low cost, scientists and engineers tried to produce multiple products by combining units. For instance, a conventional power generation system produces power and transfers waste heat energy to the environment. In a cogeneration system, this waste heat is utilized to supply heating or cooling effect. In trigeneration systems, power is produced, and waste heat is used for the heating and cooling effect. In an integrated district energy system, power is produced then waste heat is utilized again and again to generate some other useful products such as hydrogen, fresh water, hot water, drying, and chemicals.

8.3 Advantages of geothermal energy–based district systems

In order for district heating plants to become a major alternative to present or future domestic heating plants, they must supply important advantages to both the society in which they operate and to the users who purchase thermal power from the plants. Also, they must supply serious societal advantages if federal, state, or local governments are to present the financial and/or institutional assistance that is implied for successful and sustainable development. Some important benefits of geothermal district energy system applications can be defined:

1. Generally, geothermal energy fields are local power resources that can decrease the demand for imported carbon-based fuels.
2. Geothermal energy is a sustainable and safe power resource for district energy applications.
3. Geothermal energy sources have positive effects on the climate and on the environment by decreasing the carbon-based fuel combustion rate.

4. Geothermal energy can be economically competitive with conventional resources.
5. Geothermal district energy systems can work continuously, without limitations imposed by weather, unlike other alternative energy resources.
6. Geothermal energy has inherent storage potential and is the best suitable technology for base-load production.

8.3.1 Advantages to society

To expand the use of district energy systems, it is useful to provide information about the benefits and advantages of these systems to the community where these systems will be used. It should be kept in mind that society can help in the establishment of these systems if their contribution to the country's economy and the environment [6] are clearly explained. The characteristics showing the comparative analysis of these systems with other energy-generating products can assist in the public's acceptance of these systems [7]. Also, if the employment opportunities to be provided to the society during the installation of these systems can be presented clearly with appropriate analyses, these systems can encourage a potential unity throughout the society.

District energy systems are more meaningful especially when they meet the needs of the society with their renewable energy sources. If a region has a geothermal energy source, the utilization of these systems in connection with the geothermal energy source can provide potential contributions to society, such as increasing the level of performance of the system and having positive environmental effects in the region. Using the renewable energy resources of the region also contributes positively to society when the use of regional energy resources leads to positive contributions to commercial activities. Dependency on foreign countries decreases with the use of local domestic resources, and the country's economy is strengthened by the use of domestic energy resources.

Perhaps the most important contribution of district energy systems is their interaction with the environment. In general, their greatest potential contribution is in reducing the emission of gases to the air because district energy generation systems can control their outputs, including gas emissions, from the system. In addition, these systems reduce the need for space heating and cooling production systems. In this way, the level of harmful emission gases such as sulfur dioxide, which has harmful effects on the environment, decreases. District energy systems exhibit higher performance levels than those of equivalent personal heating and cooling systems. The use of renewable energy sources as energy sources for these systems also reduces emission levels.

The impact of district energy systems on society can vary depending on the level of utilization of the energy system. If incentive policies for these systems increase, they will naturally be able to take on a greater place in the social economy due to their reduced environmental impact. At the same time, increasing support for performance improvement studies on these systems may contribute to making more effective use of these systems.

Since district energy systems can meet the heating and cooling needs of a region's heating and cooling system, integrating these systems into newly established living spaces can contribute to qualifying the living spaces as Green Residential Locations. At the same time, these systems can benefit from renewable energy sources in providing heating and cooling to these settlements, which can greatly contribute to the economic development of the region. The increasing popularity of district energy systems in a region may also guide studies on industrial energy production systems in the region.

District energy generation systems attract many investors thanks to both their performance and their positive impact on the environment, so the investment cost allocated to these systems can reach huge amounts. At the same time, high amounts of tax revenue can be obtained from these systems. District energy generation systems are also an important source of employment in processes related to system installation, development, integrated operation, and expansion of the region. At the same time, the equipment needed for the installation, development, integration and growth of these systems, as well as employment for supplier companies, can be created.

8.3.2 Community advantages

Since a district energy system has many advantages over personal heating and cooling systems, it may be more attractive for a community. The cost of using a district energy system is more economical than that of personal heating and cooling systems, and its low emission values attract the attention of society. If the installation of a district energy system can be integrated into residential areas, these systems can provide cleaner, reliable, and sustainable useful output for communities. Using renewable energy sources in the region will bring together regional development. At the same time, since regional energy systems enable the use of energy resources in the regions where communities live, these systems reduce dependency on foreign countries and provide potential protection against adverse changes in the trade policies of other countries.

District energy systems can make the settlement areas where they are established more attractive for people. For this reason, as the demands for these settlements increases, employment activities may arise in various sectors with the transformation of existing residential areas and the establishment of new residential areas. At the same time, district energy systems operated by the community can be an important source of income for them.

8.3.3 Customer advantages

District energy systems bring many benefits to the regions where they are installed. Particularly, the people, businesses, and customers in the region where these systems are installed benefit most from these systems. Along with the benefits for other community members, customers benefit from having a stable energy supply.

Since district energy systems include different subsystems, many sectors deal with these systems. The energy sources used in these systems may differ according to the system to be installed. For this reason, industrial companies can work on different district energy systems in the same region. These industrial partnerships can generate income by performing multiple cost and performance improvements with multiple energy sources within the energy system. At the same time, with these energy systems, they can gain different sources of energy by obtaining a source such as hydrogen. District energy systems require less maintenance and capital expenditure than systems with personal heating and cooling. There are almost no problems in terms of security in district energy systems.

The customer's demand for additional heating and cooling can be met with district energy systems without incurring high costs [8]. Particularly, low operating and maintenance costs play an important role in the demand for these systems. At the same time, these systems are preferred because their performance is higher than that of the individual systems. In addition, district energy systems free up almost 80% of the area that would be used for customers' personal systems. In this way, a new area is gained for different applications that can bring income to customers. Backup capacity can also be added to these systems to meet the demands of customers continuously. Thanks to this backup capacity, situations such as seasonal changes and increasing need levels can be met easily.

Since production in district energy systems takes place at a central point, reliability increases. These benefits can increase the potential for district energy systems to be preferred over individual systems. The potential impact of district heating and cooling systems on the solution of environmental problems, in general, is shown in Fig. 8.5.

8.4 District heating

The residential areas' surroundings offer both challenges and opportunities relating to power demand, production, and performance. In this respect, a district heating process is one of the oldest ways to utilize geothermal resources, as well as the key technology for the success of energy security, sustainable trends at the global level, economic development, and environmental protection [9]. Classic thermal power supplies and their temperature ranges are demonstrated in Fig. 8.6.

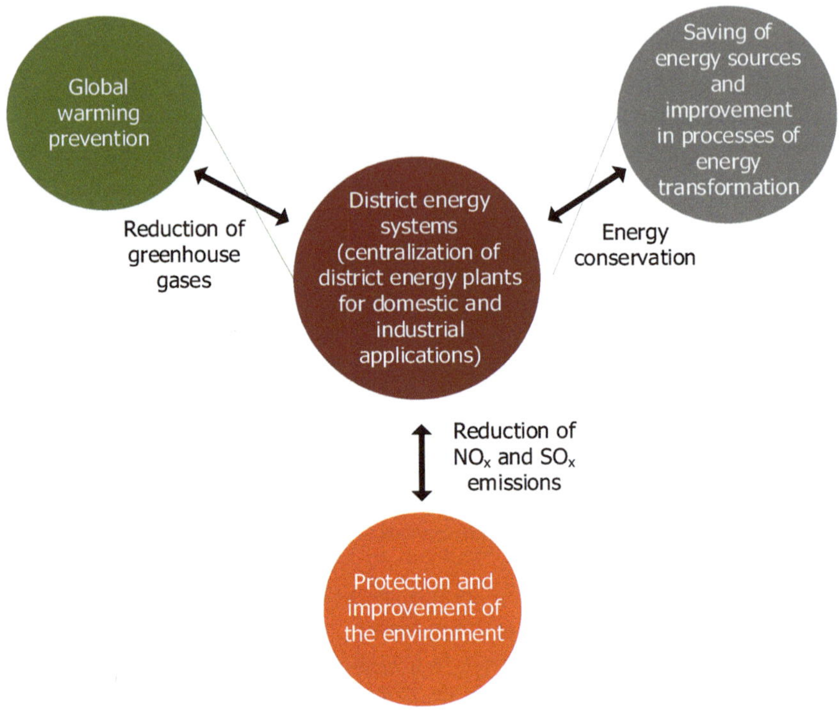

FIGURE 8.5 The positive effect of district energy technologies on environmental problems.

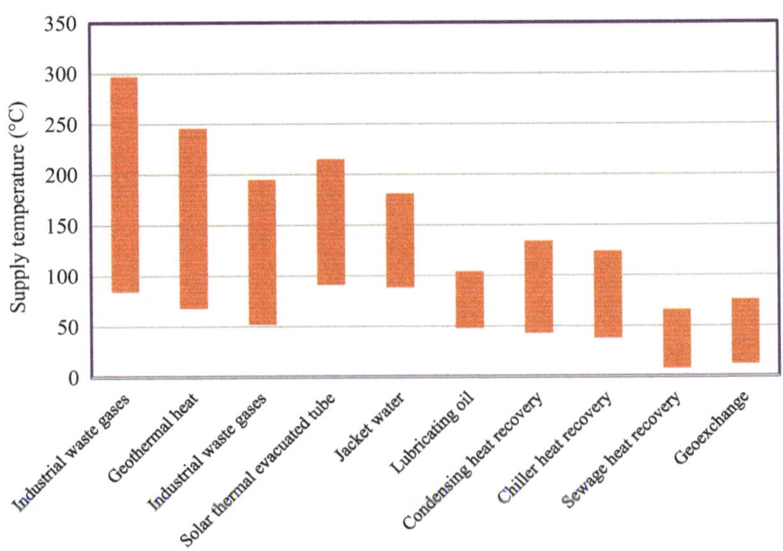

FIGURE 8.6 Energy supply and quality for district heating systems. *Data from [10].*

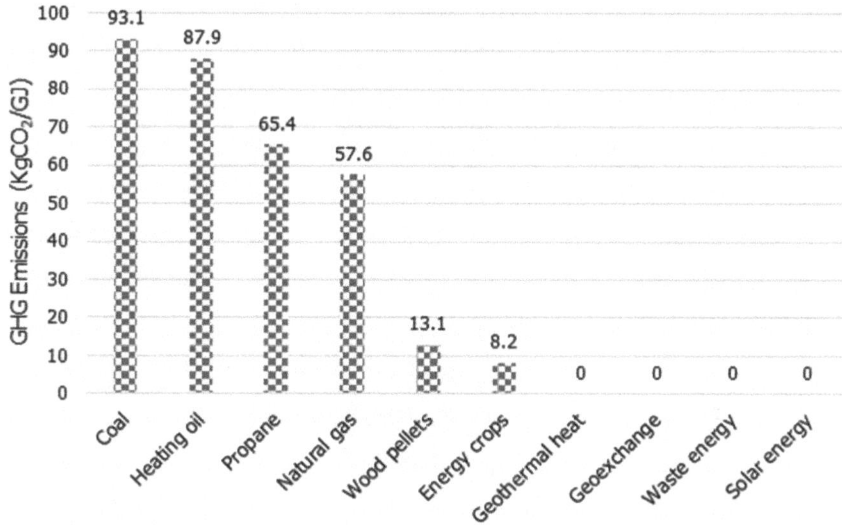

FIGURE 8.7 Greenhouse gaseous emissions of some fuels. *Data from [11].*

Greenhouse gas emissions of some fuels used for district heating plants are illustrated in Fig. 8.7, [11]. As seen from this figure, the renewable energy–based district heating plant is the major technology based on the environmental design viewpoint. To supply space heating to various users from a single well or multiple wells, district heating plants must be modeled. In addition to being an adequate and valid system, district heating has contributed to important emission reduction in many regions [12].

Extensive studies have been conducted on many energy generation systems related to DHS. These systems are generally designed according to the type of energy source that should be used for useful output production. Various features, such as energy production capacity and working fluid type, are also used to categorize these energy systems. The general types of DHS are shown in Fig. 8.8. As can be seen from the figure, these systems are divided into categories according to the energy source used.

Conventional geothermal energy–based district plants consist of wells, circulating pumps, transmission and distribution systems, adequate heat exchangers, auxiliary heating systems, and thermal energy storage processes for continuously heating the working fluid. For the independent user, the plant is identical or similar to that utilized in individual plants (forced air or radiant). All components in this heating process are equally significant, from geowater production well to the building heating systems, and all of them must be planned with maximum care. District heating plants distribute

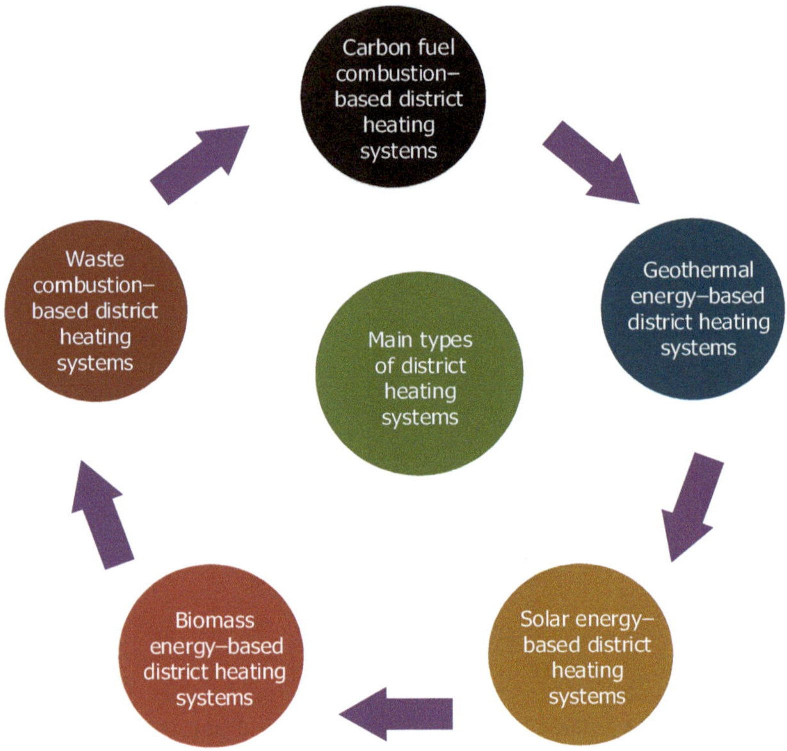

FIGURE 8.8 The main types of district heating systems.

thermal power from the main geothermal resources to residential, commercial, and/or industrial users for use in the building—water, air, process, heating, etc. This thermal power can be distributed by using steam or hot working fluid lines. Therefore thermal power comes from a distribution plant rather than being produced onsite at each plant [13]. The district energy system should meet certain design requirements:

1. The design of the district heating plant should have lower construction, material, and maintenance costs.
2. The district heating plant should have lower long-term operating costs.
3. The district heating plant must effectively supply consumer requirements.
4. The district heating plant should maximize the number of consumer links to improve performance.
5. The district heating plant should consider the need for future capacity.

A general flowchart of the geothermal energy–based district heating system, which can be divided into two processes as a central ventilation plant and space heating plant, is illustrated in Fig. 8.9. For the first type of heating

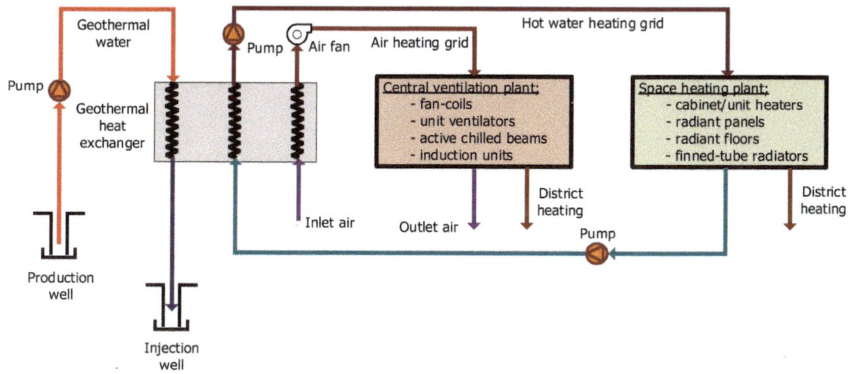

FIGURE 8.9 General flowchart of a geothermal energy–based district heating plant.

plant, the ambient air enters the geothermal heat exchanger, and after that, the heated air goes to the central ventilation plant, which consists of the fan coils, unit ventilators, active chilled beams, and induction units. In the second type of geothermal heating plant, geothermal energy is converted into hot water or steam that is distributed through the central system for space heating plants, which can be divided into a cabinet or unit heaters, radiant panels, radiant floors, and finned-tube radiators. In addition, the piping systems of the geothermal district plants should be divided into two groups: (1) piping in the central plant operation area and (2) piping in the distribution grid. These piping systems must distribute refrigerant working fluid, steam, heating/chilled/condensed water, and condensate drainage and return to and from subprocesses of the district system as directly, silently, and economically as reasonable. Also, if the temperature difference between the working fluid and ambient is very great, the distribution piping must be insulated to increase the performance of the district system.

8.4.1 Case study 8.1

A comprehensive geothermal energy–based district heating system is given in Fig. 8.10. In this district heating system, as can be seen from this figure, there are four production wells. There is also a reinjection well to ensure the sustainability of the energy sources. District heating consists of three basic parts. The first part of the system is the cycle in which heating is produced. The second part is the cycle in which the generated heat is dissipated. The third part is the cycle in which the distributed heat is consumed.

There are four production wells, a reinjection well, and a central heating station in the cycle, which allows the production of heat in the district heating system. First, the geothermal fluid is transferred to the central heating

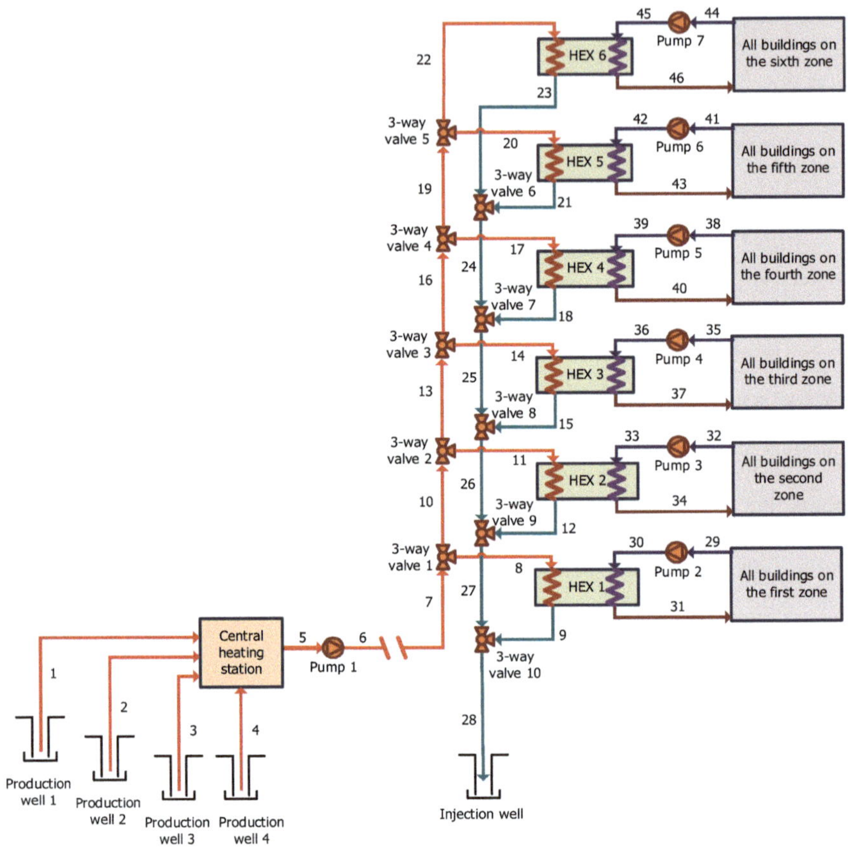

FIGURE 8.10 Schematic diagram of the geothermal energy–based district heating system.

station with flows 1, 2, 3, and 4. While this transfer is taking place, the pressure of the underground geothermal fluid is used.

There are six heat exchangers, six cycle pumps, 10 valves, and central heating station pumps in the cycle, which ensure the distribution of heat in the district heating system. The geothermal fluid pressure coming from the central heating station is sent to pump 1 with flow 5 to increase the pressure. Geothermal fluid, whose pressure is increased, is transferred to six heat exchangers with flow 6. Valves in the heat distribution system control the geothermal fluid. Adjustments can be made in the system according to the heat requirement.

The cycle in which the distributed heat is consumed depends on users' systems or tools.

In order to perform a thermodynamic analysis, the mass, energy, entropy, and exergy balance equations for geothermal energy–based integrated system components must be given. For this reason, the mass, energy, entropy, and exergy balance equations for the central heating station are defined:

$$\dot{m}_1 + \dot{m}_2 + \dot{m}_3 + \dot{m}_4 = \dot{m}_5$$

$$\dot{m}_1 h_1 + \dot{m}_2 h_2 + \dot{m}_3 h_3 + \dot{m}_4 h_4 + \dot{Q}_{L,chs} = \dot{m}_5 h_5$$

$$\dot{m}_1 s_1 + \dot{m}_2 s_2 + \dot{m}_3 s_3 + \dot{m}_4 s_4 + \dot{Q}_{L,chs}/T_{chs} + \dot{S}_g^{chs} = \dot{m}_5 s_5$$

$$\dot{m}_1 ex_1 + \dot{m}_2 ex_2 + \dot{m}_3 ex_3 + \dot{m}_4 ex_4 + \dot{E}_{L,chs}^{Q} = \dot{m}_5 ex_5 + \dot{E}x_d^{HEX1}$$

The mass, energy, entropy, and exergy balance equations for the pump 1 are defined:

$$\dot{m}_5 = \dot{m}_6$$

$$\dot{m}_5 h_5 + \dot{W}_{P1} = \dot{m}_6 h_6$$

$$\dot{m}_5 s_5 + \dot{S}_g^{P1} = \dot{m}_6 s_6$$

$$\dot{m}_5 ex_5 + \dot{E}x_{P1}^{W} = \dot{m}_6 ex_6 + \dot{E}x_d^{P1}$$

The mass, energy, entropy, and exergy balance equations for the three-way valve 1 can be defined:

$$\dot{m}_7 = \dot{m}_8 + \dot{m}_{10}$$

$$\dot{m}_7 h_7 = \dot{m}_8 h_8 + \dot{m}_{10} h_{10}$$

$$\dot{m}_7 s_7 + \dot{S}_g^{3wv1} = \dot{m}_8 s_8 + \dot{m}_{10} s_{10}$$

$$\dot{m}_7 ex_7 = \dot{m}_8 ex_8 + \dot{m}_{10} ex_{10} + \dot{E}x_d^{3wv1}$$

The mass, energy, entropy, and exergy balance equations for the three-way valve 6 can be written:

$$\dot{m}_{21} + \dot{m}_{23} = \dot{m}_{24}$$

$$\dot{m}_{21} h_{21} + \dot{m}_{23} h_{23} = \dot{m}_{24} h_{24}$$

$$\dot{m}_{21} s_{21} + \dot{m}_{23} s_{23} + \dot{S}_g^{3wvl6} = \dot{m}_{24} s_{24}$$

$$\dot{m}_{21} ex_{21} + \dot{m}_{23} ex_{23} = \dot{m}_{24} ex_{24} + \dot{E}x_d^{3wvl6}$$

The mass, energy, entropy, and exergy balance equations for the HEX 1 can be expressed:

$$\dot{m}_8 = \dot{m}_9; \dot{m}_{30} = \dot{m}_{31}$$

$$\dot{m}_8 h_8 + \dot{m}_{30} h_{30} = \dot{m}_9 h_9 + \dot{m}_{31} h_{31}$$

$$\dot{m}_8 s_8 + \dot{m}_{30} s_{30} + \dot{S}_g^{HEX1} = \dot{m}_9 s_9 + \dot{m}_{31} s_{31}$$

$$\dot{m}_8 ex_8 + \dot{m}_{30} ex_{30} = \dot{m}_9 ex_9 + \dot{m}_{31} ex_{31} + \dot{E}x_d^{HEX1}$$

The mass, energy, entropy, and exergy balance equations for the HEX 6 can be expressed:

$$\dot{m}_{22} = \dot{m}_{23}; \dot{m}_{45} = \dot{m}_{46}$$

$$\dot{m}_{22} h_{22} + \dot{m}_{45} h_{45} = \dot{m}_{23} h_{23} + \dot{m}_{46} h_{46}$$

$$\dot{m}_{22} s_{22} + \dot{m}_{45} s_{45} + \dot{S}_g^{HEX6} = \dot{m}_{23} s_{23} + \dot{m}_{46} s_{46}$$

$$\dot{m}_{22} ex_{22} + \dot{m}_{45} ex_{45} = \dot{m}_{23} ex_{23} + \dot{m}_{46} ex_{46} + \dot{E}x_d^{HEX6}$$

The exergy destruction equations for the pump, heat exchanger (HEX), and district system can be defined:

$$\dot{E}x_{D,P} = \dot{E}x_P^W - \left(\dot{E}x_{out,P} - \dot{E}x_{in,P}\right)$$

$$\dot{E}x_{D,HEX} = \dot{E}x_{out,HEX} - \dot{E}x_{in,HEX}$$

$$\dot{E}x_{D,DS} = \sum \dot{E}x_{D,P} + \sum \dot{E}x_{D,HEX}$$

Also, the exergy efficiency equations for the pump, HEX, and district system can be defined:

$$\psi_P = \frac{\left(\dot{E}x_{out,P} - \dot{E}x_{in,P}\right)}{\dot{E}x_P^W}$$

and

$$\psi_{HEX} = \frac{\dot{E}x_{out,HEX}}{\dot{E}x_{in,HEX}}$$

as well

$$\psi_{DS} = \frac{\dot{E}x_{useful,HEX}}{\sum \dot{E}x_{in,PW} - \dot{E}x_{out,IW}}$$

Here, the *PW* and *IW* show the production well and injection well, respectively. Furthermore, Lee [14] has offered a novel indicator, namely specific exergy index (*SExI*) for better classification and evaluation of geothermal energy sources:

$$SExI = \frac{h_{gw} - 273.16 x s_{gw}}{1192}$$

where h_{gw} is the geothermal water-specific enthalpy, and s_{gw} is the geothermal water-specific entropy. This indicator classifies the geothermal sources utilizing the following criteria:

1. $SExI < 0.05$ for low-quality geothermal sources;
2. $0.05 \leq SExI < 0.5$ for medium-quality geothermal sources;
3. $SExI \geq 0.5$ for high-quality geothermal sources.

For this case study, the specific exergy index of the geothermal energy−based district heating system is calculated as 0.0512. Based on the *SExI* criteria, this geothermal source can be classified as a medium-quality geothermal source. The fluid type, mass flow rate (kg/s), temperature (°C), pressure (kPa), specific enthalpy (kJ/kg), specific entropy (kJ/kg K), and exergy rate (kW) for each state of geothermal energy−based district heating system are given in Table 8.1. As given in this table, the reference temperature and reference pressure are taken as 2.5°C and 101.3 kPa, respectively. For geothermal fluid, the thermodynamic properties of water for the EES software program are used. By doing so, any possible impacts of salts and incondensable gases that might be present in the geothermal water are neglected. Also, pressure losses due to liquid flow friction are neglected.

The exergy destruction rates of the HEXs and pumps in the district heating system are given in Fig. 8.11. It can be seen from this graph that six heat exchangers and pumps in the heat distribution cycle have different performance behaviors. Exergy destruction rates of HEX 1, HEX 2, HEX 3, HEX 4, HEX 5, and HEX 6 are 249.6, 270.5, 286.2, 310, 238.4, and 243.3 kW, respectively. The exergy destruction rates of pump 1, pump 2, pump 3, pump 4, pump 5, pump 6, and pump 7 are calculated as 357.6, 61.8, 58.5, 54.26, 55.59, 41.45, and 41.16, respectively.

The exergy efficiencies of the HEXs and pumps in the district heating system are given in Fig. 8.12. It can be seen from this figure that the seven HEXs and pumps in the heat distribution cycle have different performance characteristics. The exergy efficiencies of these HEXs are computed as 0.8082, 0.7644, 0.7491, 0.647, 0.8193, and 0.7817, respectively. The exergy efficiencies of these pumps are computed as 0.4133, 0.3765, 0.3937, 0.3662, 0.3893, 0.3928, and 0.3962, respectively.

The performance curves given in Fig. 8.13 show how the reference temperature affects the energy efficiency, exergy efficiency, and exergy destruction rate of the geothermal energy−based district heating system. To examine this effect of the reference temperature, the reference temperature is gradually increased from −20°C to 20°C. In general, considering the performance curves of the district heating system, when the reference temperature is increased gradually, the energy efficiency of the system remains constant. The exergy efficiency of the system, on the other hand, increases due to the gradual increase of the ambient temperature up to 0°C and decreases after this point. Also, the exergy destruction rate of the geothermal energy−based

TABLE 8.1 State point thermodynamic data for the geothermal energy–based district heating system.

State no.	Fluid type	Mass flow rate $\dot{m}$ (kg/s)	Temperature T (°C)	Pressure P (kPa)	Specific enthalpy h (kJ/kg)	Specific entropy s (kJ/kg K)	Exergy rate $\dot{E}x$ (kW)
0	–	–	2.5	101.3	10.61	0.03829	–
1	GW[a]	100	99	180	414.9	1.296	5793
2	GW	40	96	120	402.2	1.261	2185
3	GW	40	98	200	410.7	1.284	2275
4	GW	45	93	80	389.6	1.227	2314
5	GW	175	97.09	95	406.8	1.274	9761
6	GW	175	97.79	800	410.3	1.281	10013
7	GW	175	95.09	500	398.7	1.251	9455
8	GW	29.75	95.09	500	398.7	1.251	1607
9	GW	29.75	53.09	500	222.7	0.7433	530.3
10	GW	145.3	95.09	500	398.7	1.251	7848
11	GW	30.63	95.09	500	398.7	1.251	1655
12	GW	30.63	51.09	500	214.3	0.7176	506.8
13	GW	114.6	95.09	500	398.7	1.251	6193
14	GW	31.5	95.09	500	398.7	1.251	1702
15	GW	31.5	53.09	500	222.7	0.7433	561.5
16	GW	83.13	95.09	500	398.7	1.251	4491

17	GW	33.25	95.09	500	398.7	1.251	1796
18	GW	33.25	52.09	500	218.5	0.7304	571.3
19	GW	49.88	95.09	500	398.7	1.251	2695
20	GW	24.96	95.09	500	398.7	1.251	1348
21	GW	24.96	50.09	500	210.1	0.7047	397.4
22	GW	24.92	95.09	500	398.7	1.251	1346
23	GW	24.92	48.09	500	201.8	0.6787	366.6
24	GW	49.91	72.59	500	304.2	0.9861	1622
25	GW	83.16	64.39	500	269.9	0.8857	2150
26	GW	114.7	61.29	500	256.9	0.8471	2696
27	GW	145.3	59.14	500	247.9	0.8201	3190
28	GW	175	58.11	500	243.7	0.8071	3715
29	W[b]	90.76	47	330	197.1	0.6646	1262
30	W	90.76	47.2	645	198.2	0.6671	1301
31	W	90.76	61	645	255.9	0.8434	2128
32	W	85.48	45	310	188.7	0.6384	1089
33	W	85.48	45.2	645	189.8	0.6409	1127
34	W	85.48	61	645	255.9	0.8434	2004
35	W	78.94	44	310	184.5	0.6252	961.8

(Continued)

TABLE 8.1 (Continued)

State no.	Fluid type	Mass flow rate $\dot{m}$ (kg/s)	Temperature T (°C)	Pressure P (kPa)	Specific enthalpy h (kJ/kg)	Specific entropy s (kJ/kg K)	Exergy rate $\dot{E}x$ (kW)
36	W	78.94	44.2	645	185.6	0.6277	996.7
37	W	78.94	61	645	255.9	0.8434	1851
38	W	80.51	43	310	180.3	0.612	937.2
39	W	80.51	43.2	645	181.4	0.6145	972.7
40	W	80.51	61	645	255.9	0.8434	1888
41	W	59.87	42	300	176.1	0.5988	664.5
42	W	59.87	42.2	645	177.3	0.6013	691.3
43	W	59.87	61	645	255.9	0.8434	1404
44	W	59.28	41	290	171.9	0.5855	626.5
45	W	59.28	41.2	645	173.1	0.588	653.5
46	W	59.28	61	645	255.9	0.8434	1390

[a] *GW = geothermal water.*
[b] *W = water.*

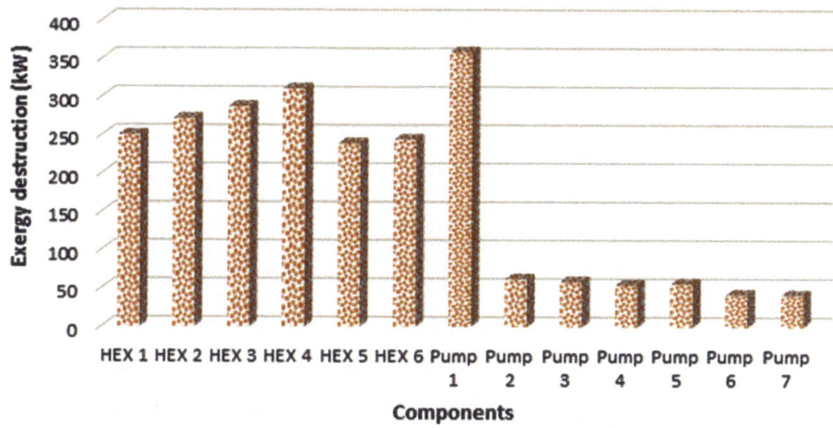

FIGURE 8.11 Exergy destruction rates of heat exchangers and pumps in the geothermal energy–based district heating system.

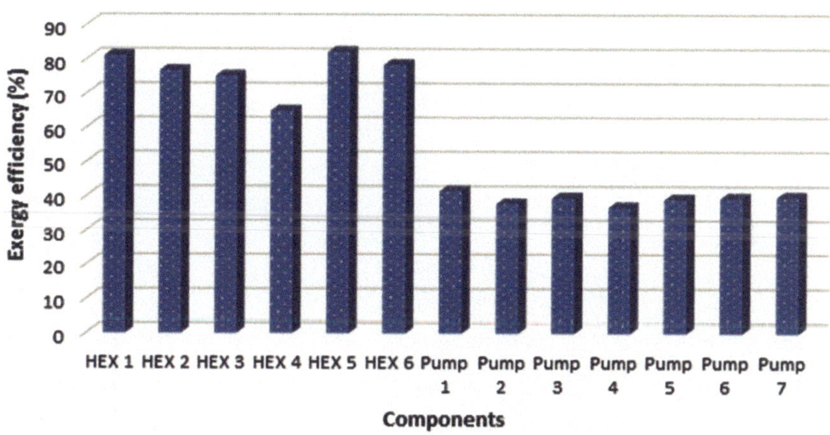

FIGURE 8.12 Exergy efficiencies of heat exchangers and pumps in the geothermal energy–based district heating system.

district heating system increases depending on the gradual increase in the reference temperature. When the reference temperature values are −20°C, −4°C, 0°C, and 20°C, the exergy destruction rates of the district heating system are 2083, 2215, 2248, and 2412 kW, respectively. For these temperature values, the energy efficiencies of the district heating system are equal. At these reference temperature values, the energy efficiency of the system is 0.6552. In addition, when the reference temperature values are −20°C, −4°C, 0°C, 4°C, and 20°C, the exergy efficiencies of the district heating system are 0.5314, 0.5543, 0.5611, 0.5529, and 0.496, respectively. As can be understood from these exergy efficiency values, the exergy efficiency of

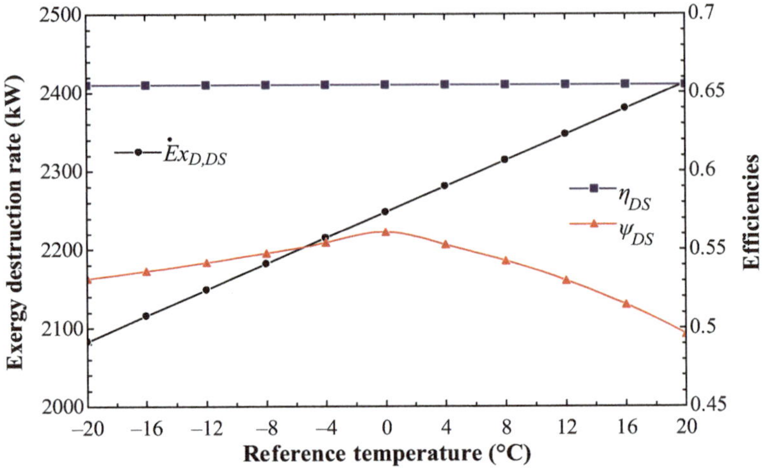

FIGURE 8.13 Effect of reference temperature on the energy and exergy efficiencies and exergy destruction rate of the geothermal energy–based district heating system.

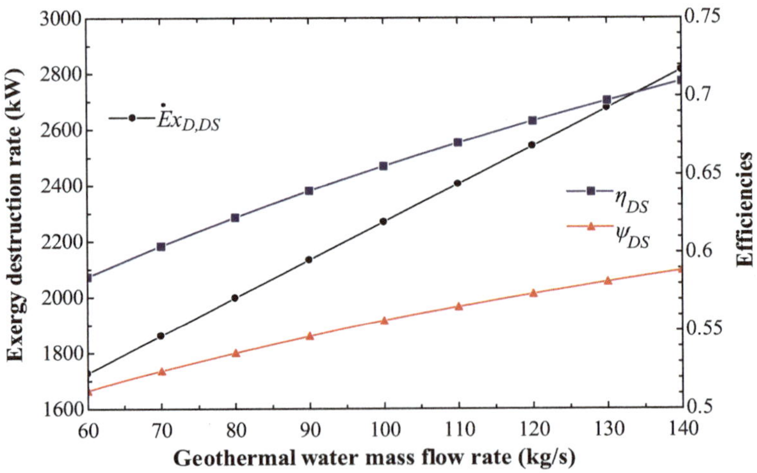

FIGURE 8.14 Effect of geothermal water mass flow rate on the energy and exergy efficiencies and exergy destruction rate of the geothermal energy–based district heating system.

the system, on the other hand, increases due to the gradual increase of the ambient temperature up to 0°C and decreases after this point.

The performance characteristics given in Fig. 8.14 show how the geothermal water mass flow rate has an effect on the energy efficiency, exergy efficiency and exergy destruction rate of the geothermal energy–based district heating system. To assess this effect of the geothermal water mass flow rate, the rate is gradually increased from 60 to 140 kg/s. In general, considering the performance characteristics of the district heating system, when the geothermal water mass flow rate is increased gradually, the energy efficiency, exergy efficiency, and exergy destruction rate of the system increase. When the geothermal water mass flow rate values are 60, 80, 100, and 140 kg/s, exergy destruction rates of the district heating system are 1727, 1997, 2268, and 2813 kW, respectively. For these geothermal water mass flow rate values, the energy efficiencies of the district heating system are 0.5845, 0.6223, 0.6552, and 0.7094. In addition, for these geothermal water mass flow rate values, the exergy efficiencies of the district heating system are 0.5116, 0.5358, 0.5562, and 0.5886, respectively. As can be seen from these values, the exergy efficiency, energy efficiency, and exergy destruction rates of the system increase in response to the gradual increase in geothermal water mass flow rate.

The performance graphs given in Fig. 8.15 show how the geothermal water temperature directs the energy efficiency, exergy efficiency, and exergy destruction rate of the geothermal energy–based district heating system. To assess this effect of the geothermal water temperature, the temperature is gradually increased from 70°C to 100°C. In general, considering the performance graphs of the district heating system, when the geothermal water temperature is increased gradually, while the energy and exergy efficiency of the system decrease, the exergy destruction rate of the system increases. When the geothermal water temperature values are 70°C, 80°C, 90°C, and 100°C, the exergy destruction rates of the district heating system are 1292, 1637, 1973, and 2301 kW, respectively. For these geothermal water temperature values, the energy efficiencies of the district heating

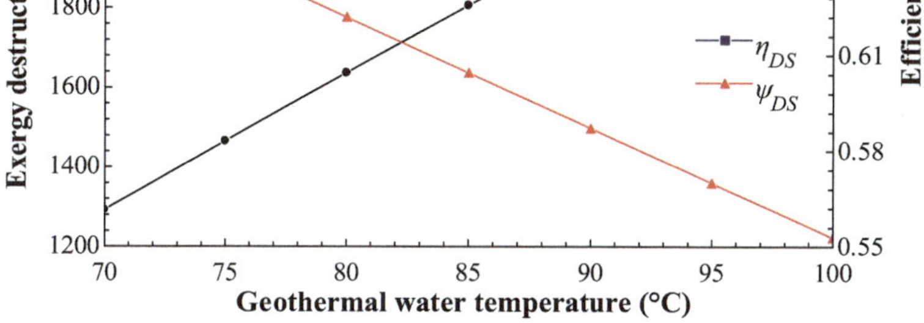

FIGURE 8.15 Effect of geothermal water temperature on the energy and exergy efficiencies and exergy destruction rate of the geothermal energy–based district heating system.

system are 0.6931, 0.6796, 0.6665, and 0.6539, respectively. In addition, for these geothermal water temperature values, the exergy efficiencies of the district heating system are 0.6563, 0.6219, 0.5871, and 0.5528, respectively.

8.5 District cooling

Similar to the geothermal energy–based district heating plant, the district cooling plant has three main plant components, such as chilled water generation system, distribution piping system, and energy transfer terminals. The major cooling subsystems containing the heat injector, chillers, pumps, chemical treatment, controls, and thermal energy rejection plant are housed in the chilled water generation system [15]. To give some useful parameters for cooling systems, different chiller performance ranges with several prime drivers are shown in Table 8.2.

General flowcharts of geothermal energy–based cooling plants with vapor compression refrigeration and an absorption and adsorption cooling cycle are illustrated in Fig. 8.16A and B, respectively. The vapor compression refrigeration cycle uses electrical energy to produce the cooling effect. As given in Fig. 8.16A, electrical energy can be produced by using

TABLE 8.2 Different chiller performance ranges with several prime drivers (based on data taken from [16]).

Cooling category	Performance level	Capacity range (ton)
Electrical power centrifugal (classic compressor)	0.52–0.7 kW/ton (COP 4.7–6.75)	500 to >1500
Electrical power centrifugal (classic dual compressor)	0.52–0.7 kW/ton (COP 4.7–6.75)	1500 to >4000
Electrical power centrifugal (industrial compressor)	0.52–0.7 kW/ton (COP 4.7–6.75)	2500 to >5500
Gaseous reciprocating engine driven centrifugal	(COP 1.5–1.9)	100 to >3000
Steam-driven centrifugal	(COP 1.2–1.8)	100 to > + 4000
Hot water driven absorption chiller (single-effect)	(COP 0.55–0.70)	<60 to >3250
Steam absorption chiller (single-effect)	(COP 0.60–0.8)	<60 to >3250
Direct fired (double-effect) absorption chiller	(COP 0.85–1.30)	<100 to >3250

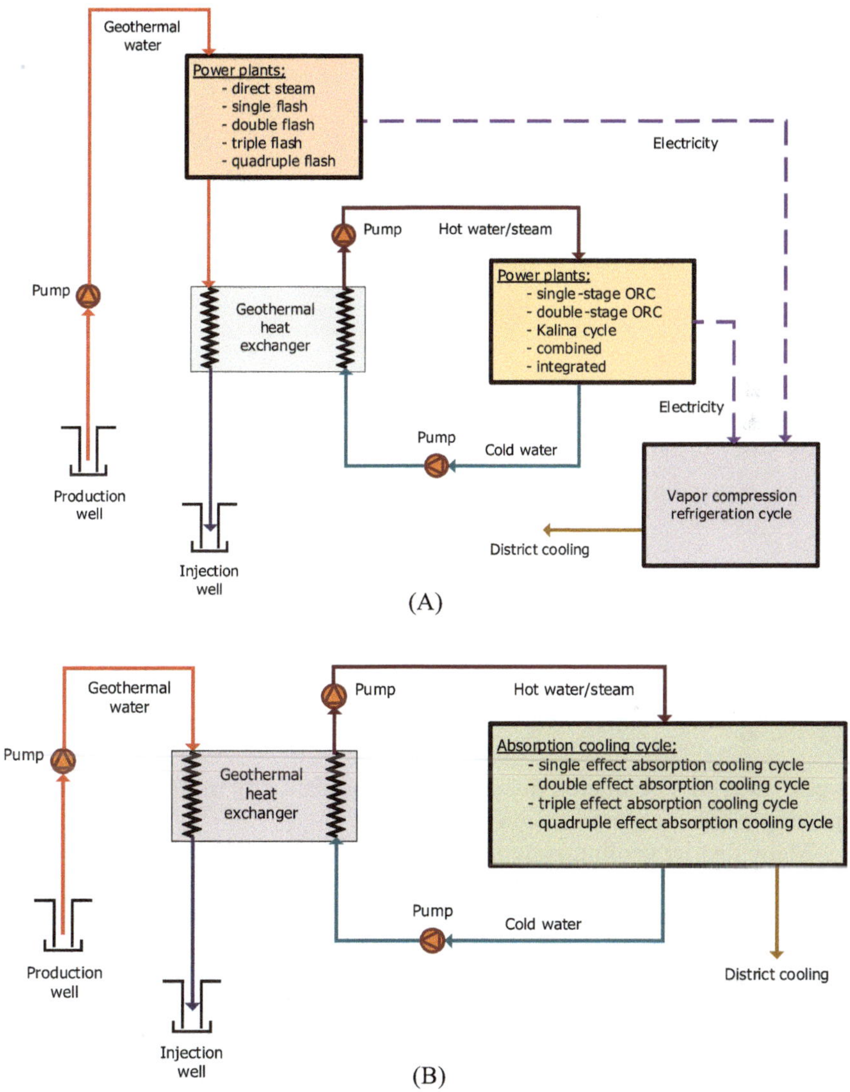

FIGURE 8.16 General flowchart of geothermal energy–based cooling plants with (A) vapor compression refrigeration and (B) absorption and adsorption cooling cycle.

geothermal energy in two ways: (1) direct utilization of geothermal water (direct steam, single-flash, double-flash, triple-flash, quadruple-flash, etc.) and (2) with secondary cycle (single-stage organic Rankine cycle (ORC), double-stage ORC, Kalina cycle, combined, integrated, etc.). The absorption cooling cycle utilizes thermal energy by using geothermal energy. As given in Fig. 8.16B, the absorption cooling cycle can be divided into four

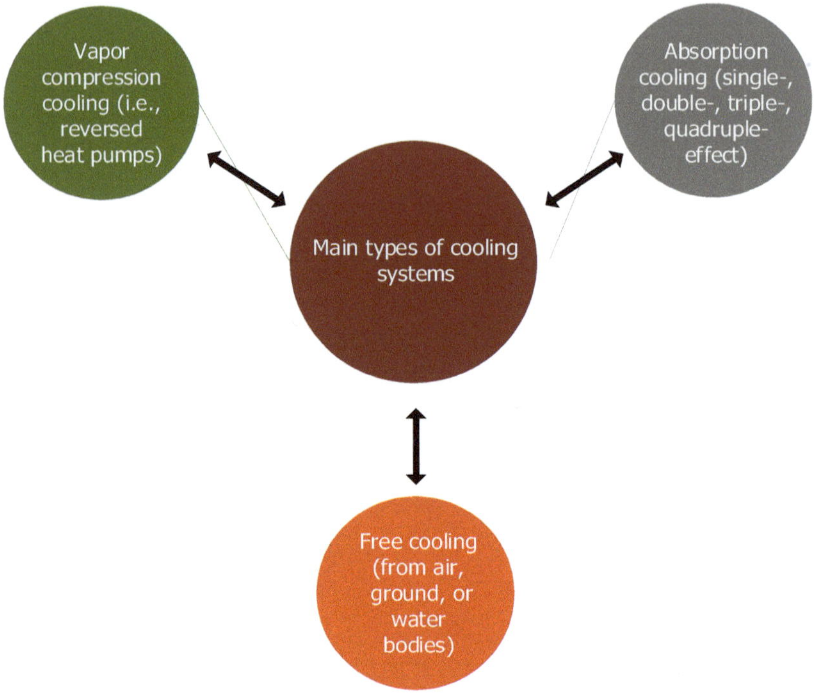

FIGURE 8.17 Main types of district cooling systems.

types, such as (1) single-effect, (2) double-effect, (3) triple-effect, and (4) quadruple-effect absorption cooling cycle.

Many district cooling systems are available to provide the district cooling product. The most common district cooling systems are given in Fig. 8.17. These systems can perform their functions with various energy sources. The efficiencies of these systems may vary depending on the performance achieved with energy sources.

8.5.1 Case study 8.2

A geothermal energy–based district cooling system is clearly given in Fig. 8.18. This cooling system consists of two subsystems: the vapor compression refrigeration system (VCRS) and the double-effect absorption refrigeration system. The system starts to work by transferring geothermal fluid to the system with flow 1 first. In order to adjust the pressure of the geothermal fluid coming into the system with the number 1 flow, the geothermal fluid is transferred to the flash chamber with the number 1 flow. This pressure setting is especially important for the separator where the fluid will be transferred in the next step. Because the necessary temperature and

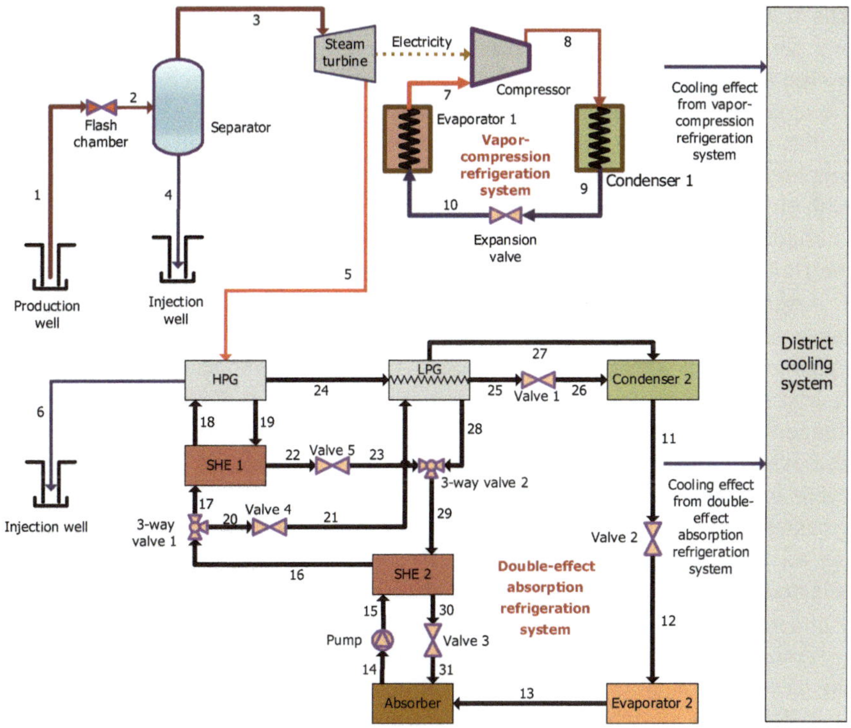

FIGURE 8.18 Schematic diagram of the geothermal energy–based district cooling system.

pressure values must be provided for the components in the system to operate cleanly. The working fluid, whose pressure has been reduced, is transferred to the separator with the flow 2. The vapor phase–rich part of the geothermal fluid in the liquid–vapor mixture is obtained in the separator. Steam-rich geothermal fluid is sent to the turbine with flow 3. The part where the liquid phase in the liquid–vapor mixture is high is reinjected to the ground with flow 4 to ensure sustainability in energy sources. When the geothermal fluid rich in the vapor phase from the separator comes to the turbine for electricity generation, it expands between streams 3 and 5. In this way, electricity is produced from the geothermal fluid. The electricity produced here is also used in the vapor compression refrigeration cycle to meet the energy requirement required for the compressor. The geothermal fluid coming out of the steam turbine is transferred to the high-pressure generator with flow 5. With this generator, energy transfer to the working fluid in the double-action absorption cooling cycle takes place. The geothermal fluid used in energy transfer in the high-pressure generator is then reinjected to the ground with flow 6 to ensure sustainability in energy sources.

In the vapor compression refrigeration cycle, the refrigerant at low temperature and low pressure is transferred to the compressor with flow 7.

The refrigerant is compressed in the compressor to increase its temperature and pressure. The energy required for compression in the compressor is met by the electricity produced in the steam turbine. Different compressor types are available for compression. These compressors can be selected according to the cost and performance analyses of the system. Fluid with increased pressure and temperature from the compressor is transferred to condenser 1 with flow 8. Condenser 1 functions here as a heat exchanger. Working fluid is condensed in the condenser. In the condensation process, the pressure of the fluid remains almost constant. The fluid pressure exiting from condenser 1 is transferred to the expansion valve with flow 9 to be adjusted. Here, the pressure of the fluid is reduced. The refrigerant, with reduced pressure, is transferred to evaporator 1 with flow 10. The temperature of the refrigerant is lower than the environment in which it is located. Taking advantage of this temperature difference, the refrigerant absorbs heat from its environment and evaporates in the evaporator. In this way, the vapor compression cooling cycle is completed, and the cooling output is obtained. There are different types of evaporators, and the appropriate evaporators can be selected according to the cost and performance analyses of the system. The key parameters selected for the geothermal energy–based integrated plant is illustrated in Table 8.3.

Another subsystem where cooling output is obtained in the district cooling system is a double-effect absorption cooling system. First, the strong fluid from the absorber comes to the pump with flow 14. The strong fluid with increased pressure is transferred to solution heat exchanger (SHE) 2 with flow 15. It is then sent to three-way valve 1 with flow 16 from SHE 2. Here, some of the strong solutions are transferred to SHE 1 with flow 17, and the other is transferred to valve 4 with flow 20. The pressure of the solution is reduced with valve 4. The low-pressure solution is transferred to the LP generator with flow 21. The high-pressure solution from SHE 1 is transferred to the high-pressure generator with flow 18. In high pressure generator (HPG), an energy transfer takes place between the solution and the geothermal fluid. The high-pressure medium concentration solution in HPG is transferred to SHE 1 with flow 19. This high-pressure medium concentration solution from SHE 1 is transferred to valve 5 with flow 22 to reduce the pressure. The low-pressure medium concentration solution is then sent to three-way valve 2 with flow 23. Here, this solution is mixed with a low concentration solution from LPG with flow 28. This solution is then transferred to SHE 2 with flow 29. From here, it is sent to valve 5 with flow 30 to adjust the pressure. This low-pressure solution is transferred back to the absorber with flow 31. In HPG, saturated steam for water vapor production is provided as the heat source that is transferred directly to LPG after it is separated from HPG. The low-pressure steam in the LPG is transferred to condenser 2 by flow 27 to be condensed. The high-pressure steam in LPG is sent to valve 1 with flow 25 to reduce the pressure. The pressurized steam is

TABLE 8.3 Key parameters selected for the geothermal energy–based district cooling system.

Parameters	Values
Reference temperature, T_o	25°C
Reference pressure, P_o	101.3 kPa
Geofluid source temperature, T_1	210°C
Geofluid source pressure, P_1	2316 kPa
Geofluid mass flow rate, $\dot{m}_1$	140 kg/s
Isentropic efficiency of the turbine, η_{Tur}	0.80
Separator 1 inlet temperature, T_2	175°C
Separator 1 inlet pressure, P_2	772 kPa
Fraction of vapor at separator 1 outlet	0.16
Turbine inlet temperature, T_3	175°C
Turbine inlet pressure, P_3	772 kPa
Turbine outlet temperature, T_5	135°C
Turbine outlet pressure, P_5	80 kPa
Isentropic efficiency of the compressor, η_{Cmp}	80%
Condenser 1 inlet pressure, P_8	1223 kPa
Condenser 1 inlet temperature, T_8	62.74°C
Expansion valve inlet pressure, P_9	1223 kPa
Expansion valve inlet temperature, T_9	47.02°C
Evaporator 1 inlet pressure, P_{10}	143 kPa
Evaporator 1 inlet temperature, T_{10}	−18.28°C
Compressor inlet pressure, P_7	143 kPa
Compressor inlet temperature, T_7	−18.28°C
High-pressure generator inlet temperature, T_5	130°C
High-pressure generator inlet temperature, T_6	108°C
Working fluid of double-effect absorption cooling system	H_2O-LiBr

transferred to condenser 2 with flow 26. In order to reduce the pressure of the refrigerant condensed in condenser 2, the refrigerant is transferred to valve 2 with flow 11. The refrigerant from valve 2 is sent to the evaporator 2 with flow 12 to be evaporated. The cooling output is obtained by using heat from the environment with the evaporator. The evaporated cooler is transferred to the absorber by flow 13. In this way, the double-effect absorption cooling cycle is completed.

To conduct a thermodynamic analysis, the balance equations for geothermal energy–based multigeneration system components must be given. For this reason, the mass, energy, entropy and exergy balance equations for the flash chamber are defined:

$$\dot{m}_1 = \dot{m}_2$$

$$\dot{m}_1 h_1 = \dot{m}_2 h_2$$

$$\dot{m}_1 s_1 + \dot{S}_g^{fc} = \dot{m}_2 s_2$$

$$\dot{m}_1 ex_1 = \dot{m}_2 ex_2 + \dot{E}x_d^{fc}$$

The mass, energy, entropy, and exergy balance equations for the separator can be expressible:

$$\dot{m}_2 = \dot{m}_3 + \dot{m}_4$$

$$\dot{m}_2 h_2 = \dot{m}_3 h_3 + \dot{m}_4 h_4$$

$$\dot{m}_2 s_2 + \dot{S}_g^{Sep} = \dot{m}_3 s_3 + \dot{m}_4 s_4$$

$$\dot{m}_2 ex_2 = \dot{m}_3 ex_3 + \dot{m}_4 ex_4 + \dot{E}x_d^{Sep}$$

The mass, energy, entropy, and exergy balance equations for the steam turbine can be expressed:

$$\dot{m}_3 = \dot{m}_5$$

$$\dot{m}_3 h_3 = \dot{m}_5 h_5 + \dot{W}_{ST}$$

$$\dot{m}_3 s_3 + \dot{S}_g^{ST} = \dot{m}_5 s_5$$

$$\dot{m}_3 ex_3 = \dot{m}_5 ex_5 + \dot{E}x_{ST}^{W} + \dot{E}x_d^{ST}$$

The mass, energy, entropy, and exergy balance equations for the compressor are defined:

$$\dot{m}_7 = \dot{m}_8$$

$$\dot{m}_7 h_7 + \dot{W}_{Comp} = \dot{m}_8 h_8$$

$$\dot{m}_7 s_7 + \dot{S}_g^{Comp} = \dot{m}_8 s_8$$

$$\dot{m}_7 ex_7 + \dot{E}x_{Comp}^{W} = \dot{m}_8 ex_8 + \dot{E}x_d^{Comp}$$

The mass, energy, entropy, and exergy balance equations for the condenser 1 can be expressed:

$$\dot{m}_8 = \dot{m}_9$$

$$\dot{m}_8 h_8 = \dot{m}_9 h_9 + \dot{Q}_{Con1}$$

$$\dot{m}_8 s_8 + \dot{S}_g^{Con1} = \dot{Q}_{Con1}/T_{Con1} + \dot{m}_9 s_9$$

$$\dot{m}_8 ex_8 = \dot{m}_9 ex_9 + \dot{E}x_{Con1}^Q + \dot{E}x_d^{Con1}$$

The mass, energy, entropy, and exergy balance equations for the expansion valve can be expressed:

$$\dot{m}_9 = \dot{m}_{10}$$

$$\dot{m}_9 h_9 = \dot{m}_{10} h_{10}$$

$$\dot{m}_9 s_9 + \dot{S}_g^{ev} = \dot{m}_{10} s_{10}$$

$$\dot{m}_9 ex_9 = \dot{m}_{10} ex_{10} + \dot{E}x_d^{ev}$$

The mass, energy, entropy, and exergy balance equations for the evaporator 1 are expressed:

$$\dot{m}_{10} = \dot{m}_7$$

$$\dot{m}_{10} h_{10} + \dot{Q}_{Eva1} = \dot{m}_7 h_7$$

$$\dot{m}_{10} s_{10} + \dot{Q}_{Eva1}/T_{Eva1} + \dot{S}_g^{Eva1} = \dot{m}_7 s_7$$

$$\dot{m}_{10} ex_{10} + \dot{E}x_{Eva1}^Q = \dot{m}_7 ex_7 + + \dot{E}x_d^{Eva1}$$

The mass, energy, entropy, and exergy balance equations for the high-pressure generator can be expressed:

$$\dot{m}_5 = \dot{m}_6; \dot{m}_{18} = \dot{m}_{19} + \dot{m}_{24}$$

$$\dot{m}_5 h_5 + \dot{m}_{18} h_{18} = \dot{m}_6 h_6 + \dot{m}_{19} h_{19} + \dot{m}_{24} h_{24}$$

$$\dot{m}_5 s_5 + \dot{m}_{18} s_{18} + \dot{S}_g^{HPG} = \dot{m}_6 s_6 + \dot{m}_{19} s_{19} + \dot{m}_{24} s_{24}$$

$$\dot{m}_5 ex_5 + \dot{m}_{18} ex_{18} = \dot{m}_6 ex_6 + \dot{m}_{19} ex_{19} + \dot{m}_{24} ex_{24} + \dot{E}x_d^{HPG}$$

The mass, energy, entropy, and exergy balance equations for the low-pressure generator are defined:

$$\dot{m}_{24} = \dot{m}_{27}; \dot{m}_{21} = \dot{m}_{25} + \dot{m}_{28}$$

$$\dot{m}_{21} h_{21} + \dot{m}_{24} h_{24} = \dot{m}_{25} h_{25} + \dot{m}_{27} h_{27} + \dot{m}_{28} h_{28}$$

$$\dot{m}_{21}s_{21} + \dot{m}_{24}s_{24} + \dot{S}_g^{LPG} = \dot{m}_{25}s_{25} + \dot{m}_{27}s_{27} + \dot{m}_{28}s_{28}$$

$$\dot{m}_{21}ex_{21} + \dot{m}_{24}ex_{24} = \dot{m}_{25}ex_{25} + \dot{m}_{27}ex_{27} + \dot{m}_{28}ex_{28} + \dot{E}x_d^{LPG}$$

The mass, energy, entropy, and exergy balance equations for the SHE 1 can be expressed:

$$\dot{m}_{17} = \dot{m}_{18}; \dot{m}_{19} = \dot{m}_{22}$$

$$\dot{m}_{17}h_{17} + \dot{m}_{19}h_{19} = \dot{m}_{18}h_{18} + \dot{m}_{22}h_{22}$$

$$\dot{m}_{17}s_{17} + \dot{m}_{19}s_{19} + \dot{S}_g^{SHE1} = \dot{m}_{18}s_{18} + \dot{m}_{22}s_{22}$$

$$\dot{m}_{17}ex_{17} + \dot{m}_{19}ex_{19} = \dot{m}_{18}ex_{18} + \dot{m}_{22}ex_{22} + \dot{E}x_d^{SHE1}$$

The mass, energy, entropy, and exergy balance equations for the three-way valve 1 can be defined:

$$\dot{m}_{16} = \dot{m}_{17} + \dot{m}_{20}$$

$$\dot{m}_{16}h_{16} = \dot{m}_{17}h_{17} + \dot{m}_{20}h_{20}$$

$$\dot{m}_{16}s_{16} + \dot{S}_g^{3wv1} = \dot{m}_{17}s_{17} + \dot{m}_{20}s_{20}$$

$$\dot{m}_{16}ex_{16} = \dot{m}_{17}ex_{17} + \dot{m}_{20}ex_{20} + \dot{E}x_d^{3wv1}$$

The energy and exergy efficiencies for subplants and the whole plant are given to make a performance assessment. The energy and exergy efficiencies for the geothermal power cycle can be defined:

$$\eta_{GPC} = \frac{\dot{W}_{Tur} + \dot{Q}_{HPG}}{\dot{m}_1(h_1 - h_6)}$$

and

$$\psi_{GPC} = \frac{\dot{E}x_{Tur}^W + \dot{E}x_{HPG}^Q}{\dot{m}_1(ex_1 - ex_6)}$$

The energy and exergy efficiencies for the VCRS can be written:

$$\eta_{VCRS} = \frac{\dot{Q}_{Cooling,VCRS}}{\dot{W}_{Comp}}$$

and

$$\psi_{VCRS} = \frac{\dot{E}x_{Cooling,VCRS}^Q}{\dot{E}x_{Comp}^W}$$

The energy and exergy efficiencies for the double-effect absorption cooling system are defined:

$$\eta_{DEACS} = \frac{\dot{Q}_{Cooling,DEACS}}{(\dot{m}_5 h_5 - \dot{m}_6 h_6) + \dot{W}_P}$$

and

$$\psi_{DEACS} = \frac{\dot{E}x^Q_{Cooling,DEACS}}{(\dot{m}_5 ex_5 - \dot{m}_6 ex_6) + \dot{E}x^W_P}$$

The energy and exergy efficiencies for the overall system can be written:

$$\eta_{CS} = \frac{\dot{W}_{Net} + \dot{Q}_{Cooling,VCRS} + \dot{Q}_{Cooling,DEACS}}{\dot{m}_1(h_1 - h_6)}$$

and

$$\psi_{CS} = \frac{\dot{E}x^W_{Net} + \dot{E}x^Q_{Cooling,VCRS} + \dot{E}x^Q_{Cooling,DEACS}}{\dot{m}_1(ex_1 - ex_6)}$$

Net energy production of the geothermal energy–based district cooling system can be defined:

$$\dot{W}_{Net} = \dot{W}_{Tur} - \dot{W}_{Pump} - \dot{W}_{Compressor}$$

The thermodynamic assessment results of the geothermal energy–based district cooling system are illustrated in Table 8.4. The useful outputs of the geothermal energy–based district cooling plant are shown in Table 8.5. As

TABLE 8.4 Thermodynamic assessment results of the geothermal energy–based district cooling system.

Subplants/whole plant	Energetic efficiency (%)	Exergetic efficiency (%)	Exergy destruction rate (kW)	Exergy destruction ratio (%)
Geothermal power cycle	48.93	56.79	8375	27.04
Vapor compression refrigeration system	32.18	28.54	16,658	53.79
Double-effect absorption cooling system	22.49	18.73	5936	19.17
Cooling system	42.65	46.81	30,969	100

TABLE 8.5 Geothermal energy–based district cooling system outputs.

Plant outputs	Values (kW)
The power production rate of the geothermal power cycle, $\dot{W}_{Tur}$	10,267
Cooling producing rate from vapor compression refrigeration system, $\dot{Q}_{Cooling,VCRS}$	18,278
Cooling producing rate from double-effect absorption cooling system, $\dot{Q}_{Cooling,DEACS}$	7659

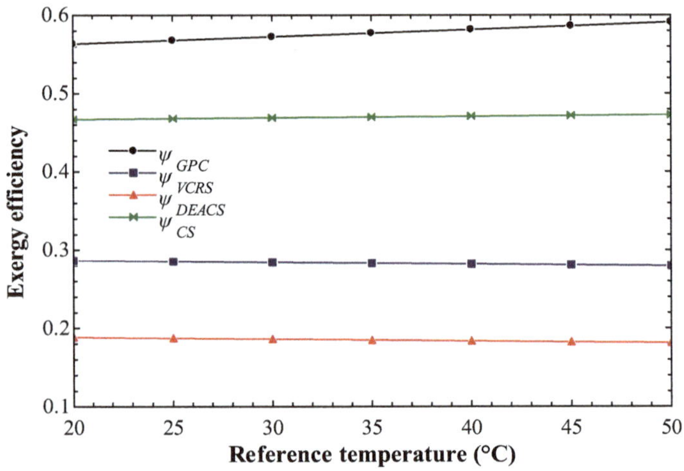

FIGURE 8.19 Effect of reference temperature on the exergy efficiency of the geothermal energy–based district cooling system and its subsystems.

given in this table, the power production rate of the geothermal power cycle is calculated as 10,267 kW, the cooling production rate from the VCRS and the double-effect absorption cooling system are computed as 182,878 and 7659 kW.

The performance curves given in Fig. 8.19 show the effect of the reference temperature on the exergy efficiency of the geothermal energy–based cooling system and its subsystems. To examine this effect of the reference temperature, the temperature is gradually increased from 20°C to 50°C. In general, considering the exergy efficiency performance curves of the geothermal energy–based cooling system and its subsystems, when the reference temperature is increased gradually, while exergy efficiencies of a geothermal power cycle (GPC) and a cooling system (CS) increase, exergy efficiencies of VCRS and double-effect absorption cooling system (DEACS) decrease. When the reference temperature value is 20°C, the exergy efficiencies of the GPC, VCRS, DEACS, and CS are 0.5633, 0.2865, 0.1884, and 0.4671, respectively. Also, when the reference temperature value is 35°C, the exergy efficiencies of the GPC, VCRS, DEACS, and CS are 0.577, 0.2831, 0.185, and 0.4699, respectively. At the maximum reference temperature value, the exergy efficiencies of GPC, VCRS, DEACS, and CS are 0.5909, 0.2797, 0.1817, and 0.4727, respectively. As can be understood from these values, when the reference temperature is increased gradually, the exergy efficiency of the geothermal energy–based cooling system increases.

The graphs given in Fig. 8.20 show the effect of the reference temperature on the power and cooling production rates of the geothermal energy–based cooling system. In general, considering the performance curves of the geothermal energy–based cooling system, when the reference temperature is increased gradually, the production of these useful outputs increases. When the reference temperature value is 25°C, the power generation of the turbine in the cooling system is 10,267 kW. Also, at this reference temperature value, the cooling production rates of VCRS and DEACS are 18,278 and 7659 kW, respectively. When the reference temperature value is 35°C, the power generation of the turbine in the cooling system is 10,514 kW. Also, at this reference temperature value, the cooling production rates of VCRS and DEACS are 18,497 and 7782 kW, respectively. At the maximum reference temperature value, the power generation of the turbine in the cooling system is 10,897 kW. Also, at this reference temperature value, the cooling production rates of VCRS and DEACS are 18,832 and 7970 kW, respectively. As can be seen from these values of the cooling system's

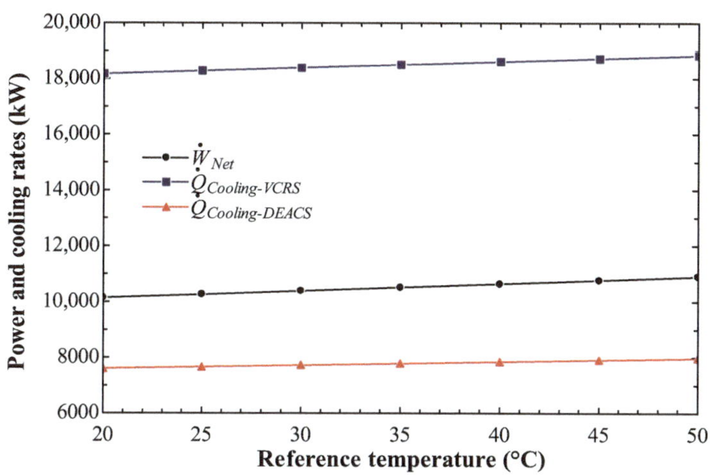

FIGURE 8.20 Effects of reference temperature on the power and cooling production rate of the geothermal energy–based district cooling system and its subsystems.

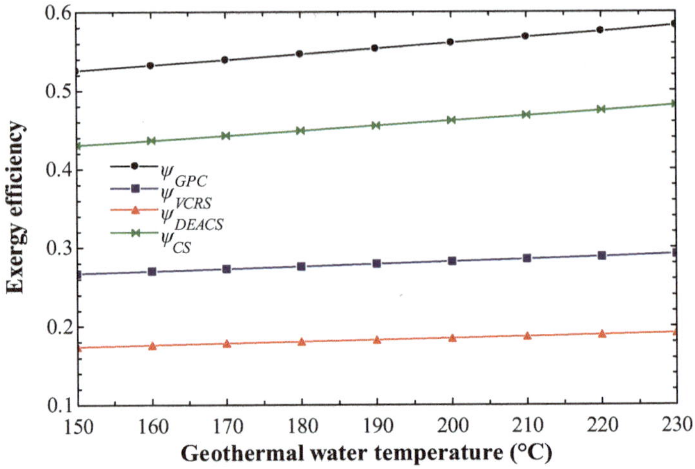

FIGURE 8.21 Influences of geothermal water temperature on the exergy efficiency of the geothermal energy–based district cooling system and its subsystems.

outputs, when the reference temperature is increased gradually, the power generation and cooling production rates of the cooling plant increase.

The performance characteristics given in Fig. 8.21 show the effect of the geothermal water temperature on the exergy efficiency of the geothermal energy–based cooling system and its subsystems. To assess the effect of the reference temperature, the geothermal water temperature is gradually increased from 150°C to 230°C. In general, considering the exergy efficiency performance characteristics of the geothermal energy–based cooling system and its subsystems, when the geothermal water temperature is increased gradually, the exergy efficiencies of GPC, VCRS, DEACS, and CS increase. When the geothermal water temperature value is 160°C, the exergy efficiencies of the GPC, VCRS, DEACS, and CS are 0.5323, 0.2702, 0.1764, and 0.4366, respectively. Also, when the geothermal water temperature value is 180°C, the exergy efficiencies of the GPC, VCRS, DEACS, and CS are 0.5463, 0.2761, 0.1807, and 0.4489, respectively. At the maximum geothermal water temperature value, the exergy efficiencies of GPC, VCRS, DEACS and CS are 0.5827, 0.2917, 0.1918, and 0.4812, respectively.

The performance graphs given in Fig. 8.22 show the effect of the geothermal water temperature on the power and cooling production rate of the geothermal energy–based cooling system. Considering the performance graphs of the geothermal energy–based cooling system, when the geothermal water temperature is increased gradually, the production of these useful outputs increases. When the geothermal water temperature value is 150°C, the power generation of the turbine in the cooling system is 9224 kW. Also, at this geothermal water temperature value, the cooling production rates of

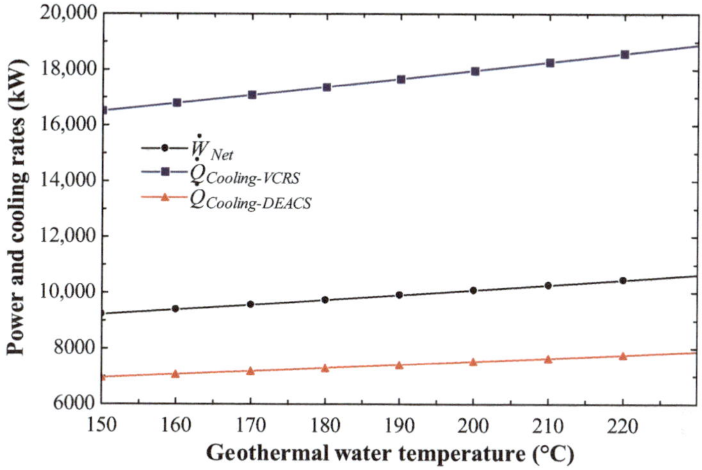

FIGURE 8.22 Impact of geothermal water temperature on the power and cooling production rate of the geothermal energy–based district cooling system and its subsystems.

VCRS and DEACS are 16,519 and 6963 kW, respectively. When the geothermal water temperature value is 200°C, the power generation of the turbine in the cooling system is 10,085 kW. Also, at this geothermal water temperature value, the cooling production rates of VCRS and DEACS are 17,972 kW and 7538 kW, respectively. At the maximum geothermal water temperature, the power generation of the turbine in the cooling system is 10,639 kW. Also, at this geothermal water temperature value, the cooling production rates of VCRS and DEACS are 18,904 and 7906 kW, respectively. As can be understood from these values of the cooling system's outputs, when the geothermal water temperature is increased gradually, the power generation and cooling production rates of the cooling plant increase.

The exergy efficiency curves given in Fig. 8.23 show the effect of the geothermal water mass flow rate on the exergy efficiency of the geothermal energy–based cooling system and its subsystems. To examines the effect of the geothermal water mass flow rate, the rate is gradually increased from 100 to 180 kg/s. Considering the exergy efficiency performance characteristics of the geothermal energy–based cooling system and its subsystems, when the geothermal water mass flow rate is increased gradually, the exergy efficiencies of GPC, VCRS, DEACS, and CS increase. When the geothermal water mass flow rate value is 110 kg/s, the exergy efficiencies of the GPC, VCRS, DEACS, and CS are 0.5479, 0.277, 0.1812, and 0.4503, respectively. Also, when the geothermal water mass flow rate value is 150°C, the exergy efficiencies of the GPC, VCRS, DEACS, and CS are 00.5747, 0.2882, 0.1893, and 0.4741, respectively. In addition, at the maximum geothermal

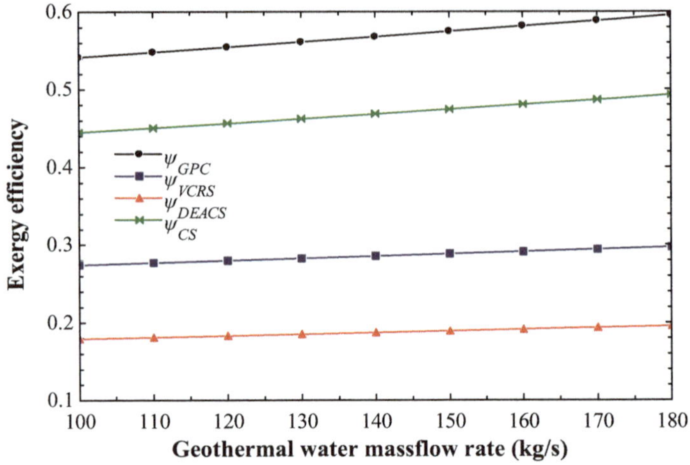

FIGURE 8.23 Effect of geothermal water mass flow rate on the exergy efficiency of the geothermal energy–based district cooling system and its subsystems.

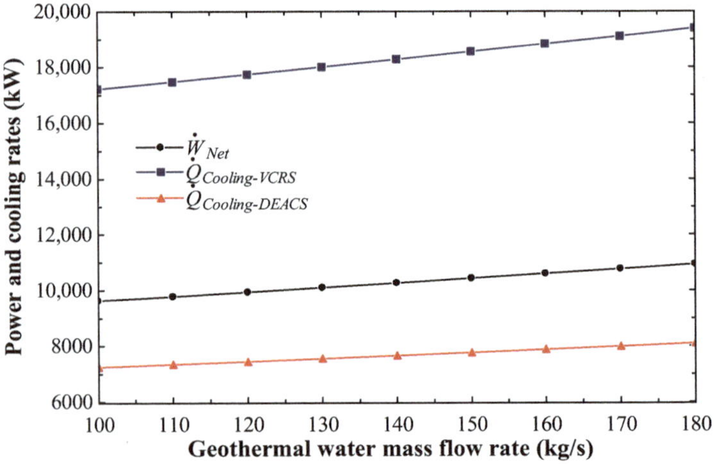

FIGURE 8.24 Effect of geothermal water mass flow rate on the power and cooling production rates of the geothermal energy–based district cooling system.

water mass flow rate value, these exergy efficiencies of GPC, VCRS, DEACS, and CS are 0.5956, 0.2969, 0.1956, and 0.4929, respectively.

The performance curves of the geothermal energy–based cooling system given in Fig. 8.24 show the effect of the geothermal water mass flow rate on the power and cooling production rates of the geothermal energy–based cooling system. In general, considering the performance characteristics of the geothermal energy–based cooling system, when the geothermal water mass flow rate is increased gradually, the production of these useful outputs increases. When the geothermal water mass flow rate value is 100 kg/s, the power generation of the turbine in the cooling system is 9635 kW. Also, at this geothermal water mass flow rate value, the cooling production rates of VCRS and DEACS are 17,221 and 7244 kW, respectively. When the geothermal water mass flow rate value is 160 kg/s, the power generation of the turbine in the cooling system is 10,598 kW. Also, at this geothermal water mass flow rate value, the cooling production rates of VCRS and DEACS are 18,830 and 7874 kW, respectively. In addition, when the geothermal water mass flow rate value is 180 kg/s, the power generation of the turbine in the cooling system is 10,940 kW. Also, at this geothermal water mass flow rate value, cooling production rates of VCRS and DEACS are 19,399 and 8096 kW, respectively. As can be understood from these values of the cooling system's outputs, when the geothermal water mass flow rate is increased gradually, the power generation and cooling production rates of the cooling plant increase.

8.6 Combined district heating and cooling plants

In the residential energy sector, the space heating, space cooling, and water heating demands are the fundamental reason for end use thermal power consumption connected with thermal power need. The district heating and cooling plants have been effectively presented as power supply plants in urban fields for the aim of saving thermal energy and also application area, limiting air pollution, or preventing residential area disaster. District heating and cooling plants deliver heating and cooling from a central facility to residential, commercial, and industrial users for space heating/cooling applications, freshwater heating, process heat production, cooking, and humidification. Also, the district heating and cooling plants intensively produce cold/hot water and steam, utilized primarily for the central plant, and supplies them to all subsystems within a certain region. Heat energy is generally delivered in the form of hot water or steam for district heating applications and chilled water for district cooling applications. As technology for distributing power and heating and cooling energy, district power plants have been shown to increase thermal power utilization performance, decrease harmful emissions, accommodate a diversity of energy resources, and capitalize on flexibility of size and scale.

The general flowchart of a geothermal energy–based heating and cooling system with (1) vapor compression refrigeration and (2) an absorption and adsorption cooling cycle is illustrated in Fig. 8.25. Utilizing recovered thermal power for reheating applications at the district standard is a general utilization for lower-grade heating aims. It should notably decrease the total power utilization of buildings that usually have simultaneous heating and cooling supplies.

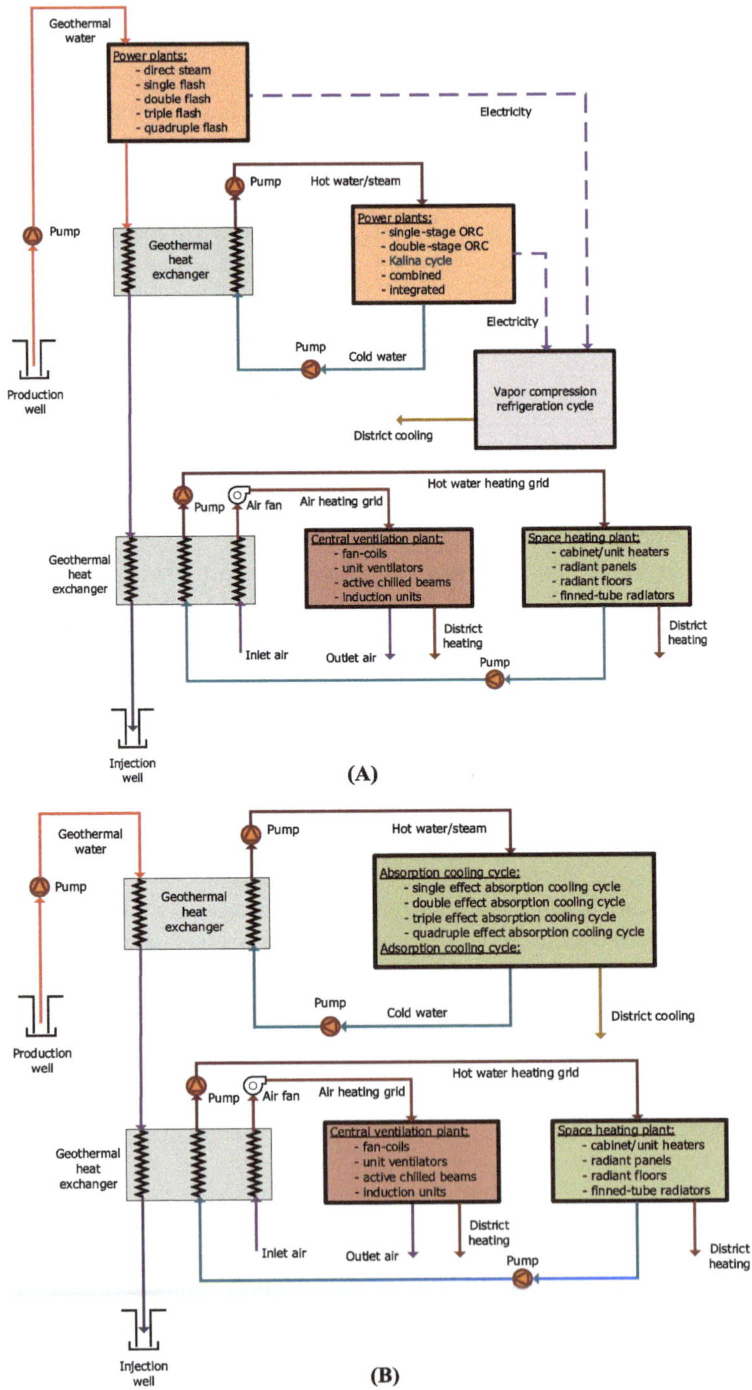

FIGURE 8.25 General flowchart of a geothermal energy–based heating and cooling plants with (A) vapor compression refrigeration and (B) absorption and adsorption cooling cycle.

8.7 Cogeneration-based district energy plants

Geothermal energy–based cogeneration systems simultaneously generate electrical energy and beneficial thermal power from the geothermal power source. On the other hand, cogeneration plants are the simultaneous production of two useful outputs in a single machine. Cogeneration generally implies the generation of power and useful thermal energy [17]. As a result, it is commonly defined as a CHP plant. Cogeneration offers a favorable and useful method, especially because its performance is generally much greater than that of separate systems for the cooutputs [18]. Cogeneration plants can work at application performances greater than those reached when thermal and electrical energy are generated in separate systems and with probably separate energy resources, for this reason, contributing to a more developed understanding of effective building solutions. District energy plants and CHP systems, that is, cogeneration systems for power and heat, are generally supplementary techniques. District energy plants can be especially useful when combined with cogeneration systems for power and thermal energy. A basic diagram of the cogeneration-based district energy plant is illustrated in Fig. 8.26. Choosing the power plant depends on the electrical and heat energy demands of the end users and on the simultaneous relationship of these demands. The characteristic of recovered heat energy from the geothermal heat exchanger is the second main factor in choosing the power plant.

8.8 Integrated district energy plants

District energy systems have many advantages in terms of both performance and environmental impact. Because system efficiency is high and emission values are low compared to individual systems, this characteristic increases interest in these systems. Therefore many comprehensive studies have been carried out in this field [19].

An integrated energy system can be defined as the system that provides useful output by integrating different subsystems [20]. Therefore the integrated district energy system can be defined as the system that generates useful output by integrating different subsystems in a certain area and that meets the energy needs of the region with this useful output. When these integrated district energy systems generate energy with renewable energy sources, a significant decrease occurs in the level of gases emitted in the air. The general flowchart of the geothermal energy–based integrated district energy system is given in Fig. 8.27. As can be seen from the figure, many useful outputs are obtained from this system. Also, subsystems that can be used to obtain useful product outputs from this system are included in the flowchart. The fact that these subsystems are included in the flow diagram explains these systems better.

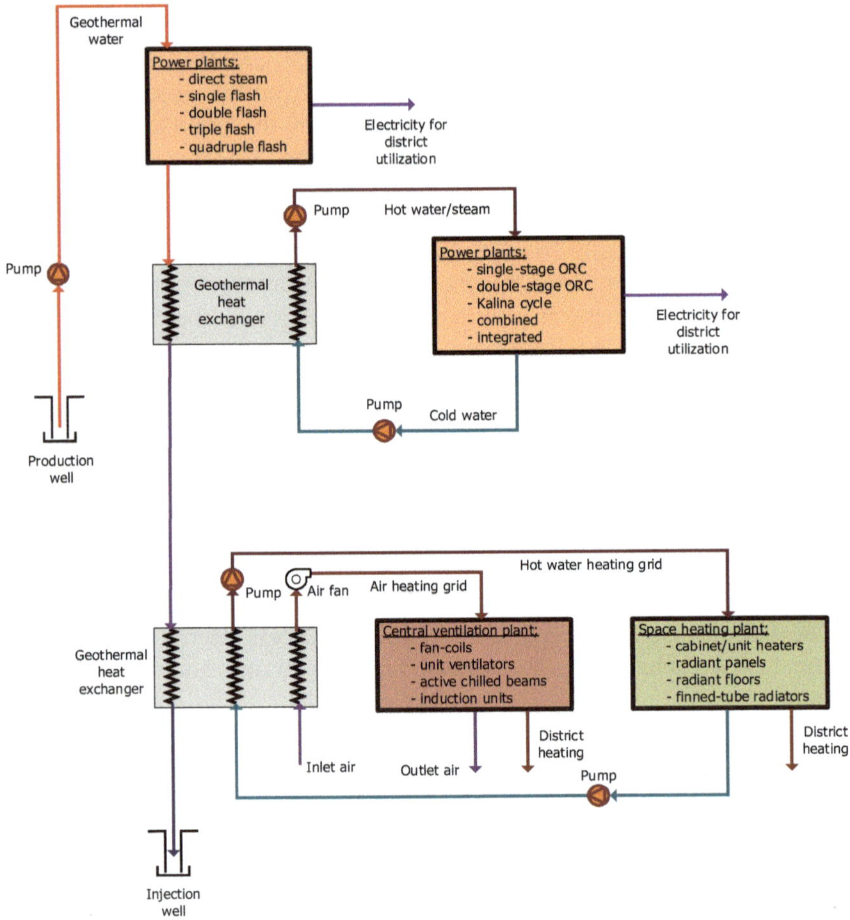

FIGURE 8.26 General flowchart of a geothermal energy–based cogeneration district energy plant.

Electricity, heating, cooling, hot water, and drying useful products can be obtained from the integrated district energy system. In the production of these products, geothermal fluid passes through different subsystems, producing useful output. As seen in Fig. 8.27, after the geothermal fluid enters the integrated system, it first enters the subsystem, which produces electricity with the vapor phase of the geothermal fluid. For the first subsystem, five different subsystems can be used for electricity generation according to the thermal properties of the geothermal fluid. The subsystems that can be used here are as follows: direct steam subplant, single-flash subplant, double-flash subplant, triple-flash subplant, quadruple-flash subplant. Theoretical explanations about these systems are given in the previous sections comprehensively. In order to obtain useful products, the subsystem to be most efficient

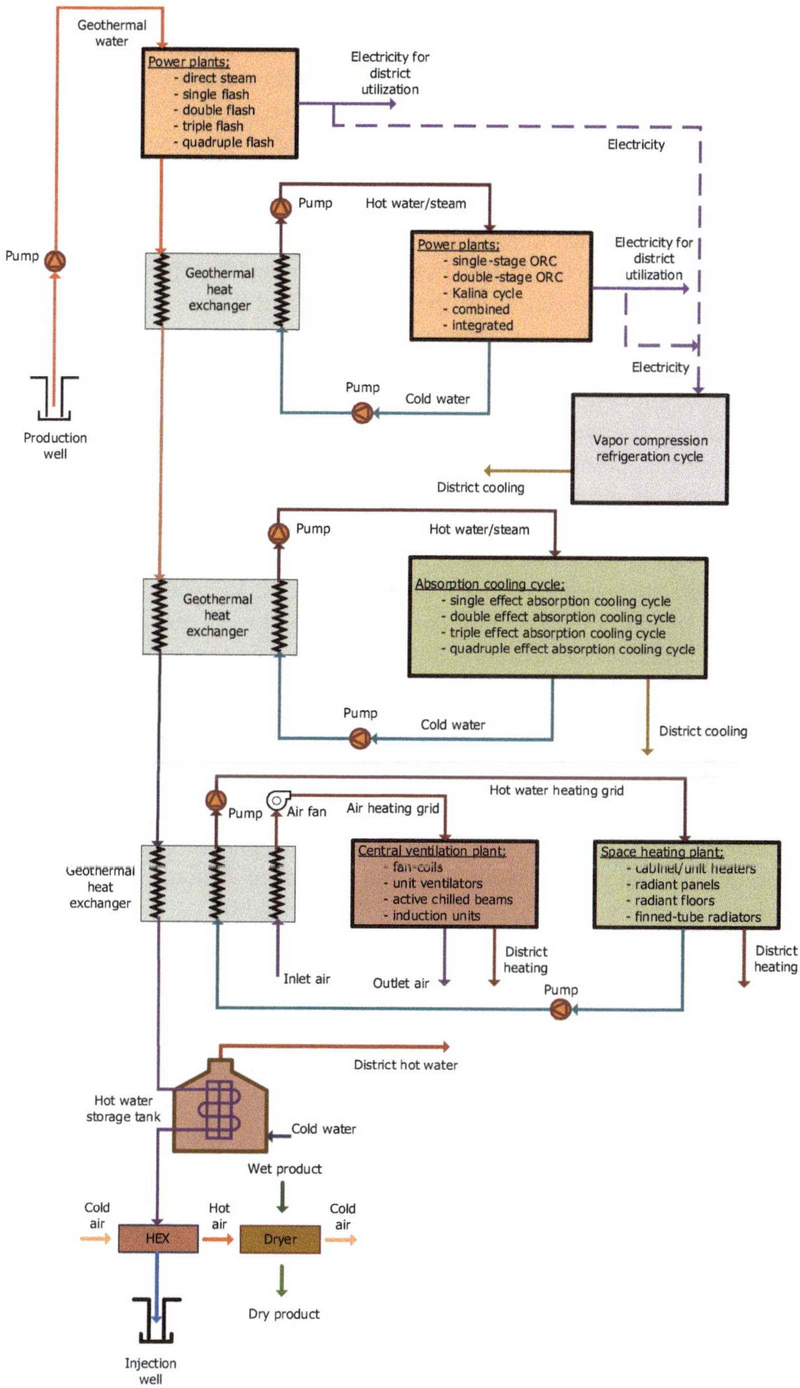

FIGURE 8.27 General flowchart of a geothermal energy–based integrated district energy plant.

in terms of system performance should be chosen since the subsystem to be selected depends on the properties of the geothermal fluid such as temperature and pressure. In addition, the integration of the subsystem to be selected into other systems should be considered. Because, as can be seen from the diagram in the figure, some of the electricity produced here is used as an energy source in another cycle, which can affect both the performance of the subsystem and system performance.

After the geothermal fluid leaves the first subsystem, transfer to another subsystem is required for useful output production. Thermal energy in the geothermal fluid is transferred to the working fluid in the second subsystem with the help of a heat exchanger so that the second subsystem can operate. Different subsystems are available to produce useful output from the subsystem here. The subsystems that can be used here are as follows: a single-stage ORC subplant, double-stage ORC subplant, combined subplant, and integrated subplant. Comprehensive descriptions of these systems have been given in detail in the previous sections. In order to achieve the desired performance from these subsystems, the effect of the working fluid used in these subsystems is great. For this reason, many studies on the selection of working fluid are carried out in these subsystems. In order to benefit from the second subsystem efficiently, both the selection of the subsystem here and the selection of the working fluid to be used in the subsystem must be carried out accurately and clearly. In addition, it is seen that the integrated subsystem is also included in the second subsystem. This shows that an integrated system can take place as a subsystem in another integrated system. As seen from the flowchart in the figure, some of the electricity produced here is used in the vapor compression refrigeration cycle, together with some of the electricity produced in the first subsystem. This can affect the performance of both the second subsystem and the entire system. For this reason, this situation should be taken into consideration while performing a performance analysis of both the subsystem and the whole system. If necessary, it should be evaluated whether the electricity produced in these two subsystems is suitable for producing another useful output and whether system performance should be improved.

The geothermal fluid coming from the second subsystem is transferred to the third subsystem for the production of cooling output. Here, thanks to the heat exchanger, the thermal energy of the geothermal fluid is transferred to the working fluid used in this subsystem. There are cycles for this subsystem that can produce cooling output. The subsystems that can be used here are a single-effect absorption cooling cycle, double-effect absorption cooling cycle, triple-effect absorption cooling cycle, and quadruple-effect absorption cooling cycle. Comprehensive theoretical explanations regarding these systems have been given in detail in previous sections. In order to obtain the desired efficiency from this subsystem, the working fluid to be used in this subsystem must be selected correctly. Many studies have been carried out on the working fluids used in these cycles. Based on these studies, the working fluid that will perform best for this subsystem can be selected.

The geothermal fluid coming from the third subsystem is transferred to the fourth subsystem to produce the heating output. Again, thanks to the heat exchanger, both space heating and ventilation are performed with the thermal energy of the geothermal fluid. As seen from the flowchart in Fig. 8.10, thermal energy in the geothermal fluid is transferred to both the air and water used in space heating. Air heating is carried out by heat transfer between the air entering the heat exchanger and the geothermal fluid. This hot air interacts with the central ventilation subplant. These subsystems can be fan coil, unit ventilator, active chilled beam, or induction unit. Hot water to be used in space heating is obtained by heat transfer between the geothermal fluid and water, thanks to a heat exchanger like air. This hot water interacts with the space heating subplant. These subsystems can be cabinet or unit heaters, radiant panels, radiant panels, or finned-tube radiators. The design of the components used in these subsystems can affect the performance of these subsystems since the production of heating output is realized thanks to these subsystems. For this reason, in order to realize the desired performance from this subsystem, these components must be either designed well or supplied well.

The geothermal fluid coming out of the fourth subsystem is transferred to the fifth and sixth subsystems, respectively, for hot water and drying output. In order to achieve the desired performance from these subsystems, the thermal energy of the geothermal fluid must be at a sufficient level because most of the thermal energy of the geothermal fluid is consumed by the four different subsystems before these subsystems produce beneficial output. In the performance analysis of the subsystems where hot water and drying outputs are obtained, these factors should be taken into consideration. The geothermal fluid coming out of the sixth subsystem is transferred back to the soil to ensure sustainability. In this way, the cycle of the integrated district energy system is completed.

8.8.1 Case study 8.3

A geothermal energy–based integrated district system is clearly given in Fig. 8.28. This integrated district system produces five beneficial outputs: electricity, heating, hot water, cooling, and dry product. The system starts working by transferring the geothermal fluid to the system with the number 1 flow. The pressure of the geothermal fluid entering the system must be adjusted before entering the separator. For this reason, the geothermal fluid entering the system is first transferred to flash chamber 1 with flow 1. The geothermal fluid, whose pressure has been adjusted, is then transferred to separator 1 by flow 2. The part of the geothermal fluid, where the vapor phase density is high, is separated in the separator and transferred to the high-pressure turbine with flow 3. The part where the liquid phase of the

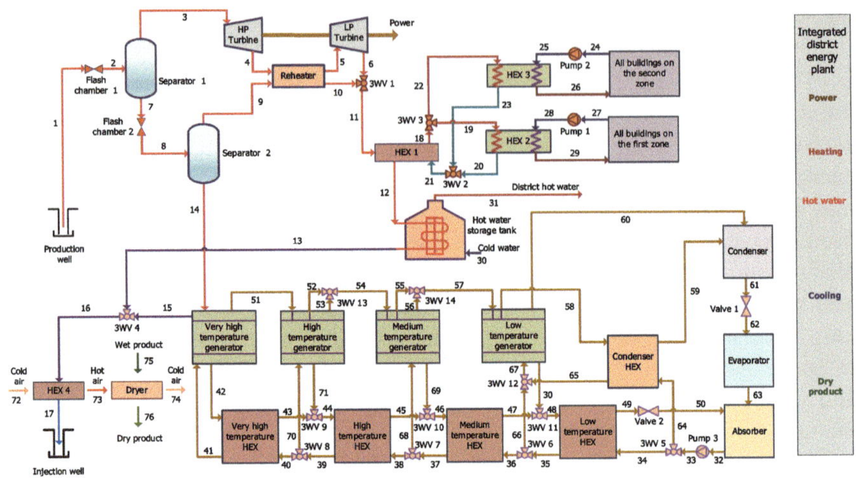

FIGURE 8.28 Schematic diagram of the geothermal energy–based integrated district system.

geothermal fluid is dense is sent to flash chamber 2 with flow 7 to adjust the pressure again. From here, it is transferred to separator 2 with flow 8. In separator 2, the part of the geothermal fluid with high vapor phase density is sent to the reheater with flow 9. The geothermal fluid part with a high density of the liquid phase in the reheater is transferred to the very high-temperature generator (VHTG) with flow 5. In the high-pressure turbine, power output is provided by expanding the geothermal fluid between flows 3 and 4. The low-pressure geothermal fluid coming from the high-pressure turbine is transferred to the reheater with flow 4. In the reheater, the fluid from the high-pressure turbine is reheated with the fluid from separator 2. This geothermal fluid is then transferred to the low-pressure turbine with flow 5. In the low-pressure turbine, power output is provided again by expanding the geothermal mixture between flows 5 and 6. Fluid from the low-pressure turbine is then sent to heat exchanger 1 with flow 11 to meet the energy need in the cycle in which the heating output is obtained. The geothermal fluid coming out of HEX 1 is then transferred to the hot water storage tank, where flow 12 will be provided with hot water output. From here, it is sent to three-way valve 4 with flow 13. This fluid coming to three-way valve 4 is mixed with the fluid from the very high-temperature generator with the number 15 flow. The geothermal fluid here is also transferred to HEX 4 with flow 16 to meet the energy need in the cycle where the drying output is provided. The geothermal fluid that supplies the energy need in HEX 4 is then reinjected to the ground with flow 17 to ensure sustainability in energy sources.

The energy requirement in the cycle in which the heating output in the system is obtained is provided by the energy transfer in HEX 1. For this HEX, energy transfer takes place between the geothermal fluid and the working fluid in the cycle in which the heating output is obtained. This working fluid is then transferred to three-way valve 3 with flow 18. With this heating cycle, the heating need for two different zones is met. For this reason, the working fluid moves after two-way valve 3 with two different flows. A part of the working fluid is sent to HEX 2 with flow 19 to meet the energy need for zone 1. The other part of the working fluid is sent to HEX 3 with flow 22 to meet the energy needs of the second zone. In these heat exchangers, the working fluid that meets the energy need is sent to three-way valve 2 with flows 20 and 23. This working fluid is then transferred back to HEX 1. The heating output is provided to the zones by the transfer of energy between the working fluid in HEX 2 and HEX 3 and the fluids used in district heating. The energetic streams are transferred to the regions with streams 26 and 29. Fluid coming from the region is transferred to the pumps with flows 24 and 27 to increase the pressure. These high-pressure fluids are also transferred to heat exchangers with flows 25 and 28 again. In this way, the cycle in which the heating output is provided is completed.

In the subsystem where hot water output is obtained, cold water is first sent to the hot water storage tank with flow 30. Here, thanks to the energy of the geothermal fluid coming from HEX 1 with flow 12, regional hot water output is provided. This hot water meets the need for hot water with flow 31 when needed.

For this plant, the subsystem where cooling output is provided is a quadruple-effect absorption cooling system. Ammonia–water is used as a working fluid for this subsystem. In this subsystem, the strong solution from the absorber with flow 41 is heated in a very high-temperature generator. The energy requirement in the very high-temperature generator is provided by the geothermal fluid coming with flow 14. The high concentration working fluid in a very high-temperature generator is transferred to the high-temperature generator (HTG) with flow 51. The weak solution in the very high-temperature generator is sent to the very high-temperature heat exchanger with flow 42. This weak solution provides heat for a very high-temperature heat exchanger. This weak solution is then transferred to the three-way valve 9 with flow 43. This weak solution and the weak solution coming with stream 71 combine in three-way valve 9. The combined weak solutions are then transferred to the high-temperature heat exchanger with flow 44. The weak solution in the high-temperature heat exchanger releases heat. The weak solution in the high-temperature heat exchanger is then transferred to three-way valve 10 with flow 45. This weak solution and the weak solution coming with flow 69 combine in three-way valve-10. The combined weak solutions are then transferred to the medium-temperature heat exchanger with flow 46. The weak solution in the medium temperature heat exchanger releases heat. The weak solution in the medium temperature heat exchanger is then transferred to three-way valve 11 with flow 47. This weak solution and the weak solution coming with flow 32 combine at three-way valve 11. The combined weak solutions are then transferred to the low-temperature heat exchanger with flow 48. The weak solution in the low-temperature heat exchanger releases heat. In the low-temperature generator (LTG), the strong solution coming with flow 34 is heated. The weak solution in the low-temperature heat exchanger is then transferred to valve 2 with flow 49. The pressure and temperature of the weak solution are reduced, and the weak solution is transferred from valve 2 to the absorber with flow 50. Refrigerant vapor in the very high-temperature generator is sent to the high-temperature generator with flow 51. This steam heats the strong solution from the high-temperature heat exchanger with flow 44. Later, this refrigerant is transferred from the high-temperature generator as flows 52 and 54 as ammonia and water. These two streams are combined with stream 54 and sent to the medium-temperature generator (MTG). These then heat the strong solution from the medium-temperature exchanger with flow 68. These two streams, which are then combined, are transferred from the medium-temperature generator as flows 55 and 56 as ammonia and water. These two streams are combined with stream 57 and sent to the low-temperature generator. These then heat the strong solution, which comes with a low flow 67. These two streams, combined, are transferred from the low-temperature generator as flows 58 and 61 as ammonia and water. Fluid coming with stream 61 is transferred to the condenser. In order to heat the solution coming from pump 3 with flow 64, the flow coming with flow 58 is transferred to the condenser heat exchanger. The heated solution is transferred to three-way valve 12 with flow 65. Ammonia–water vapor coming from condenser heat exchanger with flow 59 is transferred to the condenser. Ammonia–water vapor releases heat here. The fluid in the condenser is then transferred to valve 1 by flow 61. The pressure-adjusted fluid is transferred to the evaporator with flow 62. Evaporation with heat in the environment is carried out in the evaporator. In this way, the cooling output is obtained. The heated mixture is transferred to the absorber by flow 63. The mixture in the absorber releases heat, and this mixture is transferred to pump 3 with flow 32 in the liquid phase. In this way, the quadruple-effect absorption cooling system is completed.

In the subsystem where drying output is obtained, cold air is first sent to the HEX-4 with flow 72. Here, thanks to the energy of the geothermal fluid coming from three-way valve 4 with flow 16, hot air output is provided with flow 73. Then, this hot air meets the energy need for a dryer. In this way, the drying output provided with flow 76 is provided. The key parameters selected for the geothermal energy–based integrated district system are given in Table 8.6.

The mass, energy, entropy, and exergy balance equations for the flash chamber 1 are defined:

$$\dot{m}_1 = \dot{m}_2$$

$$\dot{m}_1 h_1 = \dot{m}_2 h_2$$

$$\dot{m}_1 s_1 + \dot{S}_g^{fc1} = \dot{m}_2 s_2$$

$$\dot{m}_1 ex_1 = \dot{m}_2 ex_2 + \dot{E}x_d^{fc1}$$

The mass, energy, entropy, and exergy balance equations for the separator 1 can be expressed:

$$\dot{m}_2 = \dot{m}_3 + \dot{m}_7$$

$$\dot{m}_2 h_2 = \dot{m}_3 h_3 + \dot{m}_7 h_7$$

$$\dot{m}_2 s_2 + \dot{S}_g^{Sep1} = \dot{m}_3 s_3 + \dot{m}_7 s_7$$

$$\dot{m}_2 ex_2 = \dot{m}_3 ex_3 + \dot{m}_7 ex_7 + \dot{E}x_d^{Sep1}$$

The mass, energy, entropy, and exergy balance equations for the high-pressure turbine can be expressed:

$$\dot{m}_3 = \dot{m}_4$$

$$\dot{m}_3 h_3 = \dot{m}_4 h_4 + \dot{W}_{HPT}$$

$$\dot{m}_3 s_3 + \dot{S}_g^{HPT} = \dot{m}_4 s_4$$

$$\dot{m}_3 ex_3 = \dot{m}_4 ex_4 + \dot{E}x_{HPT}^{W} + \dot{E}x_d^{HPT}$$

The mass, energy, entropy, and exergy balance equations for the reheater can be expressed:

$$\dot{m}_4 = \dot{m}_5; \dot{m}_9 = \dot{m}_{10}$$

$$\dot{m}_4 h_4 + \dot{m}_9 h_9 = \dot{m}_5 h_5 + \dot{m}_{10} h_{10}$$

$$\dot{m}_4 s_4 + \dot{m}_9 s_9 + \dot{S}_g^{RH} = \dot{m}_5 s_5 + \dot{m}_{10} s_{10}$$

$$\dot{m}_4 ex_4 + \dot{m}_9 ex_9 = \dot{m}_5 ex_5 + \dot{m}_{10} ex_{10} + \dot{E}x_d^{RH}$$

The mass, energy, entropy, and exergy balance equations for the three-way valve 1 can be written:

$$\dot{m}_6 + \dot{m}_{10} = \dot{m}_{11}$$

$$\dot{m}_6 h_6 + \dot{m}_{10} h_{10} = \dot{m}_{11} h_{11}$$

$$\dot{m}_6 s_6 + \dot{m}_{10} s_{10} + \dot{S}_g^{3wvl1} = \dot{m}_{11} s_{11}$$

$$\dot{m}_6 ex_6 + \dot{m}_{10} ex_{10} = \dot{m}_{11} ex_{11} + \dot{E}x_d^{3wvl1}$$

TABLE 8.6 Key parameters selected for the geothermal energy–based integrated district system.

Parameters	Values
Reference temperature, T_o	20°C
Reference pressure, P_o	101.3 kPa
Geofluid source temperature, T_1	218°C
Geofluid source pressure, P_1	2316 kPa
Geofluid mass flow rate, $\dot{m}_1$	140 kg/s
Isentropic efficiency of the turbine, η_{Tur}	0.80
Separator 1 inlet temperature, T_2	186.4°C
Separator 1 inlet pressure, P_2	1158 kPa
Separator 2 inlet temperature, T_8	157.5°C
Separator 2 inlet pressure, P_8	579 kPa
Fraction of vapor at separator 1 outlet	0.16
HP turbine inlet temperature, T_3	186.4°C
HP turbine inlet pressure, P_3	1158 kPa
HP turbine outlet temperature, T_4	157.5°C
HP turbine outlet pressure, P_4	579 kPa
LP turbine inlet temperature, T_5	157.5°C
LP turbine inlet pressure, P_5	579 kPa
LP turbine outlet temperature, T_6	65.72°C
LP turbine outlet pressure, P_6	24 kPa
HEX 1 outlet temperature, T_{18}	60.72°C
HEX 1 outlet pressure, P_{18}	250 kPa
Hot water temperature, T_{31}	45.24°C
VHTG inlet temperature, T_{14}	157.5°C
VHTG inlet pressure, P_{14}	579 kPa
VHTG temperature, T_{VHTG}	97.4°C
HTG temperature, T_{HTG}	30.3°C
MTG temperature, T_{MTG}	12.4°C
LTG temperature, T_{LTG}	3.1°C
Condenser temperature, T_{Con}	−30.5°C
Evaporator temperature, T_{Eva}	−13.4°C

(Continued)

TABLE 8.6 (Continued)

Parameters	Values
Absorber temperature, T_{Abs}	6.6°C
Working fluid of quadruple-effect absorption cooling	Ammonia–water
HEX 4 inlet temperature, T_{16}	110.63°C
Geofluid reinjection temperature, T_{17}	68.7°C

The mass, energy, entropy, and exergy balance equations for the HEX 1 can be defined:

$$\dot{m}_{11} = \dot{m}_{12}; \dot{m}_{21} = \dot{m}_{18}$$

$$\dot{m}_{11}h_{11} + \dot{m}_{21}h_{21} = \dot{m}_{12}h_{12} + \dot{m}_{18}h_{18}$$

$$\dot{m}_{11}s_{11} + \dot{m}_{21}s_{21} + \dot{S}_g^{HEX1} = \dot{m}_{12}s_{12} + \dot{m}_{18}s_{18}$$

$$\dot{m}_{11}ex_{11} + \dot{m}_{21}ex_{21} = \dot{m}_{12}ex_{12} + \dot{m}_{18}ex_{18} + \dot{E}x_d^{HEX1}$$

The mass, energy, entropy, and exergy balance equations for the pump 1 can be defined:

$$\dot{m}_{27} = \dot{m}_{28}$$

$$\dot{m}_{27}h_{27} + \dot{W}_{P1} = \dot{m}_{28}h_{28}$$

$$\dot{m}_{27}s_{27} + \dot{S}_g^{P1} = \dot{m}_{28}s_{28}$$

$$\dot{m}_{27}ex_{27} + \dot{E}x_{P1}^{W} = \dot{m}_{28}ex_{28} + \dot{E}x_d^{P1}$$

The mass, energy, entropy, and exergy balance equations for the hot water storage tank can be defined:

$$\dot{m}_{12} = \dot{m}_{13}; \dot{m}_{30} = \dot{m}_{31}$$

$$\dot{m}_{12}h_{12} + \dot{m}_{30}h_{30} = \dot{m}_{13}h_{13} + \dot{m}_{31}h_{31}$$

$$\dot{m}_{12}s_{12} + \dot{m}_{30}s_{30} + \dot{S}_g^{hwst} = \dot{m}_{13}s_{13} + \dot{m}_{31}s_{31}$$

$$\dot{m}_{12}ex_{12} + \dot{m}_{30}ex_{30} = \dot{m}_{13}ex_{13} + \dot{m}_{31}ex_{31} + \dot{E}x_d^{hwst}$$

The mass, energy, entropy, and exergy balance equations for the very high-temperature generator are defined:

$$\dot{m}_{14} = \dot{m}_{15}; \dot{m}_{41} = \dot{m}_{42} + \dot{m}_{51}$$

$$\dot{m}_{14}h_{14} + \dot{m}_{41}h_{41} = \dot{m}_{15}h_{15} + \dot{m}_{42}h_{42} + \dot{m}_{51}h_{51}$$

$$\dot{m}_{14}s_{14} + \dot{m}_{41}s_{41} + \dot{S}_g^{VHTG} = \dot{m}_{15}s_{15} + \dot{m}_{42}s_{42} + \dot{m}_{51}s_{51}$$

$$\dot{m}_{14}ex_{14} + \dot{m}_{41}ex_{41} = \dot{m}_{15}ex_{15} + \dot{m}_{42}ex_{42} + \dot{m}_{51}ex_{51} + \dot{E}x_d^{VHTG}$$

The mass, energy, entropy, and exergy balance equations for the high-temperature generator are written:

$$\dot{m}_{51} = \dot{m}_{52}; \dot{m}_{70} = \dot{m}_{53} + \dot{m}_{71}$$

$$\dot{m}_{51}h_{51} + \dot{m}_{70}h_{70} = \dot{m}_{52}h_{52} + \dot{m}_{53}h_{53} + \dot{m}_{71}h_{71}$$

$$\dot{m}_{51}s_{51} + \dot{m}_{70}s_{70} + \dot{S}_g^{HTG} = \dot{m}_{52}s_{52} + \dot{m}_{53}s_{53} + \dot{m}_{71}s_{71}$$

$$\dot{m}_{51}ex_{51} + \dot{m}_{70}ex_{70} = \dot{m}_{52}ex_{52} + \dot{m}_{53}ex_{53} + \dot{m}_{71}ex_{71} + \dot{E}x_d^{HTG}$$

The mass, energy, entropy, and exergy balance equations for the medium-temperature generator can be defined:

$$\dot{m}_{54} = \dot{m}_{55}; \dot{m}_{68} = \dot{m}_{56} + \dot{m}_{69}$$

$$\dot{m}_{54}h_{54} + \dot{m}_{68}h_{68} = \dot{m}_{55}h_{55} + \dot{m}_{56}h_{56} + \dot{m}_{69}h_{69}$$

$$\dot{m}_{54}s_{54} + \dot{m}_{68}s_{58} + \dot{S}_g^{MTG} = \dot{m}_{55}s_{55} + \dot{m}_{56}s_{56} + \dot{m}_{69}s_{69}$$

$$\dot{m}_{54}ex_{54} + \dot{m}_{68}ex_{68} = \dot{m}_{55}ex_{55} + \dot{m}_{56}ex_{56} + \dot{m}_{69}ex_{69} + \dot{E}x_d^{MTG}$$

The mass, energy, entropy, and exergy balance equations for the low-temperature generator are described:

$$\dot{m}_{57} = \dot{m}_{58}; \dot{m}_{67} = \dot{m}_{30} + \dot{m}_{60}$$

$$\dot{m}_{57}h_{57} + \dot{m}_{67}h_{67} = \dot{m}_{30}h_{30} + \dot{m}_{58}h_{58} + \dot{m}_{60}h_{60}$$

$$\dot{m}_{57}s_{57} + \dot{m}_{67}s_{67} + \dot{S}_g^{LTG} = \dot{m}_{30}s_{30} + \dot{m}_{58}s_{58} + \dot{m}_{60}s_{60}$$

$$\dot{m}_{57}ex_{57} + \dot{m}_{67}ex_{67} = \dot{m}_{30}ex_{30} + \dot{m}_{58}ex_{58} + \dot{m}_{60}ex_{60} + \dot{E}x_d^{LTG}$$

The mass, energy, entropy, and exergy balance equations for the very high-temperature HEX can be written:

$$\dot{m}_{40} = \dot{m}_{41}; \dot{m}_{42} = \dot{m}_{43}$$

$$\dot{m}_{40}h_{40} + \dot{m}_{42}h_{42} = \dot{m}_{41}h_{41} + \dot{m}_{43}h_{43}$$

$$\dot{m}_{40}s_{40} + \dot{m}_{42}s_{42} + \dot{S}_g^{VHTHEX} = \dot{m}_{41}s_{41} + \dot{m}_{43}s_{43}$$

$$\dot{m}_{40}ex_{40} + \dot{m}_{42}ex_{42} = \dot{m}_{41}ex_{41} + \dot{m}_{43}ex_{43} + \dot{E}x_d^{VHTHEX}$$

The mass, energy, entropy, and exergy balance equations for the high-temperature HEX can be defined:

$$\dot{m}_{38} = \dot{m}_{39}; \dot{m}_{44} = \dot{m}_{45}$$

$$\dot{m}_{38}h_{38} + \dot{m}_{44}h_{44} = \dot{m}_{39}h_{39} + \dot{m}_{45}h_{45}$$

$$\dot{m}_{38}s_{38} + \dot{m}_{44}s_{44} + \dot{S}_{g}^{HTHEX} = \dot{m}_{39}s_{39} + \dot{m}_{45}s_{45}$$

$$\dot{m}_{38}ex_{38} + \dot{m}_{44}ex_{44} = \dot{m}_{39}ex_{39} + \dot{m}_{45}ex_{45} + \dot{E}x_{d}^{HTHEX}$$

The mass, energy, entropy, and exergy balance equations for the medium-temperature HEX can be written:

$$\dot{m}_{36} = \dot{m}_{37}; \dot{m}_{46} = \dot{m}_{47}$$

$$\dot{m}_{36}h_{36} + \dot{m}_{46}h_{46} = \dot{m}_{37}h_{37} + \dot{m}_{47}h_{47}$$

$$\dot{m}_{36}s_{36} + \dot{m}_{46}s_{46} + \dot{S}_{g}^{MTHEX} = \dot{m}_{37}s_{37} + \dot{m}_{47}s_{47}$$

$$\dot{m}_{36}ex_{36} + \dot{m}_{46}ex_{46} = \dot{m}_{37}ex_{37} + \dot{m}_{47}ex_{47} + \dot{E}x_{d}^{MTHEX}$$

The mass, energy, entropy, and exergy balance equations for the low-temperature HEX can be described:

$$\dot{m}_{34} = \dot{m}_{35}; \dot{m}_{48} = \dot{m}_{49}$$

$$\dot{m}_{34}h_{34} + \dot{m}_{48}h_{48} = \dot{m}_{35}h_{35} + \dot{m}_{49}h_{49}$$

$$\dot{m}_{34}s_{34} + \dot{m}_{48}s_{48} + \dot{S}_{g}^{LTHEX} = \dot{m}_{35}s_{35} + \dot{m}_{49}s_{49}$$

$$\dot{m}_{34}ex_{34} + \dot{m}_{48}ex_{48} = \dot{m}_{35}ex_{35} + \dot{m}_{49}ex_{49} + \dot{E}x_{d}^{LTHEX}$$

The mass, energy, entropy, and exergy balance equations for the dryer can be expressed:

$$\dot{m}_{73} = \dot{m}_{74}; \dot{m}_{75} = \dot{m}_{76}$$

$$\dot{m}_{73}h_{73} + \dot{m}_{75}h_{75} = \dot{m}_{74}h_{74} + \dot{m}_{76}h_{76}$$

$$\dot{m}_{73}s_{73} + \dot{m}_{75}s_{75} + \dot{S}_{g}^{dry} = \dot{m}_{74}s_{74} + \dot{m}_{76}s_{76}$$

$$\dot{m}_{73}ex_{73} + \dot{m}_{75}ex_{75} = \dot{m}_{74}ex_{74} + \dot{m}_{76}ex_{76} + \dot{E}x_{d}^{dry}$$

The energy and exergy efficiencies for the geothermal power cycle can be written:

$$\eta_{GPC} = \frac{\dot{W}_{HPT} + \dot{W}_{LPT} + \dot{Q}_{HEX1} + \dot{Q}_{HWST} + \dot{Q}_{VHTG} + \dot{Q}_{HEX4}}{\dot{m}_{1}(h_{1} - h_{15})}$$

and

$$\psi_{GPC} = \frac{\dot{E}x^W_{HPT} + \dot{E}x^W_{LPT} + \dot{E}x^Q_{HEX1} + \dot{E}x^Q_{HWST} + \dot{E}x^Q_{VHTG} + \dot{E}x^Q_{HEX4}}{\dot{m}_1(ex_1 - ex_{15})}$$

The energy and exergy efficiencies for the district heating system can be defined:

$$\eta_{DHS} = \frac{\dot{Q}_{Heating}}{\dot{m}_9(h_9 - h_{10}) + \dot{W}_{P1} + \dot{W}_{P2}}$$

and

$$\psi_{DHS} = \frac{\dot{E}x^Q_{Heating}}{\dot{m}_9(ex_9 - ex_{10}) + \dot{E}x^W_{P1} + \dot{E}x^W_{P2}}$$

The energy and exergy efficiencies for the hot water production system can be written:

$$\eta_{HWPS} = \frac{\dot{Q}_{Hotwater}}{\dot{m}_{10}(h_{10} - h_{11})}$$

and

$$\psi_{HWPS} = \frac{\dot{E}x^Q_{Hotwater}}{\dot{m}_{10}(ex_{10} - ex_{11})}$$

The energy and exergy efficiencies for the quadruple-effect absorption cooling system can be defined:

$$\eta_{QEACS} = \frac{\dot{Q}_{Cooling}}{(\dot{m}_{12}h_{12} - \dot{m}_{13}h_{13}) + \dot{W}_{P3}}$$

and

$$\psi_{QEACS} = \frac{\dot{E}x^Q_{Cooling}}{(\dot{m}_{12}ex_{12} - \dot{m}_{13}ex_{13}) + \dot{E}x^W_{P3}}$$

The energy and exergy efficiencies for the drying system can be written:

$$\eta_{DS} = \frac{\dot{Q}_{Drying}}{(\dot{m}_{71}h_{71} - \dot{m}_{72}h_{72})}$$

and

$$\psi_{DS} = \frac{\dot{E}x^Q_{Drying}}{(\dot{m}_{71}ex_{71} - \dot{m}_{72}ex_{72})}$$

The energy and exergy efficiencies for the overall system can be defined:

$$\eta_{OS} = \frac{\dot{W}_{Net} + \dot{Q}_{Heating} + \dot{Q}_{Hotwater} + \dot{Q}_{Cooling} + \dot{Q}_{Drying}}{(\dot{m}_1 h_1 - \dot{m}_{15} h_{15})}$$

and

$$\psi_{OS} = \frac{\dot{E}x^{W}_{Net} + \dot{E}x^{Q}_{Heating} + \dot{E}x^{Q}_{Hotwater} + \dot{E}x^{Q}_{Cooling} + \dot{E}x^{Q}_{Drying}}{(\dot{m}_1 ex_1 - \dot{m}_{15} ex_{15})}$$

Net energy production of the geothermal energy–based integrated district system can be defined:

$$\dot{W}_{Net} = \dot{W}_{HPT} + \dot{W}_{LPT} - \sum \dot{W}_{Pump}$$

The energetic efficiency, exergetic efficiency, exergy destruction rate, and exergy destruction ratio of the geothermal energy–based integrated district system and its subsystems are illustrated in Table 8.7. The useful outputs of the geothermal energy–based integrated district system are shown in Table 8.8. As given in this table, the power production rates from the HP turbine and low-pressure (LP) turbine are calculated as 1293 and 7645 kW, respectively. In addition, the heating production rate, hot water production rate, cooling production rate, and drying production rate for the geothermal energy–based integrated district system are computed as 12,076, 9367, 7324, and 8548 kW, respectively.

TABLE 8.7 **Thermodynamic assessment results of the geothermal energy–based integrated district system.**

Subplants/whole plant	Energetic efficiency (%)	Exergetic efficiency (%)	Exergy destruction rate (kW)	Exergy destruction ratio (%)
Geothermal power cycle (GPC)	54.83	58.69	16,351	34.69
District heating system (DHS)	67.26	61.84	9873	20.94
Hot water production system (HWPS)	80.13	77.55	7624	16.17
Quadruple-effect absorption cooling system (QEACS)	24.14	20.62	5905	12.53
Drying system (DS)	72.91	68.07	7386	15.67
Whole system (WS)	63.58	59.45	47,139	100

TABLE 8.8 Geothermal energy–based integrated district system outputs.

Plant outputs	Values (kW)
Power production rate by the HP turbine, $\dot{W}_{HPT}$	1293
Power production rate by the LP turbine, $\dot{W}_{LPT}$	7645
Heating producing rate, $\dot{Q}_{Heating}$	12,076
Hot water producing rate, $\dot{Q}_{Hotwater}$	9367
Cooling producing rate, $\dot{Q}_{Cooling}$	7324
Drying producing rate, $\dot{Q}_{Drying}$	8548

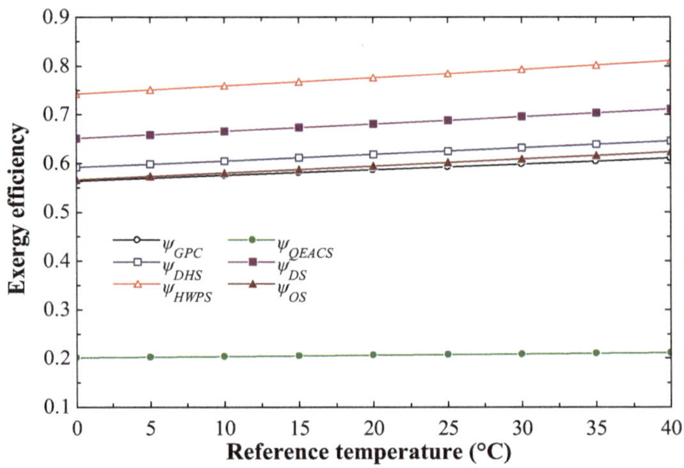

FIGURE 8.29 Effect of reference temperature on the exergy efficiency of the geothermal energy–based integrated district system and its subsystems.

The exergy efficiency characteristics given in Fig. 8.29 show the effect of the reference temperature on the exergy efficiency of the geothermal energy–based integrated district system and its subsystems. To examine the effect of the reference temperature, the temperature is gradually increased from 0°C to 40°C. In general, considering the exergy efficiency characteristics of the geothermal energy–based integrated district system and its subsystems, when the reference temperature is increased gradually, the exergy efficiencies of the geothermal energy–based integrated district system and all its subsystems increase. When the reference temperature value is 10°C, the exergy efficiencies of the GPC, DHS, HWPS, quadruple-effect absorption cooling (QEACS), drying system (DS), and WS are 0.5753, 0.605, 0.7587, 0.2037, 0.6659, and 0.5804, respectively. Also, when the reference

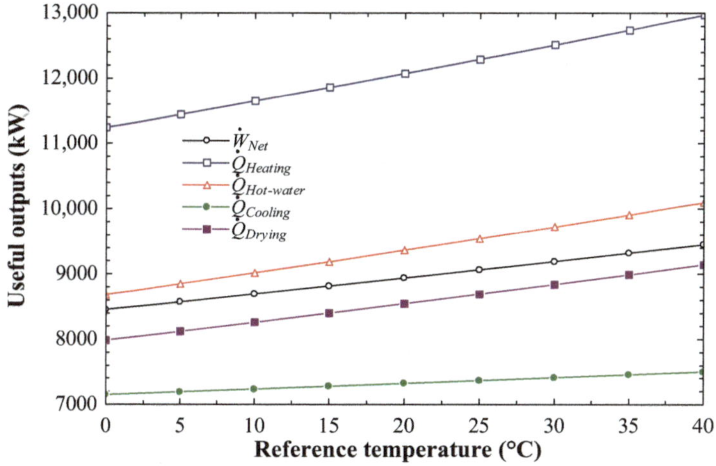

FIGURE 8.30 Effect of reference temperature on the power and cooling production rates of the geothermal energy–based integrated district system.

temperature value is 25°C, exergy efficiencies of the GPC, DHS, HWPS, QEACS, DS, and WS are 0.5927, 0.6252, 0.784, 0.2074, 0.6881, and 0.6016, respectively. At the maximum reference temperature value, the exergy efficiencies are 0.6107, 0.646, 0.8101, 0.2111, 0.7111, and 0.6235, respectively. As can be seen from these exergy efficiency values, when the reference temperature is increased gradually, the exergy efficiencies of the geothermal energy–based integrated district system and its subsystems increase.

The graphs given in Fig. 8.30 show the effect of the reference temperature on useful outputs of the geothermal energy–based integrated district system. In general, considering the performance useful output curves of the geothermal energy–based integrated district system, when the reference temperature is increased gradually, the production of these useful outputs increases. When the reference temperature values are 10°C, 25°C, and 40°C, the power generations of the geothermal energy–based integrated district system are 8692, 9063, and 9449 kW. Also, at these reference temperature values, heating outputs of the geothermal energy–based integrated district system are 11,652, 12,293, and 12,969 kW, respectively. For these reference temperature values, hot water outputs of the geothermal energy–based integrated district system are 9020, 9544, and 10,099 kW, respectively. Also, at these reference temperature values, the cooling outputs of the geothermal energy–based integrated district system are 7236, 7367, and 7501 kW, respectively. In addition, for these reference temperature values, drying outputs of the geothermal energy–based integrated district system are 8264, 8693, and 9144 kW, respectively. As can be seen from all these useful output values of the geothermal energy–based integrated district system, when the

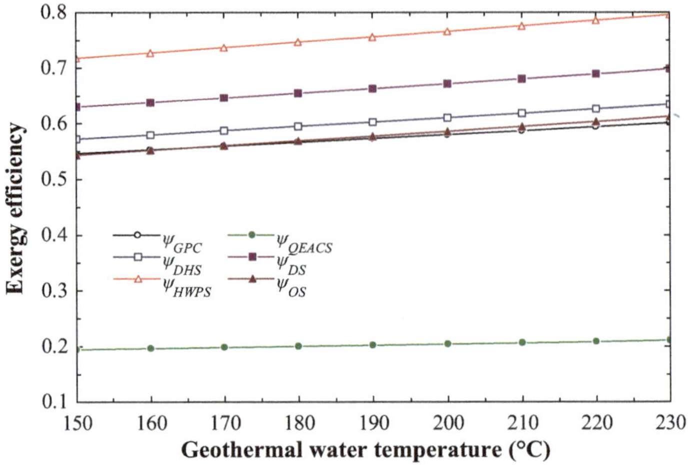

FIGURE 8.31 Impact of geothermal water temperature on the exergy efficiency of the geothermal energy–based integrated district system and its subsystems.

reference temperature is increased gradually, all useful outputs of the geothermal energy–based integrated district system increase.

The exergy efficiency curves given in Fig. 8.31 show the effect of geothermal water temperature on the exergy efficiency of the geothermal energy–based integrated district system and its subsystems. To consider this effect of the geothermal water temperature, the temperature is gradually increased from 150°C to 230°C. Considering the exergy efficiency characteristics of the geothermal energy–based integrated district system and its subsystems, when the geothermal water temperature is increased gradually, exergy efficiencies of the geothermal energy–based integrated district system and all its subsystems increase. When the geothermal water temperature value is 160°C, the exergy efficiencies of the GPC, DHS, HWPS, QEACS, DS, and WS are 0.5529, 0.5797, 0.727, 0.1961, 0.6381, and 0.5518, respectively. Also, when the geothermal water temperature value is 200°C, the exergy efficiencies of the GPC, DHS, HWPS, QEACS, DS, and WS are 0.5799, 0.6104, 0.7655, 0.2041, 0.6719, and 0.5857, respectively. When the geothermal water temperature value is 230°C, the exergy efficiencies are 0.601, 0.6345, 0.7957, 0.2103, 0.6985, and 0.6124, respectively. As can be understood from these exergy efficiency values of the system and its subsystems, when the geothermal water temperature is increased gradually, the exergy efficiencies of the geothermal energy–based integrated district system and its subsystems increase.

The beneficial output curves given in Fig. 8.32 show the effect of the geothermal water temperature on useful outputs of the geothermal energy–based integrated district system. In general, considering the useful output performance curves of the geothermal energy–based integrated district

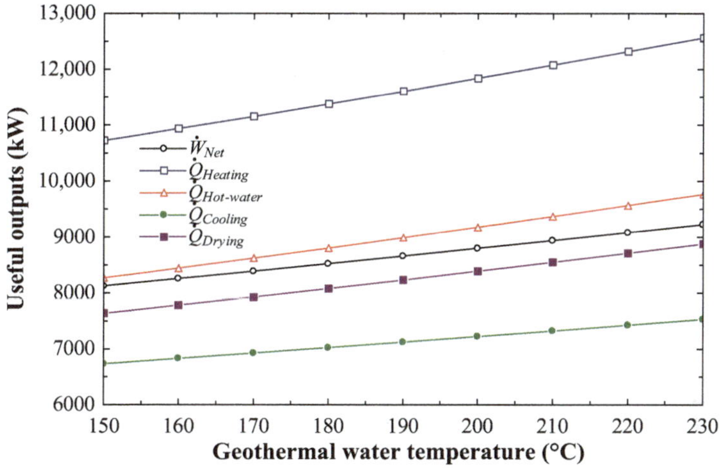

FIGURE 8.32 Impact of geothermal water temperature on the power and cooling production rate of the geothermal energy–based integrated district system.

system, when the geothermal water temperature is increased gradually, the production of these useful outputs increases. When the geothermal water temperature values are 150°C, 180°C, and 230°C, power generations of the geothermal energy–based integrated district system are 8126, 8522, and 9226 kW. At these geothermal water temperature values, the heating outputs of the geothermal energy–based integrated district system are 10,723, 11,379, and 12,563 kW, respectively. For these geothermal water temperature values, hot water outputs of the geothermal energy–based integrated district system are 8268, 8800, and 9764 kW, respectively. At these geothermal water temperature values, the cooling outputs of the geothermal energy–based integrated district system are 6737, 7024, and 7530 kW, respectively. In addition, for these geothermal water temperature values, drying outputs of the geothermal energy–based integrated district system are 7635, 8078, and 8875 kW, respectively. As can be seen from all these beneficial output values of the geothermal energy–based integrated district system, when the geothermal water temperature is increased gradually, all beneficial products of the system increase.

The exergy efficiency characteristics given in Fig. 8.33 show the effect of the geothermal water mass flow rate on the exergy efficiency of the geothermal energy–based integrated district system and its subsystems. To examine the effect of the geothermal water mass flow rate, the geothermal water mass flow rate is gradually increased from 100 to 180 kg/s. Considering the exergy efficiency characteristics of the geothermal energy–based integrated district system and its subsystems, when the geothermal water mass flow rate is increased gradually, the exergy efficiencies of the geothermal energy–based integrated district system and its all subsystems

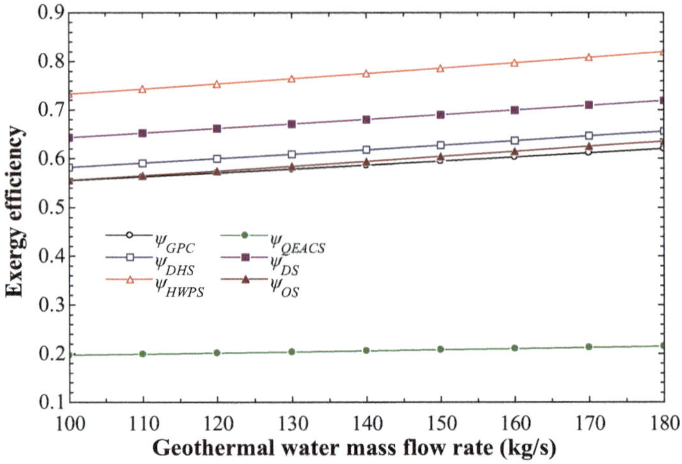

FIGURE 8.33 Effect of geothermal water mass flow rate on the exergy efficiency of the geothermal energy–based integrated district system and its subsystems.

increase. When the geothermal water mass flow rate value is 110 kg/s, the exergy efficiencies of the GPC, DHS, HWPS, QEACS, DS, and WS are 0.5629, 0.5913, 0.7438, 0.1995, 0.6528, and 0.5651, respectively. Also, when the geothermal water mass flow rate value is 160 kg/s, the exergy efficiencies of the GPC, DHS, HWPS, QEACS, DS, and WS are 0.6034, 0.637, 0.7973, 0.2107, 0.6998, and 0.6148, respectively. In case where geothermal water mass flow rate value is 180 kg/s, these exergy efficiencies are 0.6204, 0.6563, 0.8198, 0.2154, 0.7196, and 0.6359, respectively. As can be understood from these exergy efficiency values, when the geothermal water mass flow rate is increased gradually, exergy efficiencies of the geothermal energy–based integrated district system and its subsystems increase.

The performance curves of the useful outputs given in Fig. 8.34 show the effect of the geothermal water mass flow rate on the useful outputs of the geothermal energy–based integrated district system. In general, considering the useful output performance curves of the geothermal energy–based integrated district system, when the geothermal water mass flow rate is increased gradually, the production of these useful outputs increases. When the geothermal water mass flow rate values are 110, 160, and 180 kg/s, power generations of the geothermal energy–based integrated district system are 8472, 9262, and 9599 kW. At these geothermal water mass flow rate values, the heating outputs of the geothermal energy–based integrated district system are 11,312, 12,613, and 13,174 kW, respectively. For these geothermal water mass flow rate values, hot water outputs of the geothermal energy–based integrated district system are 8749, 9802, and 10,258 kW, respectively. Also, at these geothermal water mass flow rate values, the cooling outputs of the geothermal energy–based integrated district system are 6942, 7590, and

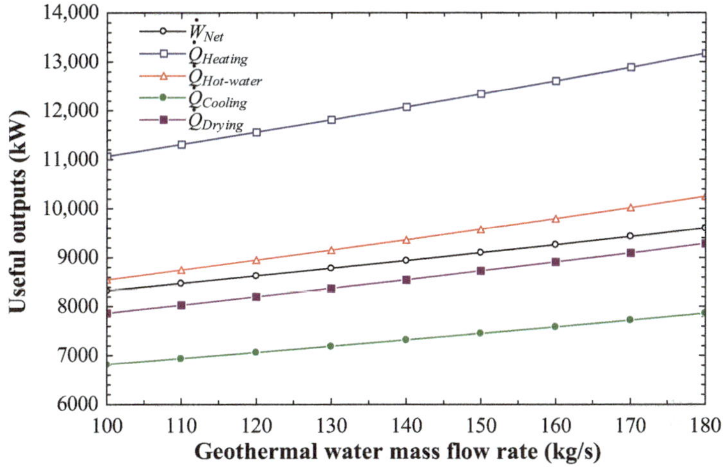

FIGURE 8.34 Effect of geothermal water mass flow rate on the power and cooling production rates of the geothermal energy–based integrated district system.

7865 kW, respectively. In addition, for these geothermal water mass flow rate values, the drying outputs of the geothermal energy–based integrated district system are 8031, 8910, and 9288 kW, respectively. As can be seen from all these beneficial output values of the geothermal energy–based integrated district system, when the geothermal water mass flow rate is increased gradually, all beneficial outputs of the geothermal energy–based integrated district system increase.

8.9 Closing remarks

In this section, geothermal energy–based district systems are explained in detail. Also, the advantages of these systems are discussed in terms of society, community, and consumer. Then these systems are evaluated in five sections as district heating plants, district cooling plants, combined district heating and cooling plants, cogeneration-based district energy plants, and integrated district energy plants. Case studies are given for these five geothermal-based energy systems. Comprehensive thermodynamic and performance analyses of these modeled systems have been carried out. Beneficial products obtained from geothermal energy–based systems are given clearly in the tables. The comparative performance characteristics of these systems and their subsystems, depending on different system indicators such as the geothermal fluid flow rate, are given in the graphics. Considering the performance analyses performed, the performance characteristics of the systems change depending on the useful product obtained from geothermal energy–based systems. This shows that systems that take full advantage of the potential of the geothermal energy source perform better. Performance evaluations of the modeled systems can be seen clearly in the tables given. The changes in the performance of the modeled systems and subsystems based on different system indicators are presented in the graphs.

Nomenclature

A	Area (m^2)
E	Energy (kJ)
e	Specific energy (kJ/kg)
$\dot{E}$	Energy rate (kW)
ex	Specific exergy (kJ/kg)
$\dot{E}x$	Exergy rate (kW)
$\dot{E}x_d$	Exergy destruction rate (kW)
$\dot{E}x^Q$	Exergy transfer rate associated with heat transfer (kW)
$\dot{E}x^W$	Exergy transfer rate associated with work (kW)
h	Specific enthalpy (kJ/kg)
H	Enthalpy (kJ)
m	Mass (kg)
$\dot{m}$	Mass flow rate (kg/s)
P	Pressure (kPa)
q	Specific heat transfer (kJ/kg)
Q	Heat (kJ)
$\dot{q}$	Specific heat transfer rate (kW/kg)
$\dot{Q}$	Heat rate (kW)
s	Specific entropy (kJ/kgK)
S	Entropy (kJ/K)
$\dot{S}$	Entropy rate (kW/K)
T	Temperature (K)
u	Internal energy (kJ/kg)
W	Work (kJ)
$\dot{w}$	Specific work rate (kW/kg)
$\dot{W}$	Work rate (kW)

Greek letters

Δ	Change in variable
$\in_{HEX}$	HEX efficiency factor
η	Energy efficiency
ψ	Exergy efficiency

Subscript

a	Air
abs	Absorber
AC	Air compressor
Cmp	Compressor
$cooling$	Cooling load

Con	Condenser
D	Destruction
e	Exit condition
Ej	Ejector
en	Energy
Erd	Energy recovery device
Eva	Evaporator
Ev	Expansion valve
ex	Exergy
f	Fuel
fls	Flashing
g	Generation
Gen	Generator
heating	Heating load
HP	High pressure
l	Liquid
LP	Low pressure
i	Inlet condition
mdm	Membrane distillation module
mine	Mineralizer
MP	Medium pressure
MR	Moisture remover
p	Pump
pr	Particulate remover
pst	Post-treatment
pt	Pretreatment
rhs	Radiator heating system
RO	Reserve osmosis
Sep	Separator
ST	Steam turbine
tot	Total
Tur	Turbine
Vl	Valve
wf	Working fluid
1...74	State numbers
0	Ambient or reference condition

Superscripts

.	Rate
Ch	Chemical

Acronyms

ACS	Absorption cooling system
BHE	Borehole heat exchanger
CHP	Combined heat and power
COP	Coefficient of performance

DEACS	Double-effect absorption cooling system
DFGP	Double-flash geothermal plant
DHS	District heating system
DS	Drying system
DSGP	Direct steam geothermal plant
EES	Engineering Equation Solver
GHE	Ground heat exchanger
GPS	Geothermal power system
GSHP	Ground source heat pump
HEX	Heat exchanger
HGHE	Horizontal ground heat exchanger
HTSE	High-temperature steam electrolysis
KC	Kalina cycle
KCGP	Kalina cycle geothermal plant
LCA	Life cycle assessment
LCI	Life cycle inventory
ORC	Organic Rankine cycle
PEM	Proton exchange membrane
QEACS	Quadruple-effect absorption cooling system
QFGP	Quadruple-flash geothermal plant
SEACS	Single-effect absorption cooling system
SExI	Specific exergy index
SFGP	Single-flash geothermal plant
SS	Single-stage
TEACS	Triple-effect absorption cooling system
TFGP	Triple-flash geothermal plant

References

[1] M. Ozturk, I. Dincer, Geothermal energy conversion, Comprehensive Energy Systems, Elsevier, 2018, pp. 474–544.

[2] I. Dincer, M.A. Rosen, Exergy Analysis of Heating, Refrigerating, and Air: Conditioning Methods and Applications, Elsevier, New York, 2015.

[3] R.G. Bloomquist, Geothermal district energy system analysis, design, and development, European Summer School on Geothermal Energy Applications, The International Geothermal Training Centre of the University of Oradea, Oradea, 2001, pp. 213–253.

[4] IEA, International Energy Agency, Global Energy Review 2020, OECD/IEA, Paris, 2020.

[5] Y.E. Yuksel, M. Ozturk, I. Dincer, Performance assessment of a solar tower based multigeneration system with thermal energy storage, Energy Storage 1 (4) (2019) 1–19.

[6] M. Asim, S. Saleem, M. Imran, M.K.H. Leung, S.A. Hussain, L.S. Miró, et al., Thermo-economic and environmental analysis of integrating renewable energy sources in a district heating and cooling network, Energy Efficiency 13 (2020) 79–100.

[7] I. Dincer, Environmental impacts of energy, Energy Policy 27 (14) (1999) 845–854.

[8] M. Ozturk, Energy and exergy analysis of a combined ground source heat pump system, Appl. Therm. Eng. 73 (1) (2014) 362–370.

[9] M.A. Rosen, S. Koohi-Fayegh, Geothermal Energy: Sustainable Heating and Cooling Using the Ground, Wiley, Chichester, 2017.

[10] M. Spurr, Economic and design optimisation in integrating renewable energy and waste heat with district energy systems. In: Proceedings of the ExCo Meeting on Presentation of the Status Report, New York City, United States, FVB Energy, September 19–21, 2013.
[11] R. Samson, Switchgrass for bioheat in Canada. Resource efficient agricultural production (REAP), <http://www.reap-canada.com>, 2008 (accessed May 2020).
[12] T.A.H. Ratlamwala, I. Dincer, Energy management in district energy systems, Comprehensive Energy Systems, Elsevier, 2018.
[13] M.S. Owen, (Ed.), ASHRAE Handbook, HVAC Systems and Equipment, 2016.
[14] K.C. Lee, Classification of geothermal resources by exergy, Geothermics 30 (2001) 431–442.
[15] Y.E. Yuksel, M. Ozturk, Thermodynamic and thermoeconomic analyses of a geothermal energy based integrated system for hydrogen production, Int. J. Hydrog. Energy 42 (4) (2017) 2530–2546.
[16] G. Phetteplace, S. Abdullah, J. Andrepont, D. Bahnfleth, A. Ghani, V. Meyer, et al., District Cooling Guide. American Society of Heating, Refrigerating, and Air-Conditioning Engineers, ASHRAE, Atlanta, GA, 2013.
[17] S. Nielsen, B. Möller, Excess heat production of future net zero energy buildings within district heating areas in Denmark, Energy 48 (1) (2012) 23–31.
[18] Y.E. Yuksel, M. Ozturk, I. Dincer, Analysis and performance assessment of a combined geothermal power-based hydrogen production and liquefaction system, Int. J. Hydrog. Energy 43 (22) (2018) 10268–10280.
[19] Y.E. Yuksel, M. Ozturk, I. Dincer, Development of a geothermal-based integrated plant for generating clean hydrogen and other useful commodities, J. Energy Resour. Technol. 142 (9) (2020) 1–13.
[20] Y.E. Yuksel, M. Ozturk, Energy and exergy analysis of renewable energy sources-based integrated system for multi-generation application, Int. J. Exergy 22 (3) (2017) 250–278.

Study questions and problems

8.1. The most general utilization of geothermal power sources is electricity production. Some different utilizations are process heating, district heating, district cooling, greenhouse heating, and heating for fish farming. From a thermodynamic analysis viewpoint and also considering the quality of energy, explain which of these utilizations you recommend most.

8.2. Which is a more effective way of increasing the performance of the geothermal district heating system considered: the energy analysis method or exergy analysis method?

8.3. Identify the sources of exergy losses in the geothermal energy–based district system considered and propose methods for reducing or minimizing them.

8.4. How are the energy and exergy efficiencies of geothermal energy–based district heating and cooling plants defined?

8.5. In geothermal district heating plants, the temperature difference between the geothermal energy source and the supply temperature of the district heating distribution network significantly affects the exergy loss rate. Explain this condition based on exergetic analysis.

8.6. What is the difference between a combined or integrated geothermal energy–based district plant?

8.7. Can exergy analysis be used in the design of a geothermal district energy system? Explain.

8.8. How can one use the results of exergy analysis of a geothermal energy–based district system to improve system efficiency?

8.9. Which use of a geothermal energy source at 155°C is better from an energy and exergy point of view?
1. District heating
2. District cooling
3. Power generation

Explain. What would your answer be if the geothermal resource is at 125°C?

8.10. How can a geothermal source at 180°C be utilized for a cooling application? How can you express the exergetic performance of such a cooling plant?

8.11. How are the energetic and exergetic performances of geothermal district heating plants described? Which definition is more valuable to you if you are a customer of geothermal district heat? Which definition is more valuable to you if you are an engineer trying to improve the effectiveness of the district plant?

8.12. Identify the primary reasons for exergy destruction in the geothermal energy–based district heating and cooling plants and propose methods for decreasing or minimizing them.

8.13. Geothermal energy sources should be classified based on the resource temperature or resource exergy. Which classification is suitable if the geothermal energy source is to be utilized for:
1. district heating?
2. district cooling?
3. power production?

Explain.

8.14. How do you explain the difference between the energetic and exergetic performances of the geothermal energy–based district heating and cooling plant considered in this chapter?

8.15. Compare the energetic and exergetic performances of a geothermal energy–based power plant. Which one is greater?

8.16. What is the impact of reference temperature on the energy and exergy efficiencies of an integrated geothermal plant?

8.17. Obtain a published article on energy and exergy analyses of geothermal energy–based district plants. Using the working data provided in the study, perform the detailed energy and exergy analyses of the plant, and compare your results to those in the original article. Also, investigate the effect of varying important operating parameters on the plant energetic and exergetic effectiveness.

8.18. Perform a case study of the geothermal energy–based district heating system shown in Fig. 8.10. Study the variation in energy and exergy efficiencies of the overall system by varying the following operating conditions:

1. Geothermal water source temperature (T5) from 90°C to 150°C
2. Mass flow rates of geothermal water from 150 to 250 kg/s. Consider the ambient pressure to be 101 kPa.

8.19. Perform a case study of the geothermal energy–based integrated system for multigeneration shown in Fig. 8.28. Study the variation in energy and exergy efficiencies of the overall system by varying the following operating conditions:

1. The geothermal water source temperature changes from 200°C to 290°C.
2. The mass flow rates of geothermal water change from 150 to 250 kg/s. Consider the ambient pressure to be 101 kPa.

8.20. Is exergy analysis more useful for geothermal energy–based integrated systems or for geothermal district energy systems plants? Explain.

Chapter 9

Future directions

Comprehensive examples present a comparative effectiveness evaluation of the single-flash, double-flash, and hybrid geothermal power application plants. Double- and multiple-flash geothermal energy systems show an advantage over those of single-flash processes. The hybrid plant can be used in both single- and multiple-flash power applications by integrating an organic Rankine cycle (ORC) with a suitable working fluid. Technical and system design challenges have been also discussed. Single-flash power systems are the most commonly utilized system configuration, which makes it very challenging to integrate a second flashing option for enhanced plant efficiency.

The high power demand for conventional electrolysis plants and high temperature demand for pure thermochemical processes might make geothermal power utilization for hydrogen generation infeasible. The hybrid thermochemical process contains at least one thermochemical step with an electrolysis step. It is reasonable to generate hydrogen by using less electrical power than that of conventional electrolysis at lower temperatures than those of pure thermochemical processes. Hence the chemical heat pumps can be used to increase the temperature of geothermal working water by means of some subsequent reactions. A sufficient increase in the temperature of heat transfer from the geothermal working fluid to the thermochemical water splitting process might be achieved by the chemical heat pump application, which increases the temperature of the heat supplied via a cyclic process driven by mechanical or electrical work.

A schematic view of progress in geothermal energy-based systems is shown in Fig. 9.1. Basic geothermal energy systems can be evaluated as a direct steam geothermal power plant, single-flash steam geothermal power system, double-flash steam geothermal power system, triple-flash steam geothermal power system, quadruple-flash steam geothermal power system, single-stage ORC geothermal power system, single-stage ORC geothermal power system with two turbines, double-stage ORC geothermal power system, Kalina cycle geothermal power system and combined flash/binary. Generally, the basic geothermal energy systems produce heat and power in one process and can provide efficiency along with operational, environmental, and financial benefits. In this book, geothermal energy—based district-

Geothermal Energy Systems. DOI: https://doi.org/10.1016/B978-0-12-820775-8.00007-6

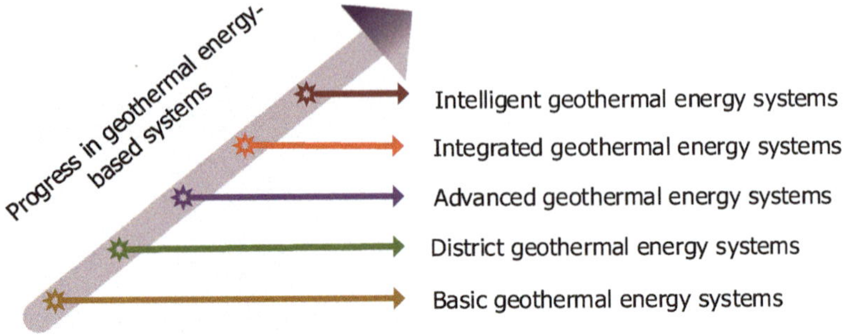

FIGURE 9.1 Schematic view of progress in geothermal energy–based systems.

energy systems are evaluated in five sections as district heating plants, district cooling plants, combined district heating and cooling plants, cogeneration-based district energy plants, and integrated district energy plants. Geothermal energy–based district systems have an important potential and role in meeting energy needs depending on developing technology. Also, the design of advanced geothermal energy systems to generate electricity, hydrogen, hot water, cold water, and freshwater plays a significant role in developed and underdeveloping countries. Advanced geothermal energy systems can be evaluated as multistaged direct systems, multiflashing systems, multistaged binary systems, multiflashing binary systems, and combined/integrated systems. Integrated geothermal energy systems are basically systems where different subsystems can perform their functions in harmony with one another. The potential of integrated geothermal energy plants varies according to the beneficial products obtained from them. The design and modeling of intelligent geothermal energy systems are among the increasingly popular areas of study. Adjusting the performance of the energy generation system according to the desired situation is important for the sustainability of energy resources. This performance setting is among the applications that can be achieved with intelligent geothermal energy systems. Making geothermal energy production systems more equipped with developing technology can enable these systems to have a better potential in energy generation systems.

An investigation of renewable power–based multigeneration ways for generating outputs such as electricity, heating, hot and freshwater, drying, cooling, hydrogen, ammonia, etc. from renewable power resources, including geothermal power, and a discussion of their benefits are very important to design more environmentally benign integrated systems. Such multigeneration methods clearly lead to improved plant effectiveness and to reduced adverse environmental effects. First, single-generation (power production only) and cogeneration systems (generally power and heat production) can be compared in terms of energy utilization performance, exergetic performance, greenhouse gaseous emissions mitigation, and payback period. It can be found that cogeneration increases the mitigation of greenhouse gaseous emissions about two to four times and that the payback time decreased about 2.8 times with respect to the single-generation condition. In addition, the results also illustrated that geothermal energy–based integrated plants for multigeneration help increase both energetic and exergetic performances, reduce cost and environmental effects, and increase sustainability. A schematic diagram of integrated systems with geothermal energy is illustrated in Fig. 9.2. Several potential applications of integrated geothermal energy utilization are, among others:

1. geothermal heat pumps,
2. space and district heating,
3. greenhouse heating,
4. aquacultural heating,
5. agricultural drying,
6. industrial process heating,
7. bathing and swimming,
8. cooling and snow melting,

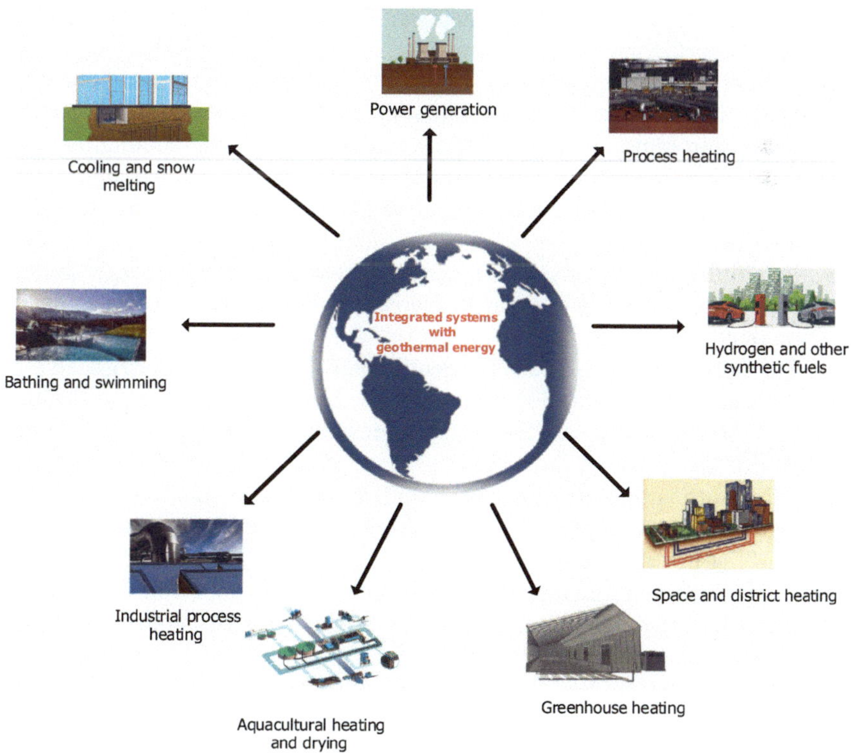

FIGURE 9.2 Schematic view of integrated systems with geothermal energy.

9. power generation,
10. process heating,
11. hydrogen and other synthetic fuels.

The following statuses for geothermal energy systems show the worldwide status of technological progress and the different main research areas for geothermal energy–based applications in the future.

1. The benefits of district energy should be defined more clearly to decrease carbon-based energy source utilization for domestic applications.
2. The major challenges to district energy plants that are primarily modified for the utilization of geothermal energy sources should be assessed future priorities as defined by the countries themselves.
3. Energy autonomy and system resilience of geothermal energy–based integrated plants are very important, and therefore design investigations and developed technologies should be performed to further real-world geothermal energy plants in order to demonstrate how they can support all the design characteristics and to improve achievement in them and hence expand their applications for domestic and industrial areas.
4. Integrated planning policy and regulations should be provided for designing highly efficient, low-cost, and large-scale geothermal energy technologies.
5. In the presented combined geothermal energy process model for multigeneration aims, such as electricity, hydrogen, heating-cooling, hot and freshwater production, and ammonia, the required capital investment and working cost for the desired generation capacities per day should be considered.
6. Flexibility and the role of technologies for geothermal energy sources should be enhanced in order to increase energy performance and useful outputs.
7. The policy and regulations for geothermal energy–based integrated systems should be more definitely described in order to show the position of these technologies for future renewable energy situations in the world.
8. Urban growth centers and land utilization design must be taken into account for the application of geothermal district energy systems, especially large-scale design aims.
9. Local economic development opportunities, depending on the geothermal energy technologies, can be investigated for reducing fossil-based source consumption.
10. Integrated system performance based on exergy analysis—especially for geothermal energy, this should be used instead of energetic performance—must be defined to make it more user-friendly and attractive.

11. The geothermal power-based ORC cycles, triple flash, binary cycle, combined or hybrid plant, and Kalina process are effective ways to produce power from low-grade temperature heat sources such as geothermal energy resources. But the required capital investment and working cost and life cycle assessment for the desired generation capacities per day should be further considered based on the design indicators of the installation area. The conclusions can be analyzed and compared so that a suitable combined plant can be suggested for the intended working conditions.
12. Future growth and trends in related areas of geothermal energy technologies should be considered in the design of novel plants for domestic and industrial applications.
13. The impacts on the total working cost of several alternative fuels instead of hydrogen energy for the geothermal power–based combined plant are suggested for evaluation.
14. The produced hydrogen from a geothermal energy–based system can be stored in gaseous, liquid, or solid form. Because of its low-density, gaseous hydrogen needs large volumes for storage applications, for this reason necessitating compression and extremely low temperatures in order to convert hydrogen to a cryogenic liquid form or combinations with other materials for solid storage. For liquid storage, the geothermal energy–based hydrogen liquefaction plants can be designed and investigated by using energetic and exergetic analyses.
15. Considering the increasing fossil energy consumption rate, environmental pollution rate, and plant and human health issues, community consultation and education for the role of renewable energy technologies in sustainable development strategies will be very important in the near future.
16. The development of new approaches in the design and analysis of district heating systems is important for the performance efficiency of these systems. Therefore the increase of long-term incentive policies for future studies on these systems will increase the potential of these systems in energy production.
17. Designing district energy systems with certain modeling approaches will increase the systems' performances. Particularly, the parameters that affect the performance of the systems should be defined so that they can be adjusted later during the modeling of the system. Thus the improvements desired in the future will be more implementable since the parameters affecting system performance are certain.
18. Increasing the level of utilization of the local energy resources of the region where the district energy systems are intended to be established will be beneficial for the economic development of the region as well as for continuous response to the energy demanded by the people of the region.

19. District energy systems will also contribute to future employment activities in the intended regions because activities such as the installation, maintenance, and development of these systems will continue to depend on one another.
20. The availability of renewable energy sources in the region where the district energy systems will be installed is an important potential for these systems. Considering the benefits of renewable energy sources, investments in both these systems and the incentives for these systems will increase.
21. If information activities on district energy systems are increased in both educational institutions and society, interest in these systems will increase. Thus these systems will be more accepted by both society and educational institutions.
22. Comparative analysis of district energy systems with respect to other energy generation systems will show more clearly the potential in these systems. For this reason, institutions and governments that can generate income with district energy systems will be interested in these analyses and will invest large amounts of research and development activities for these analyses. This will both create a sector and employ it in this field.
23. In the future, new tools and techniques will emerge to improve the performance of central energy systems. For these tools and techniques, organizations interested in these systems, doing research, or earning income will receive services from these organizations for these tools and techniques. This reveals the future potential of these techniques and tools.
24. To ensure confidence in the research on central energy systems, studies should also be carried out on real systems. Thus the need to establish energy systems in an experimental environment will arise because theoretical studies must be carried out and tested on systems in an experimental environment before the real systems can provide commercial income. Afterward, the tested studies should be applied to the systems providing commercial income, and these systems should work more efficiently.

The main utilization of geothermal energy covers a wide range of applications, such as residential heating-cooling and/or domestic hot water supply, aquaculture, greenhouse heating, swimming pools and balneology, industrial heating process, heat pumps, and electricity generation. But for higher geothermal plant effectiveness, decreased thermal energy losses and wastes, decreased working costs, decreased harmful gas emissions, better utilization of geothermal resources, multiple production ways, and increased reliability, geothermal power–based combined plants depending on the local availability of sources can be designed and operated. The outputs from this book can assist designers in developing more energy-efficient geothermal energy plants in combined arrangements. Geothermal power could be integrated with different alternative energy resources, such as solar, biomass, wind, etc., in the actual cycles, depending on their local availability, for the production of different beneficial outputs.

Generally, when the potential of central energy systems is fully used, the potential of these systems in future energy production will be great. For this reason, these systems will make great contributions to the consumer, industry, the government, and the social economy. Increasing studies on renewable energy–based central energy systems will also contribute to the sustainability of energy resources. At the same time, benefits will be obtained in these areas with studies on areas such as efficiency, environmental impact, and reliability. In addition, geothermal power may be further used in other energy forms such as hydrogen generation, absorption refrigeration and chilling, and/or several other cycles that utilize power to produce various forms of energy. Geothermal energy–based hydrogen generation, which basically utilizes geothermal power for hydrogen generation, for instance, appears to be an environmentally conscious and sustainable way for countries with abundant geothermal power sources. Hydrogen generation from geothermal sources should be achieved by utilizing direct hydrogen generation, electrolysis, hybrid thermochemical processes, or geothermal energy–based multigeneration plants.

Finally, it could be mentioned that geothermal energy is one of the most abundant renewable energy resources, without showing any intermittence as in solar, wind, and biomass applications. Considering 90% of the Earth is hotter than 100°C, further examinations are needed on geothermal power applications and usage for many aims, ranging from direct utilization of power generation, as a sustainable solution to the world's challenges on power needs and logical alternatives to carbon-based sources usage. The world has a long history in the usage of geothermal power, and further increases make this renewable source more feasible not only for geothermal heat but also for electricity generation. Geothermal energy–based integrated systems for multigeneration help to increase both energetic and exergetic performances, reduce cost and environmental effects, and increase sustainability. Hence, it is also of significance to use geothermal power resources more efficiently by integrating these resources into such mechanisms that may generate more than one energy form. Finally, improving the geothermal power–based multigeneration effectiveness would reduce greenhouse gaseous emissions and other harmful environmental effects and enhance sustainability.

Index

Note: Page numbers followed by "*f*" and "*t*" refer to figures and tables, respectively.

C

D

F

G

H

I

K

L

M

N

O

P

Q

FSC
www.fsc.org
MIX
Papier | Fördert gute Waldnutzung
FSC® C083411

Kolibri 360 GmbH
Ferdinand-Jühlke-Straße 7
99095 Erfurt, Deutschland
produktsicherheit@kolibri360.de